Lecture Notes on Mathematical Modelling in the Life Sciences

Series Editors:
Angela Stevens
Michael C. Mackey

For further volumes:
http://www.springer.com/series/10049

Urszula Ledzewicz • Heinz Schättler
Avner Friedman • Eugene Kashdan
Editors

Mathematical Methods and Models in Biomedicine

Editors
Urszula Ledzewicz
Southern Illinois University
Department of Mathematics
and Statistics
Edwardsville, IL 62026
USA

Heinz Schättler
Washington University
Department of Electrical
and Systems Engineering
St. Louis, MO 63130
USA

Avner Friedman
Ohio State University
Mathematical Biosciences Institute
100 Mathematics Bldg.
231W. 18th Avenue
Columbus, OH 43210-1292
USA

Eugene Kashdan
Tel Aviv University
Ramat Aviv
69978 Tel Aviv
Israel

ISSN 2193-4789
ISSN 2193-4797 (electronic)
ISBN 978-1-4614-4177-9
ISBN 978-1-4614-4178-6 (eBook)
DOI 10.1007/978-1-4614-4178-6
Springer New York Heidelberg Dordrecht London

Library of Congress Control Number: 2012945002

Mathematics Subject Classification: 92C30, 92C45, 92C50, 92D25, 92D30

Printed on acid-free paper

Springer is part of Springer Science+Business Media (www.springer.com)

Preface

Mathematical biomedicine is a fast developing interdisciplinary field of research that connects the natural and exact sciences in an attempt to respond to the challenges raised by biology and medicine. Biomedical research covers such diverse areas as the study of disease progression and treatment, drug development, and the analysis of epidemic trends and patterns, to name just a few. Mathematical methods in biomedicine therefore come from a correspondingly large number of fields of research that include mathematical physics, computational methods, control theory, combinatorics, statistics, and many more.

As in any research aimed to solve "real-world" problems, mathematical models in biomedicine must be grounded on experimental data. The information and model parameters usually come from in vitro or in vivo experiments, but are only rarely based on patient data. Unfortunately, quite often data are not sufficient to allow for an accurate and reliable parameter estimation. In the study of an infectious disease, the rate of infection typically varies between locations and from one time period to another. In other cases, the underlying medical processes are not fully understood and mathematical models become a tool to suggest biologically testable hypotheses.

There exist a large number of mathematical methods and procedures that can be brought in to meet the modeling and simulation challenges of understanding and predicting biological processes. This book is not meant to be a textbook or a monograph dedicated to the systematic study of any specific disease or a family of diseases. Instead, it is meant to illustrate the breadth and depth of research opportunities that exist in the general field of mathematical biomedicine. The book consists of five parts which can be read independently of each other, but are arranged to give the reader a broader picture of specific research topics and the mathematical tools that are being applied in its modeling and analysis.

There is a slight emphasis on issues related to cancer. One part not only deals with *cancer modeling* and addresses important topics like cancer stem cells, but it also focuses on specific cancers such as glioblastoma multiforme, an especially aggressive form of brain tumor. This naturally leads to the question of *treatment* and to a range of issues from modeling of gene therapies to pharmacokinetics of drugs to questions of how to optimize cancer treatment protocols. The latter topic is addressed not only for traditional treatment approaches such as chemotherapy or radiotherapy, but also for more novel procedures that involve anti-angiogenic treatments which aim at blocking the migration of new blood vessels and capillaries to the tumor. More generally, the role of the *blood vessels* and their dynamics is presented in the context of cancer (tumor-angiogenesis) and both vascular diseases resulting in blood vessel abnormalities such as aneurysm.

Another important aspect in the treatment of any disease is the understanding of the role that the *immune system* plays in its prevention and cure. In this book, the important concepts of immuno-dominance and viral epitopes are considered. Improving our understanding of these aspects helps in the study of specific diseases. One chapter presents an up-to-date mathematical study of one of the most devastating diseases of recent times, namely HIV/AIDS. Infectious diseases can be analyzed from many angles that range from the study of the spread of the disease in the human body to the spread of the infection in the population. The latter is the subject of *epidemiology* and in the book seasonal effects of epidemiological models are described which are important for diseases such as malaria, cholera, or tuberculosis.

All these topics will be studied using a palette of mathematical tools ranging from discrete cellular automata to cell population-based models described by ordinary differential equations to nonlinear partial differential equations representing complex time- and space-dependent continuous processes. The authors employ both stochastic and deterministic methods to analyze biological phenomena in various temporal and spatial settings. Despite the wide range of topics covered, we readily acknowledge that the book does not in any way contain a comprehensive overview of mathematical methods and models used in modern biomedicine. Rather, the chapters highlight some topics that we believe are at the same time both in the mainstream and on the cutting edge of biomedical research. The book provides surveys as well as suggestions for possible future research directions by including open questions and new challenges. We hope that it will be of interest to both the mathematical and the biomedical communities and to both researchers working in the field and those who might consider entering it. Much of the material is presented in a way that allows young researchers and graduate students to use it as a starting point for their own work. We believe that the book will achieve its goal if, on the one hand, it raises the awareness of the vast array of topics awaiting researchers in the field of biomedicine and, on the other hand, it shows the power and richness of mathematical methods and tools that are being used to address the challenging problems in this field.

Finally, we also would like to thank all contributors to this volume and the editors at Springer, especially Achi Dosanjh and Donna Chernyk, who have been so helpful throughout the production process.

Edwardsville, Illinois, USA — Urszula Ledzewicz
St. Louis, Missouri, USA — Heinz Schättler
Columbus, Ohio, USA — Avner Friedman
Tel Aviv, Israel — Eugene Kashdan

Contents

Contributors

Alexandra Agranovich Department of Mathematics and Gonda, Brain Research Center, Bar-Ilan University, Ramat Gan, Israel

Zvia Agur Institute for Medical BioMathematics, 10 Hate'ena St., Bene Ataroth, Israel

Frédérique Billy INRIA BANG team, BP 105, Rocquencourt, France, and Université Pierre et Marie Curie, Paris, France

Vincenzo Capasso Department of Mathematics, University of Milan and CIMAB, Milan, Italy

Jean Clairambault INRIA BANG team, BP 105, Rocquencourt, France, and Université Pierre et Marie Curie, Paris, France

Antonio Fasano Dipartimento di Matematica "U. Dini", Universita' di Firenze, Firenze, Italy

Olivier Fercoq INRIA MAXPLUS team, CMAP, Ecole Polytechnique, Palaiseau, France

Avner Friedman Department of Mathematics, The Ohio State University, Columbus, OH, USA

Alberto M. Gambaruto Department of Mathematics and CEMAT, Instituto Superior Técnico, Technical University of Lisbon, Lisboa, Portugal

Alberto Gandolfi Istituto di Analisi dei Sistemi ed Informatica "A. Ruberti" - CNR, Viale Manzoni, Roma, Italy

Frederik Graw Theoretical Biology and Biophysics, Los Alamos National Laboratory, Los Alamos, NM, USA

Gili Hochman Institute for Medical BioMathematics, 10 Hate'ena St., Bene Ataroth, Israel

Peter S. Kim School of Mathematics and Statistics, University of Sydney, NSW, Australia

Denise E. Kirschner Department of Microbiology and Immunology, University of Michigan Medical School, Ann Arbor, MI, USA

Erik Kostelich School of Mathematical and Statistical Sciences, Arizona State University, Tempe, AZ, USA

Yuang Kuang School of Mathematical and Statistical Sciences, Arizona State University, Tempe, AZ, USA

Christian Kunert Massachusetts General Hospital, Harvard Medical School, Boston, MA, USA

Urszula Ledzewicz Department of Mathematics and Statistics, Southern Illinois University Edwardsville, Edwardsville, IL, USA

Peter P. Lee Division of Hematology, Department of Medicine, Stanford University, Stanford, CA, USA

Doron Levy Department of Mathematics & Center for Scientific Computation and Mathematical Modeling (CSCAMM), University of Maryland, College Park, MD, USA

Yoram Louzon Department of Mathematics and Gonda Brain Research Center, Bar-Ilan University, Ramat Gan, Israel

Yaakov Maman The Mina and Everard Goodman Faculty of Life Sciences, Bar-Ilan University, Ramat Gan, Israel

Simeone Marino Department of Microbiology and Immunology, University of Michigan Medical School, Ann Arbor, MI, USA

Rebeccah E. Marsh Department of Physics, University of Alberta, Edmonton, Alberta, Canada

Nikolay L. Matirosyan Neurosurgery Research, Barrow Neurological Institute, St. Joseph's Hospital and Medical Center, Phoenix, AZ, USA

Joshua McDaniel School of Mathematical and Statistical Sciences, Arizona State University, Tempe, AZ, USA

Nina Z. Moore Neurosurgery Research, Barrow Neurological Institute, St. Josephs Hospital and Medical Center, Phoenix, AZ, USA

Daniela Morale Department of Mathematics, University of Milan and CIMAB, Milan, Italy

Alexandra B. Moura Department of Mathematics and CEMAT, Instituto Superior Técnico, Technical University of Lisbon, Lisboa, Portugal

Lance L. Munn Massachusetts General Hospital, Harvard Medical School, Boston, MA, USA

John Nagy Department of Biology, Scottsdale Community College, Scottsdale, AZ, USA

Alberto d'Onofrio Department of Experimental Oncology, European Institute of Oncology, Milan, Italy

Alan S. Perelson Theoretical Biology and Biophysics, Los Alamos National Laboratory, Los Alamos, NM, USA

Mark C. Preul Neurosurgery Research, Barrow Neurological Institute, St. Josephs Hospital and Medical Center, Phoenix, AZ, USA

Susana Ramalho Department of Mathematics and CEMAT, Instituto Superior Técnico, Technical University of Lisbon, Lisboa, Portugal

Heinz Schättler Department of Electrical and Systems Engineering, Washington University, St. Louis, MO, USA

Adélia Sequeira Department of Mathematics and CEMAT, Instituto Superior Técnico, Technical University of Lisbon, Lisboa, Portugal

Tal Vider Shalit Department of Mathematics and Gonda Brain Research Center, Bar-Ilan University, Ramat Gan, Israel

Alexei Tsygvintsev Unité de mathématiques pures et appliquées, École normale supérieure de Lyon, Lyon Cedex, France

Jack A. Tuszyński Department of Physics, University of Alberta, Edmonton, AB, Canada

James Alex Tyrrell Massachusetts General Hospital, Harvard Medical School, Boston, MA, USA

Abdul-Aziz Yakubu Department of Mathematics, Howard University, Washington, DC, USA

Najat Ziyadi Department of Mathematics, Morgan State University, Baltimore, MD, USA

Part I
Immune System Modeling

Spatial Aspects of HIV Infection

Frederik Graw and Alan S. Perelson

1 Introduction

Human immunodeficiency virus type 1 (HIV-1) is one of the most and intensely studied viral pathogens in the history of science. However, despite the huge scientific effort, many aspects of HIV infection dynamics and disease pathogenesis within a host are still not understood. Mathematical modeling has helped to improve our understanding of the infection as well as the dynamics of the immune response. Fitting models to clinical data has provided estimates for the turnover rate of target cells [82, 83, 111], the lifetime of infected cells and viral particles [104, 109], as well as for the rate of viral production by infected cells [21, 44]. Most mathematical models applied to experimental data on viral infections have been formulated as systems of ordinary differential equations (ODE) [91, 101, 104].

While helpful and appropriate in many situations, ODE models simplify the actual biological processes and have some limitations. One limitation is the assumption that the interacting viral and cell populations are well mixed and homogeneously distributed. This assumption, which may be realistic for populations in blood, is not realistic for populations interacting in tissues [36]. Within a tissue virus may not be distributed homogeneously and an infected cell will interact preferentially with neighboring cells. As HIV mainly infects CD4^{+} T cells [42] which are most abundant and densely packed in secondary lymphoid organs, such as lymph nodes and the spleen, the spatial arrangement of cells might influence the infection dynamics. Furthermore, during the establishment of infection, stochastic effects influenced by spatial conditions, such as the local availability of appro-

F. Graw • A.S. Perelson (✉)
Theoretical Biology and Biophysics, Los Alamos National Laboratory,
Los Alamos, NM 87545, USA
e-mail: fgraw@lanl.gov; asp@lanl.gov

U. Ledzewicz et al., *Mathematical Methods and Models in Biomedicine*, Lecture Notes on Mathematical Modelling in the Life Sciences, DOI 10.1007/978-1-4614-4178-6_1,

priate target cells, may strongly affect the outcome [43, 47, 100]. Basic ODE models are not able to capture the spatial or stochastic aspects of infection. Thus for some purposes spatial models may be preferred to ODE models.

The utilization of specific types of mathematical models requires appropriate biological data for their justification. Previous experimental techniques allow the quantification of cell or virion populations in the blood or specific organs, making the analysis of those data mainly suitable for ODE models. Recent advances in imaging techniques allow the observation of infection processes *in vivo* on a cellular and viral level [71, 72]. Spatial models that incorporate these observations might increase our understanding of the infection process. Cellular automaton simulations and agent-based models, which treat cells or virions as individual agents, are modeling frameworks that are able to incorporate this level of detail. These kinds of models have experienced an increasing interest in the analysis of experimental data and the qualitative evaluation of intervention methods (reviewed in [7]).

This chapter aims to give an overview of the different attempts which have been undertaken so far to include spatial aspects into the analysis and mathematical models of HIV infection dynamics. In the second section, we give a short introduction into the biology and dynamics of HIV infection and highlight different processes where spatial aspects may strongly affect the dynamics. The third section briefly reviews basic viral dynamic models and the insights these models have provided. In the fourth section, we introduce attempts to add a level of spatiality to these basic ODE models by using partial differential equations (PDE) or spatial compartments. Section 5 deals with enhanced computational simulation techniques, such as cellular automata and agent-based models. We describe their application to different aspects of general viral and immune dynamics and their contribution to the understanding of HIV infection dynamics. These kinds of models rely on appropriate experimental data to provide realistic simulation environments of the biological processes. Promising experimental techniques that will help modelers produce appropriate simulation environments, as well as future directions for the mathematical modeling of HIV infection dynamics, are discussed in the last section of this chapter.

2 The Biology of HIV-1 Infection and the Importance of Space

HIV is a retrovirus containing two copies of positive sense single-stranded RNA encapsulated in a viral envelope to form a particle with an average diameter of about 120 nm [129]. Glycoprotein complexes on the surface of the viral envelope allow the virus to attach to and fuse with target cells. HIV mainly replicates in lymphocytes, predominantly $CD4^+$ T cells, which are the main target cells for HIV [42]. Besides the viral load, the concentration of $CD4^+$ T cells in the blood of patients is used as a marker for disease progression, which can be categorized into three different

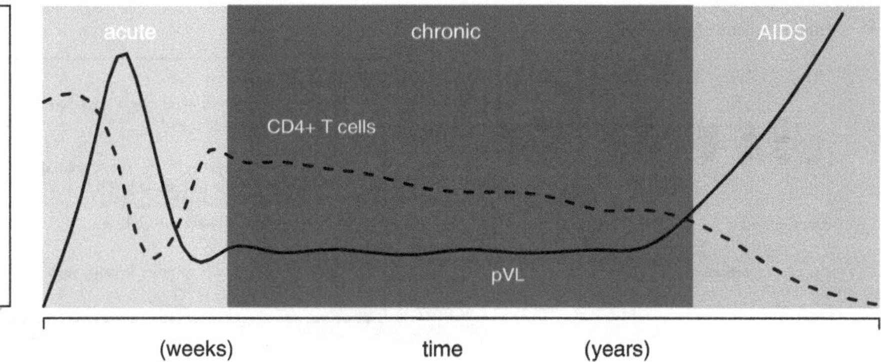

Fig. 1 Sketch of HIV disease progression. The three phases of the disease (*acute, chronic, AIDS*) characterized by the progression of the CD4$^+$ T cell count (*dashed line*) and the plasma viral load (*solid line*) are shown. The *acute* phase is characterized by an increase in the viral load and a rapid loss of CD4$^+$ T cells. After reaching a peak, the viral load declines and stabilizes on a constant level. This *chronic* phase can last for months or years with a slow decline in the CD4$^+$ T cell count. The AIDS phase is reached if the CD4$^+$ T cell count drops below a level of 200 cells/μl and the viral load increases to a value above 125 HIV-1 RNA copies/μl

stages (see Fig. 1): The *acute* infection phase, which lasts on the order of weeks, is characterized by a rapid increase in viral load and a rapid loss of CD4$^+$ T cells from the system. After reaching a peak, the viral load declines and stabilizes at a more or less constant level, called the *set-point viral load*. Reaching this level which varies among patients indicates the beginning of the *chronic* phase of the disease. The virus can persist within a patient for the duration of months or years. The terminal stage of the disease, the *AIDS* phase, is reached when the CD4$^+$ T cell count of a patient drops below 200 cells/μl and opportunistic infections occur. The factors that determine the length of the chronic phase of the disease have not been determined so far, although there exist correlations with the level of the viral load at the beginning of the chronic phase, the set-point viral load [33], and the functionality of CD8$^+$ T cell immune response [13].

Therapeutic interventions available today are not able to cure the disease. Protease inhibitors and reverse-transcriptase inhibitors interfere with the replication of the viral pathogen. Highly active antiretroviral therapy (HAART), which relies on a combination of these drugs, is able to effectively suppress viral replication in most patients, so that the viral load ultimately drops to an undetectable level. However, drug-resistant HIV mutants may exist before therapy begins or arise during therapy and diminish its effectiveness.

Spatial effects play a role throughout the entire infection process as well as the different stages of the disease (see Fig. 2).

Sexual transmission is the leading cause of HIV transmission worldwide making the female and male genital tracts the major sites of HIV invasion [47]. To successfully establish an infection, HIV has to cross the mucosal and epithelial barriers at these sites. Specific locations are more exposed for viral entry than

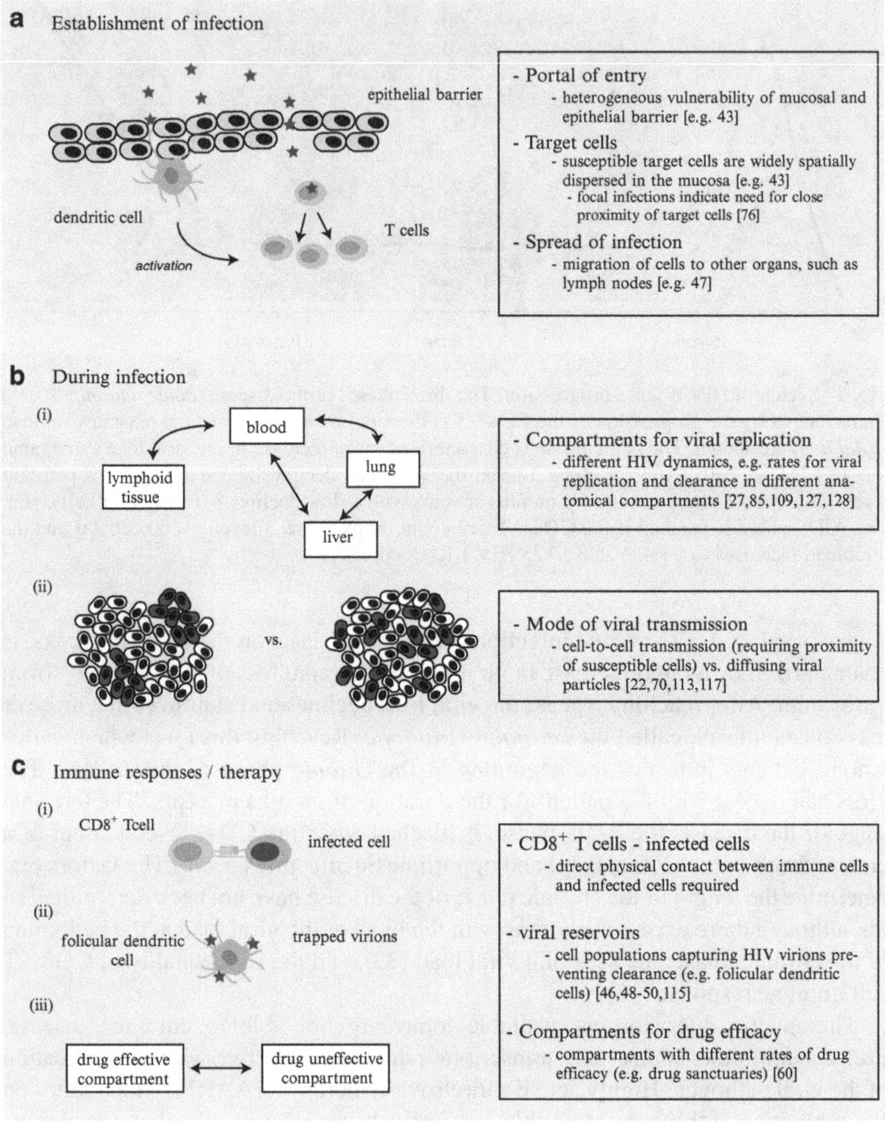

Fig. 2 Spatial aspects during different phases of HIV infection: (**a**) establishment of infection, (**b**) during the infection within a host, and (**c**) spatial aspects of immune responses and therapeutic interventions. See the main text for detailed descriptions

others [43]. Areas where the epithelium is very thin are especially vulnerable to viral invasion [47] and local physical abrasions enhance the probability of HIV to overcome the mucosal barrier. Accessibility of a sufficient pool of lymphocytes for replication at the portal of entry represents a second limitation for the establishment of infection, as these cells tend to be spatially dispersed in the cervical mucosa (see [43] and references therein). During simian immunodeficiency virus (SIV) infection of the genital mucosa of macaques only a very small fraction of the viral inoculum establishes infection and the infection is initially established in a highly focal manner [81]. Furthermore, in many cases infection is established by a single transmitted founder virus [56]. These observations indicate the blocking effect of the epithelial barrier and the need for a close proximity of susceptible target cells next to the portal of entry. Continuous viral production and sufficient expansion of the founder population of infected cells are crucial for the establishment of a systemic infection by SIV and HIV [42, 81]. The local influx of inflammatory cells into the sites of infection and the activation of nearby resting $CD4^+$ T cell populations create new target cells for the virus [43, 47] allowing the infection to get established. Furthermore, productively infected $CD4^+$ T cells and stromal dendritic cells or Langerhans cells harboring virions can emigrate into the draining lymphatic system and spread the infection to different anatomical compartments such as lymph nodes [47].

Once an infection is established, the dynamics of the cell populations involved varies between different compartments within a host. While the $CD4^+$ T cell population in mucosal tissue is massively depleted during the first weeks of infection [107], this loss is not reflected in the peripheral-blood $CD4^+$ T cell count [40]. In SIV and HIV, several studies have estimated different clearance rates of the virus in different compartments [27, 85, 127, 128]. Furthermore, some cell populations such as follicular dendritic cells in secondary lymphoid organs can act as a reservoir for HIV and maintain the infection even in the presence of host immune defenses and antiretroviral therapy [46, 48–50, 115].

The way by which HIV spreads inside a specific organ or tissue can also affect the infection dynamics. Besides the infection of cells by freely diffusing viral particles, HIV, as do several other viruses, has the ability to spread by cell-to-cell transmission [43, 114]. While free virus transmission allows the virus to spread rapidly through the blood and lymph and to infect distant tissues, the virus is vulnerable to antibody-mediated neutralization, opsonization, and phagocytosis. Besides antibodies other plasma components, such as complement, may also interact with free virions and reduce their infectivity or hasten their destruction. In contrast, cell-to-cell transmission of viral particles allows the virus to avoid exposure to plasma components, such as antibody and complement, and phagocytic cells. However, cell-to-cell transmission only allows the infection of nearby target cells, making the spread of infection dependent on the local availability of target cells. This could lead to the establishment of foci of infected cells [81] and can cause the infection to move like a traveling wave through the tissue [7]. *In vitro* HIV transmission has been observed to occur during direct cell-cell contact via so-called "virological synapses" between target T cells and infected cells [52, 53, 72] or via long-ranging nanotubes between T cells [118]. Because multiple viruses or viral genomes can be transmitted by these processes, cell-to-cell transmission has been estimated to be much more efficient than transmission via free viral particles, with estimates ranging between 10-fold and up to 18,000-fold [22, 70, 113, 117].

This finding may explain why HIV replicates more efficiently in areas densely populated with appropriate target cells in close contact to each other such as in lymphoid tissue [42, 94, 114].

The mode of transmission within a host also has an influence on the evolutionary development of the infecting viral strain. Cell-to-cell transmission could allow the virus to avoid the selective pressure mediated by the humoral immune response [22], possibly leading to foci of infected cells dominated by one viral mutant [23]. However, recent experimental observations showed that cell-to-cell transmission and free viral spread are equivalently sensitive to entry inhibitors [70]. The local clustering of infected cells dominated by one viral mutant also has implications for the effectiveness of the local immune response and therapeutic interventions.

In the same way as spatial effects play a role in the establishment and maintenance of HIV infection, space also plays a role in the immune response against the infection. To kill infected cells, antigen-specific $CD8^{+}$ T cells need to make direct physical contact with these cells. The absence of specific immune effector cells at certain spatial locations, e.g., mucosal tissue, could help explain the success of initial infection [26, 55, 108]. Similar to the immune response, therapeutic drugs are influenced by spatiality as they have to reach their targets in order to act efficiently. Due to this, drug concentration heterogeneity among different tissue compartments can facilitate the evolution of drug-resistant viral strains [60].

Overall, due to its importance for the infection and evolutionary dynamics of HIV within a host, spatial aspects should be considered while modeling and analyzing HIV infection.

3 Analyzing HIV Viral Dynamics: The Contribution of ODE Models

3.1 ODE Models of HIV Infection

The analysis of infection processes by HIV requires the consideration of virus particles, i.e., virions, and cells susceptible to infection, i.e., target cells, in which the virus can replicate and produce new virions. In mathematical terms, these interactions that occur during infection can be described by a system of ODE as formulated in Eq. (1):

$$\begin{aligned}
\frac{\mathrm{d}T}{\mathrm{d}t} &= \lambda - dT - \beta VT, \\
\frac{\mathrm{d}I}{\mathrm{d}t} &= \beta VT - \delta I, \\
\frac{\mathrm{d}V}{\mathrm{d}t} &= pI - cV.
\end{aligned} \tag{1}$$

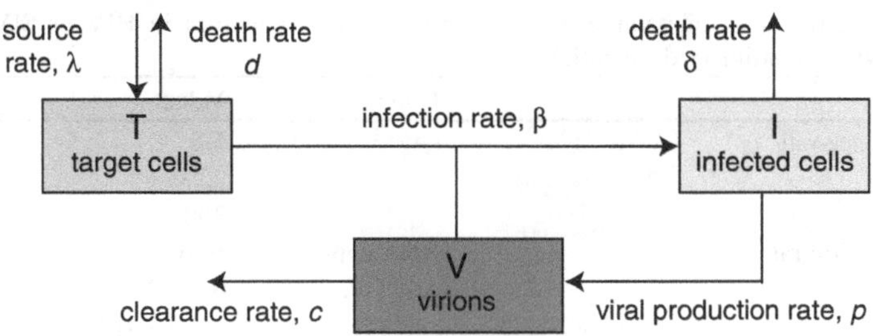

Fig. 3 Basic model of viral dynamics: target cells, T, are infected by virions, V, with a rate constant β. Infected cells, I, produce new virions at rate p. Virions and infected cells are lost with clearance rate c and death rate δ, respectively. New target cells are generated at a rate λ and target cells have an average lifetime of $1/d$. Typical values of the model parameters are given in Table 1

Equation (1) represents the basic model of viral dynamics as described by Nowak and Bangham [89], Nowak and May [91], Perelson [101], and Perelson et al. [104]. Here, T denotes the concentration of target cells, I the concentration of infected cells, and V the viral load. Target cells are created at a constant rate λ and have an average lifetime of $1/d$. Target cells can become infected by virions at a transmission rate β, and the resulting infected cells die with rate δ per cell. Productively infected cells produce new virions, which have an average lifetime of $1/c$, at rate p per cell (see Fig. 3 for a sketch of the model).

The model presented in Eq. (1) has been extended by including (1) the incorporation of an immune response, humoral as well as cell-based [91], (2) the separation between long- and short-lived infected cells [105], and (3) the consideration of an incubation period before an infected cell starts the production and release of new virions [86, 103].

The application of models based on ODEs, such as the one in Eq. (1), to experimental and clinical data has improved our knowledge about the dynamics of viral infections in several ways. The estimation of viral production and clearance rates has shown that chronic infections in which the viral load attains a set-point, such as HIV, hepatitis C virus, and human cytomegalovirus (CMV) [31, 87, 98], are maintained by high rates of viral replication and equally high rates of viral clearance [51, 89, 93, 109]. However, the estimates for the clearance rate constants vary quite substantially depending on the experimental or clinical data that are analyzed. It has been estimated that HIV virions have a very short half-life and are cleared from the circulation with an average clearance rate of $c = 23\,\text{day}^{-1}$ [109]. In contrast, the clearance rate of SIV, the monkey equivalent of HIV, was estimated to be $c = 5.5\,\text{day}^{-1}$ in lymphoid tissue [110] and $c = 300\,\text{day}^{-1}$ in the blood [128]. Partly influenced by those estimates, the estimated number of virions produced by one infected cell per day varies quite substantially as well, ranging from 100 [29, 44] to 650–3400 [110] and up to the order of $2-5\times 10^4$ [21] viral particles per infected cell per day. A lower bound for the total viral production in an infected patient was estimated to be around 10^{10} virions per day [104].

Table 1 Quantification of parameters to describe disease dynamics in HIV or SIV based on models relying on ordinary differential equations

Parameter	Unit	Value	Reference
Viral clearance rate, c	day^{-1}	5.5	[110]
		23	[109]
		300	[128] *SIV*
Viral production rate, p	RNA copies cell^{-1} day^{-1}	$\approx 10^2$	[29, 44]
		$0.7-3.4\times10^3$	[110]
		$2-5\times10^4$	[21] *SIV*
Source rate of CD4$^+$ T cells, λ	μl^{-1} day^{-1}	8	[125]
Death rate of CD4$^+$ T cells, d	day^{-1}	~0.01	[111]
Death rate of productively infected cells, δ	day^{-1}	1	[69, 104]

Values are taken from the corresponding publications

3.2 *Shortcomings of ODE Models in Spatial Situations*

As suggested by the estimates above, one can expect infection dynamics to vary among different anatomical compartments [27]. However, most of the infection occurs in solid tissue, such as in lymph nodes, the spleen or other organs in which target cells are plentiful [7, 9]. In solid tissues, local effects such as the focal release of virions from infected cells, as observed *in vivo* [44, 76], may play an important role in the infection dynamics. Once virus is released it diffuses and will preferentially infect nearby cells. In general, ODE models are not able to capture this type of spatial aspect.

In ecology and epidemiology, there is a large body of literature studying the influence of spatial structure on dynamics [28, 39, 58, 121]. The space being analyzed need not be Euclidean space but can also be a tree or more general graph representing interactions in a social network [59, 74, 88] or meta-populations [39, 58, 95]. Levin and Durrett [64] and later Webb et al. [124] analyzed the circumstances under which the results of a spatial model would differ from a mean-field model, such as that given by a simple ODE model of population dynamics which assumes a well-mixed population. For their spatial model, Webb et al. [124] examined an undirected regular network of sites which can be either empty or occupied by a susceptible, infected, or resistant individual. They found that dependent on the degree of spatial connectivity and the underlying ecological situation the prediction of the spatial and mean-field models can differ substantially given similar parameter ranges. Similar studies have been made in related fields such as statistical mechanics or genetics (reviewed in [30]).

In the following sections, we will review several methods for including spatial aspects into the modeling of infection dynamics within a host with a specific emphasis on HIV viral dynamics.

4 Getting More Spatial: From ODE to PDE

4.1 Spatial Discrete ODE Models

One way to include spatial aspects into ODE models for HIV infection dynamics on a macroscopic scale is to use compartment models. These models distinguish between different anatomical compartments in which HIV can infect cells and replicate. The published models of this type developed by De Boer and coworkers consider different sites of HIV replication and clearance, e.g., lymphoid tissue, blood, plasma and other organs such as the liver or the lung, and the interchange of viral particles among these compartments [27, 85]. Applying these models to experimental data, De Boer et al. found that the clearance rate of free virus in lymphoid organs should be rapid and that the previously estimated clearance rates measured in the blood most likely correspond to the efflux of virions from the blood to other organs [27]. By considering different compartments for viral replication and clearance, De Boer et al. were able to explain previously estimated differences in viral clearance rates as found in [109, 127, 128].

Another way to combine an ODE model with spatial aspects was introduced by Funk et al. [34]. In this case, the authors did not specifically consider different compartments and their physiological interdependence. Instead, the authors assumed a well-defined two-dimensional grid with $n \times n$ grid sites for a predefined total volume. Each grid site could represent different anatomical sites inside the host or different spatially adjoining compartments inside a specific organ, such as a lymph node. At each of the different grid sites (i, j), $i, j = 1, \ldots, n$, the concentration of target cells, $T_{i,j}$, infected cells, $I_{i,j}$, and the viral load, $V_{i,j}$, were described by ODE according to the basic model of viral dynamics in Eq. (1). Target cells and infected cells were assumed to be stationary, while virions were allowed to migrate from one grid site to a neighboring site. Virion movement is accounted for by an additional term in the virus equation [Eq. (2)], where m_V denotes the diffusion rate of free virions from site (i, j) to the eight nearest neighboring sites [34].

$$\frac{\mathrm{d}T_{i,j}}{\mathrm{d}t} = \lambda - dT_{i,j} - \beta V_{i,j} T_{i,j},$$

$$\frac{\mathrm{d}I_{i,j}}{\mathrm{d}t} = \beta V_{i,j} T_{i,j} - \delta I_{i,j},$$

$$\frac{\mathrm{d}V_{i,j}}{\mathrm{d}t} = pI_{i,j} - cV_{i,j} - \frac{m_V}{8} \sum_{i_0=i-1}^{i+1} \sum_{j_0=j-1}^{j+1} \left[V_{i,j} - V_{i_0,j_0} \right]. \tag{2}$$

Varying the degree of spatial coupling between the different sites by changing the value of the parameter m_V, the authors compared the outcome of this model to the basic population dynamics model defined by Eq. (1). They showed that given a homogeneous spatial environment, meaning that each of the different rate constants

(d,β,δ,p,c) is the same on each site of the grid, the dynamics differed from the one observed in the basic ODE model, especially early in the infection where the spatial model predicts a rapid increase of the virus load followed by a slower increase as target cells are depleted in some regions of the grid whereas the nonspatial model only shows a one-phase increase (see Fig. 2 in [34]). The authors suggest that due to this deviation parameters estimated using the basic ODE model might contain systematic errors [34].

On a heterogeneous grid, where the rate constants for each grid site are sampled randomly from a uniform distribution, spatial coupling between the different sites can change the equilibrium properties of target cells compared to the basic model of viral dynamics. However, increased spatial coupling between the sites leads to a more averaged outcome similar to the mean-field approximation given by a model with a well-mixed assumption for the different populations: If spatial coupling is high, the transport rate m_V for virions between the different sites has to be high, indicating that the average virus load in the neighborhood of a grid site has a higher influence on the equilibrium viral load at this site than more distant sites. Increased spatial coupling will lead to a smoothed viral load between the different sites ([34] and Fig. 4).

In addition to their basic model, the authors extended their model by considering an immune response targeted against infected cells. Similar to the virions, immune cells were assumed to move among the different sites of the spatial grid. Their spatially explicit model equilibrated much faster, with more damped oscillations, compared to a version of the basic model of infection dynamics that incorporated an immune response [91]. The risk that the infection persists during the invasion phase was markedly increased even for parameter regimes where the nonspatial models would predict extinction.

Similar to the consideration of spatial aspects for viral replication, viral clearance, and the dynamics of target cell populations, spatial heterogeneity can also be included when modeling the dynamics of specific immune responses. Similar to Funk et al. [34], Louzon et al. [68] studied the influence of spatial heterogeneity in a model capturing the activation and proliferation of lymphocytes. They defined a simple deterministic model considering two different populations: antigen, A, which enters the system at rate λ and is cleared at rate d_A, and lymphocytes, L, which proliferate upon activation by antigen with a constant rate r and die with death rate d_L [Eq. (3)].

$$\frac{dL}{dt} = rAL - d_L L,$$
$$\frac{dA}{dt} = \lambda - d_A A. \tag{3}$$

The authors showed that according to this ODE model the lymphocyte population will either grow exponentially or decrease to zero if the antigen concentration is either above or below a certain threshold, respectively. In a second step, the authors

1.) Viral dynamics modeled by systems of ODEs on a spatial grid for different degrees of spatial coupling between the neighboring sites

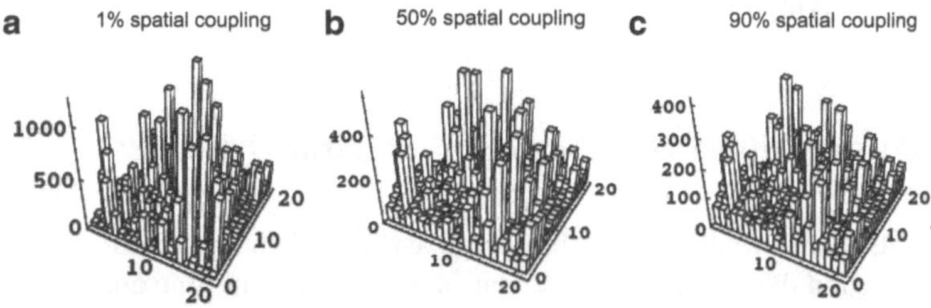

2.) Sketch of a 2D cellular automaton modeling infection in solid tissue

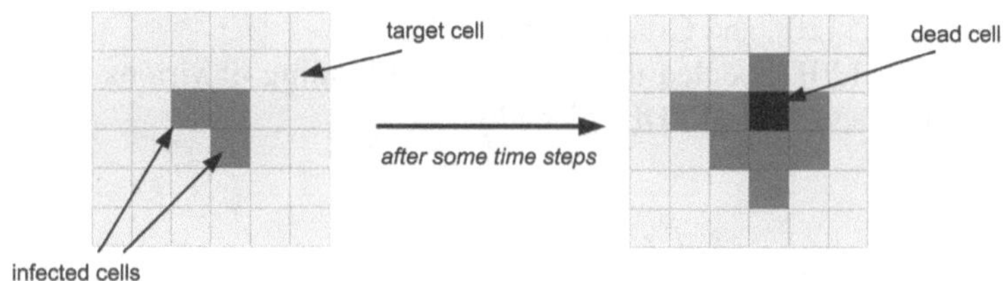

3.) Spread of infection in solid tissue modeled by a 2D cellular automaton

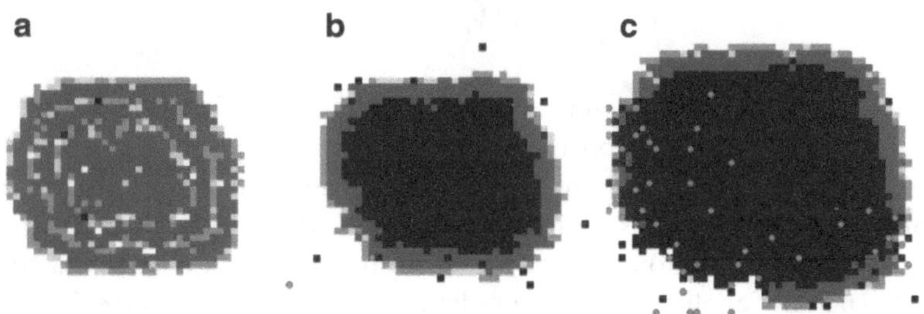

Fig. 4 Modeling spatiality in viral dynamics: (1) Spatial grid where viral dynamics on each of the different grid sites is described by a model based on ordinary differential equations [see Eq. (2)]. The pictures show the viral load at equilibrium on each of the different grid sites for different degrees of spatial coupling between the sites (Reproduced with permission from [34] © (2005) Elsevier Ltd.). (2) Sketch of the development of an infection in solid tissue modeled in a 2D cellular automaton [9]. Target cells can become infected by their infected neighboring cells that can infect other cells or die. Modeling infection by cell-to-cell transmission will lead to a wavelike pattern for the spread of the infection. (3) Snapshots of a simulation modeling spread of infection in a solid tissue with a 2D cellular automaton using different rules for the replacement of dead cells [8]: (**a**) replacement of cells independent of the location of uninfected target cells, (**b**) replacement of dead cells by proliferation of neighboring uninfected cells, (**c**) immune cells have to breach the infection wave to allow the replacement of dead cells. Cells shown in the screenshots are either uninfected (*white*), dead (*black*), infected or represent immune cells (Reproduced with permission from [7] © (2009) Elsevier Ltd.)

discretized the lymphocyte and antigen concentration in space using a regular grid as done by Funk et al. [34]. Instead of using rates, the authors calculate the probability for each reaction at each grid site and randomly perform the proliferation or death of lymphocytes according to these probabilities. Using this stochastic simulation, they observed the development of local hot spots for lymphocyte proliferation on the spatial grid even if the total antigen concentration was below the ODE predicted threshold for lymphocyte proliferation.

4.2 *Partial Differential Equations in Viral Dynamics*

Another approach to describing the spatial aspects of viral infection is to use PDE, which describe the change of cell or virion populations in time and space. Models based on PDEs eliminate the arbitrary spatial discreteness introduced by the ODE models. PDE models of viral infection dynamics have been developed by Frank [32], Strain et al. [120], and Graziano et al. [38].

Strain et al. [120] modeled the diffusion and binding of virions to target cells within a host by a system of PDEs given by

$$\frac{\partial V_{\mathrm{f}}}{\partial t} = D\Delta V_{\mathrm{f}} - \beta T V_{\mathrm{f}} - cV_{\mathrm{f}},$$
$$\frac{\partial V_{\mathrm{b}}}{\partial t} = \beta T V_{\mathrm{f}}. \tag{4}$$

Here, $V_{\mathrm{f}}(x,t)$ and $V_{\mathrm{b}}(x,t)$ denote the concentrations of free and bound virions, respectively, and $T(x,t)$ the concentration of available target cells. D is the diffusion coefficient of free virions in the system, and β and c the transmission rate and clearance rate of free virions, respectively. As done by Funk et al. [34], Strain et al. [120] used their model to compare the mean-field expectations generated by this model to spatially explicit *in silico* simulations of HIV infection (*see* Sect. 5).

Graziano et al. [38] used a finite element (FE) method to study lymphocytes and viral load as a viscous incompressible fluid occupying a two-dimensional rectangular area. Their aim was to examine the effect of the spatial distribution of T cells and the HIV viral load on HIV progression during an infection and to include the effects of therapy with a reverse transcriptase inhibitor (RTI). This FE-approach is comparable to the approach proposed by Funk et al. [34]. The modeled area is subdivided into a predefined number of subregions, the finite elements, which do not have to be of regular size. $T(x,y,t)$ and $V(x,y,t)$ are defined as the solutions for the CD4$^+$ T cell count and the viral load, respectively, for each position (x,y) in the modeled area at time point t given specific conditions for the boundaries of the area. These functions are then discretized to approximate the solution for each FE-element. The authors compared the solutions of their model to clinical data and by this determined the robustness of their model. They showed that there was no significant difference between the predicted half-life of the CD4$^+$ T cell count and the overall half-life estimated from1,500 patients. Using their approach, they showed that the speed of HIV progression depends on the initial viral distribution in the considered grid. Therapy has to inhibit viral accumulation in local regions in order to be effective.

The models proposed by Funk et al. [34] and Graziano et al. [38] aim to include spatial information into the analysis of HIV infection. However, they still rely on ordinary or PDE. PDEs are more computationally challenging than ODEs, making it more difficult to determine general behaviors of the system. Furthermore, similar to ODE models, they only provide an average or mean-field description of the system behavior [7]. Because of this, there are only a limited number of studies applying PDEs to within-host dynamics. In virus dynamics, PDE models have also been used to develop age-structured models of infection that keep track of the time a cell has been infected, i.e., its age of infection [2, 61, 86]. These models add realism in that they can account for the fact that when a cell is first infected no virus is produced until a number of steps in the viral life cycle have been completed. Thus they allow the rate of virus production to depend on the age of an infected cell. The models can also allow the death rate of an infected cell to depend on its age.

4.3 Pair Approximation

Ordinary and PDE are so-called mean-field approximations of the dynamical system that both kinds of model systems try to capture. In short, the solutions of both model systems are based on the following assumption: If our representation of space is a lattice with a total of S sites, where the status of each site $i \in S$ is denoted by σ_i, then the probability that the two neighboring sites $i, j \in S$ with status $\sigma_i\sigma_j$ change their status into $\sigma_i^{\backslash}\sigma_j^{\backslash}$ is given by

$$r_\sigma(\sigma_i\sigma_j \rightarrow \sigma_i^{\backslash}\sigma_j^{\backslash}) = r_\sigma(\sigma_i \rightarrow \sigma_i^{\backslash})r_\sigma(\sigma_j \rightarrow \sigma_j^{\backslash}). \quad (5)$$

This assumption, used in mean-field models, means that each site is considered to change its status independently of the others. However, as local interactions at different sites of the lattice might occur, this approximation might be too crude in some circumstances. The so-called pair approximation approach represents an improvement to the mean-field approximation by considering local interactions between neighboring sites. In this case the probability that the two neighboring sites $i, j \in S$ with status $\sigma_i\sigma_j$ change their status into $\sigma_i^{\backslash}\sigma_j^{\backslash}$ depends on the status of all pairs of neighboring sites in the direct neighborhood of i and j, E_{ij}:

$$r_\sigma(\sigma_i\sigma_j \rightarrow \sigma_i^{\backslash}\sigma_j^{\backslash}) = r_\sigma(\sigma_i\sigma_j \rightarrow \sigma_i^{\backslash}\sigma_j^{\backslash}|\sigma_{E_{ij}}). \quad (6)$$

The application of this kind of modeling approach to HIV infection dynamics might be relevant especially if we consider each grid site to represent a specific target cell and also include the fact that HIV can spread by cell-to-cell transmission [117]. In biology the pair approximation has so far only been applied to epidemiological and ecological problems [54, 64], and in these applications gives consistently better results than the mean-field approximation in describing the dynamics of a stochastic system.

The pair approximation can be thought of as being in between the mean-field approximation of a dynamical system and its stochastic and individual simulation as done by agent-based models, which are introduced below.

5 Cellular Automata and Agent-Based Models

Several stochastic *in silico* simulation tools have been applied to the study of viral dynamics within a host [7, 9, 18, 120, 130]. *In silico* simulations tools, such as cellular automata and agent-based models, allow one to model individual units of investigation, e.g., viruses, cells, patients, or animals, in a spatially explicit way. These types of simulation tools have been used to study an epidemic caused by a pathogen inside a population or on an agricultural farm (see, e.g., [57] and references therein).

Cellular automata and other agent-based models have three main advantages in describing complex biological systems compared to ODE or PDE models: First, the behavior of different elements in a complex biological system can be sometimes more easily described in terms of rules of interaction rather than by specific rate constants that have to be determined and that usually average the individual behavior over the whole population. Second, due to these rules, these types of models easily incorporate nonhomogeneous spatial information. Lastly, due to their implementation, cellular automata and agent-based models usually include stochasticity. In ODE and PDE-models, stochasticity needs to be explicitly implemented by assuming specific probability distributions for certain rate constants. However, the ODE and PDE models will still only represent the average dynamics of the modeled system, while in agent-based models the same starting conditions can lead to different outcomes. Agent-based models add an additional layer of complexity to the mean- field approaches by trying to mimic the underlying biological system in more detail.

5.1 Cellular Automata

Cellular automata can be considered as a specific type of agent-based model. In a cellular automaton, cells are organized in a 2D or 3D grid. Each site on the grid represents a cell or a comparable predefined unit. The position of the cell is fixed with a well-defined distance to its neighboring cells (see also Fig. 4). According to environmental conditions and predefined rules, the status of a cell changes during a simulation run. Cellular automata were first introduced by von Neumann and

Burks [123]. Since then, the mathematical properties of cellular automata have been intensively studied and are well described, e.g., [126]. Several studies have applied the concept of cellular automata to questions in the field of viral dynamics and immunity [9, 16, 18, 37, 75, 96, 97, 120, 130]. In the following we discuss some of them.

Cellular Automata and the Study of HIV Viral Dynamics

Cellular automata have been used to study HIV infection dynamics using either two-dimensional [12, 19, 120, 130] or three-dimensional representations of the modeled environment [96, 97]. In the following, we will present and discuss two studies in detail that applied two-dimensional cellular automata to the case of HIV infection.

Zorzenon dos Santos and Coutinho [130] studied the dynamics of HIV infection in a lymph node using a two-dimensional lattice. Each site on the lattice represented a $CD4^+$ T cell or a monocyte, both of which are target cells for HIV. Each cell could be in one of four different stages: (1) uninfected, (2) productively infected, (3) infected but in a final stage before the cell dies due to the action of the immune system, and (4) dead. Uninfected cells become infected dependent on the number of infected cells in the neighborhood and can be killed by an immune response after a certain number of time steps τ. The parameter τ represents the time the immune system needs to generate a target-specific immune response. With a certain probability, dead cells are replaced by uninfected cells, representing new targets which can then become infected. By starting the infection with a few infected cells, the authors found a parameter regime that recapitulates the three-phase dynamics observed in HIV infection (*acute-chronic-AIDS*) without varying the underlying parameters during the simulation. The model also showed that the infection spread in a wave form pattern through the simulated layer of cells as expected from the cell-to-cell transmission of the virus implemented in the model. Zorzenon dos Santos and Coutinho [130] also assumed that each newly infected cell carries a different lineage of the virus. Using this assumption, the authors incorporate the high mutation rate of HIV [102] into their model. Because each infected cell is assumed to carry a different viral genome the time the immune system needs to generate an immune response against an infected cell is assumed to be the same for each newly infected cell. However, this assumption might not be true. It might be also likely that due to cell-to-cell transmission clusters of infected cells are dominated by one infecting viral strain. If that is the case, then the effect of $CD8^+$ T cells, which are the immune cells responsible for recognizing and killing infected cells, might be faster in some local environments, possibly terminating the spread of the disease in this location.

In the second study, Strain et al. [120] also used a two-dimensional cellular automaton to study HIV infection dynamics in a layer of cells and compared the outcome to the results of a model formulated by ordinary differential equations [see Eq. (1)]. Similar to Zorzenon dos Santos and Coutinho [130], each site in the cellular automaton represented a T cell. Based on the diffusion rate of virions and their rate of encounter and attachment to T cells, the authors calculate the probability of a T cell to become infected. This probability depends on the distance to a virus producing cell. So, in contrast to Zorzenon dos Santos and Coutinho [130], the target cell does not have to be in the direct neighborhood of a virus producing cell in order to become infected. In the model, the authors considered the spread of the infection to be mainly driven by diffusion of free viral particles rather than cell-to-cell transmission as assumed in [130]. The initialization of the infection with one infected cell in the center of the lattice results in a radial wave of infection moving to the boundaries of the grid. Dependent on the replacement rate of dead cells, the long-term spatial pattern of the infection differs. If the replacement rate is high, virus from the wave front can diffuse back leading to a chaotic steady state with coexistence of target, infected, and dead cells. On the other hand, if the replacement rate is low, a single wavefront moves to the boundaries of the lattice and the system eventually recovers to a lattice full of target cells. In their model virus released in a burst would only spread in the direct neighborhood of the bursting cell. Because of this, infectivity depends on the concentration of T cells as neighboring cells can become infected more easily if they are tightly packed. Compared to their model, the authors found that a spatially averaged model, such as an ODE model [Eq. (1)], would overestimate viral infectiousness by more than an order of magnitude [120]. In the infected steady state of their cellular automaton, the authors found that for most parameter values about half of the $CD4^+$ T cells were infected, which is higher than actual observations from clinical data suggest [41, 116]. However, this might be due to the fact that their model neglects any type of immune response, such as those generated by $CD8^+$ T cells, which might interfere with the spatio-temporal dynamics of the infection. Furthermore, they assume the $CD4^+$ T cells, i.e., the target cells for HIV, are fixed in space. This is not a realistic assumption as T cells are highly motile [73, 77–80]. Clearly, if both virus and infected T cells move the infection dynamics will be affected in ways that have not been studied.

Cellular Automata and the Modeling of Infections in Solid Tissue

Beauchemin [8] developed a two-dimensional cellular automaton to study the effect of viral infection spread in a solid tissue [8, 9]. Instead of HIV, the Beauchemin model was applied to influenza A virus. Each site of the grid represented a target cell that can become infected by influenza at a rate dependent on the number of infected cells in the direct neighborhood of the site. Additionally, immune cells, which are able to recognize and kill infected cells, move through the modeled tissue environment. While in the two studies presented above, infection was initialized by a single infected cell, Beauchemin [8] investigated how different distributions of infected cells affect the spread of the infection in the tissue. Beauchemin [8] distinguished between infected cells randomly distributed over the grid, arranged in small isolated clusters or as a single large cluster. As infected cells can only infect their immediate neighbors, the initial distribution of infected cells has a large effect on the dynamics of the infection. If an infected cell is already part of a large cluster of infected cells then most of its neighbors are already infected. This leads to a lower effective infection rate than if infection were initiated with single infected cells where all the neighbors are infectable. However, the effective infection rate can be kept high if one allows for occasional jumps of the infection to areas where most of the cells are still uninfected. With the same argument, the type of replacement of dead cells influences the infection dynamics. A global regeneration rule in which a dead cell is replaced by an uninfected cell placed randomly on the grid, as assumed in the basic ODE model [Eq. (1)], would allow for the appearance of new uninfected cells in completely infected areas which would increase the number of infected cells. On the other hand, if the replacement of dead cells depends on the proliferation of uninfected cells in the direct neighborhood of a dead cell, the infection propagates in thin circular waves as this regeneration rule limits the growth of the infection by starving it of target cells. In this scenario, the replacement of dead cells left behind by the propagating wave of infection can only occur if immune cells detect the infection and breach the infection wave. Beauchemin [8] could show that the fraction of dead cells over time generated by this model closely agreed with experimentally observed curves [8, 14] in contrast to the result obtained by an ODE model where the replacement only depends on the number of uninfected cells and not on their location.

5.2 Agent-Based Models

Cellular automata represent a specific type of agent-based model where cells, which can change their status, are positioned at spatially fixed sites. However, the simulation of cells does not need to be constrained on a regular grid with cells of regular shape. Rather, each cell can be followed as an individual agent. As advanced experimental techniques, such as two-photon microscopy, provide additional insights into the dynamics of infection processes in vivo and computational power increases, the simulation environments described above have become more and more detailed. Bogle and Dunbar [15] used a simulation environment where cells move on a grid to simulate T-cell motility in a lymph node. The authors show that this simulation environment was able to reproduce experimental observations. Graw and Regoes [37] used a similar approach to study how the killing of infected cells by cytotoxic $CD8^+$ T cells in lymphoid organs is influenced by spatial aspects and what type of mathematical models should be used to analyze experimental data. In the following, we present several novel types of agent-based models that have been applied to the study of HIV infection dynamics and other aspects in immunology.

Agent-Based Models for HIV

Lin and Shuai [65] extended the modeling approach originally presented by Zorzenon dos Santos and Coutinho [130] to study the interplay between HIV disease dynamics and the immune response. In their simulations, they include HIV virus particles, $CD4^+$ T cells not only as the main target cell for HIV but also as helper cells for the effective action of B cells, and $CD8^+$ T cells as part of the immune response. $CD8^+$ T cells are immune cells that are able to detect and kill HIV infected cells, thereby limiting the number of virus producing cells. B cells produce HIV-specific antibodies that can bind diffusing HIV virions, making them unable to enter and infect further target cells. In their study, the authors use a two-dimensional grid with periodic boundary conditions where each grid site can be occupied by several cells but only one of each cell type. Each of the different elements, HIV virions, $CD4^+$ T cells and $CD8^+$ T cells, is considered to be motile and can change their position during a time step of the simulation. Additionally, the authors modeled HIV virions as a binary string and allow for HIV mutation. Similar to [130], the authors found a parameter regime that was able to reproduce the three-phase dynamics of HIV infection observed in individual patients. Based on their results, they postulate that the viral mutation rate is largely responsible for the HIV dynamics. Increased viral diversity leads to instability of the system and, after overcoming a certain threshold, to the inability of the immune system to control the viral growth. This observation is in line with previous studies which suggest that the antigenic diversity of the within host HIV virus population is the cause and not a consequence of the development of AIDS [90, 92].

The simulation framework presented by Lin and Shuai [65] represents a detailed model of the different populations involved in HIV infection. However, despite the increased complexity, the model of Lin and Shuai [65] has some shortcomings that might influence the validity of their results. For example, they do not distinguish between the volume occupied by one cell or one virion and assume the same diffusion dynamics for both types despite their different sizes. Without experimental validation of the chosen parameter regimes, the results of such models have to be taken with care.

Agent-Based Models and Immunity

Agent-based models have not only been applied to the study of HIV infection dynamics but also to many other aspects of the immune system affected by spatiality. While the models presented previously mainly determined the population dynamics of infected and uninfected target cells given spatial structure, recent studies model and analyze the actual motility and migration dynamics of individual immune cells in detail [10, 15]. As two-photon microscopy allows one to observe immune processes on a single cell level *in vivo*, these models can be validated by experimental observations. They also provide important insights into the activation dynamics of virus specific immune responses. Application of these realistic modeling environments to the case of HIV infection might improve our understanding of the early processes of HIV infection where migration of cells plays an important role (*see* Sect. 2).

A spatial simulation of T cell motion based on experimental observations was presented by Beltman et al. [10]. The authors used a cellular Potts model to describe T cells and used it to study the activation of naïve T cells by their cognate antigen in a lymph node. In the cellular Potts model presented by Beltman et al. [10] T cells are represented by several sites on a grid. While the volume of the cell is kept constant, the shape of the cell can vary during a simulation. Interaction of cells in a densely packed environment causes the movement of cells in certain directions. Beltman et al. [10] considered T cells and dendritic cells in their model, as well as the stromal cell environment within a lymph node. By adjusting the rules with regard to the movement direction of simulated T cells, the authors were able to reproduce motility characteristics consistent with those experimentally observed for T cells *in vivo* [73, 77–80]. Without explicit modeling, they were also able to observe the organization of small T cell streams along fibers of the stromal network in their simulation which is in line with experimental observations.

Bogle and Dunbar [16] studied the activation and proliferation of motile T cells in a lymph node. In contrast to their previous approach [15] and the other spatial models presented so far, the authors allow for a dynamic simulation environment: The size of the simulated deep cortical unit of a lymph node is allowed to expand and contract to account for the in- and outflow of T cells as well as their proliferation. The authors show that their modeling platform can reproduce realistic motility characteristics of T cells and their response to antigen.

One of the most ambitious projects to simulate immune dynamics and interactions is the `ImmunoGrid` - project funded by the European Commission [45, 99]. The project was established in 2006 as part of the `virtual physiological human` (VPN) initiative. The aim is to develop an *in silico* model of the entire immune system which could be applied to clinical practice in order to make personalized suggestions for therapy and interventions in different types of diseases such as cancer. `ImmunoGrid` is still in its infancy facing challenges that include the computational infrastructure as well as the complexity of the immune system itself (intermediate report in [45]). However, some studies done as part of this project have already generated results. Baldazzi et al. [6] and Castiglione et al. [20] investigated the effect of HAART on HIV disease dynamics using an agent-based model. As HAART is associated with side effects and requires a high level of discipline since patients need to take drugs regularly for years, designing therapeutic regimes is an area of interest. Baldazzi et al. [6] use an agent-based model to determine effective therapeutic regimes for HAART, even allowing interruption of treatment. The efficacy of therapy is allowed to vary during the simulation due to the rise of drug-resistant viral strains. In Castiglione et al. [20], the authors use this framework to find an optimal regime for HAART with the help of a genetic algorithm. Therapy was applied to virtual patients for a period of 6 months starting 7.5 years after infection. After therapy was terminated, simulated patients were challenged with an opportunistic infection. The authors compared their determined optimal therapy strategy with interruption of treatment to other regimes including the "ideal" scenario where HAART is applied to the patient each day without interruption. They found that after opportunistic infection, the survival of patients who had the "ideal" therapy was 1.14 times more efficient than those receiving the optimal therapy with interruptions. However, the optimal strategy they found with their genetic algorithm uses approximately 40% less drug than the full therapy.

The application of agent-based models to problems in infection and immunity is still in its infancy. The appropriateness of each model and simulation environment can only be validated given appropriate experimental observations. Unfortunately, some models and especially those that follow a large number of different cell populations [16, 66] require a large number of parameters, many of which have not been estimated from experimental data so far. As more interest develops in building quantitative models and with the development of new experimental techniques (*see* Sect. 6) this gap hopefully will be filled. This will allow the implementation of enhanced simulation tools and even the aims of the `ImmunoGrid`-project might be reached.

6 Conclusion and Discussion

Mathematical models based on ordinary differential equations have taught us much about the HIV infection dynamics (reviewed in [101]). These kinds of models helped to quantify rates at which HIV replicates, rates at which target cells, such as $CD4^+$ T cells, proliferate, and HIV virions or infected cells are cleared from the system. Variation in parameter estimates obtained using different models as well as advances in experimental techniques suggest that the simplistic view provided by ODE models, although sometimes useful, may need to be reconsidered in other situations. For example, the well-mixed assumption of interacting cell populations as assumed in the basic ODE models for virus dynamics [89, 91, 101, 104] has to be revised in order to model infection in tissues.

In Sect. 2 of this chapter, we outlined the importance of spatial factors throughout the different stages of the HIV infection process. Novel experiments broadened our view of the first steps of HIV infection [43]. However, there is a growing interest in obtaining a better understanding of the way in which HIV establishes infection, particularly with regard to its interplay with the different immune responses and the local tissue structure [43]. The mucosal barrier of the genital tract, the main portal of HIV entry, shows heterogeneous vulnerability to infection [43]. Proximity and constant supply of susceptible target cells close to the site of viral entry is another important factor determining the successful establishment of an infection [43, 47]. But what is the success rate of viral entry? How many target cells are needed to establish an infection at the site of entry? How do they have to be distributed? And in the end, the ultimate goal is, how the establishment of HIV infection can be prevented. While it is generally possible to formulate the different factors mentioned above in terms of stochastic events incorporated into an ODE model, determining the corresponding rate distributions provides a heavy challenge for the modeler given the current experimental data. Advances in experimental techniques, such as two-photon microscopy, have made it possible to visualize the dynamic behavior of immune cells in living tissue such as lymph nodes and the spleen [1, 17, 67, 119]. Other techniques besides two-photon microscopy such as photoactivated localization microscopy or structured illumination microscopy allow the visualization of immune cell activation processes even on a subcellular level [5]. The interaction of HIV-1 proteins with host cell molecules can be well addressed using novel fluorescent imaging techniques [25]. Agent-based models can be used to simply mimic the observed behavior of immune cells and virions by this kind of experiments [10, 15]. Using the resulting simulation frameworks and varying factors such as susceptible cell density or vulnerability of mucosal and epithelial barriers make these kinds of models suitable to address the questions mentioned above.

Spatially explicit models, such as agent-based models, will also help us to address the question of under what conditions infected cells occur in clusters. What role does cell-to-cell transmission play compared to diffusion of viral particles? Does it vary during different stages of infection? These analyses can reveal whether in a lymph node densely packed with susceptible target cells for HIV specific immune responses might play a more important role for the containment of infection than, for example, in blood where susceptible cells are more widely dispersed. The detailed spatial structure of specific organs might also explain the different HIV clearance dynamics observed in lymphoid tissue, blood, plasma, lung, and liver [27]. It might help to explain why the $CD4^+$ T cell population in mucosal tissue is massively depleted during the first weeks of infection [107], while this loss is not reflected in the peripheral blood [40].

Another advantage, and need, for agent-based models is the study of the evolution of HIV while spreading through the host. Although current evidence indicates that HIV infection can be founded by a single viral strain [112], the high replication rate of HIV leads to tremendous variability of HIV viral strains inside a single patient later during infection [62]. This diversity challenges the responding immune system, leaving it unable to control viral replication at some point. Agent-based models including spatial information allow one to follow individual cells and virions, making it possible to observe the development of clusters of infected cells containing the same viral variant. These models can be used to study the evolutionary pressure of different kinds of immune responses (innate, cell-mediated, and humoral) on HIV and, hence, on the overall disease progression and development.

The spatial models presented in the previous sections, which have already been applied to study HIV infection dynamics, have analyzed how HIV dynamics differ in different compartments [27], how HIV infection spreads in solid tissue, and how spatial patterns of infected cells change in relation to cell turnover [120, 130]. They showed that viral infectiousness can be much lower than estimated previously if spatial structure is considered [120]. However, there is still a way to go to include all the different aspects from the establishment of infection to the activation of immune responses.

In summary, we would propose that spatially explicit models should be used:

- To model the processes involved in the early stages of HIV infection in order to determine important factors for the establishment of infection, and to test and propose appropriate intervention methods.
- To understand the importance of cell-to-cell transmission of HIV relative to that of infection by cell-free virus.
- To explain the different HIV infection dynamics in different anatomical compartments and to reveal the importance of specific compartments during the progress of infection.
- To develop the "full picture" of the HIV disease dynamics by combining each detailed aspect of the infection and immune processes as it is intended by the `ImmunoGrid`-project [45, 99].

Advances in computational and mathematical techniques have made it possible to simulate complex biological processes in detail. This will allow us to address the computational challenging tasks listed above. However, all these simulation studies need to be accompanied and validated by experimental data. The already mentioned two-photon microscopy technique has made it possible to investigate infection and immune processes on a viral and cellular level in living tissue. It has revealed the motility characteristics of naïve and activated T cells inside a lymph node [73, 77–80] and in the thymus [63, 106], and the importance of the stromal network of fibroblastic reticular cells for the motility and proper activation of T cells [3, 4, 35, 84]. Two-photon imaging has been applied to study the interaction between various hosts and pathogens (reviewed in [24]), and resulting experimental data have already been used to enhance and to corroborate computational models of the immune system [10, 37]. There are several other promising techniques which might enhance our view of immune dynamics in the near future [5]. However, despite these advances in observation technology, the newly gained data have to be accompanied by careful mathematical analysis [11]. Although *in vivo* imaging techniques give us an impression about the dynamics of immune cell interactions in living tissue, artifacts might interfere with the robustness of the conclusions one can make. For example, as two-photon microscopy only allows one to observe a small fraction of the lymph node over time, fast moving cells move out of the field of observation more readily than slow moving cells, which can bias estimates of the mean motility and residence times [11].

Nevertheless, sophisticated computational models supported by experimental data can help to resolve some of the "known unknowns" in HIV disease dynamics that are currently preventing the development of a successful vaccine against this threatening disease [122].

Acknowledgements Portions of this work were done under the auspices of the U.S. Department of Energy under contract DE-AC52-06NA25396 and supported by the Center for HIV/AIDS Vaccine Immunology and NIH grants AI028433 and OD010095.

References

1. Allen, C.D., Okada, T., Tang, H.L., Cyster, J.G.: Imaging of germinal center selection events during affinity maturation. Science **315**, 528–531 (2007)
2. Althaus, C.L., De Vos, A.S., De Boer, R.J.: Reassessing the human immunodeficiency virus type 1 life cycle through age-structured modeling: Life span of infected cells, viral generation time, and basic reproductive number, R0. J. Virol. **83**, 7659–7667 (2009)
3. Bajenoff, M., Egen, J.G., Koo, L.Y., Laugier, J.P., Brau, F., Glaichenhaus, N., Germain, R.N.: Stromal cell networks regulate lymphocyte entry, migration, and territoriality in lymph nodes. Immunity **25**, 989–1001 (2006)
4. Bajenoff, M., Glaichenhaus, N., Germain, R.N.: Fibroblastic reticular cells guide T lymphocyte entry into and migration within the splenic T cell zone. J. Immunol. **181**, 3947–3954 (2008)
5. Balagopalan, L., Sherman, E., Barr, V.A., Samelson, L.E.: Imaging techniques for assaying lymphocyte activation in action. Nat. Rev. Immunol. **11**, 21–33 (2011)
6. Baldazzi, V., Castiglione, F., Bernaschi, M.: An enhanced agent based model of the immune system response. Cell. Immunol. **244**, 77–79 (2006)
7. Bauer, A.L., Beauchemin, C.A., Perelson, A.S.: Agent-based modeling of host-pathogen systems: The successes and challenges. Inf. Sci. **179**, 1379–1389 (2009)
8. Beauchemin, C.: Probing the effects of the well-mixed assumption on viral infection dynamics. J. Theor. Biol. **242**, 464–477 (2006)
9. Beauchemin, C., Samuel, J., Tuszynski, J.: A simple cellular automaton model for influenza A viral infections. J. Theor. Biol. **232**, 223–234 (2005)
10. Beltman, J.B., Maree, A.F., Lynch, J.N., Miller, M.J., de Boer, R.J.: Lymph node topology dictates T cell migration behavior. J. Exp. Med. **204**, 771–780 (2007)
11. Beltman, J.B., Maree, A.F., de Boer, R.J.: Analysing immune cell migration. Nat. Rev. Immunol. **9**, 789–798 (2009)
12. Bernaschi, M., Castiglione, F.: Selection of escape mutants from immune recognition during HIV infection. Immunol. Cell Biol. **80**, 307–313 (2002)
13. Betts, M.R., Nason, M.C., West, S.M., De Rosa, S.C., Migueles, S.A., Abraham, J., Lederman, M.M., Benito, J.M., Goepfert, P.A., Connors, M., Roederer, M., Koup, R.A.: HIV nonprogressors preferentially maintain highly functional HIV-specific CD8+ T cells. Blood **107**, 4781–4789 (2006)
14. Bocharov, G.A., Romanyukha, A.A.: Mathematical model of antiviral immune response. III. Influenza A virus infection. J. Theor. Biol. **167**, 323–360 (1994)
15. Bogle, G., Dunbar, P.R.: Simulating T-cell motility in the lymph node paracortex with a packed lattice geometry. Immunol. Cell Biol. **86**, 676–687 (2008)
16. Bogle, G., Dunbar, P.R.: Agent-based simulation of T-cell activation and proliferation within a lymph node. Immunol. Cell Biol. **88**, 172–179 (2010)
17. Bousso, P., Bhakta, N.R., Lewis, R.S., Robey, E.: Dynamics of thymocyte-stromal cell interactions visualized by two-photon microscopy. Science **296**, 1876–1880 (2002)
18. Burkhead, E.G., Hawkins, J.M., Molinek, D.K.: A dynamical study of a cellular automata model of the spread of HIV in a lymph node. Bull. Math. Biol. **71**, 25–74 (2009)
19. Castiglione, F., Poccia, F., D'Offizi, G., Bernaschi, M.: Mutation, fitness, viral diversity, and predictive markers of disease progression in a computational model of HIV type 1 infection. AIDS Res. Hum. Retroviruses **20**, 1314–1323 (2004)
20. Castiglione, F., Pappalardo, F., Bernaschi, M., Motta, S.: Optimization of HAART with genetic algorithms and agent-based models of HIV infection. Bioinformatics **23**, 3350–3355 (2007)
21. Chen, H.Y., Di Mascio, M., Perelson, A.S., Ho, D.D., Zhang, L.: Determination of virus burst size in vivo using a single-cycle SIV in rhesus macaques. Proc. Natl. Acad. Sci. U.S.A. **104**, 19079–19084 (2007)

22. Chen, P., Hubner, W., Spinelli, M.A., Chen, B.K.: Predominant mode of human immunodeficiency virus transfer between T cells is mediated by sustained Env-dependent neutralization-resistant virological synapses. J. Virol. **81**, 12582–12595 (2007)
23. Cheynier, R., Henrichwark, S., Hadida, F., Pelletier, E., Oksenhendler, E., Autran, B., Wain-Hobson, S.: HIV and T cell expansion in splenic white pulps is accompanied by infiltration of HIV-specific cytotoxic T lymphocytes. Cell **78**, 373–387 (1994)
24. Coombes, J.L., Robey, E.A.: Dynamic imaging of host-pathogen interactions in vivo. Nat. Rev. Immunol. **10**, 353–364 (2010)
25. Danielson, C.M., Hope, T.J.: Imaging of HIV/host protein interactions. Curr. Top. Microbiol. Immunol. **339**, 103–123 (2009)
26. Davenport, M.P., Ribeiro, R.M., Perelson, A.S.: Kinetics of virus-specific CD8+ T cells and the control of human immunodeficiency virus infection. J. Virol. **78**, 10096–10103 (2004)
27. De Boer, R.J., Ribeiro, R.M., Perelson, A.S.: Current estimates for HIV-1 production imply rapid viral clearance in lymphoid tissues. PLoS Comput. Biol. **6**, e1000906 (2010)
28. Dieckmann, U., Law, R., Metz, J. (eds.): The Geometry of Ecological Interactions: Simplifying Spatial Complexity. Cambridge University Press, Cambridge (2000)
29. Dimitrov, D.S., Hillman, K., Manischewitz, J., Blumenthal, R., Golding, H.: Kinetics of soluble CD4 binding to cells expressing human immunodeficiency virus type 1 envelope glycoprotein. J. Virol. **66**, 132–138 (1992)
30. Durrett, R.: Stochastic spatial models. SIAM Rev. **41**, 677–718 (1999)
31. Emery, V.C., Cope, A.V., Bowen, E.F., Gor, D., Griffiths, P.D.: The dynamics of human cytomegalovirus replication in vivo. J. Exp. Med. **190**, 177–182 (1999)
32. Frank, S.A.: Within-host spatial dynamics of viruses and defective interfering particles. J. Theor. Biol. **206**, 279–290 (2000)
33. Fraser, C., Hollingsworth, T.D., Chapman, R., de Wolf, F., Hanage, W.P.: Variation in HIV-1 set-point viral load: Epidemiological analysis and an evolutionary hypothesis. Proc. Natl. Acad. Sci. U.S.A. **104**, 17441–17446 (2007)
34. Funk, G.A., Jansen, V.A., Bonhoeffer, S., Killingback, T.: Spatial models of virus-immune dynamics. J. Theor. Biol. **233**, 221–236 (2005)
35. Germain, R.N., Bajenoff, M., Castellino, F., Chieppa, M., Egen, J.G., Huang, A.Y., Ishii, M., Koo, L.Y., Qi, H.: Making friends in out-of-the-way places: How cells of the immune system get together and how they conduct their business as revealed by intravital imaging. Immunol. Rev. **221**, 163–181 (2008)
36. Gratton, S., Cheynier, R., Dumaurier, M.J., Oksenhendler, E., Wain-Hobson, S.: Highly restricted spread of HIV-1 and multiply infected cells within splenic germinal centers. Proc. Natl. Acad. Sci. U.S.A. **97**, 14566–14571 (2000)
37. Graw, F., Regoes, R.R.: Investigating CTL mediated killing with a 3D cellular automaton. PLoS Comput. Biol. **5**, e1000466 (2009)
38. Graziano, F.M., Kettoola, S.Y., Munshower, J.M., Stapleton, J.T., Towfic, G.J.: Effect of spatial distribution of T-Cells and HIV load on HIV progression. Bioinformatics **24**, 855–860 (2008)
39. Grenfell, B.T., Bolker, B.M., Kleczkowski, A.: Seasonality and extinction in chaotic metapopulations. Proc. Sci. Bio **259**, 97–103 (1995)
40. Grossman, Z., Meier-Schellersheim, M., Paul, W.E., Picker, L.J.: Pathogenesis of HIV infection: What the virus spares is as important as what it destroys. Nat. Med. **12**, 289–295 (2006)
41. Groux, H., Torpier, G., Monte, D., Mouton, Y., Capron, A., Ameisen, J.C.: Activation-induced death by apoptosis in CD4+ T cells from human immunodeficiency virus-infected asymptomatic individuals. J. Exp. Med. **175**, 331–340 (1992)
42. Haase, A.T.: Perils at mucosal front lines for HIV and SIV and their hosts. Nat. Rev. Immunol. **5**, 783–792 (2005)
43. Haase, A.T.: Targeting early infection to prevent HIV-1 mucosal transmission. Nature **464**, 217–223 (2010)

44. Haase, A.T., Henry, K., Zupancic, M., Sedgewick, G., Faust, R.A., Melroe, H., Cavert, W., Gebhard, K., Staskus, K., Zhang, Z.Q., Dailey, P.J., Balfour, H.H., Erice, A., Perelson, A.S.: Quantitative image analysis of HIV-1 infection in lymphoid tissue. Science **274**, 985–989 (1996)
45. Halling-Brown, M., Pappalardo, F., Rapin, N., Zhang, P., Alemani, D., Emerson, A., Castiglione, F., Duroux, P., Pennisi, M., Miotto, O., Churchill, D., Rossi, E., Moss, D.S., Sansom, C.E., Bernaschi, M., Lefranc, M.P., Brunak, S., Lund, O., Motta, S., Lollini, P.L., Murgo, A., Palladini, A., Basford, K.E., Brusic, V., Shepherd, A.J.: ImmunoGrid: Towards agent-based simulations of the human immune system at a natural scale. Philos. Transact. A Math. Phys. Eng. Sci. **368**, 2799–2815 (2010)
46. Heath, S.L., Tew, J.G., Tew, J.G., Szakal, A.K., Burton, G.F.: Follicular dendritic cells and human immunodeficiency virus infectivity. Nature **377**, 740–744 (1995)
47. Hladik, F., McElrath, M.J.: Setting the stage: Host invasion by HIV. Nat. Rev. Immunol. **8**, 447–457 (2008)
48. Hlavacek, W.S., Wofsy, C., Perelson, A.S.: Dissociation of HIV-1 from follicular dendritic cells during HAART: Mathematical analysis. Proc. Natl. Acad. Sci. U.S.A. **96**, 14681–14686 (1999)
49. Hlavacek, W.S., Stilianakis, N.I., Notermans, D.W., Danner, S.A., Perelson, A.S.: Influence of follicular dendritic cells on decay of HIV during antiretroviral therapy. Proc. Natl. Acad. Sci. U.S.A. **97**, 10966–10971 (2000)
50. Hlavacek, W.S., Stilianakis, N.I., Perelson, A.S.: Influence of follicular dendritic cells on HIV dynamics. Philos. Trans. R. Soc. Lond. B Biol. Sci. **355**, 1051–1058 (2000)
51. Ho, D.D., Neumann, A.U., Perelson, A.S., Chen, W., Leonard, J.M., Markowitz, M.: Rapid turnover of plasma virions and CD4 lymphocytes in HIV-1 infection. Nature **373**, 123–126 (1995)
52. Hübner, W., McNerney, G.P., Chen, P., Dale, B.M., Gordon, R.E., Chuang, F.Y., Li, X.D., Asmuth, D.M., Huser, T., Chen, B.K.: Quantitative 3D video microscopy of HIV transfer across T cell virological synapses. Science **323**, 1743–1747 (2009)
53. Jolly, C., Kashefi, K., Hollinshead, M., Sattentau, Q.J.: HIV-1 cell to cell transfer across an Env-induced, actin-dependent synapse. J. Exp. Med. **199**, 283–293 (2004)
54. Joo, J., Lebowitz, J.L.: Pair approximation of the stochastic susceptible-infected-recovered-susceptible epidemic model on the hypercubic lattice. Phys. Rev. E Stat. Nonlin. Soft Matter Phys. **70**, 036114 (2004)
55. Kaul, R., Plummer, F.A., Kimani, J., Dong, T., Kiama, P., Rostron, T., Njagi, E., MacDonald, K.S., Bwayo, J.J., McMichael, A.J., Rowland-Jones, S.L.: HIV-1-specific mucosal CD8+ lymphocyte responses in the cervix of HIV-1-resistant prostitutes in Nairobi. J. Immunol. **164**, 1602–1611 (2000)
56. Keele, B.F., Li, H., Learn, G.H., Hraber, P., Giorgi, E.E., Grayson, T., Sun, C., Chen, Y., Yeh, W.W., Letvin, N.L., Mascola, J.R., Nabel, G.J., Haynes, B.F., Bhattacharya, T., Perelson, A.S., Korber, B.T., Hahn, B.H., Shaw, G.M.: Low-dose rectal inoculation of rhesus macaques by SIVsmE660 or SIVmac251 recapitulates human mucosal infection by HIV-1. J. Exp. Med. **206**, 1117–1134 (2009)
57. Keeling, M.J., Rohani, P.: Modeling Infectious Diseases in Humans and Animals. Princeton University Press, Princeton (2008)
58. Keeling, M.J., Rand, D.A., Morris, A.J.: Correlation models for childhood epidemics. Proc. Biol. Sci. **264**, 1149–1156 (1997)
59. Keeling, M.J., Woolhouse, M.E., May, R.M., Davies, G., Grenfell, B.T.: Modelling vaccination strategies against foot-and-mouth disease. Nature **421**, 136–142 (2003)
60. Kepler, T.B., Perelson, A.S.: Drug concentration heterogeneity facilitates the evolution of drug resistance. Proc. Natl. Acad. Sci. U.S.A. **95**, 11514–11519 (1998)
61. Kirschner, D., Webb, G.F.: A model for treatment strategy in the chemotherapy of AIDS. Bull. Math. Biol. **58**, 376–390 (1996)
62. Korber, B.T., Letvin, N.L., Haynes, B.F.: T-cell vaccine strategies for human immunodeficiency virus, the virus with a thousand faces. J. Virol. **83**, 8300–8314 (2009)

63. Ladi, E., Herzmark, P., Robey, E.: In situ imaging of the mouse thymus using 2-photon microscopy. J. Vis. Exp.11 e652. doi:10.3791/652 (2008)
64. Levin, S.A., Durrett, R.: From individuals to epidemics. Philos. Trans. R. Soc. Lond. B Biol. Sci. **351**, 1615–1621 (1996)
65. Lin, H., Shuai, J.W.: A stochastic spatial model of HIV dynamics with an asymmetric battle between the virus and the immune system. New J. Phy. **12**, 043051 (2010)
66. Linderman, J.J., Riggs, T., Pande, M., Miller, M., Marino, S., Kirschner, D.E.: Characterizing the dynamics of CD4+ T cell priming within a lymph node. J. Immunol. **184**, 2873–2885 (2010)
67. Lindquist, R.L., Shakhar, G., Dudziak, D., Wardemann, H., Eisenreich, T., Dustin, M.L., Nussenzweig, M.C.: Visualizing dendritic cell networks in vivo. Nat. Immunol. **5**, 1243–1250 (2004)
68. Louzoun, Y., Solomon, S., Atlan, H., Cohen, I.R.: Modeling complexity in biology. Physica A. **297**, 242–252 (2001)
69. Markowitz, M., Louie, M., Hurley, A., Sun, E., Di Mascio, M., Perelson, A.S., Ho, D.D.: A novel antiviral intervention results in more accurate assessment of human immunodeficiency virus type 1 replication dynamics and T-cell decay in vivo. J. Virol. **77**, 5037–5038 (2003)
70. Martin, N., Welsch, S., Jolly, C., Briggs, J.A., Vaux, D., Sattentau, Q.J.: Virological synapse-mediated spread of human immunodeficiency virus type 1 between T cells is sensitive to entry inhibition. J. Virol. **84**, 3516–3527 (2010)
71. McDonald, D., Vodicka, M.A., Lucero, G., Svitkina, T.M., Borisy, G.G., Emerman, M., Hope, T.J.: Visualization of the intracellular behavior of HIV in living cells. J. Cell Biol. **159**, 441–452 (2002)
72. McDonald, D., Wu, L., Bohks, S.M., KewalRamani, V.N., Unutmaz, D., Hope, T.J.: Recruitment of HIV and its receptors to dendritic cell-T cell junctions. Science **300**, 1295–1297 (2003)
73. Mempel, T.R., Henrickson, S.E., Von Andrian, U.H.: T-cell priming by dendritic cells in lymph nodes occurs in three distinct phases. Nature **427**, 154–159 (2004)
74. Meyers, L.A., Pourbohloul, B., Newman, M.E., Skowronski, D.M., Brunham, R.C.: Network theory and SARS: Predicting outbreak diversity. J. Theor. Biol. **232**, 71–81 (2005)
75. Mielke, A., Pandey, R.B.: A computer simulation study of cell population in a fuzzy interaction model for mutating HIV. Physica A **251**, 430–438 (1998)
76. Miller, S.E., Levenson, R.M., Aldridge, C., Hester, S., Kenan, D.J., Howell, D.N.: Identification of focal viral infections by confocal microscopy for subsequent ultrastructural analysis. Ultrastruct. Pathol. **21**, 183–193 (1997)
77. Miller, M.J., Wei, S.H., Parker, I., Cahalan, M.D.: Two-photon imaging of lymphocyte motility and antigen response in intact lymph node. Science **296**, 1869–1873 (2002)
78. Miller, M.J., Wei, S.H., Cahalan, M.D., Parker, I.: Autonomous T cell trafficking examined in vivo with intravital two-photon microscopy. Proc. Natl. Acad. Sci. U.S.A. **100**, 2604–2609 (2003)
79. Miller, M.J., Hejazi, A.S., Wei, S.H., Cahalan, M.D., Parker, I.: T cell repertoire scanning is promoted by dynamic dendritic cell behavior and random T cell motility in the lymph node. Proc. Natl. Acad. Sci. U.S.A. **101**, 998–1003 (2004)
80. Miller, M.J., Safrina, O., Parker, I., Cahalan, M.D.: Imaging the single cell dynamics of CD4+ T cell activation by dendritic cells in lymph nodes. J. Exp. Med. **200**, 847–856 (2004)
81. Miller, C.J., Li, Q., Abel, K., Kim, E.Y., Ma, Z.M., Wietgrefe, S., La Franco-Scheuch, L., Compton, L., Duan, L., Shore, M.D., Zupancic, M., Busch, M., Carlis, J., Wolinsky, S., Wolinksy, S., Haase, A.T.: Propagation and dissemination of infection after vaginal transmission of simian immunodeficiency virus. J. Virol. **79**, 9217–9227 (2005)
82. Mohri, H., Bonhoeffer, S., Monard, S., Perelson, A.S., Ho, D.D.: Rapid turnover of T lymphocytes in SIV-infected rhesus macaques. Science **279**, 1223–1227 (1998)
83. Mohri, H., Perelson, A.S., Tung, K., Ribeiro, R.M., Ramratnam, B., Markowitz, M., Kost, R., Hurley, A., Weinberger, L., Cesar, D., Hellerstein, M.K., Ho, D.D.: Increased turnover of

T lymphocytes in HIV-1 infection and its reduction by antiretroviral therapy. J. Exp. Med. **194**, 1277–1287 (2001)
84. Mueller, S.N., Germain, R.N.: Stromal cell contributions to the homeostasis and functionality of the immune system. Nat. Rev. Immunol. **9**, 618–629 (2009)
85. Müller, V., Maree, A.F., De Boer, R.J.: Release of virus from lymphoid tissue affects human immunodeficiency virus type 1 and hepatitis C virus kinetics in the blood. J. Virol. **75**, 2597–2603 (2001)
86. Nelson, P.W., Gilchrist, M.A., Coombs, D., Hyman, J.M., Perelson, A.S.: An age-structured model of HIV infection that allows for variations in the production rate of viral particles and the death rate of productively infected cells. Math. Biosci. Eng. **1**, 267–288 (2004)
87. Neumann, A.U., Lam, N.P., Dahari, H., Gretch, D.R., Wiley, T.E., Layden, T.J., Perelson, A.S.: Hepatitis C viral dynamics in vivo and the antiviral efficacy of interferon-alpha therapy. Science **282**, 103–107 (1998)
88. Newman, M.E.: Threshold effects for two pathogens spreading on a network. Phys. Rev. Lett. **95**, 108701 (2005)
89. Nowak, M.A., Bangham, C.R.: Population dynamics of immune responses to persistent viruses. Science **272**, 74–79 (1996)
90. Nowak, M.A., May, R.M.: Mathematical biology of HIV infections: Antigenic variation and diversity threshold. Math. Biosci. **106**, 1–21 (1991)
91. Nowak, M.A., May, R.M.: Virus dynamics: Mathematical Principles of Immunology and Virology. Oxford University Press, Oxford (2000)
92. Nowak, M.A., Anderson, R.M., McLean, A.R., Wolfs, T.F., Goudsmit, J., May, R.M.: Antigenic diversity thresholds and the development of AIDS. Science **254**, 963–969 (1991)
93. Nowak, M.A., Bonhoeffer, S., Hill, A.M., Boehme, R., Thomas, H.C., McDade, H.: Viral dynamics in hepatitis B virus infection. Proc. Natl. Acad. Sci. U.S.A. **93**, 4398–4402 (1996)
94. Ostrowski, M.A., Krakauer, D.C., Li, Y., Justement, S.J., Learn, G., Ehler, L.A., Stanley, S.K., Nowak, M., Fauci, A.S.: Effect of immune activation on the dynamics of human immunodeficiency virus replication and on the distribution of viral quasispecies. J. Virol. **72**, 7772–7784 (1998)
95. Ovaskainen, O., Cornell, S.J.: Asymptotically exact analysis of stochastic metapopulation dynamics with explicit spatial structure. Theor. Popul. Biol. **69**, 13–33 (2006)
96. Pandey, R.B.: Cellular automata approach to interacting cellular network models for the dynamics of cell population in an early HIV infection. Physica A **179**, 442–470 (1991)
97. Pandey, R.B., Stauffer, D.: Metastability with probabilistic cellular automata in an HIV infection. J. Stat. Phys. **61**, 235–240 (1990)
98. Pantaleo, G., Graziosi, C., Fauci, A.S.: The role of lymphoid organs in the pathogenesis of HIV infection. Semin. Immunol. **5**, 157–163 (1993)
99. Pappalardo, F., Halling-Brown, M.D., Rapin, N., Zhang, P., Alemani, D., Emerson, A., Paci, P., Duroux, P., Pennisi, M., Palladini, A., Miotto, O., Churchill, D., Rossi, E., Shepherd, A.J., Moss, D.S., Castiglione, F., Bernaschi, M., Lefranc, M.P., Brunak, S., Motta, S., Lollini, P.L., Basford, K.E., Brusic, V.: ImmunoGrid, an integrative environment for large-scale simulation of the immune system for vaccine discovery, design and optimization. Brief. Bioinformatics **10**, 330–340 (2009)
100. Pearson, J.E., Krapivsky, P., Perelson, A.S.: Stochastic theory of early viral infection: Continuous versus burst production of virions. PLoS Comput. Biol. **7**, e1001058 (2011)
101. Perelson, A.S.: Modelling viral and immune system dynamics. Nat. Rev. Immunol. **2**, 28–36 (2002)
102. Perelson, A.S., Ribeiro, R.M.: Estimating drug efficacy and viral dynamic parameters: HIV and HCV. Stat. Med. **27**, 4647–4657 (2008)
103. Perelson, A.S., Kirschner, D.E., De Boer, R.: Dynamics of HIV infection of CD4+ T cells. Math. Biosci. **114**, 81–125 (1993)
104. Perelson, A.S., Neumann, A.U., Markowitz, M., Leonard, J.M., Ho, D.D.: HIV-1 dynamics in vivo: Virion clearance rate, infected cell life-span, and viral generation time. Science **271**, 1582–1586 (1996)

105. Perelson, A.S., Essunger, P., Cao, Y., Vesanen, M., Hurley, A., Saksela, K., Markowitz, M., Ho, D.D.: Decay characteristics of HIV-1-infected compartments during combination therapy. Nature **387**, 188–191 (1997)
106. Phee, H., Dzhagalov, I., Mollenauer, M., Wang, Y., Irvine, D.J., Robey, E., Weiss, A.: Regulation of thymocyte positive selection and motility by GIT2. Nat. Immunol. **11**, 503–511 (2010)
107. Picker, L.J., Watkins, D.I.: HIV pathogenesis: The first cut is the deepest. Nat. Immunol. **6**, 430–432 (2005)
108. Pope, M., Haase, A.T.: Transmission, acute HIV-1 infection and the quest for strategies to prevent infection. Nat. Med. **9**, 847–852 (2003)
109. Ramratnam, B., Bonhoeffer, S., Binley, J., Hurley, A., Zhang, L., Mittler, J.E., Markowitz, M., Moore, J.P., Perelson, A.S., Ho, D.D.: Rapid production and clearance of HIV-1 and hepatitis C virus assessed by large volume plasma apheresis. Lancet **354**, 1782–1785 (1999)
110. Reilly, C., Wietgrefe, S., Sedgewick, G., Haase, A.: Determination of simian immunodeficiency virus production by infected activated and resting cells. AIDS **21**, 163–168 (2007)
111. Ribeiro, R.M., Mohri, H., Ho, D.D., Perelson, A.S.: In vivo dynamics of T cell activation, proliferation and death in HIV-1 infection: Why are CD4+ but not CD8+ T cells depleted? Proc. Natl. Acad. Sci. U.S.A. **99**, 15572–15577 (2002)
112. Salazar-Gonzalez, J.F., Salazar, M.G., Keele, B.F., Learn, G.H., Giorgi, E.E., Li, H., Decker, J.M., Wang, S., Baalwa, J., Kraus, M.H., Parrish, N.F., Shaw, K.S., Guffey, M.B., Bar, K.J., Davis, K.L., Ochsenbauer-Jambor, C., Kappes, J.C., Saag, M.S., Cohen, M.S., Mulenga, J., Derdeyn, C.A., Allen, S., Hunter, E., Markowitz, M., Hraber, P., Perelson, A.S., Bhattacharya, T., Haynes, B.F., Korber, B.T., Hahn, B.H., Shaw, G.M.: Genetic identity, biological phenotype, and evolutionary pathways of transmitted/founder viruses in acute and early HIV-1 infection. J. Exp. Med. **206**, 1273–1289 (2009)
113. Sato, H., Orenstein, J., Dimitrov, D., Martin, M.: Cell-to-cell spread of HIV-1 occurs within minutes and may not involve the participation of virus particles. Virology **186**, 712–724 (1992)
114. Sattentau, Q.: Avoiding the void: Cell-to-cell spread of human viruses. Nat. Rev. Microbiol. **6**, 815–826 (2008)
115. Schacker, T., Little, S., Connick, E., Gebhard-Mitchell, K., Zhang, Z.Q., Krieger, J., Pryor, J., Havlir, D., Wong, J.K., Richman, D., Corey, L., Haase, A.T.: Rapid accumulation of human immunodeficiency virus (HIV) in lymphatic tissue reservoirs during acute and early HIV infection: Implications for timing of antiretroviral therapy. J. Infect. Dis. **181**, 354–357 (2000)
116. Schacker, T., Little, S., Connick, E., Gebhard, K., Zhang, Z.Q., Krieger, J., Pryor, J., Havlir, D., Wong, J.K., Schooley, R.T., Richman, D., Corey, L., Haase, A.T.: Productive infection of T cells in lymphoid tissues during primary and early human immunodeficiency virus infection. J. Infect. Dis. **183**, 555–562 (2001)
117. Sourisseau, M., Sol-Foulon, N., Porrot, F., Blanchet, F., Schwartz, O.: Inefficient human immunodeficiency virus replication in mobile lymphocytes. J. Virol. **81**, 1000–1012 (2007)
118. Sowinski, S., Jolly, C., Berninghausen, O., Purbhoo, M.A., Chauveau, A., Kohler, K., Oddos, S., Eissmann, P., Brodsky, F.M., Hopkins, C., Onfelt, B., Sattentau, Q., Davis, D.M.: Membrane nanotubes physically connect T cells over long distances presenting a novel route for HIV-1 transmission. Nat. Cell Biol. **10**, 211–219 (2008)
119. Stoll, S., Delon, J., Brotz, T.M., Germain, R.N.: Dynamic imaging of T cell-dendritic cell interactions in lymph nodes. Science **296**, 1873–1876 (2002)
120. Strain, M.C., Richman, D.D., Wong, J.K., Levine, H.: Spatiotemporal dynamics of HIV propagation. J. Theor. Biol. **218**, 85–96 (2002)
121. Tilman, D., Kareiva, P.: Spatial Ecology. Princeton University Press, Princeton, NJ (1997)
122. Virgin, H.W., Walker, B.D.: Immunology and the elusive AIDS vaccine. Nature **464**, 224–231 (2010)
123. von Neumann, J., Burks, A.W.: Theory of Self Reproducing Cellular Automata. University of Illinois Press, Urbana and London (1966)

124. Webb, S.D., Keeling, M.J., Boots, M.: Host-parasite interactions between the local and the mean-field: How and when does spatial population structure matter? J. Theor. Biol. **249**, 140–152 (2007)
125. Wei, X., Ghosh, S.K., Taylor, M.E., Johnson, V.A., Emini, E.A., Deutsch, P., Lifson, J.D., Bonhoeffer, S., Nowak, M.A., Hahn, B.H.: Viral dynamics in human immunodeficiency virus type 1 infection. Nature **373**, 117–122 (1995)
126. Wolfram, S.: Theory and applications of cellular automata. Advanced Series on Complex Systems. World Scientific Publication, Singapore (1986)
127. Zhang, L., Dailey, P.J., He, T., Gettie, A., Bonhoeffer, S., Perelson, A.S., Ho, D.D.: Rapid clearance of simian immunodeficiency virus particles from plasma of rhesus macaques. J. Virol. **73**, 855–860 (1999)
128. Zhang, L., Dailey, P.J., Gettie, A., Blanchard, J., Ho, D.D.: The liver is a major organ for clearing simian immunodeficiency virus in rhesus monkeys. J. Virol. **76**, 5271–5273 (2002)
129. Zhu, P., Chertova, E., Bess, J., Lifson, J.D., Arthur, L.O., Liu, J., Taylor, K.A., Roux, K.H.: Electron tomography analysis of envelope glycoprotein trimers on HIV and simian immunodeficiency virus virions. Proc. Natl. Acad. Sci. U.S.A. **100**, 15812–15817 (2003)
130. Zorzenon dos Santos, R.M., Coutinho, S.: Dynamics of HIV infection: A cellular automata approach. Phys. Rev. Lett. **87**, 168102 (2001)

Basic Principles in Modeling Adaptive Regulation and Immunodominance

Peter S. Kim, Peter P. Lee, and Doron Levy

1 Introduction

In this chapter we overview our recent work on mathematical models for the regulation of the primary immune response to viral infections and immunodominance. The primary immune response to a viral infection can be very rapid, yet transient. Prior to such a response, potentially reactive T cells wait in lymph nodes until stimulated. Upon stimulation, these cells proliferate for a limited duration and then undergo apoptosis or enter dormancy as memory cells. The mechanisms that trigger the contraction of the T cell population are not well understood. Immunodominance refers to the phenomenon in which simultaneous T cell responses against multiple target epitopes organize themselves into distinct and reproducible hierarchies. In many cases, eliminating the response to the most dominant epitope allows responses to subdominant epitopes to expand more fully. Likewise, if the two most dominant epitopes are removed, then the third most dominant response may expand. The mechanisms that drive immunodominance are also not well understood.

In order to understand the processes that control the T cells expansion and contraction, Mercado et al. demonstrated experimentally that the kinetics of CD8+ T cell expansion and contraction are determined within the first day of infection

P.S. Kim (✉)
School of Mathematics and Statistics, University of Sydney, NSW 2006, Australia
e-mail: P.Kim@maths.usyd.edu.au

P.P. Lee
Division of Hematology, Department of Medicine, Stanford University, Stanford, CA 94305, USA
e-mail: ppl@stanford.edu

D. Levy
Department of Mathematics and Center for Scientific Computation and Mathematical Modeling (CSCAMM), University of Maryland, College Park, MD 20742, USA
e-mail: dlevy@math.umd.edu

U. Ledzewicz et al., *Mathematical Methods and Models in Biomedicine*, Lecture Notes on Mathematical Modelling in the Life Sciences, DOI 10.1007/978-1-4614-4178-6_2,

[20]. In another study of CD8+ T cell expansion, Kaech et al. showed that upon antigenic stimulation, naïve CD8+ T cells divide at least 7–10 times and differentiate into functional effector and memory cells even if antigen is removed [13]. An alternative experimental approach by van Stipdonk et al. also focused on CD8+ T cell stimulation [31]. They showed that naïve CD8+ T cells become activated after only 2 h of exposure to mature antigen-presenting cells (APCs). After activation, these T cells divided and differentiated into effector and eventually memory cells without a need for further antigenic stimulation. In a subsequent paper, they observed that naïve CD8+ T cells that have been stimulated for 20 h were able to carry out extensive proliferation and cytotoxic activity, characteristic of a fully developed immune response [30]. They proposed that the fate of a T cell response is governed by a "cell-instrinsic developmental program" that is set even before the first cell division takes place.

A couple of mathematical models of the T cell proliferation program have been developed in parallel to these experiments. Antia et al. devised a mathematical model to investigate whether the program is completely specified by the initial encounter with antigen or whether it can be subsequently modified by the amount of antigen present [1]. Their results favor the second paradigm in which the T cell population briefly expands in response to the amount of antigen present before committing to a fixed program. Wodarz and Thomsen [32] developed a mathematical model to find the optimal fixed program that could respond effectively to a wide variety of infections. They concluded that the 7–10 divisions observed experimentally represented such an optimum.

All together the experimental and mathematical modeling papers propose a general paradigm for T cell expansion, which can be stated as follows: upon stimulation, T cells enter a minimal developmental program of about 7–10 divisions that is followed by a period of antigen-dependent proliferation that terminates after a certain time or after a certain number of cell divisions.

While the precise mechanisms of immunodominance are not well understood, the majority of experimental and theoretical works agree on some form of T cell competition [4, 11, 14, 15, 22, 23]. The two most prevailing theories on the matter are that either T cells passively compete for a limited resource, most likely access to APCs, or that T cells actively suppress the development of other T cells.

Our approach to deriving a mathematical model of immunodominance is based on extending the adaptive regulation model to consider the case of multiple, simultaneous T cell responses. This point of view implies that immunodominance may occur as a natural result of the iTreg-mediated contraction of the T cell response proposed in [17].

Several mathematical models for immunodominance have been developed in the literature. Here we mention several key works and refer to [16] for a more complete overview of these works and their results. De Boer and Perelson show that for each target epitope only the T cell clone with the highest affinity will survive long-term T cell competition [7, 8]. Nowak develops a mathematical model and predicts that for an antigenically homogeneous virus population, the immune response will ultimately be directed against only one epitope, a situation known as complete immunodominance [21]. De Boer et al. formulate a mathematical model to analyze experimental measurements of the CD8+ T cell response to lymphocytic choriomeningitis virus [10]. The response consists of one immunodominant response and one subdominant response against different epitopes. De Boer et al. propose that differences in growth rate and recruitment times of different T cell populations can account for immunodominance. Antia et al. also formulate a model in which multiple epitope-specific T cell populations undergo a brief period of expansion in response to antigen, followed by a period of antigen-independent proliferation and contraction [1]. Handel and Antia develop a mathematical model to explain the shift in the immunodominance hierarchy between the primary and secondary responses to influenza A [12]. Scherer et al. present an alternative mathematical model in which the down-modulation of antigen-presentation leads to long-term coexistence of T cell responses [25]. A related work by Scherer et al. is an agent-based model to understand whether T cells compete for nonspecific stimuli, such as access to the surface of APCs, or for specific stimuli, such as MHC:epitope complexes [27].

Our main goal in developing mathematical models for the primary immune response and immunodominance is to identify at least some of the main mechanisms by which the primary immune response is regulated and by which immunodominance emerges. After carefully studying other approaches, we developed mathematical models that are based on the following basic principles:

1. The primary immune response should be adaptively controlled. This adaptive process can work in combination with any proliferation preprograms.
2. The adaptive control is conducted by regulatory cells. The number of regulatory cells cannot be directly proportional to the total number of effector cells as the body has no way of keeping track of this number. Instead, the process should depend only on the dynamics of individual cells.
3. Immunodominance is a by-product of adaptive regulation. Adaptive regulatory cells, which are created in an epitope-specific way, can then regulate the system in a nonspecific fashion.

When it comes to adaptive regulation, our main observation in [17] was that the preprogram paradigm as is, is inconsistent with the experimental data of Badovinac et al. [2], which showed that a 10,000-fold difference in antigen-specific naïve T cell concentrations only led to a 13-fold higher peak in the effector response. Any mechanism that relies only on a preprogrammed cell division must scale linearly with the precursor frequencies. This led us to derive a mathematical model that is based on adaptive regulatory T cells (iTregs). Our hypothesis is that T cell responses are adaptively regulated in a process that results from the dynamics of immune cells that interact based on relatively simple rules. Following the same line of thought, our model of immunodominance from [16] is based on considering immunodominance as a by-product of the regulated T cell contraction. It is sufficient to add a single rule to the model of adaptive regulation to explain immunodominance.

The model in [16] represents an "extended" model that divides T cells into helper (CD4)+ and killer (CD8+) subpopulations and considers interactions among helper T cells, killer T cells, and iTregs. This model has the advantage of presenting a more encompassing view of immune interactions; however, its complexity obfuscates the key feedback loop that drives the expansion and contraction of a primary T cell response and the development of immunodominance.

To gain insight into regulatory mechanisms, we will begin by presenting a simplified model that elucidates the key feedback loop, while still capturing the qualitative behavior of the extended model from [16]. The primary simplification is that the helper and killer subpopulations are considered as a collective population of effector T cells, since the expansion and contraction of helper and killer T cells occur roughly in parallel [9]. In this manner, the simplified model focuses on the negative feedback between the effector and regulatory T cell populations. Using the simplified model, we will discuss insights that are difficult to obtain using the extended model.

The structure of this chapter is as follows. In Sect. 2 we present our mathematical models of adaptive regulatory T cell-mediated contraction. First, in Sect. 2.1 we present the simplified model of adaptive regulation, a model that does not separate helper and killer T cells. This model is taken from [17]. The model is extended to include helper T cells in Sect. 2.2. Mathematical models of immunodominance are described in Sect. 3. The models of immunodominance are based on the models of adaptive regulation and follow the same pattern of presentation: we start in Sect. 3.1 with the basic model of immunodominance that does not include helper T cells. This model is the original model we proposed, a model that has not yet been published. We then continue in Sect. 3.2 with the extended model that includes helper T cells. This model was published in [16]. Numerical results are given in Sect. 4. We show some results for the adaptive regulation model, results that demonstrate the robustness of the system to small perturbations in the precursor frequencies. We then continue in Sect. 4.2 with simulations of both immunodominance models, focusing on results that were obtained with the simple model. A discussion and concluding remarks are provided in Sect. 5.

2 Mathematical Model of Adaptive Regulatory T cell-Mediated Contraction

In this section we overview our models for adaptive regulatory T cell-mediated contraction. We start in Sect. 2.1 with the basic adaptive regulation model we introduced in [17]. This model is based on the hypothesis that primary response may be governed by a feedback control system involving adaptive regulatory cells (iTregs) rather than by intrinsic, intracellular feedback mechanisms.

In Sect. 2.2 we extend the adaptive regulation model of Sect. 2.1 to a more comprehensive model that includes helper T cells and positive growth signals. While this model better adheres to the biology when compared with the basic model in Sect. 2.1, the basic principle that enables that contraction of the immune response remains the same: adaptive regulatory cells provide the required negative feedback to generate the desired contraction.

1) Migration of APCs to lymph node

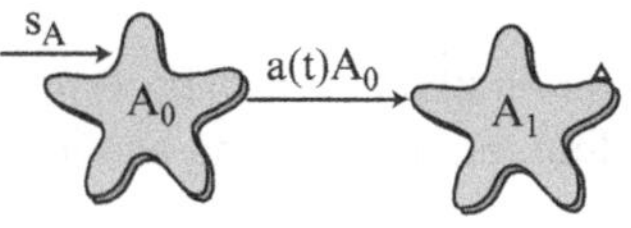

2) Initial T cell activation

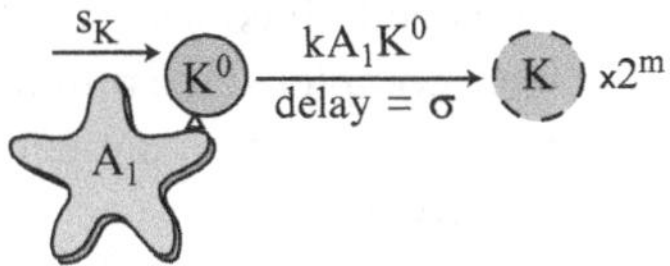

3) Antigen-dependent proliferation

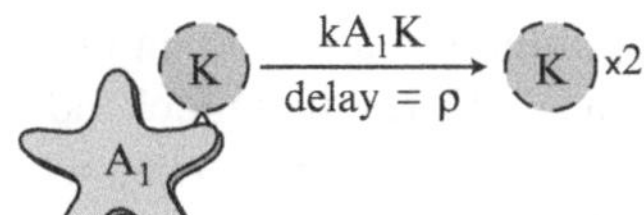

4) Effector cells differentiate into iTregs

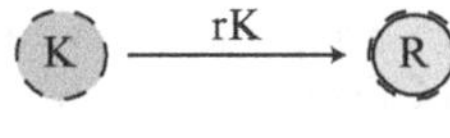

5) iTregs suppress effector cells

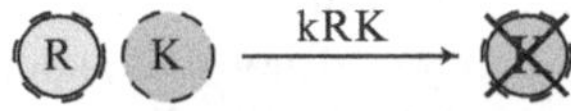

Fig. 1 Diagram of the iTreg model. (1) Immature APCs pick up antigen at the site of infection at a time-dependent rate $a(t)$. These APCs mature and migrate to the lymph node. (2) Mature antigen-bearing APCs present antigen to naïve T cells causing them to activate and enter the minimal developmental program of m divisions. (3) Effector cells that have completed the minimal program continue to divide upon further interaction with mature APCs. (4) Effector cells differentiate into iTregs at rate r. (5) The iTregs suppress effector cells. Although not indicated, each cell in the diagram has a natural death rate

In both models, the number of regulatory cells is dynamically controlled. A certain ratio of the effector cells are converted into regulatory cells. Such a process is postulated to be controlled on the level of the individual cell possibly even in a probabilistic way. There is no need for a central control of the number of regulatory cells that depends on the total number of effector cells. Such a mechanism would be biologically irrelevant. The precise means by which some of the effector cells turn into regulatory cells is irrelevant for the present work. It is possible that asymmetric differentiation is involved, or perhaps effector cells actually change their trait. In any event, all that matters is that a certain fraction of the effector cells will eventually turn into regulatory cells, due to a local process on the level of the individual cell.

2.1 *Mathematical Model of Adaptive Regulation: Feedback Loop*

We start with the simple model for monoclonal T cell responses taken from [17]. This model can be summarized in five steps (illustrated in Fig. 1):

1. APCs mature, present relevant target antigen, and migrate from the site of infection to the draining lymph node.

2. In the lymph node, APCs activate naïve T cells that enter a minimal developmental program of m cell divisions.
3. T cells that have completed the minimal developmental program become effector cells that keep dividing in an antigen-dependent manner as long as they are not suppressed by iTregs.
4. Effector cells differentiate into iTregs at a constant rate.
5. The iTregs suppress effector cells upon interaction.

For convenience, we group the entire T cell population into one unit consisting of both CD4+ and CD8+ T cells. This assumption simplifies the model and focuses on the feedback loop between effector cells and iTregs. This simplification does not capture the heterogeneous roles of CD4+ and CD8+ T cells in driving and regulating the overall T cell response. In particular, CD4+ T cells are the primary secreters of the cytokine interleukin-2 (IL-2), which drives T cell proliferation. In addition, nonregulatory CD4+ T cells are the major, if not only, source of iTregs generated in the periphery [24]. On the other hand, CD8+ T cells proliferate more rapidly and extensively than CD4+ T cells and also exhibit cytotoxic activity [9]. To capture these differences, we develop a more extensive model that includes separate CD4+ and CD8+ subpopulations in Sect. 2.2.

In addition, we assume that iTregs do not undergo further proliferation after differentiating from effector T cells. As with the previous assumption, this simplification also allows the model to focus on the feedback loop between effector cells and iTregs without incorporating an additional positive stimulation of iTreg via APCs. We also remove this simplification in the extended model of Sect. 2.2.

The T cell dynamics in the model are based on the concept of antigen-independent T cell proliferation and contraction. Various experiments have shown that the during a primary CD8+ T cell response, T cell kinetics are determined early on (after approximately 24 h of stimulation) [20], T cell expansion and differentiation are antigen-independent after initial exposure (approximately 20 h of stimulation) [30], and T cells divide at least 7–10 times after stimulation even if antigen is removed [13]. Similar results have been found for CD4+ T cells [33]. These results along with other related studies have led to the notion of *antigen-independent T cell program*. The main principle is that following initial stimulation, the primary T cell response is governed by an independent *program* that is insensitive to the nature and duration of subsequent antigen stimulation. The implication is that T cells somehow regulate themselves during a primary response without feedback from the antigen source. Since in this chapter we only consider immunodominance during a primary T cell response, we model T cell dynamics from the perspective of an antigen-independent, self-regulating process. Other examples of mathematical models of antigen-independent primary T cell response dynamics can be found in Antia et al. and Wodarz et al. [1,32].

The mathematical model corresponding to Fig. 1 is formulated as the following system of delayed differential equations (DDEs):

$$\dot{A}_0(t) = s_{\mathrm{A}} - d_0 A_0(t) - a(t) A_0(t), \qquad (1)$$

$$\dot{A}_1(t) = a(t)A_0(t) - d_1A_1(t), \tag{2}$$

$$\dot{K^0}(t) = s_K - \delta_0 K^0(t) - kA_1(t)K^0(t), \tag{3}$$

$$\begin{aligned}\dot{K}(t) = {} & 2^m kA_1(t-\sigma)K^0(t-\sigma) - kA_1(t)K(t) + 2kA_1(t-\rho)K(t-\rho) \\ & - (\delta_1 + r)K(t) - kR(t)K(t),\end{aligned} \tag{4}$$

$$\dot{R}(t) = rK(t) - \delta_1 R(t). \tag{5}$$

Here, A_0 is the concentration of APCs at the site of infection, A_1 is the concentration of APCs that have matured, started to present target antigen, and migrated to the lymph node, K^0 is the concentration of naïve T cells in the lymph node, K is the concentration of effector cells, and R is the concentration of iTregs.

Equation (1) pertains to APCs waiting at the site of infection. These cells are supplied at a constant rate s_A and die at a proportional rate d_0. Without stimulation, the population remains at its equilibrium level, s_A/d_0. The time-dependent coefficient $a(t)$ is the rate of APC stimulation from antigen at the site of infection. Equation (2) pertains to APCs that have matured, started to present relevant antigen, and migrated to the lymph node. The first term of the equation corresponds to the rate at which these APCs enter the lymph node. The second term is the natural death rate of this population.

Equation (3) pertains to naïve T cells. This population is replenished at a constant rate s_K and dies at a proportional rate δ_0. Without stimulation, the population remains at its equilibrium level, s_K/δ_0. The third term in this equation is the rate of stimulation of naïve T cells by mature APCs. The bilinear form of this term follows the law of mass action where k is the proportionality constant (or kinetic coefficient).

Equation (4) pertains to effector cells. The first term gives the rate at which activated naïve T cells enter the effector state after finishing the minimal developmental program. This term is similar to the last term of Eq. (3), except that it has an additional coefficient of 2^m and it depends on cell concentrations at time $t-\sigma$. The coefficient 2^m accounts for the increase in population of naïve T cells after m divisions, and the time delay σ is the duration of the minimal developmental program. The second term is the rate at which effector cells are stimulated by mature APCs for further division, and the third term is the rate in which cells reenter the effector population after having divided once. The fourth term is the rate that effector cells exit the population through death at rate δ_1 or differentiation into iTregs at rate r. The final term is the rate that effector cells are suppressed by iTregs. We assume that the rate of iTreg–effector interactions follows the same mass action law as APC–T cell interactions.

Equation (5) pertains to iTregs. The first term is the rate at which effector cells differentiate into iTregs, and the second term is the rate at which iTregs die. We assume that iTregs have the same death rate as effector cells.

The parameter estimates used for this model are taken from [17] and are summarized in Table 1. For the function $a(t)$, representing the rate of antigen

Table 1 Estimates for model parameters

Parameter	Description	Estimate
$A_0(0)$	Initial concentration of immature APCs	10
$K^0(0)$	Initial concentration of naïve T cells	0.04
d_0	Death/turnover rate of immature APCs	0.03
d_1	Death/turnover rate of mature APCs	0.8
δ_0	Death/turnover rate of naïve T cells	0.03
δ_1	Death/turnover rate of effector T cells	0.4
s_A	Supply rate of immature APCs	$d_0A_0(0) = 0.3$
s_K	Supply rate of naïve T cells	$\delta_0K^0(0) = 0.0012$
k	Kinetic coefficient	20
m	# of divisions in minimal developmental program	7
n	Maximum number of antigen-dependent divisions	3 to 10
ρ	Duration of one T cell division	1/3
σ	Duration of minimal developmental program	$1+(m-1)\rho = 3$
$a(t)$	Rate of APC stimulation	Eq. (6)
b	Duration of antigen availability	10
c	Level of APC stimulation	1
r	Rate of differentiation of effector cells into iTregs	0.01

Concentrations are in units of k/μL, and time is measured in days

stimulation, we assume that it starts at 0, remains positive for some time, and eventually returns to 0. To generate a smooth function for $a(t)$, we let

$$\phi(x) = \begin{cases} e^{-1/x^2}, & \text{if } x \geq 0, \\ 0, & \text{if } x < 0, \end{cases}$$

and set

$$a(t) = c\frac{\phi(t)\phi(b-t)}{\phi(b)^2}, \tag{6}$$

where $b, c > 0$. The variable t is defined such that mature APCs begin appearing in the lymph node at $t = 0$, although the infection may have begun slightly earlier. We estimate that the duration of antigen availability, b, is about 10 days. Furthermore, we estimate that the level of APC stimulation, c, is around 1. (See Fig. 2 for graphs of $a(t)$ for $b = 3$ and $b = 10$ when $c = 1$.)

2.2 *Extended Model of Adaptive Regulation: Helper and Killer T Cells*

The mathematical model presented in this section is an extension of the model in Sect. 2.1. The main extension of the model is to separate the nonregulatory

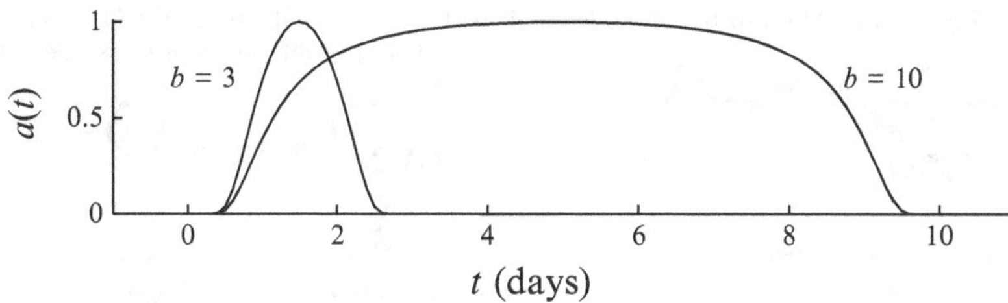

Fig. 2 Graphs of the antigen function $a(t)$ given by Eq. (6) for $b = 3$ and $b = 10$ when $c = 1$. The function $a(t)$ represents the time-dependent rate that immature APCs pick up antigen and are stimulated

T cell population into CD4+ and CD8+ T cells. CD4+ T cells are the primary producers of positive growth signal, particularly IL-2, and CD8+ T cells are the main proliferators. Furthermore, iTregs differentiate from effector CD4+ cells and suppress both effector CD4+ and CD8+ cells [24]. The extended model can be summarized in six steps (illustrated in Fig. 3):

1. APCs mature, present relevant target antigen, and migrate from the site of infection to the draining lymph node.
2. In the lymph node, APCs activate naïve CD4+ and CD8+ T cells that enter a minimal developmental program of m_1 or m_2 cell divisions, respectively.
3. Effector CD4+ and CD8+ T cells both secrete positive growth signal at different rates.
4. CD4+ and CD8+ T cells that have completed the minimal developmental program become effector cells that keep dividing as long as they are not suppressed by iTregs.
    - CD4+ T cells proliferate in response to interactions with APCs (It is assumed that CD4+ T cells produce enough IL-2 to stimulate their own growth in an autocrine loop. Hence, we do not explicitly model the secretion and consumption of IL-2 by CD4+ T cells.).
    - CD8+ T cells proliferate after consuming free positive growth signal.
5. During the immune response, some effector CD4+ T cells differentiate into iTregs.
6. The iTregs suppress effector CD4+ and CD8+ T cells and proliferate after consuming free positive growth signal.

The mathematical model corresponding to Fig. 3 is formulated as the following system of DDEs:

$$\dot{A}_0(t) = s_\mathrm{A} - d_0 A_0(t) - a(t)A_0(t), \tag{7}$$

$$\dot{A}_1(t) = a(t)A_0(t) - d_1 A_1(t), \tag{8}$$

$$\dot{H}^0(t) = s_\mathrm{H} - \delta_0 H^0(t) - kA_1(t)H^0(t), \tag{9}$$

1) Migration of APCs to lymph node

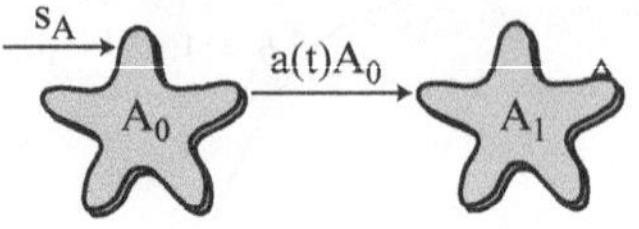

2) Initial T cell activation

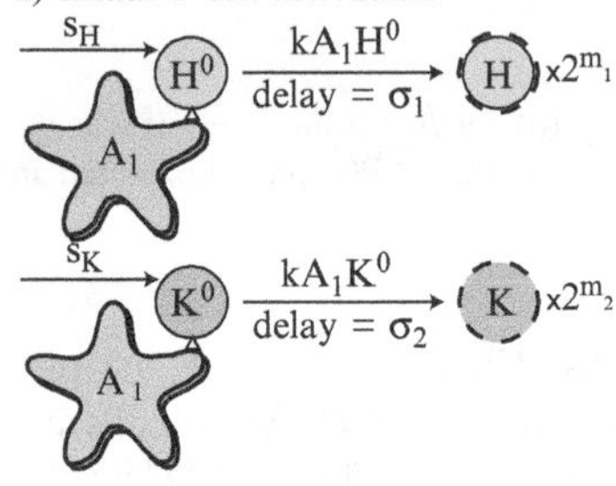

3) CD4+ and CD8+ T cells secrete positive growth signal (IL-2)

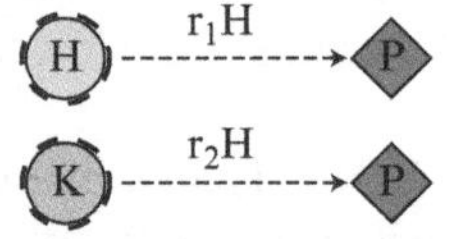

4) APC-driven proliferation of CD4+ cells. IL-2-driven proliferation of CD8+ cells & iTregs.

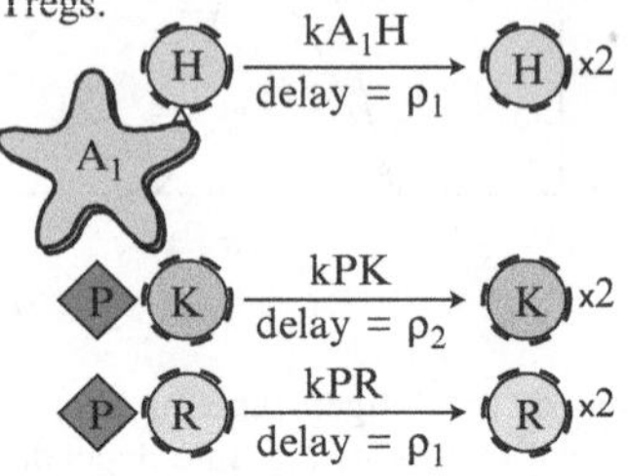

5) CD4+ T cells differentiate into iTregs

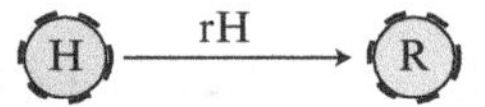

6) iTregs suppress effector T cells and consume positive growth signal.

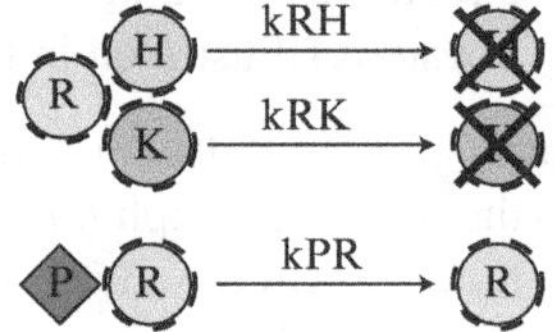

Fig. 3 Diagram of the extended adaptive iTreg model. (1) Immature APCs pick up antigen at the site of infection at a time-dependent rate $a(t)$. These APCs mature and migrate to the lymph node. (2) Mature antigen-bearing APCs present antigen to naïve CD4+ and CD8+ T cells causing them to activate and enter the minimal developmental program of m_1 and m_2 divisions, respectively. (3) Effector CD4+ and CD8+ T secrete positive growth signals at different rates. (4) CD4+ and CD8+ T cells that completed the minimal program become effector cells and continue to divide. CD4+ T cells proliferate upon further interaction with mature APCs. CD8+ T cells and iTregs proliferate after consuming positive growth signal. (5) Effector CD4+ T cells differentiate into iTregs at a constant rate. (6) The iTregs suppress effector CD4+ and CD8+ T cells. Although not indicated, each cell in the diagram has a natural death rate

$$\begin{aligned}\dot{H}(t) &= 2^{m_1}kA_1(t-\sigma_1)H^0(t-\sigma_1) - kA_1(t)H(t) + 2kA_1(t-\rho_1)H(t-\rho_1)\\ &\quad - (\delta_H + r)H(t) - kR(t)H(t),\end{aligned} \tag{10}$$

$$\dot{K}^0(t) = s_K - \delta_0 K^0(t) - kA_1(t)K^0(t), \tag{11}$$

$$\begin{aligned}\dot{K}(t) &= 2^{m_2}kA_1(t-\sigma_2)K^0(t-\sigma_2) - kP(t)K(t) + 2kP(t-\rho_2)K(t-\rho_2)\\ &\quad - \delta_K K(t) - kR(t)K(t),\end{aligned} \tag{12}$$

$$\dot{P}(t) = r_1H(t) + r_2K(t) - \delta_P P(t) - kP(t)K(t) - kP(t)R(t), \tag{13}$$

$$\dot{R}(t) = rH(t) - kP(t)R(t) + 2kP(t-\rho_1)R(t-\rho_1) - \delta_H R(t). \tag{14}$$

As in Sect. 2.1, A_0 is the concentration of APCs at the site of infection and A_1 is the concentration of APCs that have matured, started to present target antigen, and migrated to the lymph node. The variable H^0 is the concentration of naïve CD4+ killer T cells, H is the concentration of effector CD4+ cells, K^0 is the concentration of naïve CD8+ (killer) T cells, K is the concentration of effector CD8+ cells, and R is the concentration of iTregs. In addition, P is the concentration of positive growth signal (e.g., IL-2).

Equations (7) and (8) are identical to Eqs. (1) and (2) in Sect. 2.1. Equations (9) and (11) pertain to naïve CD4+ and CD8+ T cells, respectively. The CD4+ and CD8+ populations are replenished at constant rates s_H and s_K, respectively, and die at a proportional rate δ_0. The third terms in Eqs. (9) and (11) are the rates of stimulation of naïve CD4+ and CD8+ T cells by mature APCs. The bilinear form of this term follows the law of mass action where k is the proportionality constant (or kinetic coefficient). We assume that all cell–cell or cell–signal interactions follow the same law of mass action.

Equation (10) pertains to effector CD4+ cells. The first term gives the rate at which activated naïve CD4+ T cells enter the effector state after finishing the minimal developmental program of m_1 cell divisions. The time delay σ_1 is the duration of the minimal developmental program. The second term is the rate at which effector CD4+ cells are stimulated by mature APCs for further division, and the third term is the rate in which cells reenter the effector CD4+ population after having divided once. The time delay ρ_1 is the duration of one CD4+ cell division. The fourth term is the rate at which effector CD4+ cells exit the population through death at rate δ_H or differentiation into iTregs at rate r. The final term is the rate at which effector CD4+ cells are suppressed by iTregs.

Equation (12) pertains to effector CD8+ cells. The first term gives the rate at which activated naïve CD8+ T cells enter the effector state after finishing the minimal developmental program of m_2 cell divisions. The time delay σ_2 is the duration of the minimal developmental program. The second term is the rate at which effector CD8+ cells are stimulated by positive growth signal for further division, and the third term is the rate at which cells reenter the effector CD8+ population after having divided once. The time delay ρ_2 is the duration of one CD8+ cell division. The fourth term is the rate at which effector CD8+ cells die at rate δ_K. The final term is the rate at which effector CD8+ cells are suppressed by iTregs.

Equation (13) pertains to positive growth signal. The first two terms are the rates at which positive growth signal is secreted by effector CD4+ and CD8+ cells, respectively. The third term is the decay rate of positive growth signal. The fourth and fifth terms are the rates at which positive growth signal is consumed by effector CD8+ cells and iTregs, respectively.

Equation (14) pertains to iTregs. The first term is the rate at which effector CD4+ cells differentiate into iTregs The second term is the rate at which iTregs are stimulated by positive growth signal for further division, and the third term is the rate at which cells reenter the iTreg population after having divided once. The time delay ρ_1 is the duration of one CD4+ cell division. The fourth term is the rate at which iTregs die. We assume that iTregs have the same division time and death rate as CD4+ cells.

Table 2 Estimates for additional parameters used in the extended model

Param.	Description	Estimate
δ_H	Death/turnover rate of effector CD4+ T cells	0.23
δ_K	Death/turnover rate of effector CD8+ T cells	0.4
$H^0(0)$	Initial naïve CD4+ T cell concentration	see Scenario 2
$K^0(0)$	Initial naïve CD8+ T cell concentration	see Scenario 2
s_{H}	Supply rate of naïve CD4+ T cells	$\delta_0 H^0(0)$
s_{K}	Supply rate of naïve CD8+ T cells	$\delta_0 K^0(0)$
m_1	# of divisions in minimal CD4+ developmental program	2
m_2	# of divisions in minimal CD8+ developmental program	7
ρ_H	Duration of one T cell division	11/24
ρ_K	Duration of one T cell division	1/3
σ_H	Duration of min developmental program: $1+(m_H-1)\rho_H$	1.46
σ_K	Duration of min developmental program: $1+(m_K-1)\rho_K$	3
r_1	Rate of secretion of positive growth signal by CD4+ cells	100
r_2	Rate of secretion of positive growth signal by CD8+ cells	1
δ_P	Decay rate of free positive growth signal	5.5
r	Rate of differentiation of effector cells into iTregs	0.02

Other parameters are the same as those used in Table 1 for the simplified model
Concentrations are in units of k/μL, and time is measured in days

In this model, we use the same parameters as in Table 1 for the simplified model, except for those listed in Table 2. We assume that CD4+ and CD8+ T cells have halflives of 3 days and 41 h, respectively, yielding death rates of $\delta_H = 0.23$ and $\delta_K = 0.4$/day [9]. We assume that CD4+ and CD8+ populations have doubling times of 11 h and 8 h, respectively, yielding cell division rates of $\rho_H = 11/24$ and $\rho_K = 1/3$ day [9]. We do not have good estimates of the secretion rates of positive growth signal by effector T cells, hence we estimate that CD4+ and CD8+ T cells secrete growth signal at rates $r_1 = 100$ and $r_2 = 1$/day, respectively. We assume that free positive growth signal decays with a halflife of 3 h, yielding an estimate of $\delta_P =$ 5.5/day. In this model only effector CD4+ T cells can differentiate into iTregs, so the new estimate of the iTreg differentiation rate, r, must be higher than the previous estimate of $r = 0.01$/day to maintain similar dynamics. Hence, in this model, we set $r = 0.03$/day.

3 Mathematical Models of Immunodominance

In this section we explain how to expand our model for monoclonal T cell responses to polyclonal responses. We show that the expanded model automatically recreates elements of the characteristic behavior associated with immunodominance. In this way, we demonstrate that immunodominance may occur as a natural result of iTreg-mediated self-regulation of polyclonal T cell responses.

We begin in Sect. 3.1 with deriving a basic immunodominance model that demonstrated immunodominance as an extension of the basic model of adaptive regulation. We then continue in Sect. 3.2 with an extended model of immunodominance. This extended immunodominance model is the model published in [16]. The basic immunodominance model is new. While the basic model is less accurate from a biological point of view, it still captures the main principles on which the extended model is based.

3.1 A Basic Model

We extend the model from Sect. 2.1 to polyclonal T cell responses. The model includes n T cell clones that react to mature antigen-bearing APCs at different rates, k_i. The model is formulated as the following system of DDEs:

$$\dot{A}_0(t) = s_A - d_0 A_0(t) - a(t)A_0(t), \tag{15}$$

$$\dot{A}_1(t) = a(t)A_0(t) - d_1 A_1(t), \tag{16}$$

$$\dot{K}_i^0(t) = s_{K,i} - \delta_0 K_i^0(t) - k_i A_1(t) K_i^0(t), \tag{17}$$

$$\begin{aligned} \dot{K}_i(t) = {} & 2^m k_i A_1(t-\sigma) K_i^0(t-\sigma) - k_i A_1(t) K_i(t) + 2k_i A_1(t-\rho) K_i(t-\rho) \\ & - (\delta_1 + r) K_i(t) - k R_{\text{total}}(t) K_i(t), \end{aligned} \tag{18}$$

$$\dot{R}_i(t) = r K_i(t) - \delta_1 R_i(t), \tag{19}$$

where $R_{\text{total}} = \sum R_i$ and $i = 1, \ldots, n$. As before, A_0 is the concentration of immature APCs at the site of infection, and A_1 is the concentration of mature antigen-bearing APCs in the lymph node. The variables K_i^0, K_i, and R_i are the concentrations of naïve, effector, and regulatory T cells with specificity #i.

Equations (15) and (16) for the APCs are identical to Eqs. (1) and (2). Equations (17)–(19) are analogous to Eqs. (3)–(5), except that each T cell clone is supplied at a different rate $s_{N,i}$, has its own kinetic coefficient k_i, and effector cells can be suppressed by any regulatory cell, independent of their origin. The supply rate, $s_{K,i}$, of T cell clones is related to the initial concentration of that clonal population by $s_{K,i} = d_1 K_i^0(0)$. From the estimates in [17], the kinetic coefficient $k_i = p_i k_0$, where $k_0 = 40$ and p_i is the probability that T cells of the ith clone react to antigens presented on the APCs. All other parameters are taken from Table 1.

3.2 An Extended Immunodominance Model: Including the Helper T Cells

Following the basic principle of the model in Sect. 3.1, we extend the mathematical model of Sect. 2.2 to polyclonal T cell responses. The model includes n clones that react to mature antigen-bearing APCs at different rates, k_i, and is formulated as the following system of DDEs:

$$\dot{A}_0(t) = s_A - d_0 A_0(t) - a(t)A_0(t), \tag{20}$$

$$\dot{A}_1(t) = a(t)A_0(t) - d_1A_1(t), \tag{21}$$

$$\dot{H}_i^0(t) = s_{\mathrm{H,i}} - \delta_0 H_i^0(t) - k_i A_1(t) H_i^0(t), \tag{22}$$

$$\begin{aligned}\dot{H}_i(t) &= 2^{m_1} k_i A_1(t-\sigma_1) H_i^0(t-\sigma_1) - k_i A_1(t) H_i(t) + 2k_i A_1(t-\rho_1) H_i(t-\rho_1) \\ &\quad - (\delta_H + r) H_i(t) - kR_{\mathrm{total}}(t) H_i(t),\end{aligned} \tag{23}$$

$$\dot{K}_i^0(t) = s_{\mathrm{K,i}} - \delta_0 K_i^0(t) - k_i A_1(t) K_i^0(t), \tag{24}$$

$$\begin{aligned}\dot{K}_i(t) &= 2^{m_2} k_i A_1(t-\sigma_2) K_i^0(t-\sigma_2) - kP(t) K_i(t) + 2kP(t-\rho_2) K_i(t-\rho_2) \\ &\quad - \delta_K K_i(t) - kR_{\mathrm{total}}(t) K_i(t),\end{aligned} \tag{25}$$

$$\dot{P}(t) = r_1 H_{\mathrm{total}}(t) + r_2 K_{\mathrm{total}}(t) - \delta_P P(t) - kP(t) K_{\mathrm{total}}(t) - kP(t) R_{\mathrm{total}}(t), \tag{26}$$

$$\dot{R}_i(t) = rH_i(t) - kP(t) R_i(t) + 2kP(t-\rho_1) R_i(t-\rho_1) - \delta_H R_i(t). \tag{27}$$

Here $H_{\mathrm{total}} = \sum H_i$, $K_{\mathrm{total}} = \sum K_i$, and $R_{\mathrm{total}} = \sum R_i$ for $i = 1, \ldots, n$. As in Sect. 2.2, A_0 is the concentration of APCs at the site of infection and A_1 is the concentration of APCs that have matured, started to present target antigen, and migrated to the lymph node. For each clone i, the variable H_i^0 is the concentration of naïve CD4+ (helper) T cells, H_i is the concentration of effector CD4+ cells, K_i^0 is the concentration of naïve CD8+ (helper) T cells, and K_i is the concentration of effector CD8+ cells, and R_i is the concentration of iTregs. Finally, P is the concentration of positive growth signal.

Equations (20) and (21) are identical to Eqs. (7) and (8). Equations (22), (24), and (27) describe the dynamics of the naïve CD4+ T cells, naïve CD8+ T cells, and regulatory cells, respectively, for each clone i. These equations are identical to Eqs. (9), (11), and (14).

The assumption about the nonspecific suppression of the activated CD4+ and CD8+ T cells is encoded into the model in Eqs. (23) and (25). The last term in both equations shows that the suppression of the activated cells is done using the iTregs that originated from all clones.

Finally, the dynamics of the positive growth signal is proportional to the total population sizes of the activated CD4+ and CD8+ T cells, as well as the total number of iTregs in the system.

4 Results

In this section we present results obtained by simulating the mathematical model from Sects. 2 and 3. We start in Sect. 4.1 with simulations of the basic model of adaptive regulation. We focus our attention on demonstrating the robustness of the model to large variations in precursor frequencies. Additional simulations of this model can be found in [17]. In Sect. 4.2 we present simulations of the immunodominance models. Most of the simulations are of the basic model

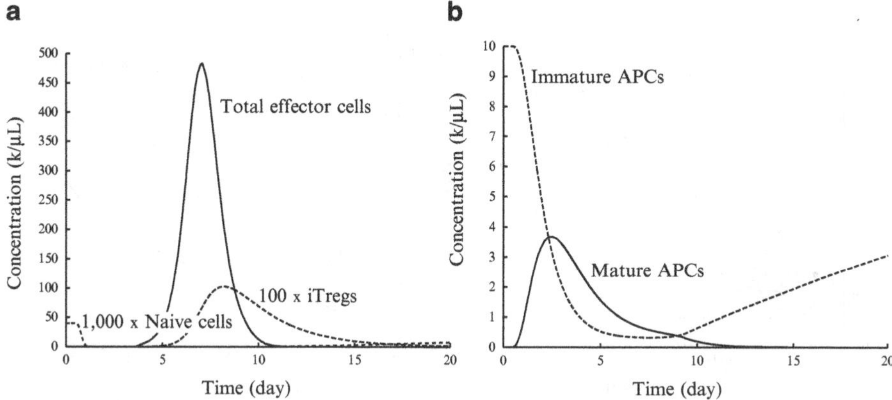

Fig. 4 Time evolution of immune cell populations. (**a**) The dynamics of naïve, effector, and regulatory T cells over 20 days. (**b**) The dynamics of immature and mature APCs

from Sect. 3.1. We provide one example of simulations of the extended model from Sect. 3.2. Additional simulations of the extended model can be found in [16].

4.1 Adaptive Regulation: Numerical Simulations

We start by numerically solving Eqs. (1)–(5). The model parameters are set according to Table 1. The simulations are done using the DDE solver "dde23" in MATLAB R2008a. The time evolution of the different cell populations is shown in Fig. 4.

It is evident from Fig. 4 that nearly all available antigen-specific naïve T cells are recruited within a day of antigen presentation, a result corroborated by the experimental data of [20]. In addition, the effector cell and mature APC populations peak at day 7.0 and day 2.5. In our model, the variable t corresponds to the time after antigen presentation begins in the lymph node. This event occurs approximately one day after infection [3]. Hence, our simulated measurements translate to T cell and APC peaks at day 8.0 and day 3.5 after infection. These results coincide well with the experimental measurements that the T cell and APC populations peak at around day 8 and day 3.2 after infection (see [3, 9]).

Figure 5a displays phase portraits of the iTreg versus the effector population for initial naïve cell concentrations of 0.0004, 0.004, 0.04, 0.4, and 4k/μL. The five curves correspond to population doublings of 18.4, 15.9, 13.6, 11.3, and 9.1, respectively, showing that every tenfold increase or decrease in precursor concentrations corresponds to approximately 2.2 fewer or 2.2 additional divisions that adjust the difference. Thus, larger initial conditions lead to larger T cell responses, but not at the level of sensitivity exhibited by the two program-based

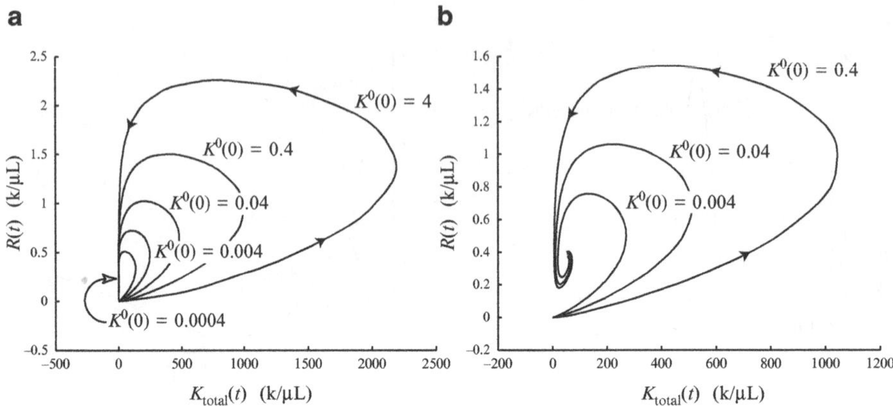

Fig. 5 Phase portraits of iTreg versus effector dynamics over 20 days. (**a**) Five different precursor frequencies, $K^0(0) = 0.0004$, 0.004, 0.04, 0.4, and 4k/μL. The curve for $K^0(0) = 0.04$ corresponds to Fig. 4. (**b**) T cell dynamics under persistent antigen stimulation, i.e., $b = 1{,}000$ days for three different precursor frequencies, $K^0(0) = 0.004$, 0.04, and 0.4k/μL

models. All phase portraits exhibit similar shapes and return to the resting state in a timely fashion. The phase portraits represent the dynamics over 20 days as in Fig. 4.

Figure 5b shows similar phase portraits as in Fig. 5a, except that the duration, b, of antigen presentation is set to 1,000 days so that antigen is chronically presented. The figure shows that the effector and iTreg populations spiral into a stable fixed point. The elongated shapes form as a result of the rapid increase in the level of antigen presentation by mature APCs over the first few days after infection before decaying to a steady level several days later. The brief burst of mature APC levels in the lymph node allows the effector concentration to expand rapidly for a brief time before being attracted to the stable fixed point.

4.2 Immunodominance: Numerical Simulations

We start by showing results that were obtained from simulating the basic immunodominance model from Sect. 3.2. We numerically simulate solutions to Eqs. (15)–(19). The numerical solution is obtained using the DDE solver "dde23" in MATLAB R2008a. We consider several scenarios of multiple T cell clones responding to the same target at once. Each T cell clone is characterized by its reactivity to target antigen, p_i, and its initial concentration, $K_i^0(0)$.

Scenario 1. (Five T cell clones, different reactivities). We consider five T cell clones that differ only in terms of their reactivities to the target antigen. For $i = 1,\ldots,5$ we set reactivity #i as $p_i = 2^{-i}$. The initial concentrations are given by $K_i^0(0) = 0.01$ k/μL, $\forall i$. We also consider cases of single knockout (SKO),

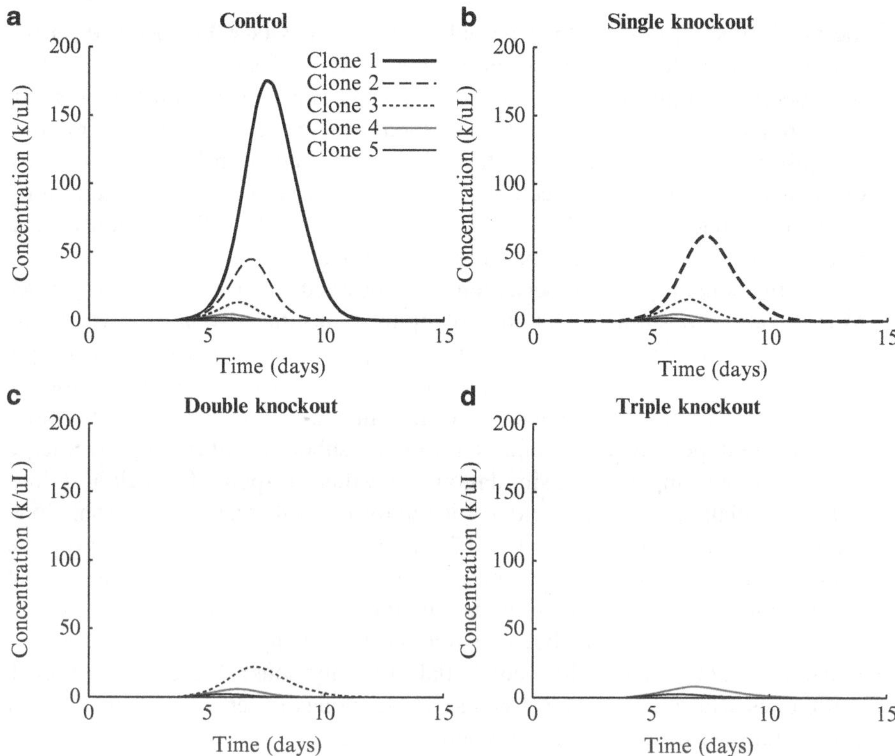

Fig. 6 Basic immunodominance model: time evolution of effector cell clones for Scenario 1. Five T cell clones are present at the same initial concentration $K_i^0(0) = 0.01$ k/μL and reactivities $p_1 = 1/2$, $p_2 = 1/4$, $p_3 = 1/8$, $p_4 = 1/16$, and $p_5 = 1/32$. All other parameters are taken from Table 1. (**a**) Control experiment: clones 1–5 all respond. (**b**) SKO: clone 1 is removed. Only clones 2–5 respond. (**c**) DKO: clones 1 and 2 are removed. Only clones 3–5 respond. (**d**) TKO: clones 1–3 are removed. Only clones 4 and 5 respond

double knockout (DKO), and triple knockout (TKO) experiments in which the T cell responses mediated by one, two, or three immunodominant T cell clones are removed. The following cases are considered:

(a) Control: No T cells are removed. Clones 1–5 all respond.
(b) SKO: clone 1 is removed. Only clones 2–5 respond.
(c) DKO: clones 1 and 2 are removed. Only clones 3–5 respond.
(d) TKO: clones 1–3 are removed. Only clones 4 and 5 respond.

Figure 6 shows the numerical simulations obtained in all four cases. As expected, we see in Fig. 6a that the five T cell clones fall into a hierarchy based on their reactivities. When the dominant clone is removed (a case shown in Fig. 6b), the second most reactive clone partially compensates, i.e., whereas the peak of the response from clone 2 is 44.85 in the control experiment, it rises to 62.39 in the SKO experiment. Similarly, when the two most dominant clones are removed the third most dominant clone partially compensates (see Fig. 6c) and so on, but the ability of less reactive T cell responses to compensate for more reactive ones decreases rapidly. In the case of the TKO experiment shown in Fig. 6d, the immune response from clone 4 is much weaker than the original immune response generated by clone 1 in the control case shown in Fig. 6a. Our study of Scenario 1 shows that T cell reactivities play a strong role in determining immunodominance hierarchies. Furthermore, less reactive T cell clones have limited ability to compensate for more reactive ones.

The phenomenon of compensation was observed experimentally by van der Most et at. who showed that loss of epitope-specific responses was almost inevitably associated with compensatory responses against subdominant epitopes. In addition, their experiments showed that noticeable compensation by a subdominant response depended on the removal of all or most of the more dominant epitopes, creating room, as it were, for subdominant epitopes to emerge [29]. In the same manner, our simulations show that a response from clone 2 does not substantially emerge until clone 1 is removed and that a response from clone 3 does not emerge until clones 1 and 2 are removed, and so on. By extension, a response against a subdominant epitope is likely not to emerge until all or most T cell clones, specific for the dominant epitope (or epitopes), are removed. The degrees of shift in hierarchy become more prominent in the following examples.

Scenario 2. (Four clones, different initial concentrations). We consider four T cell clones with the same reactivities. These clones differ only in their initial concentrations. In this case, the reactivities are set as $p_i = 1/2$, $i = 1,\ldots,4$. The initial concentrations are taken as: $K_1^0(0) = 0.04$ k/μL, $K_2^0(0) = 0.01$ k/μL, $K_3^0(0) = 2.5 \times 10^{-3}$ k/μL, $K_4^0(0) = 6.25 \times 10^{-4}$ k/μL. As before, we consider SKO, DKO, and TKO experiments. Figure 7 shows the results of the numerical simulations.

Figure 7a shows that the four T cell clones fall into a hierarchy based on their initial concentrations. Specifically, the T cell response of clone 1 starts and remains exactly four times higher than that of clone 2. Likewise, the response of clone 2 starts and remains exactly four times higher than that of clone 3, and so on. Since the reactivities p_i are identical for all four clones, the equations determined by Eqs. (15)–(19) for each clone are also identical, meaning that the four T cell responses fall into a linear relation determined by their initial conditions.

When the dominant clone is removed, the second most frequent clone compensates effectively, even though it starts with an initial concentration that is four times less than that of clone 1 (see Fig. 7b). Indeed, the T cell response for clone 2 more than doubles between the control and SKO experiments. Similarly, when the two most dominant clones are removed the third most frequent clone also compensates effectively and so on (see Fig. 6c, d).

Scenario 3. (Two clones, one with a higher reactivity and one with a higher precursor concentration). In Scenarios 1 and 2, we examined the effects of varying reactivities and initial concentrations separately. In this case we vary both parameters and consider two clones. We start by considering a possible

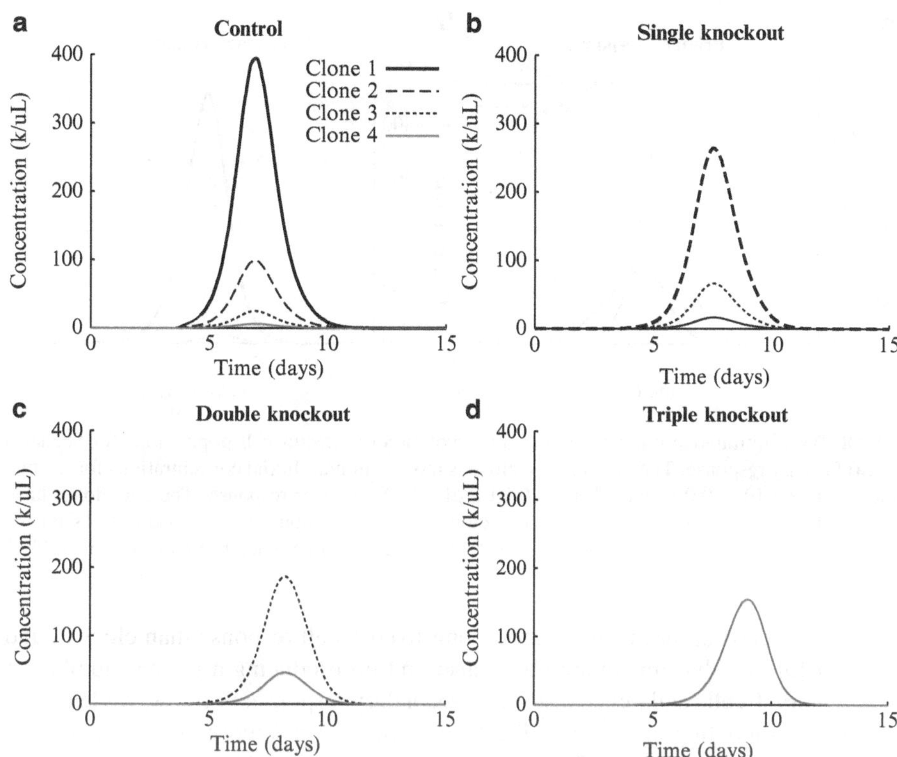

Fig. 7 Basic immunodominance model: time evolution of effector cell populations for Scenario 2. Four T cell clones are present with the same reactivity $p_i = 1/2$ and initial concentrations $K_1^0(0) = 0.04$, $K_2^0(0) = 0.01$, $K_3^0(0) = 2.5 \times 10^{-3}$, and $K_4^0(0) = 6.25 \times 10^{-4}$ k/μL. All other parameters are taken from Table 1. (**a**) Control experiment: clones 1–4 all respond. (**b**) SKO: clone 1 is removed. Only clones 2–4 respond. (**c**) DKO: clones 1 and 2 are removed. Only clones 3 and 4 respond. (**d**) TKO: clones 1–3 are removed. Only clone 4 responds

primary response in which the more reactive clone starts at a lower concentration than the less reactive clone. For our hypothetical secondary response, the initial concentrations are reversed.

1. Reactivity: $p_1 = 1/2$
   Initial concentration: $K_1^0(0) = 0.004$ (primary), 0.04 (secondary) k/μL
2. Reactivity: $p_2 = 1/4$
   Initial concentration: $K_2^0(0) = 0.04$ (primary), 0.004 (secondary) k/μL

Figure 8 shows numerical solutions for Scenario 3. In Fig. 8a we see that the clone with the higher initial concentration dominates during the primary response. Indeed, clone 2 produces a response that is about three times as high as the response of clone 1. However, by day 10, the population of clone 1 persists whereas the population of clone 2 has nearly vanished. The more reactive clone,

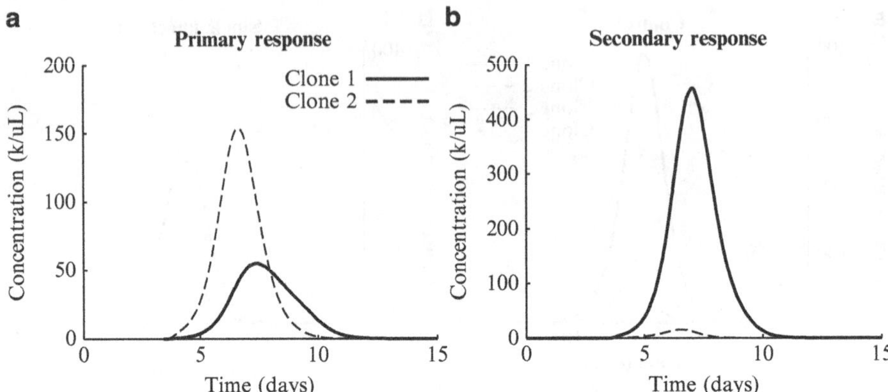

Fig. 8 Basic immunodominance model: time evolution of effector cell populations for Scenario 3. (**a**) Primary response. The less reactive clone is more common. Initial concentrations for the two clones are $K_1^0(0) = 0.004$ and $K_2^0(0) = 0.04$ k/μL. (**b**) Secondary response. The two clones have switched places, and now the more reactive clone is more common. Initial concentrations for the two clones are $K_1^0(0) = 0.04$ and $K_2^0(0) = 0.004$ k/μL. All other parameters are taken from Table 1

clone 1, ends up producing a more long-lived T cell response than clone 2, and so it follows that this clone might also end up producing a greater number of memory T cells and hence a stronger secondary response. For now, we leave the explicit modeling of memory T cell formation for a future work. Nonetheless, we see from Fig. 8a that iTreg-mediated contraction could give rise to a natural process of "collective affinity maturation" that enables the memory repertoire to select for highly reactive clones even when these clones do not produce the most dominant primary responses.

Without explicitly modeling memory T cell formation, let us suppose that between primary and secondary responses, the composition of the T cell repertoire shifts in favor of the more reactive T cell clone. In particular, suppose that for the hypothetical secondary response, the initial concentrations are reversed. Then, Fig. 8b shows that clone 1 clearly dominates the secondary response. Furthermore, both primary and secondary responses start with the same total initial concentration of T cells, but a much stronger response from clone 1 causes the combined secondary response to peak at over twice the height of the combined primary response.

For simplicity, we generated a hypothetical secondary response by switching the initial concentrations of the two T cell populations, but there is no reason to assume that initial concentrations must switch or that the total initial population must stay the same. In fact, the memory pool generated after a primary response is probably larger than the original naïve T cell pool. Yet even with this simplified view of collective affinity maturation, we see that simple shifts in the relative distribution of T cell clones may result in large differences in subsequent responses. Hence, a mechanism of immunodominance mediated by iTregs may serve as a global,

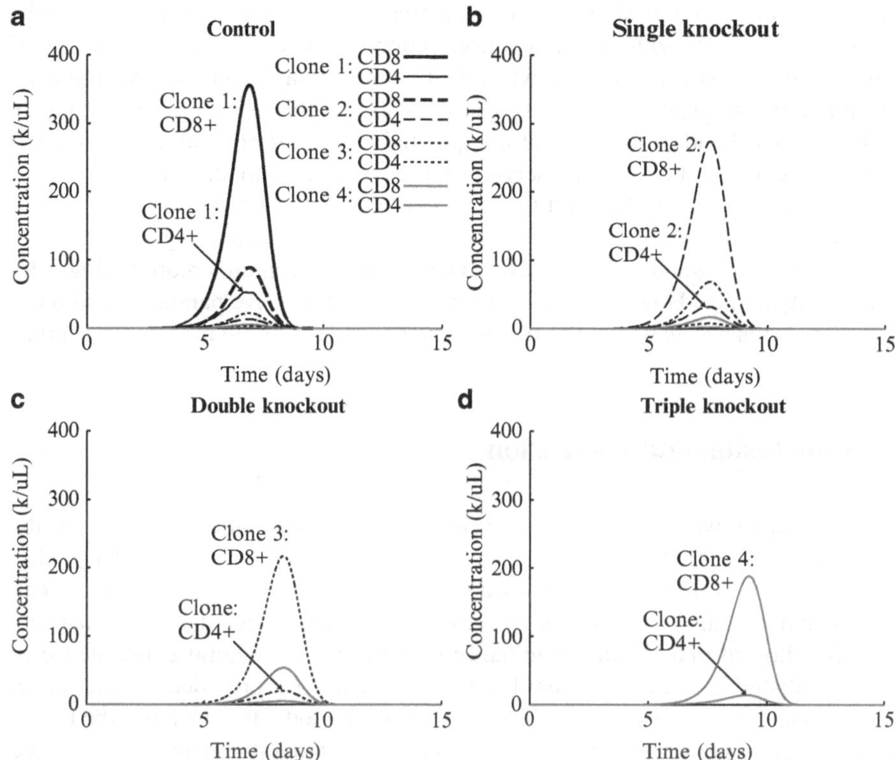

Fig. 9 Extended immunodominance model: time evolution of effector cell populations for Scenario 2. Four T cell clones are present with the same reactivity $p_i = 1/2$ and initial naïve CD8+ concentrations $K_1^0(0) = 0.04$, $K_2^0(0) = 0.01$, $K_3^0(0) = 2.5 \times 10^{-3}$, and $K_4^0(0) = 6.25 \times 10^{-4}$ k/μL. Initial naïve CD4+ concentrations are given by $H_i^0(0) = 1.5K_i^0(0)$. Parameters that are not listed in Table 2 are taken from Table 1. (**a**) Control experiment: clones 1–4 all respond. (**b**) SKO: clone 1 is removed. Only clones 2–4 respond. (**c**) DKO: clones 1 and 2 are removed. Only clones 3 and 4 respond. (**d**) TKO: clones 1–3 are removed. Only clone 4 responds

self-organizing phenomenon among simultaneous T cell responses that serves to improve the overall quality (rather than just the quantity) of the T cell repertoire.

To compare with the basic model Eqs. (1) and (2), we simulate *Scenario 2* from Sect. 4 with the extended model Eqs. (7)–(14). Following Senario 2 of the basic model, we consider four T cell clones with the same reactivities that only differ in terms of their initial concentrations. All reactivities are assumed to be identical: $p_i = 1/2$, $i = 1,\ldots,4$, and the initial concentrations of naïve CD8+ cells are taken as: $K_1^0(0) = 0.04$ k/μL, $K_2^0(0) = 0.01$ k/μL, $K_3^0(0) = 2.5 \times 10^{-3}$ k/μL, and $K_4^0(0) = 6.25 \times 10^{-4}$ k/μL. For each clone, the initial concentrations of naïve CD4+ cells are taken to be $H_i^0(0) = 1.5K_i^0(0)$, which is a typical observed proportion of CD4+ and CD8+ T cells [5].

The results of the simulation are shown in Fig. 9. In Fig. 9a we see that the four T cell clones fall into a hierarchy based on their initial concentrations. When the dominant clone is removed, the second most frequent clone compensates effectively, even though it starts with an initial concentration that is four times less than that of clone 1 (see Fig. 9b). In addition, when the two most dominant clones are removed the third most frequent clone also compensates effectively, and so on (see Fig. 9c, d).

From Figs. 9 and 7, we see that the qualitative behavior of immunodominance seen in the basic model is preserved in the extended model, which explicitly incorporates separate CD4+ and CD8+ T cell dynamics. The basic model allows us to focus on the role of negative feedback between effector and regulatory T cells in producing immunodominance. The extended model captures more biologically accurate dynamics. However, it requires more comprehensive parameter estimates and more extensive analysis. The overall characteristics of both models are similar.

5 Conclusion and Discussion

In this chapter we provided an overview of our mathematical models for the regulation of the primary T cell response and of immunodominance. Our mathematical models were constructed based on a set of basic principles. A robust T cell contraction was shown to emerge as a result of an adaptive regulatory mechanism. We also showed that immunodominance may occur as a natural consequence of iTreg-mediated T cell contraction. For both problems, we provided a basic model that does not include helper T cells and an extended model that includes the helper T cells. Our numerical simulations focused on the immunodominance models. The simulations showed that the qualitative behavior of the simple and of the extended models is identical.

The main point that we emphasized throughout the chapter is that the modeling of these biological phenomena should focus on the basic principles that control the emerging dynamics. While it is desirable that the mathematical models accurately capture the main biological ingredients, certain simplifications allow us to focus on the basic principles. A basic model that captures the desirable qualitative features can be always extended later on to reflect more accurate biology. This was the methodology we followed when developing these models.

Our models do not take into account the suppression of APCs by iTregs, although it is a known function of regulatory T cells [6]. Incorporating suppression of APCs is a direction for a future work and may partly explain why competition is only observed for epitopes presented on the same APC. In this light, considering spatial elements is another relevant extension, since regulatory T cells locally suppress cells in a contact-dependent manner, but no longer inhibit cells that have moved out of the vicinity [28]. In the context of immunodominance, regulatory (or suppressor) T cells give rise to highly localized inhibition that operates only in the context of one or a few common APCs [26]. In their mathematical models, León et al. assume that regulatory and effector cells need to be activated by APCs that are close in space and time in order to interact [18, 19]. Indeed, such localization may be necessary to prevent a regulatory response from shutting down the whole immune system.

One of the consequences of our work is that immunodominance provides a means of peripheral positive selection that may be optimal in most circumstances, since it generates highly adapted responses against specifically targeted antigens (a response that targets the most reactive clones). Such a pattern must be disadvantageous against rapidly evolving pathogens such as HIV or cancer that can evade narrow T cell responses. Hence, our model of iTreg-mediated immunodominance may have implications for improving therapy via T cell vaccinations. In particular, our model suggests a possible negative correlation between immunodominance, driven by contraction, and epitope spread, driven by expansion. In this case, the strength and timing of the iTreg response may cause a shift in T cell dynamics toward a narrower or broader response, i.e., toward immunodominance or epitope spread. From these results, we hypothesize that temporarily suppressing the de novo generation of iTregs following T cell vaccination may result in a broader T cell response than normal against multiple target epitopes, which will then make it more likely for the immune system to eliminate rapidly evolving targets that would otherwise escape immune detection.

Acknowledgements This work was supported in part by the joint NSF/NIGMS program under Grant Number DMS-0758374 and in part by Grant Number R01CA130 817 from the National Cancer Institute. The content is solely the responsibility of the authors and does not necessarily represent the official views of the National Cancer Institute or the National Institutes of Health.

References

1. Antia, R., Bergstrom, C.T., Pilyugin, S.S., Kaech, S.M., Ahmed, R.: Models of CD8+ responses: 1. What is the antigen-independent proliferation program. J. Theor. Biol. **221**(4), 585–598 (2003)
2. Badovinac, V.P., Haring, J.S., Harty, J.T.: Initial T cell receptor transgenic cell precursor frequency dictates critical aspects of the CD8(+) T cell response to infection. Immunity **26**(6), 827–841 (2007)
3. Belz, G.T., Zhang, L., Lay, M.D., Kupresanin, F., Davenport, M.P.: Killer T cells regulate antigen presentation for early expansion of memory, but not naïve, CD8+ T cell. Proc. Natl. Acad. Sci. USA **104**(15), 6341–6346 (2007)
4. Borghans, J.A., Taams, L.S., Wauben, M.H., de Boer, R.J.: Competition for antigenic sites during T cell proliferation: A mathematical interpretation of in vitro data. Proc. Natl. Acad. Sci. USA **96**(19), 10782–10787 (1999)
5. Catron, D.M., Itano, A.A., Pape, K.A., Mueller, D.L., Jenkins, M.K.: Visualizing the first 50 hr of the primary immune response to a soluble antigen. Immunity **21**(3), 341–347 (2004)
6. Chang, C.C., Ciubotariu, R., Manavalan, J.S., Yuan, J., Colovai, A.I., Piazza, F., Lederman, S., Colonna, M., Cortesini, R., Dalla-Favera, R., Suciu-Foca, N.: Tolerization of dendritic cells by T(S) cells: The crucial role of inhibitory receptors ILT3 and ILT4. Nat. Immunol. **3**(3), 237–243 (2002)
7. De Boer, R.J., Perelson, A.S.: T cell repertoires and competitive exlusion. J. Theor. Biol. **169**(4), 375–390 (1994)
8. De Boer, R.J., Perelson, A.S.: Toward a general function describing T cell proliferation. J. Theor. Biol. **175**(4), 567–576 (1995)

9. De Boer, R.J., Homann, D., Perelson, A.S.: Different dynamics of CD4+ and CD8+ T cell responses during and after acute lymphocytic choriomeningitis virus infection. J. Immunol. **171**(8), 3928–3935 (2003)
10. De Boer, R.J., Oprea, M., Antia, R., Murali-Krishna, K., Ahmed, R., Perelson, A.S.: Recruitment times, proliferation, and apoptosis rates during the CD8(+) T-cell response to lymphocytic choriomeningitis virus. J. Virol. **75**, 10663–10669 (2001)
11. Grufman, P., Wolpert, E.Z., Sandberg, J.K., Karre, K.: T cell competition for the antigen-presenting cell as a model for immunodominance in the cytotoxic T lymphocyte response against minor histocompatibility antigens. Eur. J. Immunol. **29**(7), 2197–2204 (1999)
12. Handel, A., Antia, R.: A simple mathematical model helps to explain the immunodominance of CD8 T cells in influenza A virus infections. J. Virol. **82**(16), 7768–7772 (2008)
13. Kaech, S.M., Ahmed, R.: Memory CD8+ T cell differentiation: Initial antigen encounter triggers a developmental program in naïve cells. Nat. Immunol. **2**(5), 415–422 (2001)
14. Kedl, R.M., Kappler, J.W., Marrack, P.: Epitope dominance, competition and T cell affinity maturation. Curr. Opin. Immunol. **15**(1), 120–127 (2003)
15. Kedl, R.M., Rees, W.A., Hildeman, D.A., Schaefer, B., Mitchell, T., Kappler, J., Marrack, P.: T cells compete for access to antigen-bearing antigen-presenting cells. J. Exp. Med. **192**(8), 1105–1113 (2000)
16. Kim, P.S., Lee, P.P., Levy, D.: A theory of immunodominance and adaptive regulation. Bull. Math. Biol. **73**, 1645–1665 (2011)
17. Kim, P.S., Lee, P.P., Levy, D.: Emergent group dynamics governed by regulatory cells produce a robust primary T cell response. Bull. Math. Biol. **72**, 611–644 (2010)
18. León, K., Lage, A., Carneiro, J.: How regulatory CD25+CD4+ T cells impinge on tumor immunobiology? On the existence of two alternative dynamical classes of tumors. J. Theor. Biol. **247**(1), 122–137 (2007)
19. León, K., Lage, A., Carneiro, J.: How regulatory CD25+CD4+ T cells impinge on tumor immunobiology: The differential response of tumors to therapies. J. Immunol. **179**(9), 5659–5668 (2007)
20. Mercado, R., Vijh, S., Allen, S.E., Kerksiek, K., Pilip, I.M., Pamer, E.G.: Early programming of T cell populations responding to bacterial infection. J. Immunol. **165**(12), 6833–6839 (2000)
21. Nowak, M.A.: Immune responses against multiple epitopes: A theory for immunodominance and antigenic variation. Semin. Virol. **7**, 83–92 (1996)
22. Probst, H.C., Dumrese, T., van den Broek, M.F.: Cutting edge: Competition for APC by CTLs of different specificities is not functionally important during induction of antiviral responses. J. Immunol. **168**(11), 5387–5391 (2002)
23. Roy-Proulx, G., Meunier, M.C., Lanteigne, A.M., Brochu, S., Perreault, C.: Immunodomination results from functional differences between competing CTL. Eur. J. Immunol. **31**(8), 2284–2292 (2001)
24. Sakaguchi, S., Yamaguchi, T., Nomura, T., Ono, M.: Regulatory t cells and immune tolerance. Cell **133**(5), 775–787 (2008)
25. Scherer, A., Bonhoeffer, S.: Epitope down-modulation as a mechanism for the coexistence of competing T-cells. J. Theor. Biol. **233**(3), 379–390 (2005)
26. Sercarz, E.E., Lehmann, P.V., Ametani, A., Benichou, G., Miller, A., Moudgil, K.: Dominance and crypticity of T cell antigenic determinants. Annu. Rev. Immunol. **11**, 729–766 (1993)
27. Scherer, A., Salathé, M., Bonhoeffer, S.: High epitope expression levels increase competition between T cells. PLoS Comput. Biol. **2**(8), e109 (2006)
28. Trimble, L.A., Lieberman, J.: Circulating CD8 T lymphocytes in human immunodeficiency virus-infected individuals have impaired function and downmodulate CD3 zeta, the signalling chain of the T-cell receptor complex. Blood **91**(2), 585–594 (1998)
29. van der Most, R.G., Murali-Krishna, K., Lanier, J.G., Wherry, E.J., Puglielli, M.T., Blattman, J.N., Sette, A., Ahmed, R.: Changing immunodominance patterns in antiviral CD8 T-cell responses after loss of epitope presentation or chronic antigenic stimulation. Virology **315**(1), 93–102 (2003)

30. van Stipdonk, M.J., Hardenberg, G., Bijker, M.S., Lemmens, E.E., Droin, N.M., Green, D.R., Schoenberger, S.P.: Dynamic programming of CD8+ T lymphocyte responses. Nat. Immunol. **4**(4), 361–365 (2003)
31. van Stipdonk, M.J., Lemmens, E.E., Schoenberger, S.P.: Naïve CTLs require a single brief period of antigenic stimulation for clonal expansion and differentiation. Nat. Immunol. **2**(5), 423–429 (2001)
32. Wodarz, D., Thomsen, A.R.: Effect of the CTL proliferation program on virus dynamics. Int. Immunol. **17**(9), 1269–1276 (2005)
33. Yang, Y., Kim, D., Fathman, C.G.: Regulation of programmed cell death following T cell activation in vivo. Int. Immunol. **10**(2), 175–183 (1998)

Evolutionary Principles in Viral Epitopes

Yaakov Maman, Alexandra Agranovich, Tal Vider Shalit, and Yoram Louzoun

1 Introduction: Can We Learn from Epitope Distribution on Viral Evolution?

The infection of a cell by a virus elicits a Cytotoxic T Lymphocyte (CTL) response to viral peptides presented by the Major Histocompatibility Complex class I molecules [6, 20]. Such a CTL response plays a critical role in the host's anti-viral immune response [39]. This role is suggested by studies indicating a drop of viral loads and the relief of the acute infection symptoms following the emergence of virus-specific CTLs [8], as well as by data from CTL depleted animal models [33, 41]. The CTL response is also associated with a rapid selection of viral CTL escape variants [23, 34]. In the last few years we have applied an immunomic methodology combining genomic data and multiple bioinformatic tools to study the anti-viral CTL response [5, 19, 28, 35, 38, 55–57] and found that viruses selectively mutate proteins inducing the highest danger to their survival. We here summarize these results, and propose some general conclusions regarding the selective forces affecting viruses and their human host.

In general, CTL epitopes are short peptides that can be recognized by CTLs when presented on MHC-I molecules, as will be further explained. These epitopes originate from short peptides cleaved by the proteasome [48] that can pass through the transporter associated with antigen processing (TAP) and associate non-covalently with the groove of MHC-I molecules. A large fraction of these

Y. Maman
The Mina and Everard Goodman Faculty of Life Sciences, Bar-Ilan University, Ramat Gan, 52900 Israel

A. Agranovich • T.V. Shalit • Y. Louzoun (✉)
Department of Mathematics and Gonda Brain Research Center, Bar-Ilan University, Ramat Gan 52900, Israel
e-mail: louzony@math.biu.ac.il

U. Ledzewicz et al., *Mathematical Methods and Models in Biomedicine*, Lecture Notes on Mathematical Modelling in the Life Sciences, DOI 10.1007/978-1-4614-4178-6_3,

epitopes are nine-mers (although 8-mers, 10-mers and even longer epitopes are also observed). A cleaved nine-mer is presented on an MHC-I molecule only if its affinity to the MHC molecule is sufficiently high. Normally, MHC-I molecules present fragments originated from cellular proteins. After viral infection, virus-originated peptides are presented on these molecules, enabling immune recognition of the virus [6]. We have recently developed and improved a set of bioinformatic tools to estimate all peptides within a virus that can be presented to the immune system (the CTL epitope repertoire) [54].

The human leukocyte antigen (HLA) locus, the locus that encodes the MHC molecules in human, is the most polymorphic locus in the human genome. In the class I locus HLA-A, B and C have over 2,000 reported alleles, and their number keeps increasing [47]. This large polymorphism permits a rapid selection of alleles that can respond to viral threats. On the other hand, viruses can mutate rapidly. For example, the human immunodeficiency virus (HIV) mutation rate is approximately 1.e−3–1.e−4 mutations per base pair per division [13], which is approximately one mutation per division for the entire viral genome. This high mutation rate, coupled with a short viral life cycle (24–72 h for most viruses [23]), allows viruses to modify their epitope repertoire within a short time. We here provide multiple examples to show how viruses have evolved to maximize their infectivity in the presence of threat from the host.

2 Methodology: *SIR* Score

We have defined the normalized epitope density of a protein or an organism as the size of immune repertoire (SIR) score [24, 28, 32, 35, 38, 54, 55]. The number of predicted CTL epitopes from a sequence was computed by applying a sliding window of nine amino acids, and computing for each nine-mer and its two flanking residues whether it is cleaved by the proteasome and whether it binds to the TAP channel and to a given MHC-I allele (Fig. 1). The SIR score was defined as the ratio between the computed CTL epitope density (fraction of nine-mers that were predicted to be epitopes) and the epitope density expected within the same number of random nine-mers. The random nine-mers were taken from a long random peptide built, using the amino acid distribution calculated over the sequences of all fully sequences of viruses are available, and taking into account the correlation between the frequencies of neighbouring amino acids in these viruses [55]. An average SIR score of less than 1 represents an under-presentation of epitopes, whereas an average SIR score of more than 1 represents an over-presentation of epitopes. For example, assume a hypothetical sequence of 1,008 amino acids (1,000 nine-mers) containing 15 HLA A*0201 predicted epitopes. If the average epitope density of HLA A*0201 in a large number of random proteins with an amino acid

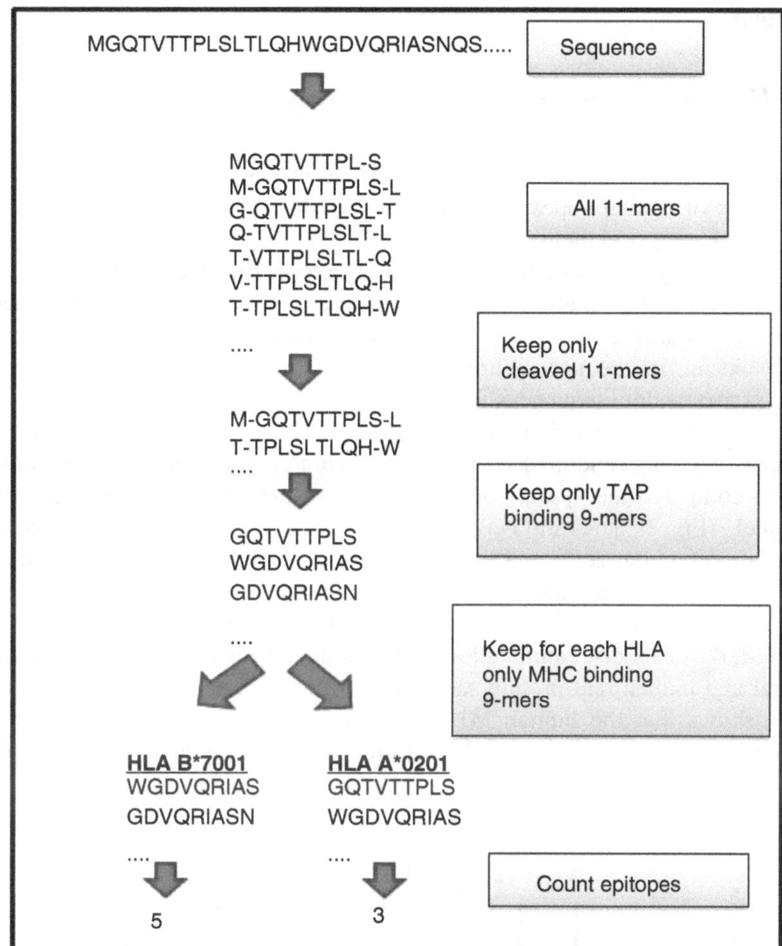

Fig. 1 Algorithm for SIR score computation. Each protein is divided into all nine-mers and the appropriate flanking regions (**a**). For each ninemer a cleavage score is computed (**b**). We compute for all nine-mers with a positive cleavage score a TAP binding score and choose only suprathreshold peptides (**c**). The MHC binding score of all TAP binding and cleaved nine-mers is computed (**d**). Nine-mers passing all these stages are defined as epitopes. We then compute the number of epitopes per protein per HLA allele (**e**). The ratio between the number of predicted epitopes and the parallel number in a random sequence with a similar amino acid distribution is defined as the SIR score

distribution typical to viruses was 0.01 (i.e., 10 epitopes in 1,000 nine-mers), then the SIR score of the sequence for HLA A*0201 would be 1.5 (15/10). The average SIR score of a protein was defined as the average of the SIR scores for each HLA allele studied, weighted by the allele's frequency in the average human population.

3 Results

3.1 Human Versus Viral Epitope Density

In order to check if the MHC allele distribution has evolved to maximize the presentation of viral epitopes, we measured the epitope density of human and viral proteins and compared them in the same allele [5].

Systematically, viral proteins express more epitopes than their human counterpart over the vast majority of alleles (Fig. 2 empty bars) ($p < 1.e-7$). The most natural mechanism for such a selection is that HLA alleles, binding residues that are over-represented in viral sequences are preferred. In order to test this hypothesis, we produced two random sequences with different amino acid distributions. We trained two distinct Markov models on all human and all viral proteins, respectively, and produced very long random sequences (1.e6 amino acids) based on these models. We then compared the epitope densities of these two random sequences in each HLA allele (Fig. 2). The ratio between the epitope density of the random sequence based on viral amino acids and the random sequence based on human amino acid is even larger than the ratio between the real viral and human protein epitope densities (Fig. 2) ($p < 0.02$ T test on the ratio between epitopes from the viral and human Markov models, compared with the ratio between computed epitopes from the viral and human real protein sequences). The difference between the Markov models shows that the human MHC system is evolving to recognize the viral amino acid distribution. The smaller difference between human and viral epitope

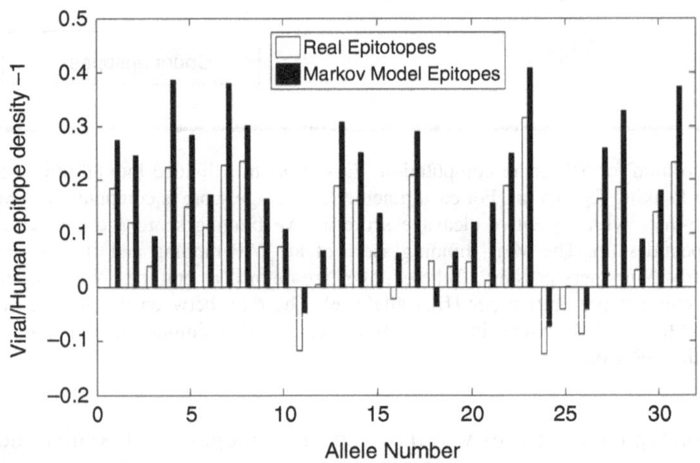

Fig. 2 Comparison of epitope density in human and viral proteins. The *white bars* represent the ratio between the relative change between the viral and human epitope densities [i.e., (viral epitope density/human epitope density) −1]. All positive values represent a higher viral epitope density. The *black bars* represent the same comparison performed on a random sequence based on human and viral amino acid distributions

numbers shows that within the broad specificity of human MHC to preferentially bind epitopes with viral amino acid distribution, viruses specifically mutate their epitopes to limit detection.

3.2 Human and Non-human Hosts

Among viruses, the attempts to mutate epitopes on human HLA alleles should only be observed in human viruses (viruses infecting a human host). We compared the epitope density in viruses infecting a human host and the epitope density in their nonhuman counterparts. In order to compare viruses from human and non human hosts, we used three groups of viruses from different families. We compared the SIR scores of human herpesviruses and the ones of non-human herpesviruses. Five human and 18 non-human herpes strains were tested on human MHC alleles [55] (Fig. 3a). The average SIR score of the human herpesviruses was lower than their non-human counterparts (T-test, $p < 0.003$). A similar result can be observed when comparing HIV I and II with the respective simian immunodeficiency virus (SIV) from which they originated [56] ($p < 0.02$). We computed the SIR score of all HIV-1, HIV-2, SIVcpz, SIVsm, and SIVmac sequences in the LANL and NCBI databases [30]. The SIVcpz, which is the ancestor of HIV-1, has a higher SIR score than average SIR score of all HIV-1 sequences. The SIR score of HIV-2 is also smaller than the SIR scores of its ancestors SIVsm and SIVmac [56] (Fig. 3b). Another virus with human strains as well as homologues infecting other species is Hepatitis B virus (HBV). Viruses similar to HBV exist, among others, in ducks and squirrels. As was the case for the herpes and the HIV, the SIR score of non-human hepatitis viruses is approximately 1, while the SIR score of the HBV is, lower than 1 (Fig. 3c) (T-test, $p < 0.01$).

More generally, when comparing the SIR scores of all human-infecting viruses, it is significantly lower than the one of non-human viruses (Fig. 3d) (T-test, $p < 1.e-7$). However, this last result should be taken with a grain of salt, since the SIR score of different virus families can significantly vary. Thus, the average over multiple families is affected by the number of available fully sequenced viruses in each family. The human papillomavirus (HPV) family (papillomaviridae) is one of the most highly sequenced human viral families. HPV have a low SIR score, reducing the average human viruses SIR score. Thus, the proper analysis is the case-by-case comparison of similar viruses, as was done in the herpes, HIV and HBV cases. In the following sections, we discuss each such virus separately.

3.3 Herpes: The Effect of Different Viral Life Stages

The herpesviridae is a large family of double-stranded DNA viruses that infect a wide range of hosts. Based on biological characteristics the herpesviruses are

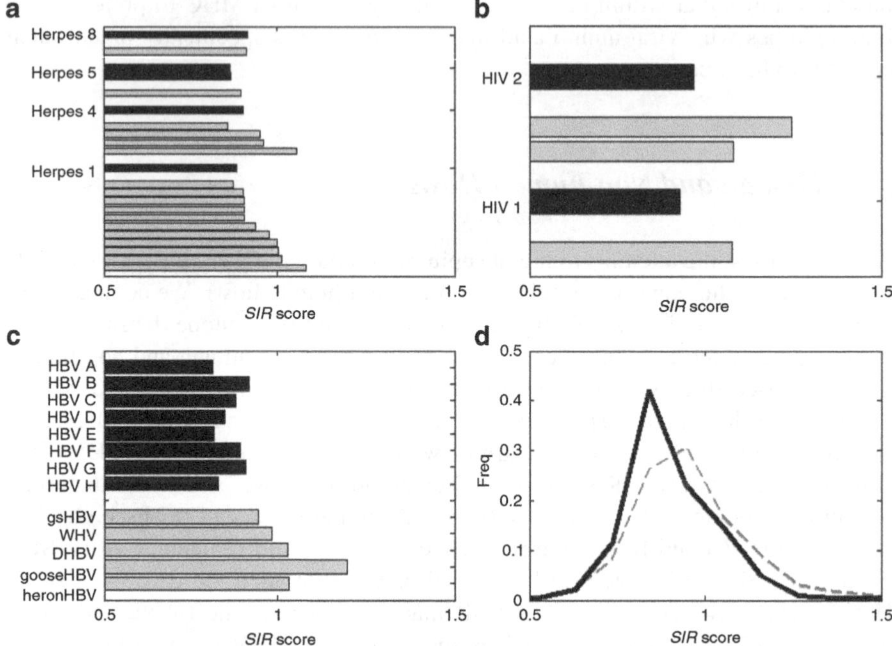

Fig. 3 The average SIR score of human vs. non human viruses. Data is shown for three viruses: herpes virus (HHV-1, HHV-4, HHV-5 and HHV-8), HIV (HIV-1 and HIV-2) and Hepatitis B (strains A-H) virus. The *black columns* represent human strains while the *grey columns* represent non-human strains. In most cases the average SIR score of the human viruses is lower than the non human viruses. The *lower right* drawing represents the SIR score distributions of all full sequenced viruses from the NCBI. In general, the SIR score of the non-human viruses was distributed around 1 while the human virus was less than 1

classified into three major subfamilies and eight herpesviruses. They have been identified as having humans as their primary host. With the exception of KSHV and HSV2, the human herpesviruses are ubiquitous, and infections with these viruses are common worldwide. The herpesvirus genome is 120–250 kbp encoding dozens of genes and several microRNAs. Within all characterized herpesvirus genomes there are conserved regions that mainly encode for structural proteins and replication enzymes, and are more conserved between members of the same subfamily than they are between subfamilies. Herpesviruses exhibit two divergent phases of infection: lytic (productive) and latent (non-productive), characterized by distinct patterns of viral gene expression.

Extensive viral gene expression characterizes productive infection which culminates in virus production and release along with cell lysis and death. In contrast, only few viral genes are expressed and no viral progeny is produced during the latent phase in which the virus genome is maintained as a circular episome in the host cell. Several physiological conditions may induce reactivation of a hidden virus, switching the latent infection into a lytic. In fact, primary host infection with herpesvirus leads to life-long latent infection with periodic reactivation of the virus. A limited set of viral genes that enable the virus to replicate are expressed during latent infection phase of gamma herpesviruses, while in the alpha subfamily, there are no latent proteins, but there are latent RNA transcripts [23]. Latency of the beta subfamily is less understood and appears to rely in host gene control and host differentiation signals. The production of the virus depends on another cascade that classifies the viral genes into three major groups according to their respective expression pattern; immediate-early (IE), early and late. The immediate early (IE) genes encode proteins that are expressed first and are regulatory in nature. These proteins control the expression of viral and cellular genes and are critical in the determination whether the infection will be abortive or productive [15].

We hypothesized that the destruction of the virus during latent or early productive infection would prevent further virus infection and dissemination. We expected to observe reduced exposure of viral proteins expressed during the latent phase of the infection. We investigated the epitope density in proteins expressed in each viral life cycle stage. Here, we focus on: herpes simplex virus type 1 (HSV1-HHV1), human cytomegalovirus (HCMV-HHV5), Epstein-Barr virus (EBV-HHV) and the Kaposi's sarcoma-associated herpesvirus (KSHV-HHV8).

We have first looked for all latent proteins. The KSHV genome encodes over 90 proteins. Eighty-two of these proteins were classified into three groups (latent, IE and lytic) based on their expression phase [15]. We expected the SIR score of the latent proteins to be lower than all others. The 77 EBV proteins were also divided into IE, latent and lytic proteins [45]. In the HSV1 there are only latent transcripts, and no viral proteins are expressed. HCMV also has latent proteins, but a list of such proteins was not found. The HCMV genes were divided into lytic and IE genes.

The SIR score of the latent proteins in both KSHV and EBV was indeed significantly lower than the SIR score for all proteins, while lytic proteins have a higher SIR score (Fig. 4).

If the lower SIR score is indeed a measure of immune-induced selection pressure to reduce the number of epitopes, other immune evasion mechanisms may be used in these high-risk proteins. One such mechanism could be self-mimicry. In other words, one would expect that the fraction of epitopes similar to self to be also higher in the latent group. This was indeed the case for both EBV and KSHV (Fig. 5). The escape from the immune system is thus obtained using both a reduction in the number of epitopes and mimicry. The number of immunogenic epitopes is however still not reduced to zero. The remaining epitopes may either evoke no immune response or develop alternative evasion mechanisms, such as MHC downregulation, or interference with the cleavage process [1, 2]. Obviously, some epitopes are left, since many of the low SIR score proteins are actually very immunogenic. The low SIR score is not evidence of low immunogenecity, but of the fact that the virus attempts to hide the epitopes of this specific protein. In fact, it may actually be the opposite. The proteins that the virus would try to hide may be the most immunogenic proteins.

Virions are also not present in the earliest phase of infection, during the expression of immediate early (IE) genes. The IE proteins are expressed during the

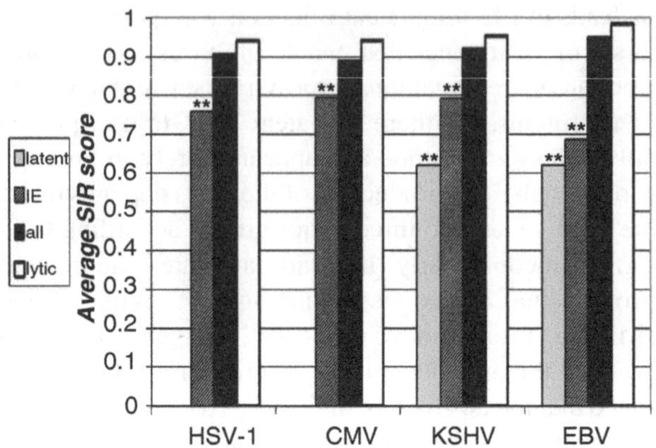

Fig. 4 The average SIR score of latent, IE and lytic proteins. Data is shown for HSV1, CMV, KSHV and EBV. Each column represents proteins highly expressed in different phases: latent, IE, all and lytic. All represents all the viral proteins and IE represent proteins highly expressed in IE phase and so on. *Two stars* represent a p value of less than 0.01 and *one star* represents a p value of less than 0.05. The latent proteins express less epitopes (per candidate nonamer) than the IE (for the KHSV, $p < 0.001$, while for the EBV the difference is not statistically different). Both the IE and the latent proteins express less than the lytic proteins ($p < 0.001$). The values are the average SIR score over all HLAs, weighted by the HLA frequency in the total human population

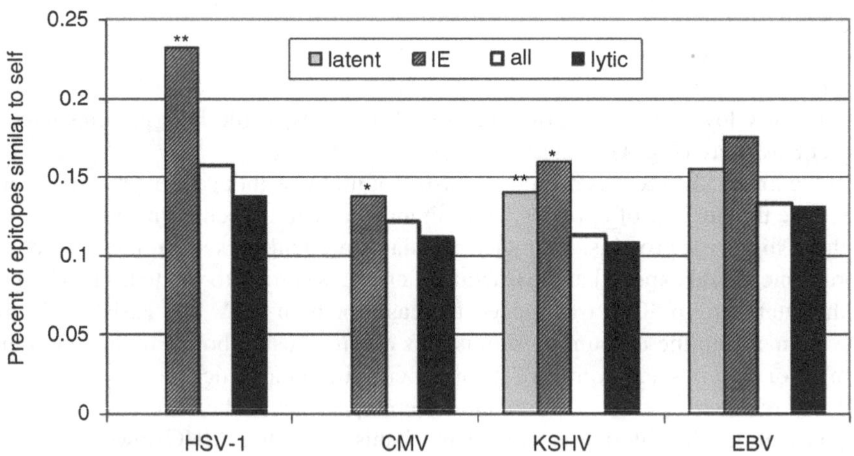

Fig. 5 The similarity of viral epitopes to self-epitopes. Data is shown for all the viruses. Each column represents proteins highly expressed in different phases: All represents all the viral proteins. Each of the other groups represents proteins highly expressed in the lytic, latent and IE phases. The IE and the latent proteins are much more similar to the selfepitopes than the lytic proteins (for HSV-1 and latent KHSV $p < 0.01$, marked with *two stars*, for CMV and IE KHSV $p < 0.05$, marked with a *star*). There is no statistical difference between similarity to self of proteins in the latent and IE groups

initial infection, having mainly regulatory functions (although they are also required for genome replication). As in the case of the latent proteins, one would expect the SIR score of the IE proteins to be low. We defined the IE proteins for the HSV1, EBV, HCMV and KSHV.

The SIR score of the IE proteins was indeed significantly lower than the SIR score of all proteins in the virus, which in turn was lower than the SIR score of lytic proteins (Fig. 4 $p < 1.e-6$). Similarly, the fraction of epitopes similar to self was higher within the IE proteins than in all other epitopes (on average). The SIR score of IE proteins and their similarity to self was intermediate between the latent and all other lytic proteins, although for KSHV there was no statistical difference between the similarity to self of latent and IE proteins. In the early phase of infection, when the stealth of the infection and viral population growth may be critical, it is probably important (from the virus point of view) to reduce the host detection, but not as much as in the latent phase. In contrast with latent proteins, the IE proteins are expressed for a short period, since they are downregulated by the early proteins. Thus, even if they are partially presented to the immune system, CTLs have a limited time to use the IE epitopes in order to destroy the virally infected cells.

3.4 HIV: A Second Example of the Same Principle

As in the herpes case, beyond the total reduced number of epitopes, we can see in HIV a non-uniform epitope density distribution among proteins. We separated the HIV-1-M(B) CTL epitope repertoire into its nine genetic components and computed the appropriate SIR score for each protein. Within the nine HIV proteins, a clear and logical hierarchy of the SIR score emerges. Regulatory proteins (Tat and Rev) have very few epitopes left already at the HIV ancestral sequence. Their SIR score is close to the basal level induced by the errors inherent to the prediction algorithms. The "late" virion associated proteins (Gag, Pol and Env) are found at the top of the list. Accessory proteins also have an intermediate SIR score (Fig. 6). This is in good agreement with the observed critical role of Tat specific CTL in stopping the acute infection stage [27] and with the differential total variability of the HIV genes [3, 7]. Epitopes from proteins that are produced during the first stages of the HIV gene replication after cell entry, such as the regulatory proteins Rev and Tat, are exposed early to CTLs within the cell. Their detection by CTLs can lead to cellular destruction, long before new virions are produced.

Epitopes originating from Gag and Env proteins are presented to CTLs only later in the infection. The detection of such epitopes may not critically impair HIV, as some virions may have the time to bud before the cell is destroyed. The same trend was observed using the sub-Saharan allele frequency (Fig. 6). One can actually observe similar results at the single allele level when looking at frequent HLA alleles (that are probably highly sampled in the non-serotyped population) (data now shown).

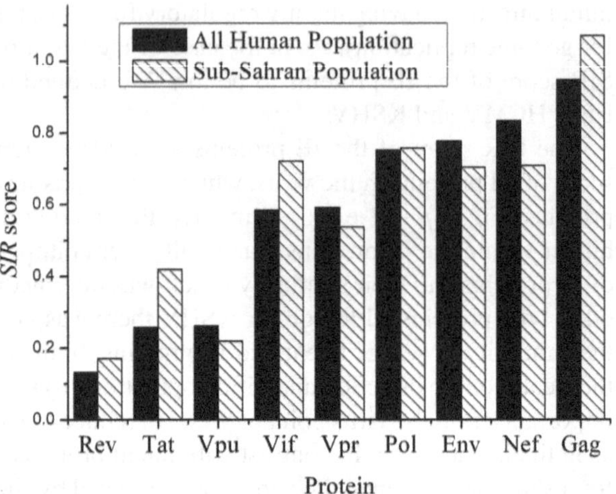

Fig. 6 Average SIR score of HIV-1 genes. The average SIR score was computed for two possible populations: the all human population and the sub-Saharan population. The genes were ordered according to their score in the all human population

The SIR results presented up to now represents the average effect on the population of mutations. These results may be affected by the HLA allele usage in the population, or by sampling effects on the HLA usage of HIV carriers.

We have thus repeated the analysis using a smaller number of sequences, where the host HLA serotyping is known. These sequences are the aggregated results of all HIV-1 clad B cohorts available in the LANL database. We computed for these HIV sequences the SIR scores in the HLA allele of their host, and used protein/HLA combinations for which at least ten sequences were available, to obtain reliable averages. As was observed in the average analysis, Tat and Rev systematically have low SIR scores, while Gag has high ones (Fig. 6). Interestingly, when comparing the SIR of HIV genes on the HLA of their host with the average SIR score of all sequences (i.e. serotyped and un-serotyped sequences) of the same gene on the same HLA allele, the weighted average of the difference is significantly negative for the regulatory genes Tat and Rev and significantly positive for Gag ($R = 0.2$, $p < 0.0001$) (Fig. 7). One can thus clearly see from multiple angles a positive selection of epitopes in Gag and a negative selection of such epitopes in regulatory genes in hosts with matching HLA.

The SIR score only provides statistical information. In order to translate these results to the properties of specific epitopes, we tracked the HLA-A*0201 Gag, Pol and Tat epitopes over the last 30 years. HLA-A*0201 is the most frequent allele, and a large number of experimental epitopes were tested on this allele. Tat has no epitopes in the ancestral sequence and only two epitopes (QPLQIVVIV and VPIAIVKSV) in all early sequences; both appear in a single sequence, and disappear over time. Some transient epitopes appear, but only briefly. Not only do the epitopes themselves disappear, but their position in the genome is highly mutated. By the end of our sampling it is completely different (only 30 % similarity). Interestingly, once the epitopes disappeared, this position remained conserved. Rev evolution is very similar to Tat, showing a systematic elimination of the epitopes.

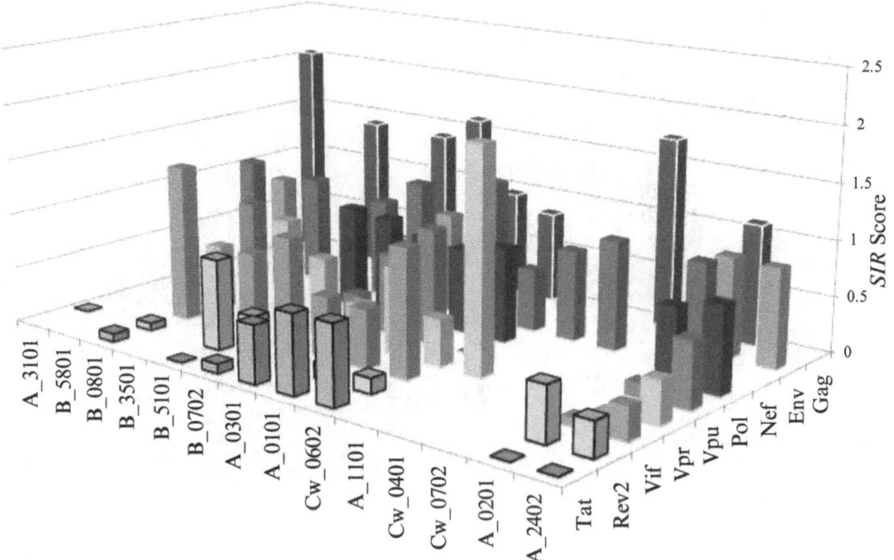

Fig. 7 The SIR score of all candidate HLA alleles for different proteins in serotyped hosts. Tat and Rev systematically have low SIR scores, while Gag has high ones. Not all proteins had measured sequences in all alleles. *Empty spots* represent protein/allele combinations with less than ten sequences

A very different trend was observed in Gag/Pol. Gag has originally a large number of computed epitopes, including the immunodominant SLYNTIATL. The original epitopes are actually highly conserved over time, and a good alignment to the original epitopes can be found in 85 % of the sequences in the last recorded period. Pol demonstrates a similar epitope evolution to Gag, showing again the very different evolution of Tat, Rev, Gag and Pol.

3.5 Early Versus Late Viral Proteins in Multiple Viruses

The main conclusion from these two examples is that early proteins are selected to present less epitopes than late ones. We have systematically tested if that was the case in the vast majority of viruses [57]. We have compared early to late proteins in 24 different viruses, mostly adenoviruses, herpesviruses and HPV. In most HLA alleles tested the average SIR score ratio of late to early proteins is higher than one. This result is valid for most viruses, as well as for the average of all viruses (all positive values in Fig. 8 represent allele/virus combination with a higher epitope density in late proteins than in early proteins). The difference between early and late proteins is significant, when comparing the average SIR over all alleles (T test, $p < 0.0001$), or when the SIR for each allele is taken into account (Anova, $p < 1.e-100$).

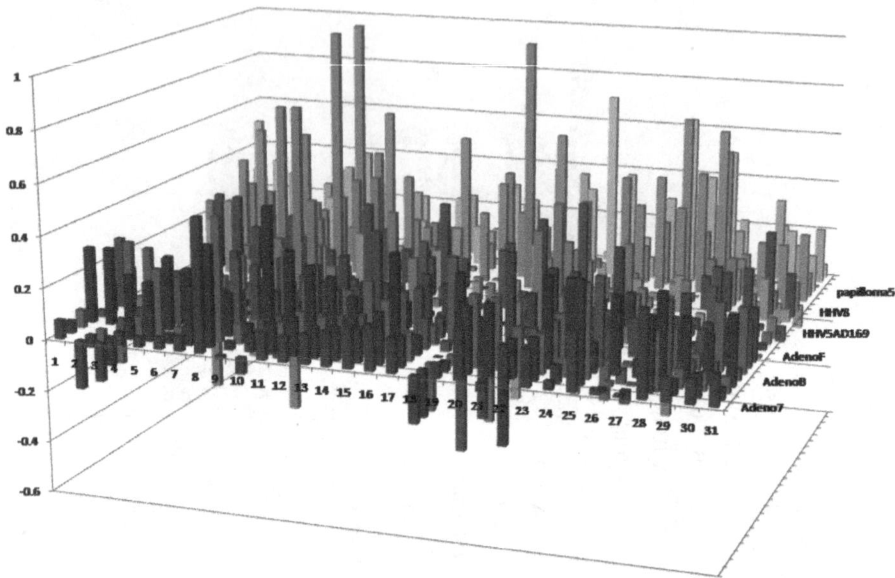

Fig. 8 SIR score of early vs. late proteins. The data is shown for 24 viruses of all candidate HLA alleles (31 alleles). Each column represents the ratio between the difference of the late and early SIR score to their $sum([SIR(late) - SIR(early)]/[SIR(late) + SIR(early)])$. For most $HLA/virus$ the ratio is more than zero, indicating a significant positive difference in the number of presented epitopes between these groups

Thus, quite systematically, late viral proteins express more epitopes than early ones. Assuming viruses would make all possible efforts to evade detection, one could assume that viruses would remove all epitopes through mutations. However, given the large number of epitopes resulting from the MHC polymorphism and the probable cost of mutations, viruses are probably limited in their attempts to reduce epitopes. Given these restrictions, our results show that most of the effort is targeted to early proteins that are probably the most dangerous for the virus.

3.6 *The Effect of Protein Density: HBV*

Another factor that might determine the strength of immune induced selection on a viral protein is its copy number. In general, it is more advantageous for the virus to remove epitopes in proteins with high copy numbers than in proteins with low copy numbers. The latter would have a low total epitope number even if each protein copy has a high epitope density. Epitope generation from degraded molecules occurs at a rate of about 1/10,000 epitope/molecule [44]. Thus if a protein has a low copy number in the cell, it will practically never produce epitopes. The effect of copy number on immune-induce selection is here tested using the HBV.

HBV, the only member of the hepadnaviridae family that infects humans, is a small, enveloped virus with a partially double-stranded circular DNA genome of 3.2 kbp. HBV has a high prevalence—with about three hundred fifty million carriers of HBV world-wide [58]. The HBV genome contains the following four protein-coding open reading frames: (a) The preCore/Core reading frame encodes for the capsid protein (Core) and for the HBeAg protein whose function is not fully clear. HBeAg is known to be secreted and is thought to have a role in the regulation of the immune response [11, 12, 18, 42, 43]. (b) In the same transcript (called pgRNA), which also acts as a template for the virus replication, there is the open reading frame for Polymerase, which has a reverse transcriptase activity [10, 16]. (c) The preS\S open reading frame encodes for Surface proteins: large, middle and small intermembrane proteins located on the ER membrane. The large Surface protein is probably the protein that interacts with the receptor on the hepatocyte membrane and has a role in the release of the virus from the cell [10, 16]. (d) The fourth open reading frame encodes for the X protein which thought to have transcription regulation activity by some studies [22, 37]. It is also proposed to have a cytosolic function as a regulator of proteasome cleavage of some proteins [49]. The entire Surface gene, the C-terminus of preCore/Core and the N-terminus of X overlap with Polymerase [17].

X can accumulate 10,000–50,000 copies per cell in WHV-hepadnavirus that infects woodchucks [14]. We assume that the protein copy number in HBV is not very different. Surface is a structural protein. It also exists in multiple copies per virion. By the same logic, the virus should attempt to hide it. Thus, simply based on the copy number, it is expected that X, Core and Surface should be subject to a more stringent pressure than Polymerase.

We first tested for a general decrease in the epitope density in HBV. We used the SIR score to evaluate the epitope density in each protein compared to the score of the same protein in non-human orthologues. The SIR baselines defined by random viruses can bias the result, since different proteins have different amino acid compositions. While the basic characters of a protein are conserved during evolution, the immunological characters are species-specific. Non-human hosts have different MHC and TAP molecules (although they share a similar proteasome) [9, 31, 40, 51]. Thus, if a specific evolutionary pressure induces epitope removal in HBV, its SIR score should be lower than the ones of non-human hepadnaviruses. The average HBV SIR score as well as the scores of the HBV Surface, Core and X proteins are indeed significantly lower than in other hepadnaviruses (T test, $p < 5e-3$, Fig. 9) (note that X is expressed exclusively in mammalian hepadnaviruses). In Polymerase, however, the SIR score of the HBV protein is similar to or even higher than in the non-human hepadnavirus protein (T test, $p < 5e-3$).Thus, evolutionary pressure seems to affect the epitope number in Core, Surface and X, but not in Polymerase. The high epitope density in Polymerase can either be due to a high fitness cost of mutations or the weak immune pressure induced by Polymerase, as shall be further discussed.

Since the above SIR score is averaged over the 33 most frequent HLA alleles, we further tested the SIR score of HBV for each allele separately and compared it

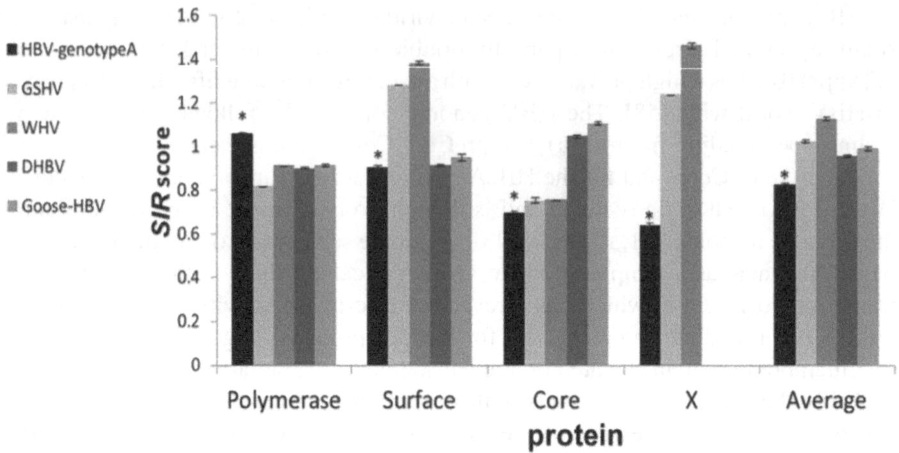

Fig. 9 Hepadnaviruses proteins SIR score. SIR score (Hepatitis B Virus (HBV), Ground Squirrel Hepatitis Virus (GSHV), WoodChuck Hepatitis Virus (WHV), Duck Hepatitis B Virus (DHBV), Goose Hepatitis B Virus) proteins. Avihepadnaviruses lack the X protein. **pvalue* << 0.005

to all other hepadnaviruses in our study (GSHV, WHV, DHBV and GHBV) for the same allele. As expected, the same pattern (namely, lower SIR score for the average of all HBV proteins as well as specifically, Core, Surface and X proteins, and higher SIR score for Polymerase in HBV compared with other hepadnaviruses) was shown in most alleles, mainly in the alleles with higher frequency in human population (T-test, $p < 0.05$).

Among the proteins found to have low SIR scores in HBV, the SIR score of Surface is relatively high while Core and X proteins have significantly lower scores (T-test, $p < 1.e - 33$). This difference can originate from differences in the immune pressure induced by these proteins or by inherent limitations on the amino acid composition. To differentiate between the two mechanisms, we tested the "neutral SIR" of each protein. We define the "neutral SIR" as the SIR score of random sequences derived from scrambling each protein. In other words, instead of scrambling full viral genomes, we scrambled each viral protein by itself 1,000 times and compared the scrambled sequences SIR score distribution to the SIR score distribution of 107 non-scrambled sequences of the same protein (data not shown). The real sequences of Core, Surface and X have a lower average SIR score than the scrambled sequences (T-test, $p < 1.e - 20$). Polymerase real sequences have higher SIR scores ($p < 1.e - 15$) than their scrambled version. Note that the differences between Core, X and Surface disappear when a comparison is performed between the real SIR score and the "neutral SIR" score. Thus, Core, Surface and X seem to induce a similar immune response.

To summarize, one can conclude from all these examples that at least two elements affect the viral immune escape attempts:

- The timing of the protein expression.
- The level of expression.

In the following section we develop a model to explain these two effects.

3.7 Mathematical Model

We have recently modelled a virus [4], as a set of different proteins expressed in a host cell. Each protein i has its own expression profile $x_i(t)$—concentration of protein i at time t—and can induce a polyclonal T cell response $T_i(t)$. Each T_i represents the number of cells in a group of clones reacting to the epitopes in a given protein. For the sake of simplicity we denote T_i as a clone, although it is not a single clone in the strict sense of the word, since it is composed of multiple clones reacting to possibly different epitopes on the same protein. Finally, we denote by p_i the number of different epitopes presented by each viral protein.

The viral T lymphocytes dynamics can be simplified by the following multidimensional ODE:

$$\frac{\mathrm{d}x}{\mathrm{d}t} = h(t), \qquad \frac{\mathrm{d}T}{\mathrm{d}t} = f(p, \{x\}, T), \tag{1}$$

where x and p are the vectors containing $x_i(t)$, p_i and T_i, $h(x)$ represents the dynamics of the viral proteins following cell entry in a given cell, and $f(p, \{x\}, T)$ represents the dynamics of T cell clones during the course of a disease. The dynamics of x are the dynamics of proteins inside an infected cell with a typical timescale of hours. These dynamics end at cell death or following budding. The dynamics of T cells evolve at a much slower timescale and span many viral life cycles. Moreover, the dynamics of T cell clones are affected by all infected cells. We denote the set of the $x_i(t)$ values in all infected cells as $\{x\}$.

If a given protein has x_i copies and p_i epitopes per copy, it presents a total of $p_i x_i$ epitopes. One can roughly estimate the total number of epitopes presented by a virus in a cell as $\sum_i p_i x_i(t)$, assuming no large differences in the affinity of each peptide. Further assuming a constant killing rate per T cell and per epitope (μ_0), the total killing rate before budding of a given virally infected cell can be estimated by

$$\mu(t) = \mu_0 \int_0^{t_{\mathrm{budding}}} \mathrm{d}t \sum_i p_i x_i T_i. \tag{2}$$

The probability that this host cell would survive for a long enough time allowing viral budding is thus:

$$P(survival) = \exp\left(-\mu_0 \int_0^{t_{\mathrm{budding}}} p_i x_i(t) T_i(t) \mathrm{d}t\right). \tag{3}$$

The virus host dynamics can be treated as an epidemic model, where each cell is susceptible. The virus can spread among the host cells if each infected cell infects on average at least one susceptible cell. This is the parallel of having an R0 value of more than 1 in the SIR model [26]. One can assume that on evolutionary timescales, viruses have evolved to maximize this survival probability. We propose to compute the distribution of p maximizing $P(survival)$. Other factors obviously affect the survival probability, but they do not depend on p and will not affect the optimization results.

The simplest distribution of p would obviously be a constant value of zero epitopes for all viral proteins. However, given the fitness cost of adaptations, a more complex picture emerges. Assuming each mutation has a multiplicative cost on the survival probability, the survival probability is

$$P(survival) = \prod_i d^{p_{i0}-p_i} \exp\left(-\mu_0 \int_0^{t_{\text{budding}}} \mathrm{d}t \sum_i p_i x_i T_i\right), \tag{4}$$

where d represents the contribution of each mutation to the death probability ($d < 1, p_{i0}$ —basal number of epitopes of protein i). The results of the optimization will be similar even if a more complex function of the mutations is assumed:

$$P(survival) = \varphi\left(\sum_i p_i\right) \exp\left(-\mu_0 \int_0^{t_{\text{budding}}} \mathrm{d}t \sum_i p_i x_i T_i\right)$$

as long as φ is a positive function with a minimum at 0. We do not consider here the cost of adaptations increasing the epitopes number.

Instead of maximizing $P(survival)$, $-log(P(survival))$ is minimized, leading to the following minimization problem:

$$L(p^*) = \min_{p_i \geq 0}\left(\mu_0 \int_0^{t_{\text{budding}}} \mathrm{d}t \sum_i p_i x_i T_i + \overline{d} \sum_i (p_{i0} - p_i)\right), \tag{5}$$

where $\overline{d} = -\ln d > 0$. This problem is formally similar to assuming a constraint on the number of epitopes ($\sum_i p_i > c$) and minimizing:

$$\mu_0 \int_0^{t_{\text{budding}}} \mathrm{d}t \sum_i p_i x_i T_i.$$

We here test the effect of two aspects of viral protein dynamics on the optimal epitope number: the expression time and the total protein copy number. We assume a simplistic model of viral protein dynamics to test these effects. We focus on the life cycle of virus in a given cell and compute the optimal epitope density distribution within this life cycle. This distribution is affected by the T cell distribution. We present two extreme cases: the first case occurs when a large number of cells are infected, and the protein expression in a given cell has no effect on the T cell population. The second case is when very few cells are infected and the T cell clones are small. In such a case, the T cell clone growth rate is determined by the activation of T cells, which is in turn determined by the epitopes presented.

In the first case, the viral entry time is different for each infected cell, and the expression time of viral proteins is not synchronized among cells. We denote this case as the asynchronous model. The second case represents the initial stages of the disease, where we assume a small number of virions seeding host cells. We assume that in this case, all cells in a given generation of infected cells are infected simultaneously. This case is denoted as the synchronous model.

General Solution

The minimization problem (5) can be translated to a positive constrained minimization problem:

$$\mu(p^*) = \min_{\sum_i p_i \geq c, p \geq 0} \mu(p). \tag{6}$$

The first constraint in Eq. (6) represents the fitness cost of mutation. In the synchronous model that represents the earliest phases of the infection, we assume that the T cells are being activated by the epitopes and that their population is low enough to ignore possible saturation terms. The asynchronous model represents the late or chronic stages of the disease, where we assume that the T cell populations are in steady state. In such a case, their concentration is determined by the average activation signal they receive. These considerations lead to the following simplified dynamics of two concerned cases:

$$\dot{x} = h(x), \qquad \dot{T}_i = p_i x_i T_i, \qquad 1 \leq i \leq n \tag{7}$$

or

$$\dot{x} = h(x), \qquad \dot{T}_i = < p_i x_i >, \qquad 1 \leq i \leq n. \tag{8}$$

The exponential growth model describes the synchronous case. The second model describes the case where the T cell populations have reached a steady state and is denoted as the asynchronous model. In all our models, we do the assumption that all virions bud before some particular time t_{budding}.

Asynchronous Model

A long time after the initial infection, the T cell clones population size can be seen as static through the life cycle of a virus and proportional to the T cell average activation rate. This activation rate is in turn proportional to the average copy number of the appropriate viral protein. The system dynamics can be described as

$$\begin{cases} \dot{x} = h(x), \\ T_i^{as} = T^{as} p_i \int\limits_0^{t_{\text{budding}}} x_i \mathrm{d}t, 1 \leq i \leq n, \end{cases} \tag{9}$$

where we name the constant concentration of T lymphocytes of type i as T_i^{as}. The infected cell clearance rate until budding can be computed by substituting the solution of Eq. (9) into Eq. (6) to get $\mu(p) = C_1 \sum_i A_i (p_i)^2$ where $A_i = g_i^2$ and $g_i = \int_0^{t_{\text{budding}}} x_i \mathrm{d}t$. We look for the set p^* satisfying

$$\mu(p^*) = \min_{\sum_i p_i \geq c, p \geq 0} \sum_i A_i (p_i)^2. \tag{10}$$

Solution of the Optimization Problem

For all i, $A_i > 0$ and the optimization problem is a constrained convex quadratic optimization problem that can be solved using the KKT (Karush–Kuhn–Tucker) theorem [25, 29], leading to a unique positive minimum [36].

The inequality constraint $\sum_i p_i \geq c$ is always active ($\sum_i p_i = c$). We can thus write $p_{i0} = c - \sum_{i \neq i0} p_i$ where i_0 is the index of some nonzero element p_i and reformulate the optimization problem (10) as follows:

$$\min_{p \geq 0} \left(\sum_{i \neq i_0} A_i (p_i)^2 + A_{i_0} \left(c - \sum_{i \neq i_0} p_i \right)^2 \right). \tag{11}$$

The Lagrangian is

$$L(p, \eta) = \left(\sum_{i \neq i_0} A_i (p_i)^2 + A_{i_0} \left(c - \sum_{i \neq i_0} p_i \right)^2 \right) - \sum_{i \neq i_0} \eta_i p_i. \tag{12}$$

Up to a constant difference, this is equivalent to the maximization problem in Eq. (5). The KKT equations are

$$\frac{\partial L}{\partial p_i} = 2A_i p_i - 2A_{i_0} \left(c - \sum_{i \neq i_0} p_i \right) - \eta_i = 0, \quad 1 \leq i \leq n \text{ and } i \neq i_0, \tag{13a}$$

$$\frac{\partial L}{\partial \eta_i} = -p_i \leq 0, \qquad 1 \leq i \leq n \text{ and } i \neq i_0, \tag{13b}$$

$$\eta_i p_i = 0, \qquad 1 \leq i \leq n \text{ and } i \neq i_0, \tag{13c}$$

$$\eta_i \geq 0, \qquad 1 \leq i \leq n \text{ and } i \neq i_0. \tag{13d}$$

From Eq. (13a) we get that

$$\eta_i = 2A_i p_i - 2A_{i_0}\left(c - \sum_{i \neq i_0} p_i\right).$$

A strictly positive solution can be obtained if $\eta_i = 0$ for all $1 \leq i \leq n$ and $i \neq i_0$, yielding

$$p_i = \frac{A_{i_0}}{A_i} p_{i_0}, \qquad 1 \leq i \leq n \quad \text{and} \quad i \neq i_0$$

that, when substituted into condition $\sum_{i=1}^{n} p_i = c$, takes the form

$$p_i = \frac{c/z}{A_i}, \tag{14}$$

where $z = \sum_{j=1}^{n} \frac{1}{A_j}$ (i.e., the number of epitopes is inversely proportional to A_i). We have proven that all p_i are strictly positive in the optimum solution [4].

Synchronous Model

In some viral infections, such as HIV, long-term reservoirs of infected cells are produced at the earliest stages of the infection [21, 46, 53]. Thus, this very early stage may be the most critical from the viral perspective. In the initial period of the infection, the T cell clone size growth can be described by a mass-action formalism [50, 52]. Furthermore, assuming the invading virions simultaneously invade a very limited number of host cell, we compute the T cell dynamics with all viral proteins synchronized, to obtain:

$$\dot{x} = h(x), \qquad \dot{T}_i = p_i x_i T_i, \quad 1 \leq i \leq n. \tag{15}$$

Equation (15) can be substituted into Eq. (6) to produce the following non-linear convex optimization problem:

$$\min_{p \geq 0, \sum_{i=1,\ldots,n} p_i \geq c} (\mu) = min \sum_{i=1,\ldots,n} T_i^{0} A_i^{p_i}, \tag{16}$$

where

$$A_i = \mathrm{e}^{g_i} \qquad \text{and} \qquad g_i = \int_0^{t_{\text{budding}}} x_i \mathrm{d}t.$$

Solution of the Optimization Problem

Equation (16) is a non-linear optimization problem and can be solved again using the KKT (Karush–Kuhn–Tucker) conditions [25, 29]. Since it is a convex optimization problem, the KKT point is the unique minimum [36]. The Lagrangian of Eq. (16) is

$$L(p,\eta,\nu) = \sum_i T_i^0 A_i^{p_i} - \nu\left(\sum_i p_i - c\right) - \sum_i \eta_i p_i. \tag{17}$$

For the entire development of KKT see in [4]. As in the asynchronous case, we can write $p_{i_0} = c - \sum_{i \neq i_0} p_i$ and formulate a new optimization problem:

$$\min_{p_i \geq 0}\left(\sum_{i \neq i_0} T_i^0 A_i^{p_i} + T_{i_0}{}^0 A_{i_0}{}^{c-\sum_{i\neq i_0} p_i}\right) \tag{18}$$

with the following solution for the active constrains:

$$\eta_i = T_i^0 A_i^{p_i} lnA_i - T_{i_0}{}^0 A_{i_0}{}^{c-\sum_{i\neq i_0} p_i} lnA_{i_0} \tag{19}$$

for $1 \leq i \leq n$ and $i \neq i_0$. The fraction of active constraints is a function of T_i^0 and A_i. For all nonzero solutions, the value of p_i is

$$p_i = ln\left(\frac{T_{i_0}{}^0 lnA_{i_0}}{T_i^0 lnA_i} A_{i_0}{}^{p_{i_0}}\right) / lnA_i. \tag{20}$$

Equation (20) can be simplified by defining

$$Z_i = \frac{T_{i_0}{}^0 lnA_{i_0}}{T_i^0 lnA_i}$$

to obtain

$$\begin{aligned} p_i &= \frac{1}{lnA_i}\left[lnZ_i + \left(c - \sum_j \frac{lnZ_j}{lnA_j}\right) \Big/ \left(\sum_j \frac{1}{lnA_j}\right)\right] \\ &= \frac{lnZ_i + C_1}{lnA_i}, \end{aligned} \tag{21}$$

where C_1 is constant.

Comparison Between Synchronous and Asynchronous Models

A few basic differences exist between the solutions of the two models. Both models are fully determined by the value of $g_i = \int_0^{t_{\text{budding}}} x_i dt$. Thus the optimal epitope distribution is not a function of the peak density of each protein or of its precise expression pattern. The only thing that affects the distribution is the integral of the copy number until the virus can bud.

However, one can already see from the general solution that the synchronous model has a sharper distribution, with some proteins having no epitopes and other having all the epitopes. In contrast in the asynchronous models, all p_i are strictly positive.

Note that in both model, either increasing the total copy number or decreasing the initial expression time will increase g_i and lead to a lower optimal epitope number as is indeed observed. An interesting question remaining is the effect of changing both factors simultaneously in opposite directions. We here checked a couple of simplified models for the protein dynamics to study these effects. Given more realistic protein expression patterns, one can simply compute g_i for each protein and assess the optimal epitope distribution. A basic conclusion from the model up to now is however that given two proteins with different values of g_i, the one with the higher g_i is expected to have less epitopes.

Numerical Results

We assume a low initial value for each protein ($x_i^0 = 0.1$) and that budding occurs after the expression of the last protein. If the saturation level is constant (we set $\lambda_i/\sigma_i = 1$), A_i is an increasing function of λ_i, and p_i is a decreasing function of λ_i (Fig. 10a). The more interesting situation is when early expressed proteins have lower protein copy numbers. In such a case the low density could have been expected to balance the risk induced by the early exposure. A simple example would be to set $\lambda_i/\sigma_i \propto 1/\lambda_i$, in other words, setting the saturation level to be inversely proportional to the expression level. In such a case, early expressed proteins would have a low saturation level. However, even in this model early expressed proteins have less epitopes in the optimal solution both in the asynchronous (Fig. 10a) and the synchronous case (Fig. 10b).

4 Conclusion and Discussion

We have here presented a systematic bioinformatics and theoretical analysis showing the effect of two main elements on viral survival: (a) the protein expression timing and (b) the protein total expression level. We have shown using the bioinformatics analysis and the theoretical analysis that indeed if either the protein copy number is low or its expression stage is late, the virus does not feel a strong pressure to reduce its epitope number.

The interesting issue is the effect of their combination. We have shown using the theoretical model that in case where these two are combined, the main element shaping the viral escape is actually the expression time and not the copy number.

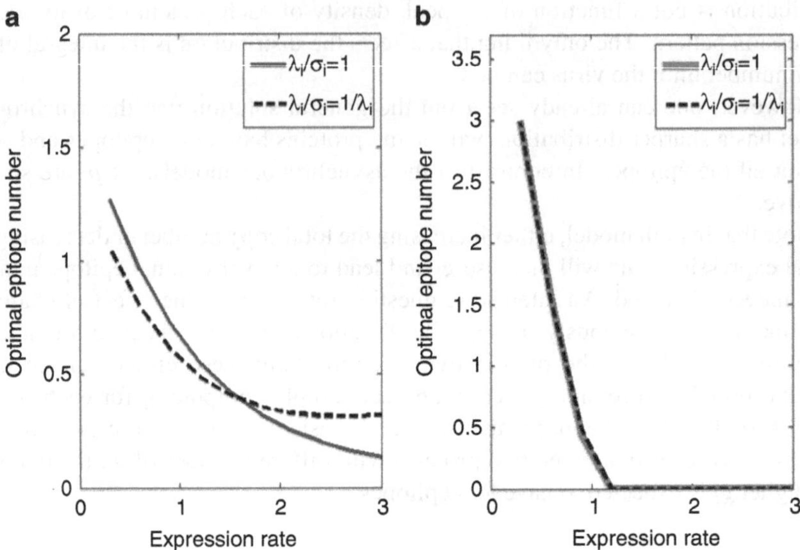

Fig. 10 Optimal number of epitopes as a function of the expression rate. (**a**) Asynchronous model. The virus budding time is normalized to 1. The early stage optimal solution is presented for a set of ten proteins, each with a different expression rate: $\lambda_i = 3 - 0.3(i-1); 1 \leq i \leq 10$. Two saturation levels are considered: a constant saturation level (*full line*): $\lambda_i/\sigma_i = 1$ and a saturation level that is inversely proportional to expression rate (*dashed line*): $\lambda_i/\sigma_i = 1/\lambda_i$. The total epitope number is $C = 5$. The initial lymphocyte clone size was arbitrarily set to $T_i^0 = 1$. The initial protein concentration was $x_i^0 = 0.1$. The optimization was solved analytically according to expression (10). In the two cases, early expressed proteins have a lower expected number of epitopes. (**b**) Synchronous model. The presented results follow the assumptions used for the asynchronous model. The results follow the trend of the asynchronous stage, with the exception that the epitope number can be zero for some of the proteins

This analysis represents the combined analysis of a large number of viruses and represents an example of real-time evolution. The conclusions from this analysis go beyond viral dynamics to the combination of elements in evolutionary computations. The proposed model can be expanded to other more complex evolutionary scenario. The main advantages of the viral dynamics analysis are the high mutation rate of viruses and their short life cycle. In contrast with other organisms, viruses may have actually reached their optimal properties.

References

1. Abendroth, A., Arvin, A.: Varicella-zoster virus immune evasion. Immunol. Rev. **168**, 143–156 (1999)
2. Abendroth, A., Arvin, A.: Immune evasion mechanisms of varicella-zoster virus. Arch. Virol. Suppl. (17), 99–107 (2001)

3. Addo, M.M., Yu, X.G., Rosenberg, E.S., Walker, B.D., Altfeld, M.: Cytotoxic t-lymphocyte (ctl) responses directed against regulatory and accessory proteins in hiv-1 infection. DNA Cell Biol. **21**(9), 671–678 (2002)
4. Agranovich, A., Vider-Shalit, T., Louzoun, Y.: Optimal viral immune surveillance evasion strategies. Theor. Popul. Biol. **80**(4), 233–243 (2011)
5. Almani, M., Raffaeli, S., Vider-Shalit, T., Tsaban, L., Fishbain, V., Louzoun, Y.: Human self-protein cd8+ t-cell epitopes are both positively and negatively selected. Eur. J. Immunol. **39**(4), 1056–1065 (2009)
6. Ambagala, A.P., Solheim, J.C., Srikumaran, S.: Viral interference with mhc class i antigen presentation pathway: The battle continues. Vet. Immunol. Immunopathol. **107**(1-2), 1–15 (2005)
7. Betts, M.R., Yusim, K., Koup, R.A.: Optimal antigens for hiv vaccines based on cd8+ t response, protein length, and sequence variability. DNA Cell Biol. **21**(9), 665–670 (2002)
8. Borrow, P., Lewicki, H., Hahn, B.H., Shaw, G.M., Oldstone, M.B.: Virus-specific cd8+ cytotoxic t-lymphocyte activity associated with control of viremia in primary human immunodeficiency virus type 1 infection. J. Virol. **68**(9), 6103–6110 (1994)
9. Burgevin, A., Saveanu, L., Kim, Y., Barilleau, E., Kotturi, M., Sette, A., van Endert, P., Peters, B.: A detailed analysis of the murine tap transporter substrate specificity. PLoS One **3**(6), e2402 (2008)
10. Cattaneo, R., Will, H., Schaller, H.: Hepatitis b virus transcription in the infected liver. EMBO J. **3**(9), 2191–2196 (1984)
11. Chang, C., Enders, G., Sprengel, R., Peters, N., Varmus, H.E., Ganem, D.: Expression of the precore region of an avian hepatitis b virus is not required for viral replication. J. Virol. **61**(10), 3322–3325 (1987)
12. Chen, H.S., Kew, M.C., Hornbuckle, W.E., Tennant, B.C., Cote, P.J., Gerin, J.L., Purcell, R.H., Miller, R.H.: The precore gene of the woodchuck hepatitis virus genome is not essential for viral replication in the natural host. J. Virol. **66**(9), 5682–5684 (1992)
13. Coffin, J.M.: Hiv population dynamics in vivo: Implications for genetic variation, pathogenesis, and therapy. Science **267**(5197), 483–489 (1995)
14. Dandri, M., Schirmacher, P., Rogler, C.E.: Woodchuck hepatitis virus x protein is present in chronically infected woodchuck liver and woodchuck hepatocellular carcinomas which are permissive for viral replication. J. Virol. **70**(8), 5246–5254 (1996)
15. Dourmishev, L.A., Dourmishev, A.L., Palmeri, D., Schwartz, R.A., Lukac, D.M.: Molecular genetics of kaposi's sarcoma-associated herpesvirus (human herpesvirus-8) epidemiology and pathogenesis. Microbiol. Mol. Biol. Rev. **67**(2), 175–212 (2003)
16. Enders, G.H., Ganem, D., Varmus, H.: Mapping the major transcripts of ground squirrel hepatitis virus: The presumptive template for reverse transcriptase is terminally redundant. Cell **42**(1), 297–308 (1985)
17. Fields, B.N., Knipe, D.M., Howley, P.M.: Fields Virology, 5th edn. Wolters Kluwer Health/Lippincott Williams & Wilkins, Philadelphia (2007)
18. Garcia, P.D., Ou, J.H., Rutter, W.J., Walter, P.: Targeting of the hepatitis b virus precore protein to the endoplasmic reticulum membrane: After signal peptide cleavage translocation can be aborted and the product released into the cytoplasm. J. Cell Biol. **106**(4), 1093–1104 (1988)
19. Ginodi, I., Vider-Shalit, T., Tsaban, L., Louzoun, Y.: Precise score for the prediction of peptides cleaved by the proteasome. Bioinformatics **24**(4), 477–483 (2008)
20. Gulzar, N., Copeland, K.F.: Cd8+ t-cells: Function and response to hiv infection. Curr. HIV Res. **2**(1), 23–37 (2004)
21. Haase, A.T.: Population biology of hiv-1 infection: Viral and cd4+ t cell demographics and dynamics in lymphatic tissues. Ann. Rev. Immunol. **17**(1), 625–656 (1999)
22. Haviv, I., Shamay, M., Doitsh, G., Shaul, Y.: Hepatitis b virus px targets tfiib in transcription coactivation. Mol. Cell Biol. **18**(3), 1562–1569 (1998)
23. Howley, P.M., Roizman, B., Straus, S.E., Martin, M.A., Griffin, D.E.: Fields Virology, 4th edn. Lippincott Williams & Wilkins (LWW), Philadelphia (2001)

24. Itzhack, R., Louzoun, Y.: Random distance dependent attachment as a model for neural network generation in the caenorhabditis elegans. Bioinformatics **26**(5), 647–652 (2010)
25. Karush, W.: Minima of functions of several variables with inequalities as side constraints. M.sc. dissertation, Univ. of Chicago (1939)
26. Kermack, W.O., McKendrick, A.G.: A contribution to the mathematical theory of epidemics. Proc. R. Soc. Lond. A **115**, 700–721 (1927)
27. Kiepiela, P., Leslie, A.J., Honeyborne, I., Ramduth, D., Thobakgale, C., Chetty, S., Rathnavalu, P., Moore, C., Pfafferott, K.J., Hilton, L., Zimbwa, P., Moore, S., Allen, T., Brander, C., Addo, M.M., Altfeld, M., James, I., Mallal, S., Bunce, M., Barber, L.D., Szinger, J., Day, C., Klenerman, P., Mullins, J., Korber, B., Coovadia, H.M., Walker, B.D., Goulder, P.J.: Dominant influence of hla-b in mediating the potential co-evolution of hiv and hla. Nature **432**(7018), 769–775 (2004)
28. Kovjazin, R., Volovitz, I., Daon, Y., Vider-Shalit, T., Azran, R., Tsaban, L., Carmon, L., Louzoun, Y.: Signal peptides and trans-membrane regions are broadly immunogenic and have high cd8+ t cell epitope densities: Implications for vaccine development. Mol. Immunol. **48**(8), 1009–1018 (2011)
29. Kuhn, H.W., Tucker, A.W.: Nonlinear programming. In: Proceedings of the Second Berkeley Symposium on Mathematical Statistics and Probability. University of California Press (1951)
30. Kuiken, C., Korber, B., Shafer, R.W.: Hiv sequence databases. AIDS Rev. **5**(1), 52–61 (2003)
31. Kumanovics, A., Takada, T., Lindahl, K.F.: Genomic organization of the mammalian mhc. Annu. Rev. Immunol. **21**, 629–657 (2003)
32. Lavi, O., Klement, E., Louzoun, Y.: Effect of vaccination in environmentally induced diseases. Bull. Math. Biol. **73**(5), 1101–1117 (2011)
33. Letvin, N.L., Schmitz, J.E., Jordan, H.L., Seth, A., Hirsch, V.M., Reimann, K.A., Kuroda, M.J.: Cytotoxic t lymphocytes specific for the simian immunodeficiency virus. Immunol. Rev. **170**, 127–134 (1999)
34. Lichterfeld, M., Yu, X.G., Le Gall, S., Altfeld, M.: Immunodominance of hiv-1-specific cd8(+) t-cell responses in acute hiv-1 infection: At the crossroads of viral and host genetics. Trends Immunol. **26**(3), 166–171 (2005)
35. Louzoun, Y., Vider, T., Weigert, M.: T-cell epitope repertoire as predicted from human and viral genomes. Mol. Immunol. **43**(6), 559–569 (2006)
36. Luenberger, D.G., Ye, Y.: Linear and nonlinear programming. International Series in Operations Research & Management Science, vol. 116, 3rd edn. Springer, New York (2008)
37. Maguire, H.F., Hoeffler, J.P., Siddiqui, A.: Hbv x protein alters the dna binding specificity of creb and atf-2 by protein-protein interactions. Science **252**(5007), 842–844 (1991)
38. Maman, Y., Blancher, A., Benichou, J., Yablonka, A., Efroni, S., Louzoun, Y.: Immune-induced evolutionary selection focused on a single reading frame in overlapping hepatitis b virus proteins. J. Virol. **85**(9), 4558–4566 (2011)
39. McMichael, A.J., Gotch, F.M., Noble, G.R., Beare, P.A.: Cytotoxic t-cell immunity to influenza. N Engl. J. Med. **309**(1), 13–17 (1983)
40. Miller, M.M., Bacon, L.D., Hala, K., Hunt, H.D., Ewald, S.J., Kaufman, J., Zoorob, R., Briles, W.E.: 2004 nomenclature for the chicken major histocompatibility (b and y) complex. Immunogenetics **56**(4), 261–279 (2004)
41. Negri, D.R., Borghi, M., Baroncelli, S., Macchia, I., Buffa, V., Sernicola, L., Leone, P., Titti, F., Cara, A.: Identification of a cytotoxic t-lymphocyte (ctl) epitope recognized by gag-specific ctls in cynomolgus monkeys infected with simian/human immunodeficiency virus. J. Gen. Virol. **87**(Pt 11), 3385–3392 (2006)
42. Ou, J.H., Laub, O., Rutter, W.J.: Hepatitis b virus gene function: The precore region targets the core antigen to cellular membranes and causes the secretion of the e antigen. Proc. Natl. Acad. Sci. USA **83**(6), 1578–1582 (1986)
43. Ou, J.H., Yeh, C.T., Yen, T.S.: Transport of hepatitis b virus precore protein into the nucleus after cleavage of its signal peptide. J. Virol. **63**(12), 5238–5243 (1989)

44. Princiotta, M.F., Finzi, D., Qian, S.B., Gibbs, J., Schuchmann, S., Buttgereit, F., Bennink, J.R., Yewdell, J.W.: Quantitating protein synthesis, degradation, and endogenous antigen processing. Immunity **18**(3), 343–354 (2003)
45. Pudney, V.A., Leese, A.M., Rickinson, A.B., Hislop, A.D.: Cd8+ immunodominance among epstein-barr virus lytic cycle antigens directly reflects the efficiency of antigen presentation in lytically infected cells. J. Exp. Med. **201**(3), 349–360 (2005)
46. Rehermann, B., Nascimbeni, M.: Immunology of hepatitis b virus and hepatitis c virus infection. Nat. Rev. Immunol. **5**(3), 215–229 (2005)
47. Robinson, J., Waller, M.J., Parham, P., de Groot, N., Bontrop, R., Kennedy, L.J., Stoehr, P., Marsh, S.G.: Imgt/hla and imgt/mhc: Sequence databases for the study of the major histocompatibility complex. Nucleic Acids Res. **31**(1), 311–314 (2003)
48. Rock, K.L., York, I.A., Saric, T., Goldberg, A.L.: Protein degradation and the generation of mhc class i-presented peptides. Adv. Immunol. **80**, 1–70 (2002)
49. Sirma, H., Weil, R., Rosmorduc, O., Urban, S., Israel, A., Kremsdorf, D., Brechot, C.: Cytosol is the prime compartment of hepatitis b virus x protein where it colocalizes with the proteasome. Oncogene **16**(16), 2051–2063 (1998)
50. Sykulev, Y., Cohen, R.J., Eisen, H.N.: The law of mass action governs antigen-stimulated cytolytic activity of cd8+ cytotoxic t lymphocytes. Proc. Natl. Acad. Sci. USA **92**(26), 11990–11992 (1995)
51. Tanaka, K., Tsurumi, C.: The 26s proteasome: Subunits and functions. Mol. Biol. Rep. **24**(1-2), 3–11 (1997)
52. Untersuchungen über die chemischen Affinitäten. Abhandlungen aus den Jahren 1864 1867 1879. Ostwald's Klassiker der exakten Wissenschaften nr. 104. 1899 Leipzig: W. Engelmann. 182. Guldberg, C.M.
53. van Rompay, K.K.A., Dailey, P.J., Tarara, R.P., Canfield, D.R., Aguirre, N.L., Cherrington, J.M., Lamy, P.D., Bischofberger, N., Pedersen, N.C., Marthas, M.L.: Early short-term 9-[2-(r)-(phosphonomethoxy) propyl] adenine treatment favorably alters the subsequent disease course in simian immunodeficiency virus-infected newborn rhesus macaques. J. Virol. **73**(4), 2947 (1999)
54. Vider-Shalit, T., Louzoun, Y.: Mhc-i prediction using a combination of t cell epitopes and mhc-i binding peptides. J. Immunol. Meth. **374**(1-2), 43–46 (2010)
55. Vider-Shalit, T., Fishbain, V., Raffaeli, S., Louzoun, Y.: Phase-dependent immune evasion of herpesviruses. J. Virol. **81**(17), 9536–9545 (2007)
56. Vider-Shalit, T., Almani, M., Sarid, R., Louzoun, Y.: The hiv hide and seek game: An immunogenomic analysis of the hiv epitope repertoire. AIDS **23**(11), 1311–1318 (2009)
57. Vider-Shalit, T., Sarid, R., Maman, K., Tsaban, L., Levi, R., Louzoun, Y.: Viruses selectively mutate their cd8+ t-cell epitopes–a large-scale immunomic analysis. Bioinformatics **25**(12), i39–44 (2009)
58. (WHO), W.H.O. http://www.who.int/mediacentre/factsheets/fs204/en/.

Part II
Blood Vessel Dynamics

A Multiscale Approach Leading to Hybrid Mathematical Models for Angiogenesis: The Role of Randomness

Vincenzo Capasso and Daniela Morale

1 Introduction

In biology and medicine we may observe a wide spectrum of formation of patterns, usually due to self-organization phenomena. This may happen at any scale; from the cellular scale of embryonic tissue formation, wound healing or tumor growth, and angiogenesis to the much larger scale of animal grouping. Patterns are usually explained in terms of a collective behavior driven by "forces," either external and/or internal, acting upon individuals (cells or organisms). In most of these organization phenomena, randomness plays a major role; here we wish to address the issue of the relevance of randomness as a key feature for producing nontrivial geometric patterns in biological structures. As working examples we offer a review of two important case studies involving angiogenesis, i.e., tumor-driven angiogenesis [7] and retina angiogenesis [8]. In both cases the reactants responsible for pattern formation are the cells organizing as a capillary network of vessels, and a family of underlying fields driving the organization, such as nutrients, growth factors, and alike [18, 19].

A fruitful approach to the mathematical description of such phenomena, suggested since long by various authors [16, 22, 26, 27, 30, 31], is based on the so-called *individual based models*, according to which the "movement" of each individual is described, embedded in the total population. This is also known as *Lagrangian approach*. Possible randomness is usually included in the motion, so that the variation in time of the (random) locations $X_N^k(t) \in \mathbb{R}^d, k = 1, \ldots, N(t)$, of individuals in a group of size $N(t)$ at time $t \geq 0$ is described by a system of stochastic differential equations driven by gradients of suitable underlying fields. On the other hand the

V. Capasso (✉) • D. Morale
Department of Mathematics, University of Milan, 20133 Milan, Italy

CIMAB (Interuniversity Centre for Mathematics Applied to Biology, Medicine and Environmental Sciences), University of Milan, 20133 Milan, Italy
e-mail: vincenzo.capasso@unimi.it; daniela.morale@unimi.it

U. Ledzewicz et al., *Mathematical Methods and Models in Biomedicine*, Lecture Notes on Mathematical Modelling in the Life Sciences, DOI 10.1007/978-1-4614-4178-6_4,

number $N(t)$ of elements in the system may be subject to a stochastic dynamics, driven itself by the mentioned underlying fields. Vice versa these fields are usually strongly coupled with the dynamics of individuals in the population. The strong coupling of the kinetic parameters of the relevant stochastic branching-and-growth of the capillary network, with the family of interacting underlying fields, is a major source of complexity from a mathematical and computational point of view. This is the reason why in literature we may find a large variety of mathematical models addressing some of the features of the angiogenic process, and still the problem of integration of all relevant features of the process is open [1,9,33,34,36,40,41].

Thus our main goal is not to provide additional models for the angiogenic phenomenon, but to address the mathematical problem of reduction of the complexity of such systems by taking advantage of its intrinsic multiscale structure; the dynamics of cells will be described at their natural scale (the microscale), while the dynamics of the underlying fields will be described at a larger scale (the macroscale).

In our specific cases, in a Lagrangian approach, endothelial cells are described by a family of stochastic processes $\{(X^k(t),Y^k(t))\}_{t\in\mathbb{R}_+}$, for $k\in\{1,\dots,N(t)\}$, where $X^k(t)$ denotes the position of the kth cell out of $N(t)$, where $N(t)$ counts the random number of cells at time $t\in\mathbb{R}_+$. $Y^k(t)$ will represent either the velocity or the type of the cell. The branching mechanism of blood vessels is modelled as a stochastic marked counting process describing the birth of endothelial cells, while capillary extensions are described by a system of a random number of stochastic differential equations; the whole network of vessels is thus obtained as the union of their individual trajectories. As a matter of mathematical treatability, for the capillary extension we have adopted a system of Itô type stochastic differential equations with additive noise, modelled by a family of independent Wiener processes; in this way we may take advantage of the results of the Itô calculus. As anticipated above, the kinetic parameters of the branching and extension stochastic processes depend upon a family of underlying fields, whose evolution will be discussed later. Alternately, one may adopt an Eulerian description of the system of cells according to which we consider their collective behavior via the corresponding empirical measure

$$\hat{Q}_N(t)=\frac{1}{N(t)}\sum_{k=1}^{N(t)}\varepsilon_{(X^k(t),Y^k(t))},$$

where ε denotes the usual Dirac measure.

As anticipated above, the underlying fields are modelled at a larger scale; they are taken as spatially continuous fields admitting spatial densities with respect to the usual Lebesgue measure; their evolution is then modelled in terms of a system of partial differential equations for such spatial densities, with reaction terms depending upon $\hat{Q}_N(t)$, the empirical spatial distribution of the cell population, hence on the evolving capillary network.

Here it comes the complexity of the whole system; since the capillary network carries a significant randomness, the evolution equations of the underlying fields are a set of random partial differential equations, leading to random kinetic parameters for the stochastic processes of branching and growth. We are thus facing a problem of double stochasticity. This is a major source of complexity which may tremendously increase as the number of cells becomes extremely large, as it may happen in many cases of real interest. Under these last circumstances, by taking into account the natural multiple scale nature of the system a *mesoscale* may be introduced, which is sufficiently small with respect to the macroscale of the underlying fields and sufficiently large with respect to typical cell size [5, 6, 27, 29]. This is made explicit in the evolution of the underlying fields. Indeed, by applying suitable (*laws of large numbers*) at the mentioned mesoscale, we may approximate the empirical distribution by a deterministic measure admitting mean spatial densities in the equations for the underlying fields, thus providing a family of deterministic underlying fields. We may then use these approximated mean fields to drive the parameters of the relevant stochastic processes describing the dynamics of the cells at the microscale. In this way only the simple stochasticity of the geometric processes of birth (branching) and growth is kept.

These kinds of models are known as *hybrid models*, since we have substituted all stochastic underlying fields by their "averaged" counterparts; most of the current literature could now be reinterpreted along these lines. However, the main scope of this chapter is to investigate the possibility that such hybrid approach may still generate a nontrivial and realistic geometric pattern of the capillary network. Unfortunately we have been able to evidence that the averaging of the underlying fields in the early stages of the vascularization process may lead to unrealistic dynamical behaviors, which miss a realistic patterning of the vasculature, as shown here by numerical simulations.

In conclusion, the two different approaches (Lagrangian and Eulerian) describe the system at different scales: the finer scale description is based on the (stochastic) behavior of individuals (microscale) and the larger scale description is based on the (continuum) behavior of densities of relevant underlying fields (macroscale).

> "The central problem is to determine how information is transferred across scales; one of the aims of the modelling is to catch the main features of the interaction at the scale of single individuals that are responsible, at a larger scale, for a more complex behavior that leads to the formation of patterns" [16].

A very similar approach has been recently proposed in [41], though the authors do not keep an explicit stochasticity at the microscale. In order to obtain a nontrivial capillary network, in their work the authors mimic the natural intrinsic spatial randomness of the phenomenon at the microscale by introducing a given spatial heterogeneity in the kinetic parameters of the extracellular matrix, which acts as a spatially heterogeneous underlying field. This same argument is widely discussed in [1]. On the contrary, in our model we do not impose any artificially superimposed heterogeneity, but we confirm that a spatial, possibly random, heterogeneity of the underlying fields is necessary in order to generate a nontrivial and realistic network. In our treatment we show that the required spatial heterogeneity is generated by the dynamics of the model itself, due to the strong coupling of the microscopic stochastic evolution of the cell with the underlying fields, which thus inherit their randomness. The advantage of using averaged quantities at the larger scales is anyway convenient, both from a theoretical point of view and for being computational affordable; the message is: do it when suitably supported by applicability of laws of large numbers.

For the time being, we have performed a heuristic derivation of a law of large numbers as the number of cells increases, showing that, when the number of cells is sufficiently large, the empirical measure $\tilde{Q}_N(t)$ can be approximated by its limit measure $Q(t)$ [6, 7, 27] which admits a spatial density with respect to the usual Lebesgue measure. We also provide the evolution equations for the relevant spatial densities associated with the limit measure. Indeed, one should check that the hybrid system is compatible with a rigorous derivation of the evolution for the vessel densities. Nonlinearities in the full model are a big difficulty in this direction. In literature, examples of derivation of limit systems from a stochastic particle dynamics may be found also in [10, 29, 39], but to the best of the knowledge of the authors, the kind of stochastic hybrid models considered here has not been studied yet.

In Sect. 2 we discuss the mathematical modelling of the stochastic interacting population when the number of individuals is finite. We consider both Lagrangian and Eulerian (discrete) descriptions. In Sects. 2.1 and 2.2, we analyze two specific cases. The first case refers to a model for stochastic tumor-induced angiogenesis. There is a widespread literature on the subject [1, 3, 11–13, 15, 23, 24, 32–34, 37]. The interested reader may refer to the introduction of [7] for a detailed description. The second model considered here is a model for retinal angiogenesis. Both models are examples of stochastic fiber processes, coupled with the continuum underlying field of a chemoattractor (indirect interaction). In Sect. 3 we study the derivation of the corresponding hybrid models, for the two working examples. Finally, in Sect. 4 we discuss the role of randomness in the proposed model via simulations.

2 Cells, Interactions and Evolution

We denote by N a parameter of scale of the process; from a modelling point of view, it may well represent the total number of cells involved in the phenomenon, in a way or another. Let $N(t) \in \mathbb{N}$ be the random number of cells entered in the dynamics by time t, $T_{\mathrm{b}}^i \in \mathbb{R}_+$, the random time of birth of the i-th cell; $T_{\mathrm{d}}^i \in \mathbb{R}_+$, the random time of death of the i-th cell. Here, cell states are described by a bivariate stochastic process $\underline{Z}^i = \{\underline{Z}^i(t)\}_{t\in[0,T]}$, such that

$$\underline{Z}^i(t) = (X^i(t), Y^i(t)), \quad i = 1, \ldots, N(t),$$

where $X^i(t) \in \mathbb{R}^d$ is the random location of the i-th cell at time $t \in \mathbb{R}_+$, and $Y^i(t) \in \mathbb{I}$ may represent a possible random characterization of the i-th cell at time $t \in \mathbb{R}_+$. Later on, we will see that $Y^i(t)$ will represent the velocity of the i-th cell, in the first model, or the type of the cell, in the second one. We model sprout extension by tracking the random network $X(t)$, given by all the trajectories $X^i(t)$, i.e.,

$$X(t) = \bigcup_{i=1}^{N(t)} \{X^i(s), T_b^i \leq s \leq \min(t, T_d^i)\}. \tag{1}$$

Thus, the random network of endothelial cells is the union of the random trajectories described by all existing cells.

The *Lagrangian description* of the system may be also given in terms of measures. Precisely, if we denote by $\mathbb{D}$ the range $\mathbb{R}^d \times \mathbb{I}$, we might describe the state of the k-th cell by

$$\varepsilon_{\underline{Z}^k(t)} = \varepsilon_{(X^k(t),Y^k(t))} \in \mathcal{M}(\mathbb{D}),$$

which is the *Dirac measure* localized at $(X^k(t),Y^k(t))$. This is such that for any $B_1 \times B_2 \in \mathcal{B}_{\mathbb{D}}$,

$$\varepsilon_{\underline{Z}^k(t)}(B_1 \times B_2) = \begin{cases} 1, & X^k(t) \in B_1, Y^k(t) \in B_2, \\ 0, & \text{otherwise,} \end{cases}$$

so that, for any sufficiently smooth $g : \mathbb{R}^d \times \mathbb{S} \to \mathbb{R}$,

$$\int_{\mathbb{R}^d} \int_I g(x,y) \varepsilon_{(X^k(t),Y^k(t))}(\mathrm{d}y \times \mathrm{d}y) = g\left(X^k(t), Y^k(t)\right).$$

According to an *Eulerian discrete approach*, the system can be described via a space-type distribution of cells by considering the global random empirical measure Q_N of the process, such that, for any $t \in \mathbb{R}_+$,

$$Q_N(t) = \frac{1}{N} \sum_{i=1}^{N(t)} \varepsilon_{(X^k(t),Y^k(t))} \in \mathcal{M}(\mathbb{D}). \tag{2}$$

We define the marginal measure $\widetilde{Q}_N$ as

$$\widetilde{Q}_N(t) = Q_N(t)(\cdot, \mathbb{I}) = \frac{1}{N} \sum_{i=1}^{N(t)} \varepsilon_{X^k(t)} \in \mathcal{M}(\mathbb{R}^d). \tag{3}$$

A key question concerns the modelling of the *interaction*; interaction among cells may be direct or indirect. Here we consider two cases of indirect interaction, i.e., the force exerted on each cell depends on an underlying field whose evolution depends on the distribution of the entire population; as a consequence the dependence of the evolution of the spatial distribution of a single individual upon the spatial distribution of the whole population is mediated by the underlying field.

We consider pairwise interaction. Thus the interaction between two locations $x, y \in \mathbb{R}^d$ is modelled by a symmetric reference kernel K_1, depending on their distance. We assume that, in a population of N cells, the interaction of two cells, out of N, located in x and y, respectively, is modelled by

$$\frac{1}{N} K_N(x-y), \quad \text{where} \quad K_N(z) = N^{\eta} K_1(N^{\eta/d} z), \tag{4}$$

which expresses the rescaling of K_1 with respect to the total member N of cells, in terms of a scaling coefficient $\eta \in [0,1]$. At a position $x \in \mathbb{R}^d$, the interaction with the population of cells at time $t \in \mathbb{R}_+$ is given by a mollified version of the marginal spatial random distribution Eq. (3)

$$\frac{1}{N} \sum_{k=1}^{N(t)} K_N(x - X^k(t)) = \left(\widetilde{Q}_N(t) * K_N\right)(x).$$

The choice of η in Eq. (4) determines the range and the strength of the influence of neighboring cells; indeed, the number of cells influencing the underlying field at a point $x \in \mathbb{R}^d$ is of the order $N^{1-\eta}$. In particular, for N increasing to infinity, the choice $\eta < 1$ let $N^{1-\eta}$ tend to infinity, allowing the applicability of a local law of large numbers. From a mathematical point of view, the use of mollifiers reduces analytical complexity; from a modelling point of view this might correspond to the range of a nonlocal interaction between cells and the relevant underlying fields. Another possible interpretation would be that a cell is not exactly a point, but it may extend as a spot in the relevant spatial domain. Altogether the parameter η defines the order of what we call here a mesoscale, i.e., $N^{-\eta}$ [6,7,28].

A general model may then appear of the following form, for any $t \geq 0$,

$$\mathrm{d}\underline{Z}^k(t) = F\left[C(\cdot,t)\right](\underline{Z}^k(t))\mathrm{d}t + \underline{\sigma}\,\mathrm{d}\underline{W}^k(t), \qquad k = 1,\ldots,N(t); \tag{5}$$

$$\frac{\partial}{\partial t} C(x,t) = Op_2(C(\cdot,t))(x) + Op_3\left[\widetilde{Q}_N(t) * K_N\right](x), \qquad x \in \mathbb{R}^d. \tag{6}$$

System (5)–(6) say that the stochastic evolution of an individual state $\underline{Z}^k(t)$ is driven by an underlying field $C(x,t)$ (such as nutrient, growth factor, and alike [18, 19]) via the operator $F\left[C(\cdot,t)\right]$ depending on the field and acting on each individual; on the other hand the evolution equation of the field $C(x,t)$ depends itself upon the structure of the system of individuals by means of an operator Op_3, which depends on the convolution $\widetilde{Q}_N(t) * K_N$ of individuals, acting at a spatial location x. In our concrete examples, reported in Sects. 2.1 and 2.2, the underlying fields are better specified as suitable biochemical substances. For simplicity, the diffusion coefficient $\underline{\sigma}$ in Eq. (5) is assumed to be a constant, in time and space, diagonal matrix with entries in $\mathbb{R}_+$. Randomness is modelled via a diagonal matrix with diagonal entries given by two independent Wiener processes $W_i^k(t), i = 1,2$. Note that also the evolution of the stochastic process $\{N(t)\}_{t \in \mathbb{R}_+}$ may depend upon the underlying field $C(t,x)$. By specializing operators $F[C], Op_i, i = 1,2,3$, we may discriminate different dynamics.

For any $g \in C_b^{2,1}(\mathbb{D} \times \mathbb{R}_+)$, Itô's formula applied to $g(\underline{Z}^k(t),t))$, for any $k = 1,\cdots,N(t)$, leads to the *Eulerian (discrete) description* via an evolution equation for the empirical measure $Q_N(t)$ [6, 27]; indeed,

$$\int_{\mathbb{D}} g(\underline{z},t) Q_N(t)(\mathrm{d}\underline{z}) = \int_{\mathbb{D}} g(\underline{z},0) Q_N(0)(\mathrm{d}\underline{z}) + \int_0^t Op_1\left(Q_N(s), C(\cdot,s), g(\cdot,s)\right) \mathrm{d}s$$
$$+ M_N\left[\underline{Z},\underline{W}\right](t), \tag{7}$$

where $M_N\left[\underline{Z},\underline{W}\right](t)$ is a zero mean martingale.

So far, we have shown how the models (5)–(6) may provide a general Lagrangian description of the evolution of the processes $\underline{Z}^k$, coupled with the evolution of a field C. On the other hand Eq. (7), again coupled with Eq. (6), is an Eulerian (discrete) description of the system, where single identities of cells are lost. Both equations may be coupled with other stochastic processes too, due for example to the evolution of $N(t)$.

In the next two sections we provide two examples of models for angiogenesis, in which we specify system (5)–(6), and then Eq. (7).

2.1 Tumor-Induced Angiogenesis

An interesting example of formation of patterns may be found in the process of tumor growth and in particular in angiogenesis. Tumor-induced angiogenesis is believed to occur when normal tissue vasculature is no longer able to support the growth of an avascular tumor. A well-known model in literature is based on the idea that endothelial cells proliferate and migrate in response to different signalling cues; in particular they move though a gap in the basement membrane and into extra cellular matrix (ECM). They secrete proteolytic enzymes, which also degrade the ECM. Migration is through to be controlled by chemotaxis, the directed cell movement up the gradient of a diffusible substance [17], a growth factor emitted by the tumor [here tumour angiogenetic factor (TAF)], and by haptotaxis, the directed cell movement along a nondiffusible substance, an adhesive gradient (here the fibronectin). TAF and fibronectin bind to specific membrane receptors of endothelial cells, activating cell migration machinery. Then cells produce a matrix degrading enzyme (MDE), which improves the attachment of the cells to fibronectin contained in the extracellular matrix. As a consequence, endothelial cells are able to exert the traction forces needed for migration. ECs subsequently respond to the TAF concentration gradients by forming sprouts, dividing and migrating towards the tumor. So, at an individual level, cells interact and perform a branching process coupled with elongation, under the stimulus of a chemical field produced by a tumor.

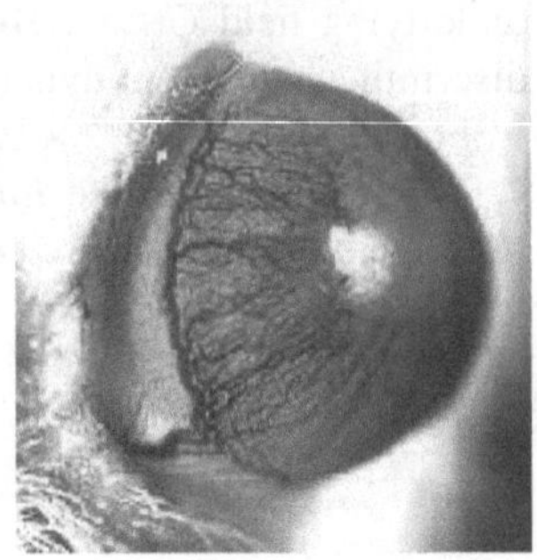

Fig. 1 Angiogenesis on a rat cornea: experiment performed at IFOM, FIRC research centre [14]

In this way formation of aggregating networks (vessels) is shown as a consequence of collective behavior. Figure 1 shows an example of real angiogenesis induced by a biochemical substance playing the role of an angiogenic tumor on a rat cornea [14].

Based on the above discussion, the main features of the process of formation of a tumour-driven vessel network are

1. Vessel branching
2. Vessel extension
   - Chemotaxis, in response to a generic TAF, released by tumour cells
   - Haptotatic migration, in response to fibronectin gradients, emerging from the extracellular matrix, and through degradation and production by endothelial cells themselves
3. Anastomosis, when a capillary tip meets an existing vessel

The initiation of sprouting from preexisting parental vessels is not considered here; in order to avoid further mathematical technicalities, we assume a given number N_0 of initial capillary sprouts; we refer to literature [23] for details on this topic.

2.1.1 Modelling the Evolution of the Capillary Network and the Coupling with the Underlying Fields

We again denote by N a parameter of scale of the process. We model the network by tracking the tip of vessels. Thus, the considered individual stochastic processes are

$$\underline{Z}^i(t) = (X^i(t), v^i(t)) \in \mathbb{R}^d \times \mathbb{R}^d,$$

which denote the location and the velocity of the tip of the i-th vessel at time t, for $i = 1, \ldots, N(t)$, respectively. $N(t)$ represents the number of tips at time t. The network of endothelial cells is described by the union of the trajectories of the tips as in Eq. (1).

Vessel Extension. The movement (extension) of tips follows a Langevin model; at any $t > T^i$ and for any $k \in \{1,\dots,N(t)\}$ we have

$$\begin{aligned} \mathrm{d}X^i(t) &= v^i(t)(1-\gamma\mathbb{I}_{X(t)}(X^i(t))\mathrm{d}t, \\ \mathrm{d}v^i(t) &= \left(-kv^i(t) + F\left(C(t,X^i(t)), f(t,X^i(t))\right)\right)\mathrm{d}t + \sigma \mathrm{d}W^i(t). \end{aligned} \quad (8)$$

The advection term includes the typical inertial component $-kv^i(t)$, while, according to a typical chemotaxis, velocity $v^i(t)$ is driven by a function F of the underlying fields, the TAF C and the fibronectin f. As in [34, 36], we take the bias depending on the TAF and the ficronectin fields

$$\begin{aligned} F\left(C(t,X^i(t)), f(t,X^i(t))\right) &= d_C(C(t,X^i(t)))\,\nabla C(t,X^i(t)) \\ &\quad + d_f(f(t,X^i(t)))\,\nabla f(t,X^i(t)). \end{aligned}$$

d_C, d_f, are turning coefficients, modelled as follows:

$$\begin{aligned} d_C(C(t,X^k(t))) &= d_1 \frac{\left|\nabla C(t,X^k(t))\right|}{(1+q_1 C(t,X^k(t))^{q_2}}, \\ d_f(f(t,X^k(t))) &= d_2 \left|\nabla f(t,X^k(t))\right|, \end{aligned}$$

with $q_1, q_2 \geq 0$. So the reorientation of cells increases as a function of the magnitude of the chemotactic, haptotactic gradient; furthermore cells become desensitized to chemotactic gradients at high attractant concentrations, as stressed in [1, 23]. In Eq. (8) the parameter γ may assume only the values 0 and 1; $\gamma = 0$ means that no impingement is considered; otherwise, for $\gamma = 1$ the phenomenon of anastomosis is taken into account (for further information see [7] and references therein).

Vessel Branching. In literature two kinds of branching have been identified, either from a tip or from a mature vessel (see, e.g., [1, 25, 34]). Here we describe explicitly the branching processes from a mathematical point of view only for the tip branching. The reader may refer to [7] for a possible mathematical modelling of the mature vessel branching. We consider, as it is always supposed, that the branching rates depend on the field of concentration of TAF. The tip vessel branching is described by a process $\Phi = \sum_i \varepsilon_{(T_i^b, X^i)}$ on the σ-algebra $\mathscr{B}_{(\mathbb{R}_+, X^i)}$, i.e., a marked counting process such that, for any measurable set $A \subseteq \mathscr{B}_{\mathbb{R}^+ \times \mathbb{R}^d}$,

$$\Phi(A) := \sum_i \varepsilon_{(T_i^b, X^i)}(A) = \mathrm{card}\{n : (T^n, Y^n) \in A\}$$

is the random variable counting those tips which are born in A. By definition $\Phi(\{0\} \times \mathbb{R}^d) = N_0$. This process Φ is characterized by the following stochastic intensity:

$$\mu(dt \times dx) = prob(\Phi(dt \times dx) = 1 | \mathscr{F}_{t^-}) = \Lambda_t(dx)\, dt, \text{ for } (t,x,s) \in \mathbb{R}_+ \times \mathbb{R}^d.$$

$\mathscr{F}_{t^-}$ denotes the σ algebra history of the process up to time $t-$ and, given a nonnegative function $h \in C_b(\mathbb{R}^d)$ and $\alpha \in \mathbb{R}_+$,

$$\Lambda_t(dx) = \alpha h(C(t,x)) \sum_{i=1}^{N(t^-)} \varepsilon_{X^i(t)}(dx). \tag{9}$$

When a tip located in x branches at time t, the initial value of the state of the new tip is taken as $(X^{N(t)+1}, v^{N(t)+1}) = (x, v_0)$, where v_0 is a non random velocity and $T^b_{N(t)+1} = t$. Given the branching process Φ, the counting process $N(t) = \Phi([0,t] \times \mathbb{R}^d)$ is a Poisson-like stochastic process, counting the total number of cells born before $t \in \mathbb{R}_+$, with intensity

$$\nu(dt) = \Lambda_t(\mathbb{R}^d)dt = \sum_{i=1}^{N(t)} h(C(t, X^i(t)))dt.$$

Evolution of the Fields. The chemotactic field $C(t,x)$ diffuses and degrades; the consumption is proportional to the extension velocities $v^i, i = 1, \ldots, N(t)$. So, for any $(t,x) \in \mathbb{R}_+ \times \mathbb{R}^d$,

$$\frac{\partial}{\partial t} C(t,x) = c_1 \delta_A(x) + c_2 \triangle C(t,x) - c_3 C(t,x) \frac{1}{N} \sum_{i=1}^{N(t)} (v^i(t) \delta_{X^i(t)} * K_N)(x). \tag{10}$$

Parameters $c_1, c_2, c_3 \in \mathbb{R}^+$ in Eq. (10) represent the rate of production of a source located in a region $A \subset \mathbb{R}^d$, modelling, e.g., a tumor mass, the diffusivity and the rate of consumption, respectively.

Fibronectin is known to be attached to the extracellular matrix and does not diffuse [2], thus the equation for fibronectin does not contain any diffusion term. As in [36], degradation of fibronectin, characterized by a coefficient β_2, depends on the concentration of MDE, produced by the cells. Hence, the concentration of fibronectin $f(x,t)$ produced by the endothelial cells at the tip evolves as

$$\frac{\partial}{\partial t} f(t,x) = \beta_1 \frac{1}{N} \sum_{i=1}^{N(t)} (\delta_{X^i(t)} * K_N)(x) - \beta_2 m(t,x) f(t,x), \tag{11}$$

for $\beta_1, \beta_2 \in \mathbb{R}_+$. The MDE, once produced with rate ν_1, diffuses locally with diffusion coefficient ε_1 and is spontaneously degraded at a rate ν_2.

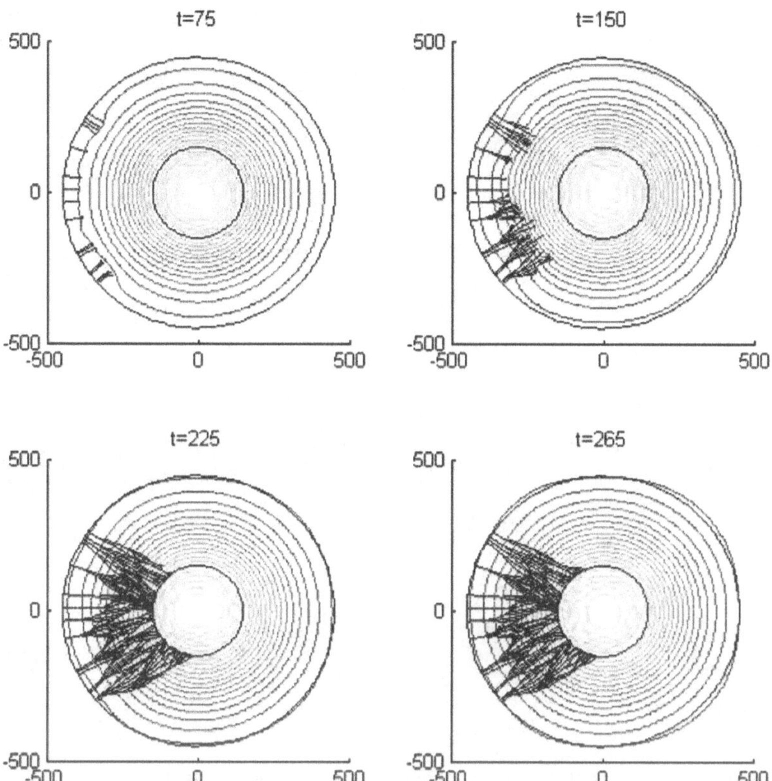

Fig. 2 The evolution of the vessel network with a tumor localized in the center and 20 initial tips

$$\frac{\partial}{\partial t}m(t,x) = \varepsilon_1 \triangle m(t,x) + \nu_1 \frac{1}{N}\sum_{i=1}^{N(t)} (\delta_{X^i(t)} * K_N)(x) - \nu_2 m(t,x). \tag{12}$$

In Eqs. (10)–(12), we consider, as discussed in Sect. 2, a mollified version of the marginal spatial random distribution, by means of the convolution with the kernel K_N, given by Eq. (4). Figures 2 and 3 show simulation results in the case the tumor, source of TAF, is located in the centre of the cornea; we consider 20 initial capillary sprouts. In such a simulation anastomosis is also taken into account. In particular Figure 2 shows four steps of the capillary network formation in response to the tumor mass, and its interaction with the underlying chemotactic field, represented by the shading contour lines in the background. Figure 3 shows different zooms that illustrate the network structure at lower scales; anastomosis is highlighted.

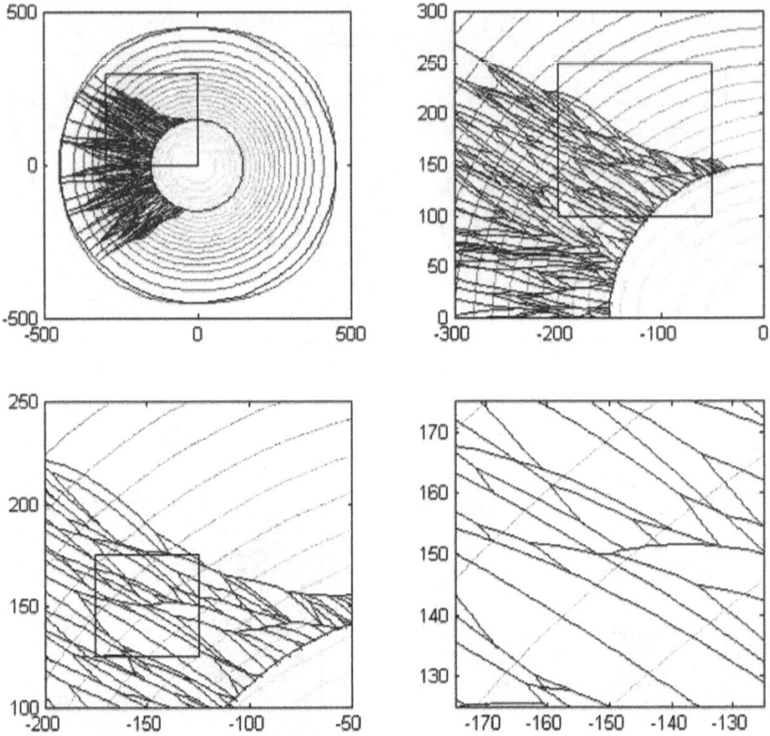

Fig. 3 Subsequent zooms of a simulation the network. Anastomosis is shown

2.1.2 Modelling the Evolution of the Empirical Measures and the Coupling with the Underlying Fields

As discussed in the general context of Sect. 2, an Eulerian description may be given via an empirical measure as Eq. (2). In this case, we may associate two fundamental random spatial measures, describing the network at time t; given a suitable scale parameter N, Q_N, the empirical measure associated with the processes $(X^k(t), v^k(t)), k = 1, \ldots, N(t)$, is given by

$$Q_N(t) = \frac{1}{N} \sum_{i=1}^{N(t)} \varepsilon_{(X^k(t), v^k(t))} \tag{13}$$

while $V_N(t)$, the empirical spatial distribution of velocities, is given by

$$V_N(t) = \frac{1}{N} \sum_{i=1}^{N(t)} v_k(t) \varepsilon_{X^k(t)} = \int_{\cdot \times \mathbb{R}^d} v \, Q_N(t)(\mathrm{d}(x,v)).$$

Furthermore, we obtain the random empirical distribution of tips $X_N(t)$ as a marginal of the measure Eq. (13)

$$X_N(t) = \frac{1}{N} \sum_{i=1}^{N(t)} \varepsilon_{X^k(t)} = Q_N(t)(\cdot \times \mathbb{R}^d). \tag{14}$$

From the Lagrangian description above, Itô's formula applied to a function $g \in C_b(\mathbb{R}^d \times \mathbb{R}^d)$ of $\underline{Z}_N^k(t)$, for any $k = 1, \ldots, N(t)$, provides the time evolution equation of the empirical measure Q_N. Indeed, by Itô's formula [4], from system (8), we get

$$\begin{aligned} \mathrm{d}g((X^k(t), v^k(t))) &= \mathrm{d}g((X^k(0), v^k(0))) + \nabla_x g((X^k(t), v^k(t)))\mathrm{d}X^k(t) \\ &\quad + \nabla_v g((X^k(t), v^k(t)))\mathrm{d}v^k(t) \\ &\quad + \frac{1}{2}\Delta_v g((X^k(t), v^k(t)))(\mathrm{d}v^k(t))^2 \\ &= \mathrm{d}g((X^k(0), v^k(0))) + \nabla_x g((X^k(t), v^k(t)))v^k(t) \\ &\quad - \nabla_v g((X^k(t), v^k(t))) \\ &\quad \times \left[kv^k(t) - F\left(C(t, X^k(t)), f(t, X^k(t))\right)\right] \mathrm{d}t \\ &\quad + \frac{\sigma^2}{2}\Delta_v g((X^k(t), v^k(t)))\mathrm{d}t \\ &\quad + \sigma \nabla_v g((X^k(t), v^k(t)))\mathrm{d}W^k(t). \end{aligned} \tag{15}$$

By summing up into Eq. (15), we may obtain evolution equations for the random measure Q_N, as follows. For any $B \in \mathscr{B}_{\mathbb{R}^d \times \mathbb{R}^d}$

$$\begin{aligned} \int_B g(x,v) Q_N(t)\mathrm{d}(x,v) &= \int_B g(x,v) Q_N(0)\mathrm{d}(x,v) \\ &\quad + \int_0^t \int_B \Big[\nabla_x g(x,v)v + g(x,v)\alpha h(C(s,x))\, \delta_{v_0}(v) \\ &\quad - \nabla_v g(x,v)\left[kv - F\left(C(t,x), f(t,x)\right)\right] \\ &\quad + \frac{\sigma^2}{2}\Delta_v g(x,v))\Big] Q_N(t)(\mathrm{d}(x,v))\mathrm{d}s + \tilde{M}_N(t), \end{aligned} \tag{16}$$

where the last term

$$\begin{aligned} \tilde{M}_N(t) &= \int_0^t \int_{\mathbb{R}^n} g(s,x)\left[\Phi(\mathrm{d}s, \mathrm{d}x) - N\alpha h(C(s,x)) X_N(s)(\mathrm{d}x)\, \mathrm{d}s\right] \\ &\quad + \int_0^t \frac{\sigma}{2N} \sum_{k=1}^{N(t)} \nabla_v g((X^k(t), v^k(t)))\mathrm{d}W^k(t) \end{aligned} \tag{17}$$

is a zero mean martingale. Equation (16) is coupled with the system of PDEs

$$\frac{\partial}{\partial t}C(t,x) = c_1\delta_A(x) + c_2\triangle C(t,x) - c_3C(t,x)\,(K_N(t) * V_N(t))(x), \tag{18}$$

$$\frac{\partial}{\partial t}f(t,x) = \beta_1(K_N * X_N(t))(x) - \beta_2\, m(t,x)f(t,x), \tag{19}$$

$$\frac{\partial}{\partial t}m(t,x) = \varepsilon_1\triangle m(t,x) + \nu_1(K_N * X_N(t))(x) - \nu_2 m(t,x). \tag{20}$$

Finally, let us notice that, since $h(x,s)$ is uniformly bounded, the process

$$Z(t) = \frac{N(t)}{N} = \langle Q_N(t), 1\rangle$$

is stochastically dominated by the process $\frac{1}{N}Y_{N\langle Q_N(0),1\rangle}$, where Y_k is a Yule process with birth rate given by $\tilde{h} = \sup h(x,s), \quad Y_k(0) = k$. This implies that

$$\lim_{n\to\infty} \sup_{N\in\mathbb{N}} prob\left(\sup_{t\leq T} \langle Q_N(t), 1\rangle \geq n\right) = 0. \tag{21}$$

In conclusion in the example presented here, the Lagrangian description of system (5)–(6), discussed in the previous section, has the form of system (8)–(12), while the Eulerian discrete description Eqs. (6)–(7) is given by system (16)–(20).

2.2 Retinal Angiogenesis

The development of the retina vasculature is preceded by an invasion of migrating specific cells called astrocytes into the retina; they spread radially and proliferate forming a cellular network that provides an initial template for the blood vessels. Astrocytes express the *vascular endothelial growth factor* (VEGF) which is a key stimulus for angiogenesis [38, 42]. The second step of the biological process consists of the angiogenesis itself. As in the previous model, specialized endothelial tip cells are identified at the leading edge of the growing vascular network [20]. They might guide the growth of vascular sprouts by detecting the attractant or repulsive external cues (VEGF) provided by astrocytes. Further experiments have shown that another type of endothelial cells is present in the angiogenic sprouts, the *stalk cells*, located in the neighborhood of tip cells. Immediately after birth, the retinal vascular system starts to develop as a sprout from the optic disc and initially forms a primitive vascular plexus which is rapidly remodelled into large and small vessels. During the first postnatal week, retinal vessels continue to extend radially over the superficial layer of the retina to form a two-dimensional vascular structure.

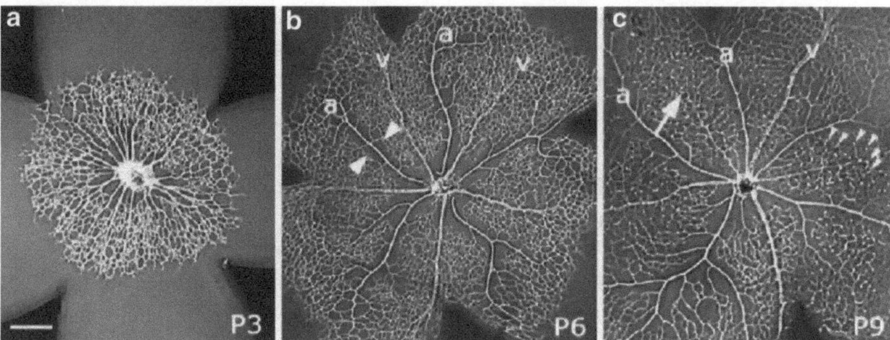

Fig. 4 Developing mouse retinal vasculature stained with an antibody against Collagen IV in retinal whole mount preparations. The relatively uniform plexus visible in 3-day-old (P3) mouse pups (**a**) expands and remodels by P6 (**b**) into a network with clearly distinguishable arteries (a) and veins (v). Arteries can be identified by their capillary free zones (*arrowheads* in **b**). At P9 (**c**) the primary plexus has reached the retinal periphery, vessels start to sprout into the retina and are visible as *white dots* (*small arrowheads*) establishing the deeper network of the retinal vasculature. Some of the veins disappear from the primary plexus (*arrow*) and relocate by a process of remodelling to the deeper plexus (not visible). Scale bar is 200 μm. Credits: [18]

A third step of the network formation is due to the vascular remodelling and maturation.

Here we mainly consider the second step, during which the sprout of the network starts due to the VEGF, and a planar network is formed, as in Fig. 4. The main features of the model are the following. As already mentioned, the dynamics involves three different *cell types*:

1. Type 1 cells: the *mural cells* which are the mature cells. They supply nutrients; when a low concentration in their neighborhoods is detected (read, angioproteins are present), they duplicate generating type 2 cells. Mural cells are subject to death, while their displacement has been considered negligible.
2. Type 2 cells: the active cells, both the specialized *endothelial tip cells* at the leading edge of the growing vascular network, and the *stalk cells*, located in the neighborhood of tip cells. They can proliferate and die and their movement is regulated by repulsive chemotaxis with respect to nutrients produced by type 1 cells and attractive chemotaxis with respect to the VEGF. When the type 2 cell population increases, these cells start to organize, converting themselves into type 1 cells. Brownian diffusion may affect their movement.
3. Type 3 cells: the dead cells. Both type 1 and 2 may become type 3 cells.

The VEGF (g) and the nutrients (e.g., oxygen) (u) activate the migration and the dynamics of endothelial cells at the microscale. We may suppose that at the macroscale such fields may be described by continuum quantities evolving in time via partial differential equations, whose parameters depend vice versa on the state of the cells themselves.

Hence, the main features of the dynamics include *branching*, due to proliferation and change of state of cells, *elongation*, due to aggregation and repulsion phenomena, *remodelling*, and finally *blood circulation*. As already mentioned, here we do not consider the latter two.

2.2.1 Modelling the Evolution of the Capillary Network and the Coupling with the Underlying Fields

We again denote by N a parameter of scale of the process. We model the network by tracking the cells building up the vessels. Let be $\mathbb{S} = \{1,2,3\}$. The considered individual stochastic processes are

$$\underline{Z}^i(t) = (X^i(t), C^i(t)) \in \mathbb{R}^d \times \mathbb{S},$$

which denote the location and the type of the i-th cell at time t, for any $i = 1,\ldots,N(t)$, respectively. $N(t)$ is the total number of cells at time t. The network of endothelial cells is described by the union of the trajectories of the tips as in Eq. (1).

Vessel Extension. We suppose that only type 2 cells are subject to the action of the underlying fields, while type 1 cells are only subject to a possible randomness and type 3 cells do not move anymore. Randomness is modelled by additive independent Wiener processes $\{W_t^i\}_{t\in\mathbb{R}_+}$. Hence, for $i = 1,\ldots,N(t)$ and $t > T_i^b$,

$$\mathrm{d}X^i(t) = \beta\, [\nabla g(X^i(t),t) - \nabla u(X^i(t),t)]\delta_{C^i(t),2}\,\mathrm{d}t + \sigma(C^i(t))\mathrm{d}W_t^i, \tag{22}$$

where

$$\sigma(C^i(t)) = \begin{cases} \sigma_j,\ C^i(t) = j, & \text{for } j = 1,2; \\ 0,\ \ C^i(t) = 3, & \end{cases} \tag{23}$$

$\sigma_1, \sigma_2 \in \mathbb{R}_+$ are diffusion coefficients, $\delta_{i,j}$ is the Kroenecker delta, and $\beta \in \mathbb{R}_+$ represents the strength of response to the drift.

Cell Proliferation. As above, proliferation is described by a branching process, modelled as a marked counting process, by means of a random measure $\Phi = \sum_i \varepsilon_{(T_i^b, X^i, C^i)}$, on $\mathscr{B}_{\mathbb{R}_+\times E\times\mathbb{S}}$. Hence, for any measurable set $A \subseteq \mathscr{B}_{\mathbb{R}_+\times\mathbb{R}^d\times\mathbb{S}}$

$$\Phi(A) := \sum_i \varepsilon_{(T_i^b, X^i, C^i)}(A) = \mathrm{card}\{i : (T_i^b, X^i, C^i) \in A\}$$

is the random variable which counts those cells which are born in A. The process Φ is characterized by the following stochastic intensity; for any $(t,x,s) \in \mathbb{R}_+ \times \mathbb{R}^d \times \mathbb{S}$

$$\Lambda_t(\mathrm{d}x \times \{s\})\,\mathrm{d}t = h(x,s) \sum_{i=1}^{N(t^-)} \varepsilon_{(X^i(t),C^i(t))}(\mathrm{d}x \times \{s\})\,\mathrm{d}t. \tag{24}$$

The type 1 cell proliferation rate is constant, while the type 2 cell proliferation rate decreases with their local density number. As a consequence the nonnegative function $h(.,s) \in C_b(\mathbb{R}^d)$, for $s \in \mathbb{S}$, may be modelled in the following way:

$$h(\underline{Z}^i(t)) = h\left((X^i(t), C^i(t))\right) = \lambda_1 \delta_{C^i(t),1} + \lambda_2 \frac{1}{1 + N(Q_N^{[2]} * K)(X^i(t))} \delta_{C^i(t),2}, \quad (25)$$

where $\lambda_1, \lambda_2 \in \mathbb{R}_+$, with $\lambda_1 \ll \lambda_2$ and K is just a regularizing kernel.

The counting process $N(t) = \Phi([0,t] \times \mathbb{R}^d \times \mathbb{S})$ has intensity given by

$$\nu(\mathrm{d}t) = \Lambda_t(\mathbb{R}^d \times \mathbb{S})\mathrm{d}t = \sum_{i=1}^{N(t)} h(\underline{Z}^i(t))\mathrm{d}t$$

Cell State Evolution. The change of state of each cell is described via a continuous time Markov chain. The associated time-dependent stochastic matrix for the i-th cell is given by the following:

$$\mathbf{M}(\underline{Z}^i(t), t) = \begin{bmatrix} -(m_{12} + m_{13}) & m_{12} & m_{13} \\ m_{21} & -(m_{21} + m_{23}) & m_{23}, \\ 0 & 0 & 0 \end{bmatrix}, \quad (26)$$

where

$$m_{hk} := m_{hk}(\underline{Z}^i(t)) = \lim_{\Delta t \to 0} \frac{prob\{C^i(t + \Delta t) = k | C^i(t) = h\}}{\Delta t}$$

are given by

$$m_{12}(x,1) = \frac{\lambda_{12}}{u(x,t)}, \quad m_{13}(x,1) = \frac{\lambda_{13}}{(Q_N^{[1]} * K)(x)}, \quad (27)$$

$$m_{21}(x,2) = \lambda_{21}(Q_N^{[2]} * K)(x), \quad m_{23}(x,2) = \lambda_{23}, \quad (28)$$

for $i, j = 1, 2, 3$, $\lambda_{ij} > 0$.

Evolution of the Fields. VEGF diffuses and naturally degrades; furthermore, as a matter of modelling simplification, we assume that VEGF is produced at the front of type 2 cells [19]. So, given a region of interest $E \subset \mathbb{R}^d$ (for $d = 2$ or 3), for any $(t,x) \in \mathbb{R}_+ \times \mathbb{R}^d$, the concentration of VEGF $g(x,t)$ is subject to the following evolution equation:

$$\frac{\partial g(x,t)}{\partial t} = -d_g g(x,t) + D_g \Delta g(x,t) + \alpha_g \frac{1}{N} \sum_{j=1}^{N(t)} \left(\varepsilon_{(\cdot, C^j(t))}(x,2) * K_N\right)(X^j(t)). \quad (29)$$

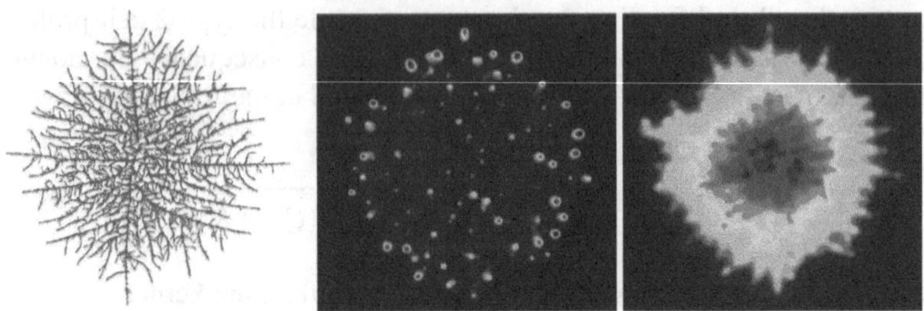

Fig. 5 Snapshot of the vascular network (*left*), the VEGF (*center*), and the nutrient (*right*) density field for the fully stochastic model

In a similar way, we assume that the nutrient diffuses, naturally degrades, and is produced by type 1 cells [19], so that, for any $(t,x) \in \mathbb{R}_+ \times \mathbb{R}^d$, the concentration of nutrient $u(x,t)$ is subject to the following evolution equation:

$$\frac{\partial u(x,t)}{\partial t} = -d_u u(x,t) + D_u \Delta u(x,t) + \alpha_u \frac{1}{N} \sum_{j=1}^{N(t)} \left(\varepsilon_{(\cdot, C^j(t))}(x,1) * K_N \right) (X^j(t)). \quad (30)$$

Parameters $d_i, D_i, \alpha_i \in \mathbb{R}^+, (i = g, u)$ in Eqs. (29) and (30) represent the rates of natural degradation, the diffusivities, and the rates of production, respectively.

By summarizing, at the finer scale of the microscale, the Lagrangian description is based on the stochastic behavior of individual cells, given by the system of stochastic differential equations (22)–(23), the branching process $\Phi_N(t)$ with rates Eqs. (9)–(25), the Markov chain of the state change Eqs. (26)–(28), coupled with the stochastic partial differential equation describing the evolution of the underlying fields Eqs. (29) and (30).

An example of the obtained results is shown in Fig. 5.

The geometrical complexity of the network is now recovered: there are frequent branchings which appear in almost every area of the domain and even several anastomoses occur. The pattern of the nutrient field u (right) reflects the structure of the stochastic network (left). The pattern of the growth factor g (center) indicates the location of type 2 cells: they are mostly concentrated on the boundary regions. The stochastic branching leading to the realistic pattern of the vessel network [Fig. 5 (left)] is a direct consequence of the stochasticity of the underlying fields [Fig. 5 (center and right)].

2.2.2 Modelling the Evolution of the Empirical Measures and the Coupling with the Underlying Fields

From the empirical measure process $Q_N \in C([0,T], \mathcal{M}(\mathbb{R}^d \times \mathbb{S}))$, where for any $t \in [0,T]$, $Q_N(t)$ is given by Eq. (2), one can derive the empirical spatial distribution of the cells of type $s \in \mathbb{S}$

$$Q_N^{[s]}(t) = Q_N(t)(\cdot \times \{s\}) = \frac{1}{N} \sum_{k \in H(s,t)} \varepsilon_{X^k(t)} \in \mathscr{M}(\mathbb{R}^d), \tag{31}$$

where $H(s,t) = \{k \in \{1,\ldots,N(t)\} : C^k(t) = s\}$. Again from Itô's formula, we obtain that the time evolution of the measure $Q_N(t)$, for any $t \in [0,T]$ and for any $B \in \mathscr{B}(\mathbb{R}^d)$ is the following

$$\begin{aligned}
\sum_{c=1}^{3} \int_B f(x,c) Q_N(t)(\mathrm{d}x,c) = & \sum_{c=1}^{3} \int_B f(x,c) Q_N(0)(\mathrm{d}x,c) \\
& + \int_0^t \left\{ \int_B \sum_{c=1}^{2} \frac{\sigma_c^2}{2} \Delta_x f(x,c)\, Q_N^{[c]}(s)(\mathrm{d}x) \right. \\
& + \int_B \beta [\nabla_x g(x) - \nabla_x u(x)] \cdot \nabla_x f(x,2)\, Q_N^{[2]}(s)(\mathrm{d}x) \\
& + \sum_{\substack{c=1 \\ j \neq c}}^{2} \int_B (f(x,j) - f(x,c))\, m_{cj}(x,c)\, Q_N^{[c]}(s)(\mathrm{d}x) \\
& \left. + \sum_{c=1}^{2} \int_B h(x,c)\, f(x,c)\, Q_N^{[c]}(s)(\mathrm{d}x) \right\} \mathrm{d}s \\
& + M_N[Q_N, W](t),
\end{aligned} \tag{32}$$

where

$$\begin{aligned}
M_N[Q_N,W](t) = & \int_0^t \sum_{c=1}^{2} \frac{1}{N} \sum_{k=1}^{N_B(s)} \sigma_c \nabla_x f(X^k(s),c) \mathrm{d}W^k(s) \delta_{c,C^k(s)} \\
& + \int_0^t \frac{1}{N} \sum_{k=1}^{N_B(s)} f(X^k(s),C^k(s)) \left[N_B(\mathrm{d}s) - \sum_{c=1}^{2} h(X^k(s),c) \delta_{c,C^k(s)}] \mathrm{d}s \right] \\
& + \int_0^t \frac{1}{N} \sum_{k=1}^{N_B(s)} \left\{ f(X^k(s),C^k(s)) \right. \\
& \left. - \sum_{j \neq C^k} [f(X^k(s),j) - f(X^k(s),C^k(s))\, m_{C^k j}(X^k(t),C^k(t)) \right\} \mathrm{d}s.
\end{aligned} \tag{33}$$

It is a zero mean martingale with respect to the natural filtration $\{\mathscr{F}_t\}_{t \in \mathbb{R}_+}$ generated by the process $\{(X^k(t),C^k(t)),N(t)\}_{t \in \mathbb{R}_+}$. Equation (32) is coupled with the system of PDEs

$$\frac{\partial g(x,t)}{\partial t} = -d_g g(x,t) + D_g \Delta g(x,t) + \alpha_g \left(Q_N^{[2]} * K_N \right)(x); \tag{34}$$

$$\frac{\partial u(x,t)}{\partial t} = -d_u u(x,t) + D_u \Delta u(x,t) + \alpha_u \left(Q_N^{[1]} * K_N \right)(x). \tag{35}$$

System (32)–(35) gives an Eulerian description of the fully stochastic retinal angiogenic process.

3 Towards Hybrid Models

Let us go back to the general Lagrangian system (5)–(6) and to the Eulerian version given by Eq. (7). The analysis and the computation of the above system require the knowledge of the evolution of all individuals up to time t. In both examples we have considered, in the detailed models, either the Lagrangian one or the Eulerian one, the evolution of the stochastic processes of branching and extension is driven by parameters which depend upon the underlying fields; since the evolution of these ones is vice versa coupled with the above stochastic processes, they are themselves stochastic; as a consequence, we are dealing at the microscale with a doubly stochastic system. A major difficulty, both analytical and computational, derives from the fact that indeed the parameters are $\{\mathscr{F}_{t^-}\}$—stochastic, i.e., their values at time $t > 0$ depend upon the actual history $\mathscr{F}_t$ of the whole system up to time t^-. The strong coupling with the fields is a source of complexity, as already discussed. Under these circumstances, a possible way to reduce complexity, which has been suggested by the authors [7] and by a large literature [35, 40], is to apply suitable laws of large numbers at the mesoscale, i.e., in suitably scaled neighborhoods of any relevant point $x \in \mathbb{R}^d$, such that, at that scale we may approximate $Q_N(t)$ by a deterministic measure (the so called mean field approximation) $Q(t)$, possibly having a density $\rho(\cdot,t)$ with respect to the usual Lebesgue measure ν^d, i.e.,

$$Q(t) = \rho(x,y,t)\nu^d. \tag{36}$$

Given that Eq. (36) is the limit measure of the sequence of the empirical measures $\{Q_N(t)\}_{t\in\mathbb{R}_+}$, we may additionally observe the following. From Eq. (4), it is clear that

$$\lim_{N\to+\infty} K_N = \delta_0,$$

where δ_0 is Dirac's delta function; as a consequence, the measure $\widetilde{Q}_N(t)$ is such that its regularized measure, defined, for any $s \in \mathbb{S}$, as

$$h_N(x,t) = (\widetilde{Q}_N(t) * K_N)(x) = \frac{1}{N^{1-\eta}} \sum_k K_1(N^{\eta/d}(x - X^k(t))), \tag{37}$$

is such that

$$\lim_{N\to+\infty} h_N(x,t) = \widetilde{\rho}(x,t) = \int_{\mathbb{I}} \rho(x,y,t)\mathrm{d}y. \tag{38}$$

Note that from Eq. (6), the interaction of the underlying field C with the neighboring cells is realized via the function h_N. The size of the spatial region of interaction is a consequence of the choice of our mesoscale, related to the value of $\eta < 1$. Indeed, Eq. (37) represents an averaging over a large number of cells, at the level of the mesoscale (given by the size of η); furthermore K_N converges to a Dirac δ-function sufficiently slow, so that we may apply a "law of large numbers" in such a way that $h_N(x,t)$ approaches the limit Eq. (38). We need to be sure that the sum in Eq. (37) is performed over a sufficiently large number of cells, which is equivalent to say that the range of K_N has not to shrink to zero too fast with respect to the increase of the number of cells.

In the limit the relevant density will satisfy a deterministic evolution equation of the type

$$\int_{\mathbb{D}} g(\underline{z},t)\rho(\underline{z},t)\mathrm{d}\underline{z} = \int_{\mathbb{D}} g(\underline{z},0)\rho(\underline{z},0)\mathrm{d}\underline{z} + \int_0^t Op_1\left(\rho(\cdot,s),\widetilde{C}(\cdot,s),g(\cdot,s)\right)\mathrm{d}s. \quad (39)$$

This approximated measure then substitutes the random measure $Q_N(t)$ in the evolution equations for the underlying fields at the macroscale, so that they are now completely deterministic

$$\frac{\partial}{\partial t}\widetilde{C}(x,t) = Op_2[\widetilde{C}(\cdot,t)](x) + Op_3\left[\widetilde{C}(\cdot,t) * \widetilde{\rho}(\cdot,t)\right](x); \quad (40)$$

they may be used to drive the kinetic parameters of the cell processes at the microscale, thus leading to simple stochastic processes

$$\mathrm{d}\underline{Z}^k(t) = F\left[\widetilde{C}(\cdot,t)\right](\underline{Z}^k(t))\mathrm{d}t + \underline{\sigma}\,\mathrm{d}\underline{W}^k(t),, \qquad k = 1,\ldots,N(t). \quad (41)$$

These kind of models are known as *hybrid models*, since they are stochastic at the level of individuals Eq. (41) and deterministic at the level of the field Eq. (40).

3.1 Tumor-Induced Angiogenesis

Given the evolution equation (16) for the empirical measure $Q_N(t)$, let us observe that due to the Doob's inequality and Eq. (21), the quadratic variation of the zero mean martingale Eq. (17) is such that

$$\mathbb{E}\left[\sup_{t\leq T}|\tilde{M}_N(t)||\mathscr{F}_0\right]^2 \leq 4C\mathbb{E}\left[|\tilde{M}_N(T)|^2|\mathscr{F}_0\right]$$

$$
\begin{aligned}
&= C_1 \mathbb{E}\left[\int_0^T \frac{1}{N^2} \sum_{i=1}^{N(t)} |g(X^k(t), v_0)|^2 \mathrm{d}t \middle| \mathscr{F}_0\right] \\
&\quad + C_2 \mathbb{E}\left[\int_0^T \frac{\sigma^2}{N^2} \sum_{i=1}^{N(t)} |\nabla_v g(x,v)|^2 \mathrm{d}t \middle| \mathscr{F}_0\right] \\
&\leq C\frac{T}{N}(\|g\|_2^2 + \|\nabla g\|_2^2)\mathbb{E}\left[\sup_{t\leq T} \langle T_N(t), 1\rangle \middle| \mathscr{F}_0\right] \\
&< C\frac{T}{N}. \qquad (42)
\end{aligned}
$$

From Eq. (42), $\mathbb{E}\big[\sup_{t\leq T} |M_N[Q_N, W](t)|\big]^2$ tends to zero, as N increases to infinity. Hence, from $\mathbb{E}[M_N(t)] = 0$, the martingale Eq. (17) vanishes in probability, i.e., uniformly in $[0,T]$,

$$
M_N[Q_N, W](t) \xrightarrow{P} 0.
$$

This is the substantial reason of the deterministic limiting behavior of the process Q_N, as N increases to infinity.

Let us consider now a heuristic derivation of a continuum dynamics, by assuming that, as $N \to \infty$, the measure $Q_N(t)$ admits a limit measure $Q(t)$, i.e., formally, we take

$$
Q_N(t)(\mathrm{d}(x,v)) \to Q_\infty(t)(\mathrm{d}(x,v)) = p(t,x,v)\mathrm{d}x\mathrm{d}v, \qquad (43)
$$

then we would have limit measures also for $T_N(t)$ and $V_N(t)$, i.e., $T_\infty(t)(\mathrm{d}x) = \tilde{p}(t,x)\mathrm{d}x$; and $V_\infty(t)(\mathrm{d}x) = w(t,x)\mathrm{d}x$, where

$$
\tilde{p}(t,x) = \int p(t,x,v)\mathrm{d}v, \quad w(t,x) = \int v\, p(t,x,v). \qquad (44)
$$

Then system (16)–(20) becomes

$$
\begin{aligned}
\int_B g(x,v)p(t,x,v)\mathrm{d}x\mathrm{d}v = \int_0^t \int_B p(s,x,v)\mathrm{d}s\,\mathrm{d}x\mathrm{d}v \Bigg[&\frac{\sigma^2}{2}\Delta_v g(x,v) \\
&+ \nabla_x g(x,v)v + g(x,v)\alpha h(\widetilde{C}(s,x))\delta_{\{v_0\}}(v) \\
&- \nabla_v g(x,v)\left[kv - F\left(\widetilde{C}(t,x), \widetilde{f}(t,x)\right)\right]\Bigg] \qquad (45)
\end{aligned}
$$

$$
\frac{\partial}{\partial t}\widetilde{C}(t,x) = c_1\delta_A(x) + d_1\triangle\widetilde{C}(t,x) - \eta\widetilde{C}(t,x)w(t,x); \qquad (46)
$$

$$
\frac{\partial}{\partial t}\widetilde{f}(t,x) = \beta\tilde{p}(x,t) - \gamma\widetilde{m}(x,t)\widetilde{f}(t,x); \qquad (47)
$$

$$
\frac{\partial}{\partial t}\widetilde{m}(t,x) = \varepsilon_1\triangle\widetilde{m}(t,x) + \nu_1\tilde{p}(x,t) - \nu_2\widetilde{m}(t,x). \qquad (48)
$$

Equation (45) may be seen as the weak form of the following partial differential equation for the density $p(t,x,v)$:

$$\frac{\partial}{\partial t}p(t,x,v) = -v\cdot\nabla_x p(t,x,v) + k\nabla_v\cdot(vp(t,x,v)) + \alpha h(\widetilde{C}(t,x))p(t,x,v_0)$$
$$-\nabla_v\cdot\left[F\left(\widetilde{C}(t,x),\widetilde{f}(t,x)\right)p(t,x,v)\right] + \frac{\sigma^2}{2}\Delta_v p(t,x,v). \qquad (49)$$

Given the limit system (45)–(48), one may recover the dynamics of the individuals coupling such a system with

$$dX^i(t) = v^i(t),$$
$$\mathrm{d}v^i(t) = -kv^i(t) + F[\widetilde{C},\widetilde{f}]\mathrm{d}t + \mathrm{d}W^i(t).$$

3.2 *Retinal Angiogenesis*

As considered above, let us perform a heuristic derivation of a continuum dynamics. Note that also in this case we may state that the martingale Eq. (33) vanishes in probability. Let us assume that, as $N\to\infty$, the measure $Q_N(t)$ admits a limit measure $Q(t)$ with a density $p(x,s,t)$, such that, for any $A\subset\mathbb{S}$,

$$Q(t)(\mathrm{d}x\times A) = \sum_{s\in A} p(x,s,t)\mathrm{d}x.$$

This is equivalent to say that for any $s\in\mathbb{S}$, the empirical measure related to the cell of type s, $Q_N^{[s]}(t)$ given by Eq. (31) converges to a measure $Q^{[s]}(t)$ such that

$$Q^{[s]}(t)(\mathrm{d}x) = p(x,s,t)\,\mathrm{d}x =: p_s(x,t)\,\mathrm{d}x. \qquad (50)$$

From Eqs. (32) and (50), together with Eqs. (29) and (30), a weak form for the time evolution of the continuum densities $p_s(x,t), s\in\mathbb{S}$, is obtained

$$\sum_{s\in\mathbb{S}}\int_B f(x,s)p_s(x,t)\mathrm{d}x = \sum_{s\in\mathbb{S}}\int_B f(x,s)p_s(x,0)\mathrm{d}x$$
$$+\int_0^t\int_B\left[\frac{\sigma_1^2}{2}\Delta_x f(x,1)p_1(x,t) + h(x,1)f(x,1)p_1(x,t)\right.$$
$$-m_{12}f(x,1)p_1(x,t) - m_{13}f(x,1)p_1(x,t)$$
$$\left.+m_{21}f(x,2)p_2(x,t)\right]\mathrm{d}x\mathrm{d}t$$

$$
\begin{aligned}
&+\int_0^t\int_B\Big[\frac{\sigma_2^2}{2}\Delta_x f(x,2)p_2(x,t)+h(x,2)f(x,2)p_2(x,t)\\
&\qquad +\beta\left(\nabla_x\tilde{g}(x,t)-\nabla_x\tilde{u}(x,t)\right)\cdot\nabla_x f(x,2)p_2(x,t)\\
&\qquad -m_{21}f(x,2)p_2(x,t)-m_{23}f(x,2)p_2(x,t)\\
&\qquad +m_{12}f(x,2)p_1(x,t)\Big]\mathrm{d}x\mathrm{d}t\\
&+\int_0^t\int_B\Big[m_{13}f(x,3)p_1(x,t)-m_{23}f(x,3)p_2(x,t)\Big]\mathrm{d}x\mathrm{d}t.
\end{aligned}
\tag{51}
$$

with

$$
m_{12}:=m_{12}(x,1)\frac{\lambda_{12}}{\tilde{u}(x,t)},\qquad m_{13}:=m_{13}(x,1)=\frac{\lambda_{13}}{(p_1(\cdot,t)*K)(x)},\tag{52}
$$

$$
m_{21}:=m_{21}(x,2)=\lambda_{21}(p_2(\cdot,t)*K)(x),\quad m_{23}:=m_{23}(x,2)=\lambda_{23},\tag{53}
$$

and, by Eq. (38), coupled with the dynamics of the (now purely deterministic) underlying fields $\tilde{g}$ and $\tilde{u}$

$$
\frac{\partial\tilde{g}(x,t)}{\partial t}=-d_g\tilde{g}(x,t)+D_g\Delta\tilde{g}(x,t)+\alpha_g p_2(x,t),\tag{54}
$$

$$
\frac{\partial\tilde{u}(x,t)}{\partial t}=-d_u\tilde{u}(x,t)+D_u\Delta\tilde{u}(x,t)+\alpha_u p_1(x,t).\tag{55}
$$

In particular, by considering the weak evolution Eq. (51) over the three subsets $B\times s\in\mathscr{B}_E\otimes\mathbb{S}$, for any $s=1,2,3$, we find the weak form of the following system of PDEs:

$$
\frac{\partial p_1(x,t)}{\partial t}=(h(x,1)-m_{12}-m_{13})p_1(x,t)+m_{21}p_2(x,t),\tag{56}
$$

$$
\begin{aligned}
\frac{\partial p_2(x,t)}{\partial t}&=\frac{\sigma_2^2}{2}\Delta p_2(x,t)-\alpha_2\nabla\Big[\left(\nabla\tilde{g}(x,t)-\nabla\tilde{u}(x,t)\right)p_2(x,t)\Big]\\
&\quad(h(x,2)-m_{21}-m_{23})p_2(x,t)+m_{12}p_1(x,t),
\end{aligned}
\tag{57}
$$

$$
\frac{\partial p_3(x,t)}{\partial t}=m_{13}p_1(x,t)+m_{23}p_2(x,t),\tag{58}
$$

coupled with Eqs. (54) and (55). System (54)–(58) represents the continuum description of the angiogenic network at the macroscale. It may be taken into account only whenever a law of large numbers may be applied. It describes the continuum densities of the different types of cells coupled with the deterministic PDEs for the underlying fields.

In this way we obtain an approximation of the driving fields which are deterministic at the macroscale. These fields now drive, at the microscale, a simply stochastic evolution for the single cells. More specifically, a typical cell k will satisfy the following system of stochastic differential equations, for any $k = 1,\dots,N(t)$:

$$\mathrm{d}Y^k(t) = \beta\,[\nabla\tilde{g}(Y^k(t),t) - \nabla\tilde{u}(Y^k(t),t)]\,\delta_{C^k(t),2}\,\mathrm{d}t + \sigma(C^k(t))dW_t^k\mathrm{d}t, \quad (59)$$

subject to a change of state governed by the matrix $M(Y^k(t),C^k(t))$, with proliferation rate

$$h\left((Y^k(t),C^k(t))\right) = \lambda_1\,\delta_{C^k(t),1} + \lambda_2\frac{1}{1+N(p_2(\cdot,t)*K)(Y^k(t))}\delta_{C^k(t),2}, \quad (60)$$

and coupled with systems (54)–(55) for the VEGF $\tilde{g}$ and the nutrient $\tilde{u}$.

In conclusion, we may describe this retinal angiogenic process in two different ways; either by a fully stochastic system, involving the individual cell processes, and the associated empirical measure process, strongly coupled with the evolution of relevant underlying fields, or as a simply stochastic system coupled with the evolution of the deterministic mean underlying field obtained upon a limit of large numbers of the empirical measure Q_N. Whenever applicable, this last approach makes the model computationally more affordable.

4 Conclusions and Discussion

The following question arises; do we lose any important feature of the dynamics of the system by considering the hybrid model, i.e., by eliminating the stochastic variability of the underlying fields? We have addressed this question by means of numerical simulations.

In this section we compare the simulation results (a) for the fully deterministic system of spatial densities of both cells and underlying fields, (b) for the hybrid system, i.e., stochastic birth and growth processes (at the microscale (cells)) driven by deterministic averaged underlying fields (at the macroscale), and (c) for the fully stochastic model of retinal anagenesis.

The Deterministic System. First of all, we consider system (54)–(58). From Fig. 6 (thanks to spherical symmetry it is sufficient to show 2D sections of the relevant distributions) it is clear that:

1. Each profile is smooth and symmetric with respect to the origin.
2. Type 2 cells show a higher density in the external regions, where the capillaries are built (up-left).
3. Type 1 cell density decreases from the origin to the frontier (up-right).

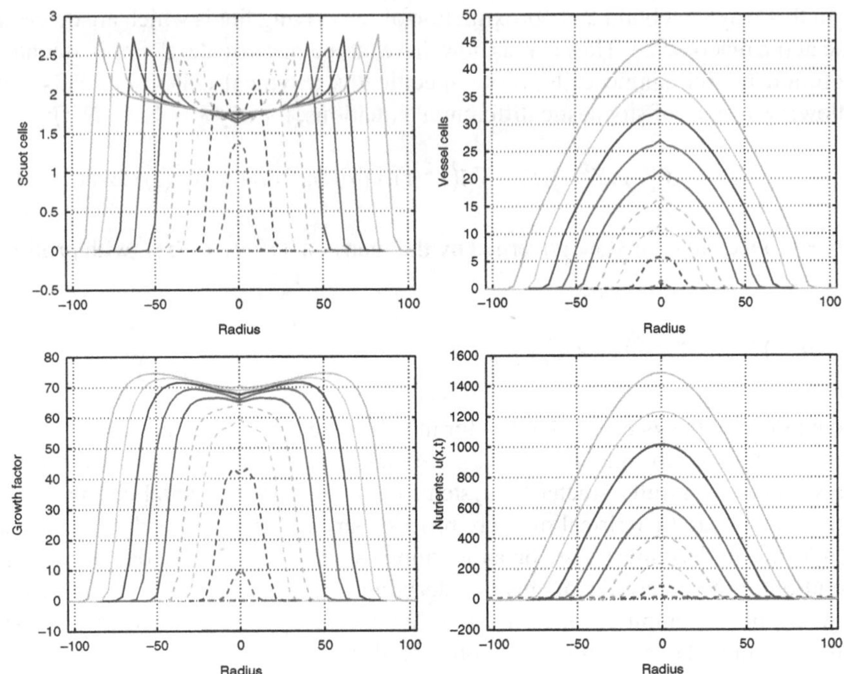

Fig. 6 Simulations of the time evolution of the profiles of type 2 cell (*up-left*), type 1 cells (*up-right*), VEGF $\tilde{g}$ (*down-left*) and nutrient $\tilde{u}$ (*down-right*) densities of the fully deterministic model

4. Nutrient $\tilde{u}$ (down-right) and growth factor $\tilde{g}$ (down-left) concentrations reflect the densities of cells by which they are produced: diffusion makes their profiles much smoother.

The time evolution shows how type 2 cells move progressively from the center to the frontier; we may also notice the expansion of the vascularization by building new vessels at the boundary region without a significant modification of the profile in the central area. In the internal part the nutrient concentration $\tilde{u}$ increases (supplied to the tissue cells) and the growth factor starts to decrease because type 2 cells have moved towards the frontier. Hence, if we were not interested in the detailed geometric structure of the capillary network, the deterministic model would be able to catch some of the relevant features of the qualitative behavior of the phenomenon under study.

The Hybrid System. Let us consider now the hybrid model in which the stochastic evolution Eqs. (59) and (60) of the system of cells $\{(X^k, C^k)\}_k$ is driven with the deterministic dynamics Eqs. (54) and (55) of the underlying fields $\tilde{g}, \tilde{u}$, obtained as a result of the mesoscale averaging.

Clearly, in these simulations, we may use the same numerical results for the underlying fields obtained above. Hence, as before, the homogenized underlying

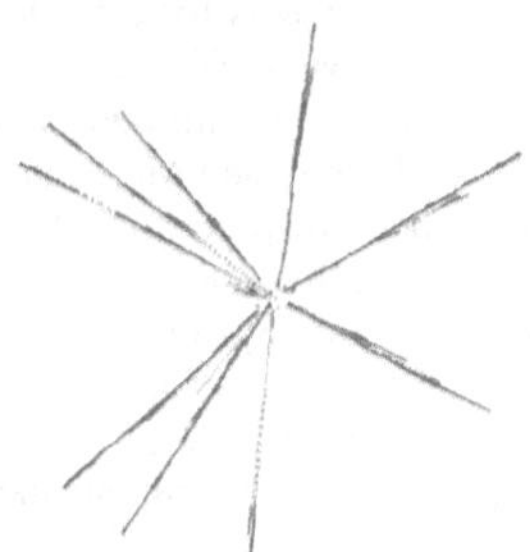
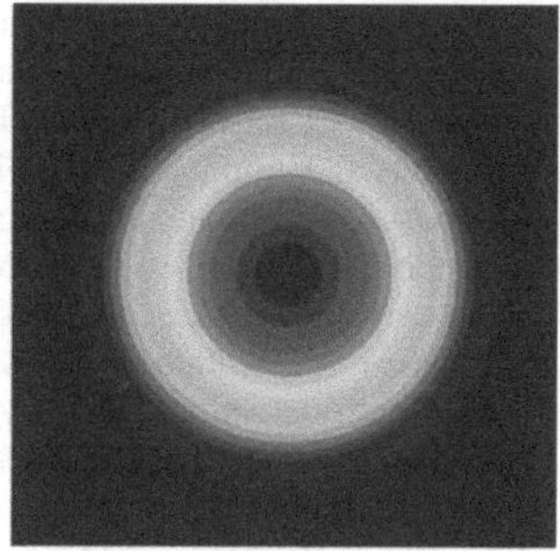

Fig. 7 Snapshot at time $T = 60$ of type 1 and 2 cells (*left*) and radially symmetric nutrient field $\tilde{u}$ (*right*) for the hybrid model

fields $\tilde{g}$ and $\tilde{u}$ exhibit spherical symmetry (Fig. 7, right). As a consequence, capillaries are built radially from the center, but there is almost no branching (Fig. 7, left). So while we gain a reduction of the computational costs with respect to the fully stochastic model, we lose the formation of a realistic vessel network.

A possible explanation may be found in the excessively reduced randomness of the system; by assuming ab initio a field dynamics subject to the purely deterministic equations (54) and (55), we have lost a significant part of the spatial heterogeneity. After all, at the beginning of the simulation, the population size of vascularizing cells is still too small, and then the adopted approximation (which requires a large N) is not valid yet.

For the fully stochastic model, we have already discussed how the geometrical complexity of the network is recovered, as in Fig. 5.

In conclusion, from a fully stochastic model, by using convergence results of martingale theory (laws of large numbers), we can derive a deterministic approximation of the density of cells to be coupled with the evolution equations of the underlying fields. Its numerical integration provides the kinetic parameters of the stochastic processes modelling branching and elongations of vessels (*hybrid model*). The warning is that, though in this way we gain a convenient reduction of the computational complexity with respect to the fully stochastic model, by averaging "ab initio" the random spatial distribution of the population of cells in the evolution equations of the underlying fields, we lose a significant part of their spatial structure due to an anomalous reduction of the otherwise unavoidable randomness.

As another possible interesting application, we may refer to embryonic an postnatal development of the nervous system, where neuronal migration and axon pathfinding are guided by extracellular cues [21].

References

1. Anderson, A.R.A., Chaplain, M.A.: Continuous and discrete mathematical models of tumour-induced angiogenesis. Bull. Math. Biol. **60**, 857–900 (1998)
2. Birdwell, C., Brasier, A., Taylor, L.: Two-dimensional peptide mapping of fibronectin from bovine aortic endothelial cells and bovine plasma. Biochem. Biophys. Res. Commun. **97**, 574–81 (1980)

3. Byrne, H., Chaplain, M.: Mathematical models for tumour angiogenesis: Numerical simulations and nonlinear wave solutions. Bull. Math. Biol. **57**, 461-486 (1995)
4. Capasso V., Bakstein D.: An Introduction to Continuous-Time Stochastic Processes - Theory, Models and Applications to Finance, Biology and Medicine. Birkhäuser, Boston (2005)
5. Capasso, V., Morale, D.: Rescaling stochastic processes: Asymptotics. In: Multiscale Problems in the Life Sciences From Microscopic to Macroscopic, Lecture Notes in Mathematics / Fondazione C.I.M.E., vol. 1940, pp. 91–146. Springer, Heidelberg (2008)
6. Capasso, V., Morale, D.: Asymptotic behavior of a system of stochastic particles subject to nonlocal interactions. Stoch. Anal. And Appl. **27**(3), 574–603 (2009)
7. Capasso, V., Morale, D.: Stochastic modelling of tumour-induced angiogenesis. J. Math. Biol. **58**, 219–233 (2009)
8. Capasso, V., Morale, D., Facchetti, G.: The role of stochasticity for a model of retinal angiogenesis. IMA J. Appl. Math. (2012) (In press)
9. Capasso, V., Micheletti, A., Morale, D.: Stochastic geometric models, and related statistical issues in tumour-induced angiogenesis. Math. Biosci. **214**, 20–31 (2008)
10. Champagnat, N., Méléard, S.: Invasion and adaptive evolution for individual-based spatially structured populations. J. Math. Biol. **55**, 147–188 (2007)
11. Chaplain, M.: The mathematical modelling of tumour angiogenesis and invasion. Acta Biotheor. **43**, 387–402 (1995)
12. Chaplain, M.A.J., Anderson, A.R.A.: Modelling the growth and form of capillary networks. In: Chaplain, M.A.J., et al. (eds.) On Growth and Form. Spatio-temporal Pattern Formation in Biology. Wiley, Chichester (1999)
13. Chaplain, M., Stuart, A.: A model mechanism for the chemotactic response of endothelial cells to tumour angiogenesis factor. IMA J. Math. Appl. Med. Biol. **10**, 149–168 (1993)
14. Corada, M., Zanetta,L., Orsenigo, F., Breviario, F., Lampugnani, M.G., Bernasconi, S., Liao, F., Hicklin, D.J., Bohlen, P., Dejana, E. : A monoclonal antibody to vascular endothelial-cadherin inhibits tumor angiogenesis without side effects on endothelial permeability. Blood **100**, 905–911 (2002)
15. Davis, B.: Reinforced random walk. Prob. Theor. Rel. Fields **84**, 203–22 (1990)
16. Durrett, R., Levin, S.A.: The importance of being discrete (and spatial). Theor. Pop. Biol. **46**, 363–394 (1994)
17. Folkman, J., Klagsbrun, M.: Angiogenic factors. Science **235**, 442–447 (1987)
18. Fruttiger, M.: Development of retinal vasculature. Angiogenesis **10**, 177–88 (2007)
19. Gariano, R.F., Gardner, T.W.: Retinal angiogenesis in development and disease. Nature **438**, 960–966 (2003)
20. Gerhardt, H., Golding, M., Fruttiger, M., Ruhrberg, C., Lundvkvist, A., Abramsson, A., Jeltsch, M., Mitchell, C., Alitao, K., Shima, D., Betsholtz, C.: VEGF guides angiogenic sprouting utilizing tip cell filopodia. J. Cell Biol. **161**, 1163–1177 (2003)
21. Guan K.L., Rao, Y.: Signalling mechanisms mediating neuronal responses to guidance cues. Nature Rev. Neurosci. **4**, 941–956 (2003) doi:10.1038/nrn1254
22. Gueron, S., Levin, S.A.: The dynamics of group formation. Math. Biosci. **128**, 243–264 (1995)
23. Levine, H.A., Sleeman, B.D., Nilsen-Hamilton, M.: Mathematical modelling of the onset of capillary formation initiating angiogenesis. J. Math. Biol. **42**, 195–238 (2001)
24. Liotta, L., Saidel, G., Kleinerman, J.: Diffusion model of tumor vascularization. Bull. Math. Biol. **39**, 117–128 (1977)
25. McDougall, S.R., Anderson, A.R.A., Chaplain, M.A.J.: Mathematical modelling of dynamic tumour-induced angiogenesis: Clinical implications and therapeutic targeting strategies. J. Theor. Biol. **241**, 564–589 (2006)
26. Meleard, S., Fernandez, B.: Asymptotic behaviour for interacting diffusion processes with space-time random birth. Bernoulli **6**, 1–21 (2000)
27. Morale, D., Capasso, V., Oelschläger, K.: An interacting particle system modelling aggregation behavior: From individuals to populations. J. Math. Biol. **50**, 49–66 (2005)
28. Oelschläger, K.: A law of large numbers for moderately interacting diffusion processes. Z. Wahrscheinlichkeitstheorie verw. Gebiete **69**, 279–322 (1985)

29. Oelschläger, K.: On the derivation of reaction-diffusion equations as lilit dynamics of systems of moderately interacting stochastic processes. Prob. Th. Rel. Fields **82**, 565–586 (1989)
30. Okubo, A.: Dynamical aspects of animal grouping: Swarms, school, flocks and herds. Adv. BioPhys. **22**, 1–94 (1986)
31. Okubo, A., Levin, S.: Diffusion and ecological problems: Modern Perspectives. Springer, Heidelberg (2002)
32. Orme, M., Chaplain, M.: A mathematical model of the first steps of tumour-related angiogenesis: Capillary sprout formation and secondary branching. IMA J. Math. Appl. Med. **14**, 189–205 (1996)
33. Plank, M.J., Sleeman, B.D.: A reinforced random walk model of tumour angiogenesis and anti-angiogenic strategies. IMA J. Math. Med. Biol. **20**, 135–181 (2003)
34. Plank, M.J., Sleeman, B.D.: Lattice and non-lattice models of tumour angiogenesis. Bull. Math. Biol. **66**(6), 1785–1819 (2004)
35. Rejniak, K.A., Anderson, A.R.A.: Hybrid models of tumor growth. Wiley Interdiscipl. Rev. Syst. Biol. Med. **3**, 115–125 (2011) doi: 10.1002/wsbm.102
36. Stéphanou, A., McDougall, S.R., Anderson, A.R.A., Chaplain, M.A.J.: Mathematical modelling of the influence of blood rheological properties upon adaptative tumour-induced angiogenesis. Math. Comput. Model. **44**, 96–123 (2006)
37. Stokes, C.L., Lauffenburger, D.A.: Analysis of the roles of microvessel endothelial cell random motility and chemotaxis in angiogenesis. J. Theor. Biol. **152**, 377–403 (1991)
38. Stone, J., Itin, A., Alon, T., Peter, J., Gnessin, H., Chan-Ling, T., Keshet, E.: Development of retinal vasculature is mediated by hypoxia induced vascular endothelial growth factor (VEGF). J. Neurosci. **15**, 4738–4747 (1995)
39. Sznitman, A.S.: Topics in propagation of chaos. Ecole d'Ete de Probabilites de Saint-Flour XIX, 1989. In: Lecture Notes in Mathematics, vol. 1464, pp. 165–251. Springer, Berlin (1991)
40. Sun, S., Wheeler, M.F., Obeyesekere, M., Patrick, C.W. Jr.: A deterministic model of growth factor-induced angiogenesis. Bull. Math. Biol. **67**, 313–337 (2005)
41. Sun, S., Wheeler, M.F., Obeyesekere, M., Patrick, C.W. Jr.: A Multiscale angiogenesis modeling using mixed finite element methods. SIAM Multiscale Model. Simul. **4**, 1137–1167 (2005)
42. West, H., Richardson, W.D., Fruttiger, M.: Stabilization of the retinal vascular network by reciprocal feedback between blood vessel and astrocytes. Development **132**, 1855–1862 (2005)

Modeling Tumor Blood Vessel Dynamics

Lance L. Munn, Christian Kunert, and J. Alex Tyrrell

1 Introduction

Tumor blood vessels are structurally abnormal and functionally inefficient, resulting in incomplete perfusion of tumor vessel networks and nonuniform delivery of chemotherapeutics to the tumor cells. Excessive production of the angiogenic growth factor VEGF (vascular endothelial growth factor) contributes to tumor vessel abnormalities, and many anti-VEGF therapies can cause remodeling or stabilization of tumor blood vessels. This remodeling resembles the process of angioadaptation previously studied in the context of normal physiology and ischemia. During angioadaptation (also know as adaptive remodeling), endothelial cells respond to blood forces to alter blood flow. Some segments dilate, while others contract, eventually producing an efficient network. Although not well understood, it is likely that adaptive remodeling depends on blood shear forces, transvascular pressure, upstream signals transmitted along the endothelium as well as growth factors such as VEGF. To provide an analytical framework for understanding these processes in the context of tumor vasculature, we have developed a mathematical model, supported by multiparameter imaging methodology, which incorporates the necessary elements for predicting the transport of nutrients and drugs throughout tumor vessels and tissue, as well as the adaptive remodeling of the blood vessel network. A better understanding of the mechanisms responsible for the network dynamics may lead to novel approaches for treating tumors or other diseases involving vascular pathologies.

Anti-angiognenic therapy induces structural and functional changes in tumor vasculature, collectively known as vascular normalization [37, 60], that can

L.L. Munn (✉) • C. Kunert • J. Alex Tyrrell
Massachusetts General Hospital, Harvard Medical School, Boston, MA 02114, USA
e-mail: lance@steele.mgh.harvard.edu; kuni@steele.mgh.harvard.edu; james@steele.mgh.harvard.edu

U. Ledzewicz et al., *Mathematical Methods and Models in Biomedicine*, Lecture Notes on Mathematical Modelling in the Life Sciences, DOI 10.1007/978-1-4614-4178-6_5,

improve the efficacy of subsequently administered chemotherapeutics [34, 38, 62, 92, 96]. The predominant effects of anti-VEGF therapies are decreased vessel leakiness (hydraulic conductivity), decreased vessel diameters and pruning of the unstable vessel network [42, 88, 97] (Fig. 1). It is thought that each of these effects can influence perfusion of the vessel network, inducing flow in regions that were previously sluggish or stagnant. Unfortunately, changes induced by anti-VEGF therapies are transient and overlapping in time, and a rational, generalized strategy for using normalization in combination with chemotherapeutics remains elusive. This is in part due to the nonlinearity in the system and the inability to distinguish the effects of decreased vessel leakiness from those due to network structural changes.

In many physiological systems, angiogenic blood vessels respond to blood forces, reorganizing locally to optimize the network globally [1, 33, 35, 36, 71]. Undergoing structural adaptation, some segments dilate while others contract; excessive contraction can even lead to complete regression of some vessel segments. A stable configuration is eventually reached and then supported and fortified by perivascular cells. In contrast, tumor blood vessels are chronically immature, probably due to the high levels of VEGF and other growth factors in the microenvironment. Interestingly, many anti-VEGF, therapies can cause maturation and stabilization of tumor blood vessels through a process resembling flow-based adaptive remodeling (Fig. 1b). Because it so closely resembles flow-based adaptation, it is possible that the central mechanism underlying tumor vessel normalization is the same: blood shear forces drive the structural remodeling. But how the blood forces, VEGF and other signals are integrated by the endothelium is not known. An appropriate mathematical model, which incorporates the necessary elements for predicting the convection and diffusion of nutrients and drugs throughout tumor vessels and tissues, as well as the adaptive remodeling of the network topology, is needed to dissect the critical determinants of vessel normalization and to predict how it might be used most effectively in combination with cytotoxic drugs.

2 Anti-angiogenic Therapies and Tumor Vessel Normalization

Because tumor expansion depends on nutrients delivered by blood vessels, Folkman proposed that preventing new vasculature would effectively control the progression of solid tumors [74]. And because VEGF is a major growth factor that supports the growth of new vasculature, it has been the target of many recent therapies [24–26, 39].

Unfortunately, the original goal of anti-angiogenic therapies to starve tumors by targeting their blood supply has been difficult to achieve. In most clinical studies, these therapies have had inconsistent effects on tumor physiology, with no long-term inhibition of tumor growth [48, 49]. Nevertheless, drugs that block angiogenic growth factors such as VEGF often have dramatic effects on the structure and function of the vascular network, normalizing the tumor blood vessels and

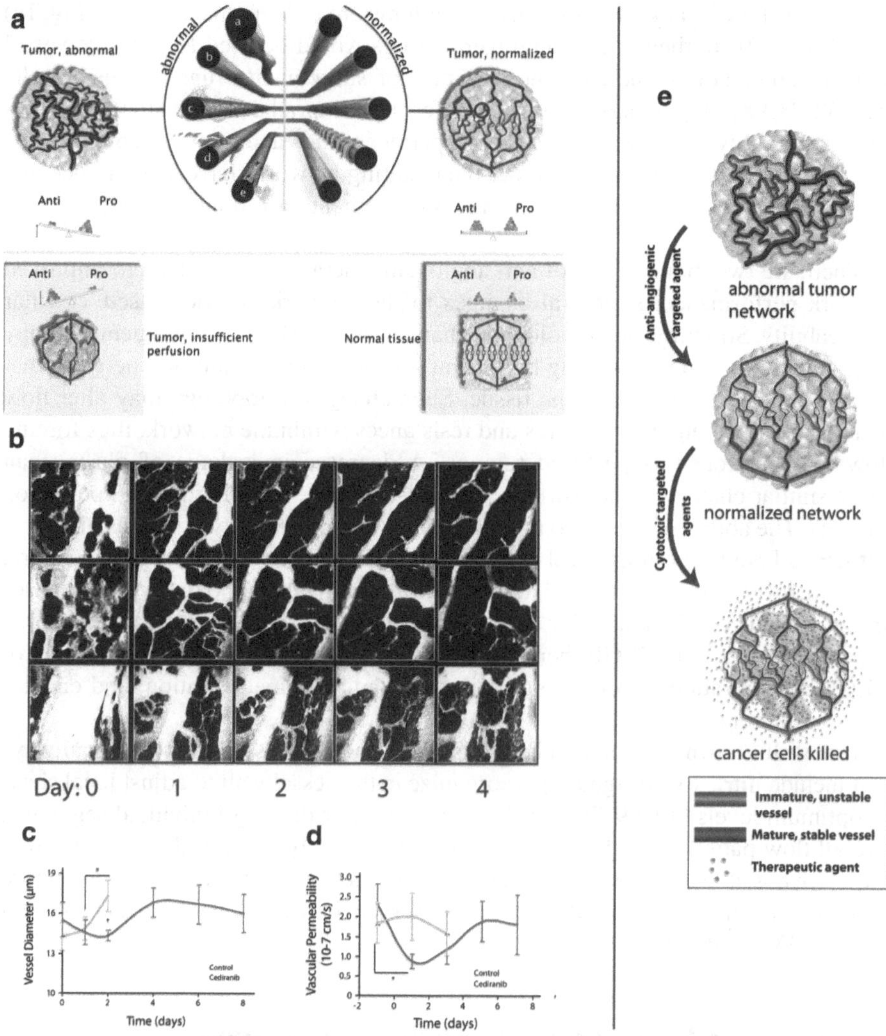

Fig. 1 Tumor vessel abnormalities and normalization. (**a**) Functional abnormalities are caused by structural abnormalities (*left*, a–e) that undergo the following changes with anti-angiogenic treatment (shown at *right*): vessel tortuosity decreases (a), insufficient basement membrane coverage increases (b), excessive and poorly organized basement membrane adopts normal morphology (c), perictye coverage increases (d), and large intercellular gaps tighten (e) (from Jain et al PPO). (**b**) Structural changes in a tumor vessel network after treatment with anti-VEGF antibody (DC101; MCaIV carcinoma in SCID mouse). Three different regions show similar changes over the 4 days after treatment, including significant diameter reduction and pruning of some segments. Glioma vessels show similar changes in diameter (**c**) as well as permeability (**d**) after anti-angiogenic therapy (from Kamoun et al. JCO, 2009). (**e**) It has been proposed that anti-angiogenic agents can make a tumor network more efficient and improve the delivery of subsequent cytotoxic chemotherapy [37]

improving the efficacy of subsequently administered chemotherapeutics (Fig. 1e) [18, 37, 60, 72]. In theory, anti-angiogenic drugs would be used to improve vessel network efficiency to increase the transport of subsequent drugs to cancer cells [4, 7, 32, 78, 98]. In practice, though, the effects are unpredictable. Anti-VEGF therapy can variably affect vessel diameters, permeabilities and network connectivity, but it is difficult to determine whether the resulting networks have actually become more efficient or whether the network improvements are transient or preserved [43, 88, 97].

There are two main effects of anti angiogenic therapy that can lead to improved network performance: structural changes to the network and decreased vascular permeability. Structural and topological changes induced by anti-angiogenic therapy (e.g., diameter changes, pruning redundant segments) may result in a network that is more evenly distributed in the tissue. Such changes in topology may alter flow patterns by changing the pressures and resistances within the network, thus forcing flow into previously unperfused regions. A decrease in leakage of plasma can cause similar changes in network pressure distribution, restoring normal perfusion patterns. The ability of VEGF to modulate vascular permeability is well established, but less is known about its involvement in structural changes in tumor vasculature. At high levels, as in untreated tumors with significant hypoxia, VEGF causes sprouting and vessel dilation, but at lower, normal levels, homeostasis is maintained. We propose that anti-VEGF therapy brings the system into this normal regime of VEGF concentrations, prevents sprouting and indiscriminate dilation, and enables shear-based remodeling.

It is well known that endothelial cells sense shear stress and, through pathways that include nitric oxide signaling, reorganize in the vessel wall to adjust local shear to optimum levels [12, 69, 76, 99]. By contracting or dilating individual segments, overall flow patterns develop that distribute flow evenly through the network. It is therefore possible that shear stress can serve as a driving force to remodel immature tumor vessel networks, similar to that observed in wound healing and skeletal muscle [33, 35, 36, 71].

3 Models of Angiogenesis, Structural Adaptation, and Normalization

Many mathematical models have been formulated and used to study blood vessel development and function [14, 15, 27, 41, 45, 56, 67, 73, 86], drug delivery to tumor tissue [5, 8–10, 63] or vessel normalization [18, 40, 53, 72, 81]. Unfortunately, there is no single comprehensive, predictive mathematical framework for studying the nonlinear relationships between vessel structure and function in real networks.

A number of elegant theoretical and experimental studies have addressed the issues of tumor angiogenesis and structural adaptation of blood vessels. Lee et al. modeled a growing tumor and a dynamically evolving blood vessel network, reproducing inhomogeneous tumor-like capillary networks [55]. This model also reproduced vessel collapse due to reduced blood flow and mechanical compression. Welter and coworkers developed a model to analyze the vascular remodeling process of an arterio-venous vessel network during tumor growth and were able to reproduce complex vascular geometry with necrotic zones and "hot spots" of increased vascular density and blood flow of varying size [95]. Szczerba and Szekely presented a simple computational model of intussusceptive angiogenesis and remodeling, predicting bifurcation formation and micro-vessel separation in a porous cellular medium [87].

Because shear-based remodeling is, at its foundation, built on a very elegant mathematical mechanism, many computational models have attempted to describe it. Hacking et al. performed one of the first detailed analyses of network adaptation, showing that a system with single inlet driven only by shear-driven diameter changes is destined to degenerate to a single flow path from inlet to outlet [31]. Using similar models and experimental systems in vivo, Pries and Secomb have addressed angioadaptation in normal physiology and ischemia. They conclude that four parameters are necessary to drive the adaptation to a stable steady state: endothelial wall shear stress, intravascular pressure, a flow-dependent metabolic stimulus, and a stimulus conducted from distal to proximal segments along vascular walls [65–68].

Directly addressing the problem of combination treatment, Ledzewicz et al. applied optimal control theory to determine desired scheduling of anti-angiogenic and cytotoxic drugs [54, 72]. Other models have been used to analyze perturbations in blood flow, such as caused by vascular obstructions. Gruionu et al. used morphometric data from vascular casts to simulate the changes in blood flow caused by obstructions and the subsequent adaptive remodeling. They found that vascular arcades can partially maintain blood flow after vascular blockage and that structural adaptation is important for modifying vessel diameters and controlling flow in this system [30].

4 Modeling Tumor Physiology

For many years we have applied the lattice-Boltzmann approach to study interactions between red and white blood cells in dynamic flow [59, 84] and the effects of vessel geometry [83] and RBC aggregation [85] on blood rheology. In addition, this approach has allowed us to analyze the relationship between plasma, intravascular hematocrit, and blood velocities in simple vessel networks [81] and blood forces on digitized vessel walls [82]. More recently, we have adapted the model to analyze flowing blood cells with realistic rest shapes and mechanical properties in three-dimensions [20–22, 61]. In other modeling work, we have used deterministic models to analyze fluid convection and pressure distribution in tumors during normalization [40]. But this approach is limited, as it does not explicitly account for blood vessel architecture or adaptive remodeling. The current model, therefore, extends our lattice-Boltzmann approach to larger domains and more complicated networks to address questions of heterogeneous perfusion and transport in relatively large volumes of tumor tissue.

5 Modeling Vascular Adaptation During Anti-angiogenic Therapy

Although many of the previous models of vascular remodeling are useful for integrating physiological information, they generally make simplifying, yet crippling, assumptions about vasculature, blood flow, and the boundary conditions. To overcome these limitations and form a better understanding of this complex physiology, we have developed a mathematical model that simulates blood flow in complex vascular networks, incorporating the necessary components for simulating vascular dynamics and function during anti-VEGF therapy. These components include:

1. Lattice-Boltzmann calculations of the full flow field within the vasculature and within the tissue
2. Diffusion and convection of soluble species such as oxygen or drugs within vessels and the tissue domain
3. Distinct and spatially-resolved vessel hydraulic conductivities and permeabilities for each species
4. Upstream signaling to control inlet vessel tone
5. Erythrocyte particles advecting in the flow and delivering oxygen with real oxygen release kinetics
6. Hypoxia-driven VEGF production and vascular remodeling driven by VEGF level and shear stress

The model decouples vascular leakiness and network structural changes, allowing determination of how each influences flow patterns and oxygen and drug delivery during anti-angiogenic therapy. An understanding of how hydraulic conductivity and network architecture act independently "or in synergy" to enhance delivery of chemotherapeutics will allow rational use of existing drugs, or the design of new ones, to improve chemotherapy.

To develop our mechanistic model of structural adaptation, we use a lattice-Boltzmann model (LBM), which solves flow and transport of blood [20, 61, 81, 83, 84], oxygen, drugs and growth factors in the tissue. Being fully local, LBM offers a distinct advantage over traditional CFD when dealing with complex boundary conditions and highly irregular geometry within the computational domain. Further, coding an LBM solver is rather simple, and the algorithm is essentially the same in either two or three dimensions. In addition, the LBM approach is imminently adaptable to parallel and distributed implementations.

Having its roots in kinetic theory, LBM also offers the advantage of studying the continuum macroscopic properties of a system from its microscopic phenomena.

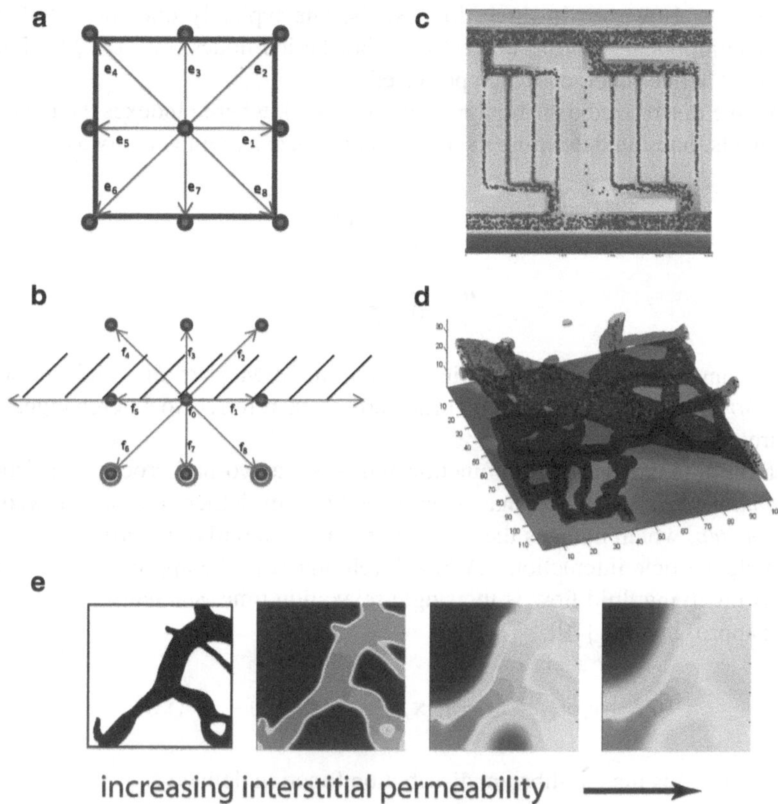

Fig. 2 Simulating vessel network dynamics. (**a**) LBM discretized velocities. (**b**) Boundary treatment in LBM. (**c**) A two-dimensional idealized network with flow from left to right in the top vessel (arteriole), from right to left in the bottom vessel (venule) and from top to bottom through the capillary bed. The particles delivering oxygen are *red*, and the color map gives the oxygen profile in the tissue. (**d**) Flow and oxygen transport from tumor vasculature in three-dimensions. The vasculature of a U87 glioma growing in a mouse was traced using techniques described in [89]. (**e**) Simulating intravascular and interstitial transport of drugs with varying diffusivity. A portion of a tumor network was traced and the model simulates flow through the network and interstitial space. By changing the bounce-back conditions at nodes outside the vessel boundary, we can simulate varying levels of interstitial diffusivity

Central to the lattice-Boltzmann method is the single-particle distribution function $f(\mathbf{p},\mathbf{x},t)$. It describes a fictive particle's state in terms of a statistical ensemble with a measurable probability of being observed with a certain position $\mathbf{x}$ and momentum $\mathbf{p}$ at time t. This distribution function is discretized to have a finite number of velocity components confined to a regular Cartesian lattice (Fig. 2a).

The choice of lattice structure is classified according to the *DnQm* scheme. Here, D specifies the dimension and Q the number of discrete velocities. In two dimensions the *D2Q9* lattice is preferred, i.e., two-dimensions with 8 velocity components linking nearest-neighbor lattice nodes plus one velocity reserved for rest particles. However, in three dimensions, one typically does not need all 26 velocity components linking nearest neighbor lattice nodes, i.e., D3Q27. Instead, the D3Q19 lattice scheme is often preferred.

From the discretized distribution function $f_i(\mathbf{x},t)$, where i indexes the $\mathbf{c}_i$ velocity components, one can determine the macroscopic density and velocity as

$$\rho_f = \sum_i f_i \quad (1)$$

$$\mathbf{u} = \frac{1}{\rho_f} \sum_i f_i \mathbf{c}_i \quad (2)$$

Importantly, the ideal gas equation of state in LBM gives the fluid pressure as $P = 1/3\rho$ and hence the two quantities are often referred to interchangeably as pressure/density.

Determining the distribution function at time t is a two-part process consisting of a *streaming step*, which propagates particles between lattice nodes, followed by a *collision step*, which updates the momentum of the underlying particles as a result of particle–particle interactions. A key development in the application of LBM to systems involving fluid flow is the single-relaxation time Bhatnagar–Gross–Krook (BGK) approximation [79]:

$$f_i \mathbf{x} + \mathbf{c}_i \delta t, t + \delta_t = f_i(\mathbf{x},t) - \frac{1}{\tau}\left(f_i(\mathbf{x},t) - f^{\mathrm{eq}}(\mathbf{x},t)\right) \quad (3)$$

where $f^{\mathrm{eq}}(\mathbf{x},t)$ is the equilibrium distribution function defined as

$$f_i^{\mathrm{eq}}(\mathbf{x},t) = w_i \rho_f \left(1 + 3\mathbf{c}_i \cdot \mathbf{u} + \frac{9}{2}(\mathbf{c}_i \cdot \mathbf{u})^2 - \frac{3}{2}\mathbf{u} \cdot \mathbf{u}\right) \quad (4)$$

where w_i are lattice weights. The collision operator is defined as a relaxation towards local equilibrium. The parameter τ controls the rate of relaxation and is directly related to the kinematic viscosity of the fluid under study:

$$\nu = \frac{1}{3}\left(\tau - \frac{1}{2}\right) \quad (5)$$

5.1 Boundary Conditions

Complex and irregular boundaries are simple to handle using LBM, making it ideal for modeling vascular networks. There are essentially two important types of boundary conditions: (a) those that involve fluid/solid interfaces at internal lattice nodes and (b) those that involve the domain boundary. To handle the first type, a zero-slip boundary condition is enforced on all fluid/wall interfaces. A particularly simple approach is to omit the collision step above at all solid nodes and instead reflect all incoming fluid packets in the reverse outgoing direction. This is referred to as the mid-plane bounce-back rule:

$$f_{\alpha}^{\text{out}}(\mathbf{x},t) = f_{-\alpha}^{\text{in}}(\mathbf{x},f^{*}) \tag{6}$$

where $\alpha, -\alpha$ correspond to opposite lattice directions, i.e., $\mathbf{c}_{\alpha}, \mathbf{c}_{-\alpha}$, and the t^{*} indicates the half-time step just prior to collision. The mid-plane bounce-back rule gets its name from the fact that the wall is actually situated half-way between adjacent fluid/solid nodes. Note, in practice, the rule works best when $\tau \approx 1$.

When a lattice node lies on the boundary of the domain, one has a problem during the streaming step, as certain fluid packets will be undefined. To illustrate, consider the 2-D example at a north wall interface in Fig. 2b. After streaming, the direction-specific densities f_6, f_7, f_8 are unknown since the implied off-lattice nodes above the boundary don't contribute during the streaming step. The simplest solution is to invoke periodic boundary conditions where outgoing fluid packets are wrapped around the domain. Additionally, one can specify Dirichlet (pressure) boundaries at the inlet and outlet of a channel. This requires solving for the three unknown fluid packets as well as the macroscopic velocity in the direction normal to the boundary. The solution involves simple algebra. Similarly, one can specify an incoming velocity and outgoing pressure (flux boundaries). Details of the solution in both 2/3-dimensions can be found in [100].

5.2 *Transvascular Convection and Diffusion*

Another attractive aspect of the LBM technique is its ability to simulate multi-component flows including a fluid and gas phase. When a gas species, e.g., oxygen, has a momentum that is negligible compared to blood plasma, it can be considered a *passive solute*, which does not impact the pressure or velocity fields of the advecting fluid. Rather, the solute is simply advected by the background fluid.

The key artifice is to introduce another distribution function that uses a simplified equilibrium distribution function:

$$g_i^{\text{eq}}(\mathbf{x},t) = w_i C_g \left(1 + 3\mathbf{c}_i \cdot \mathbf{u}\right) \tag{7}$$

where C_g analogously represents the gas concentration rather than a fluid density. The macroscopic velocity $\mathbf{u}$ is taken directly from the underlying fluid. Constant flux/concentration boundary conditions can be specified in a manner similar to the Dirichlet boundary condition described above [80].

An important aspect of LBM solute transport models is that they are known to generally solve the convection–diffusion/dispersion equation under a variety of conditions [80]. For example, in the case where the fluid has no velocity, LBM can be shown to solve the diffusion equation. Here, the diffusion coefficient of the solute is then analogous to the kinematic viscosity of the underlying fluid [80]:

$$D = \frac{1}{3}\left(\tau_g - \frac{1}{2}\right) \tag{8}$$

Note that the relaxation rate of the gas solute and underlying fluid are independent, hence the notation τ_g.

5.3 Interstitial Convection and Diffusion

LBM techniques have been applied successfully to the study of complex porous media [17]. The idea is to specify a computation domain consisting of open "fluid" nodes bounded by "solid" nodes that represent a porous medium. Though simple and effective, the lattice resolution must be on the order of the smallest pore or conduit in the material being studied. It also implicitly requires one to know the pore-space geometry, which might be infeasible. Instead, a partial bounce-back (PBB) rule can be employed to simulate a mesoscale permeability [94]:

$$f_{\alpha}^{\text{out}} = (1 - n_s) f_{\alpha}(\mathbf{x}, t) + n_s f_{-\alpha}^{\text{in}}(\mathbf{x}, t^{-}) \tag{9}$$

Note, that this revised partial bounce-back rule is similar to the mid-plane rule introduced above, except that a fraction n_s of the distribution before collision (note the t^{-}) is added to the distribution after collision. Further, when the solid fraction is unity, the partial bounce-back rule actually recovers the original mid-plane boundary condition. Similarly, for a solid fraction of zero, the standard collision rule is in effect.

By utilizing PBB, the entire computational domain becomes open to both flow and solute transport. Thus, we are able to study interstitial convection and diffusion at the mesoscale level. Further, local spatial modulation of the PBB is ideal for simulating phenomenon-like vessel wall permeability and can be used to independently control diffusion in plasma and the interstitium. Naturally, since each component of the flow has a separate distribution function, permeability can be tailored independently.

5.4 Incorporation of RBC Particles and Oxygen Release

As presented thus far, the basic LBM approach is unable to capture the physiological effects that red blood cells (RBCs) have on the transport of oxygen in blood. For example, at oxygen partial pressures typical in arterioles (20–60 mm Hg), the oxygen content of hemoglobin in a volume of blood at 30 % hematocrit is nearly two orders of magnitude greater than the dissolved oxygen in an equivalent volume of plasma [93]. Therefore, to accurately simulate oxygen delivery to tissue, a mechanism to simulate the oxygen content of hemoglobin in blood must be included. A particle suspension is a natural choice. Figure 2c shows oxygen delivery in a domain containing an idealized 2-D capillary bed: a single arteriole and single venule span the domain, connected by parallel capillaries. Plasma and RBC particles flow into the network, delivering oxygen based on local oxygen tension. The oxygen then convects and diffuses through the tissue (color map). In Fig. 2d, a 3D network was acquired from actual tumor tissue, binarized with our vessel tracing algorithm and perfused with simulated blood.

Our ongoing work has focused on highly accurate modeling of cell-cell and cell-wall interactions in blood flow. For example, important phenomena such as inertial lift forces and RBC aggregation evolve naturally from these complex models based on the underlying physics. In the current simulations, we utilize a simpler particle model and impose the same physics as external constraints. This allows us to focus on the primary problem of interest, oxygen delivery and consumption in the interstitium. Hence, we utilize an approach whereby RBCs simply advect with the underlying fluid. To describe the position $\mathbf{p}_j$ of each particle we also include a Brownian motion along with an inter-particle and wall repulsion force:

$$\mathbf{p}_j(t+\delta t) = \mathbf{p}_j(t) + \delta t \left(\mathbf{u}(\mathbf{p}_j(t)) + \mathbf{w}_\sigma + \mathbf{F}_{\text{wall}}(\mathbf{p}_j(t)) + \sum_{k \neq j} \mathbf{F}_p(\mathbf{p}_i(t), \mathbf{p}_k(t)) \right) \tag{10}$$

Here, the underlying fluid velocity $\mathbf{u}(\mathbf{p}_j(t))$ provides the advective component, and w_σ is a Gaussian white-noise process whose integration over time is equivalent to a Brownian motion. The wall force density is a nonlinear function involving the magnitude and gradient of the Euclidean distance map. This map gives the minimum distance between any point in the domain and the nearest wall point. The wall force is tuned to ensure an appropriate cell-free layer. Inter-particle forces are defined similarly over a small local neighborhood around each particle.

5.5 Metabolism

Accurately modeling the kinetics of oxygen delivery and consumption are a key aspect of the proposed model. Modeling reactive flows is straightforward using LBM and in its simplest form amounts to a modification of the collision operator to include a change in density/concentration at each time step [3, 79]:

$$f_i(\mathbf{x}+\mathbf{c}_i\delta t, t+\delta t) = f_i(\mathbf{x},t) - \frac{1}{\tau}\left(f_i(\mathbf{x},t) - f_i^{\text{eq}}(\mathbf{x},t)\right) + R_i \tag{11}$$

where $\bar{R} = \sum_i R_i$ represents the reaction rate. Key parameters like the diffusion rate and solubility of oxygen in blood/tissue, as well as the oxygen consumption rate, are reported in [52], based on studies in a rat model. Data related to the reaction rate of oxygen and hemoglobin are reported in [23, 47, 77].

5.6 Simulating VEGF Production and Drug Delivery

Simulating VEGF production and drug delivery: Adding a passive reactive solute to the model is straightforward. The difference between VEGF and oxygen is primarily in the incorporation of different boundary conditions and release mechanism. Sources of VEGF are identified as regions deemed sufficiently hypoxic within the interstitium (< 1 mm Hg). The source is added as a local increase in concentration flux of an additional distribution corresponding to VEGF. The transport of VEGF follows the same advection–diffusion through the interstitium as oxygen, but with its own, lower coefficient of diffusion. Drugs are handled in a similar way, but the source is restricted to the inlet vessels, with concentrations varying according to an imposed function. Drugs or particles of various sizes can be included by assigning different vessel permeabilities and different interstitial diffusivities (set via bounce-back rules). Figure 2e demonstrates the ability of the model to simulate varying interstitial permeabilities by adjusting bounce-back conditions for the extra-vascular nodes. In this way, we can simulate delivery of drugs of various sizes.

5.7 Distal to Proximal Conductive Signaling

Though the precise mechanism is not well understood, Pries and Secomb make a compelling argument that conductive signaling along vessel walls is biologically plausible and more importantly, critical to the stabilization of their theoretical model [65, 66, 68]. Similarly, we have found it necessary to incorporate a mechanism to conduct an upstream signal from distal, hypoxic vessels to proximal feeding arterioles. The approach is relatively simple. Regions of the vasculature that are identified as being hypoxic are first segmented. Each distinct region is able to generate an upstream signal with strength proportional to the region size and with an exponential decay based on distance upstream as measured along the vessel wall. Finally, signals from distinct regions are summed:

$$S(x) = \sum \log(\mathrm{Area_i}) \times \mathrm{e}^{-d_i(x)/c} \tag{12}$$

where $d_i(x)$ is the distance to a particular region and c is decay constant.

Propagation is achieved by using an active contour technique adopted from image processing [75]. A typical application of active contours in image processing involves monotonically growing a small initial region boundary outward with a fixed speed until the boundary intersects with an image feature like an edge. This approach requires an external rule that is then used to locally stop growth, for example, when the active contour hits a vessel wall. Refer to Appendix (C.2) for an overview of the theory and a brief mathematical development of active contours. In particular, the development of geodesic active contours with level sets is described. Using the concepts outlined in Appendix (C.2), we incorporate a classic balloon force into a speed function that drives a curve C according to the following equation of motion:

$$\partial_t C = [B(C) \times H(\mathbf{u} \cdot \mathbf{n})](-v + c\kappa)\mathbf{n} \tag{13}$$

where κ is the boundary curvature and encourages smoothness with a rate proportional to the constant c, $\mathbf{n}$ is the unit boundary normal (inward pointing), v is a constant speed (e.g., unity), and the nonlinear function $[B(C) \times H(\mathbf{u} \cdot \mathbf{n})]$ acts as a stopping force on the evolving curve. Here, the stopping function is used to enforce two separate constraints. The first is that the curve is confined to evolve within fluid nodes. This is achieved by choosing B as a binary image that is unity if a location in the computational domain is a fluid site and zero otherwise. Hence, $B(C)$ will be zero if the curve intersects with a non-fluid site in the domain. The second constraint is that the curve can only evolve in a direction opposite the flow field $\mathbf{u}$. With $H(\cdot)$ as the Heaviside function and the convention that the unit normal is inward pointing, $H(\mathbf{u} \cdot \mathbf{n}) = 0$ if the boundary is oriented $> 90^\circ$ to the direction of flow, thus ensuring that the curve only evolves upstream. By initializing the active contour to lie within a vessel, the evolution equation given above, by virtue of the balloon force, will drive the contour through branches and other convex regions, eventually settling against every point on the vessel wall that is upstream from the initial region. By further registering the epoch that each point on the active contour first reaches a vessel wall, the "distance" from the initial region can be calculated.

5.8 Vessel Remodeling

The algorithm uses a set of key parameters to control the remodeling process. Fundamentally, this includes two distinct regimes: one where remodeling is dominated by wall shear stress and the other where VEGF causes an interruption in the shear-based remodeling. In the shear-based regime, vessels will dilate/contract in an effort to achieve a nominal shear rate. As our model is based on a discrete domain, i.e., a binary image with discrete pixels, shear-based remodeling essentially follows a discrete rule $\delta r = \text{sign}(\tau - \tau_0)$. Similar to Hacking et al. this rule is designed to achieve constant shear in all vessels, which follows directly from Murray's law [31]. In contrast, when in a *moderate to high-VEGF* regime, vessels will undergo dilation in an effort to promote an increase in flow. The rate of dilation is constant, e.g., 1 pixel (1 lattice-Boltzmann unit) per iteration. In the highest VEGF regime, upstream signaling will cause additional dilations in proximal vessels. As described above, the signal varies continuously in strength as a function of distance and initial size of the hypoxic region. The signal must therefore be discretized for use in our model, which involves simple thresholding of the signal. Ultimately, our remodeling rules should lead to a convergence state, whereby all vessels exhibit nominal shear, as well as being exposed to nominal VEGF levels.

5.9 Stability and Convergence Behavior

To be sure, efforts to achieve this, or similar convergence states, have met some difficulty in the literature. As described in Hacking et al. shear-based remodeling alone is fundamentally unstable, leading to arteriovenous shunting, complete regression, or even infinite expansion [31]. Following previous approaches in the literature, we use additional metabolic signaling to promote stability during the remodeling process [65, 66, 68]. However, we have still found it necessary to impose some additional rules to achieve satisfactory convergence results.

First, our simple model does not include specific delineation of branches and segments. There is no knowledge of vascular structure or topology, and all remodeling is fully local. As a result, an ill-timed contraction of even the smallest region of a vessel can potentially shut down an entire downstream vascular bed before VEGF signaling can counteract regression. Hence, we maintain a minimum vessel radius for a fixed number of remodeling iterations, thereby delaying complete vessel regression. This is biologically feasible, since endothelial cell reorganization takes time. Further, as our modeling domain is grid-based, it exhibits a finite resolution that inherently introduces small artifacts into the flow field. As a result, our shear-based remodeling rule must be modified to remain insensitive to disturbances that lie within a narrow band around the nominal shear rate. However, we've also found it necessary at times to relax this band further in order to achieve good convergence behavior.

Another important factor in achieving good stability characteristics is the choice of flow inlet/outlet boundary conditions. As demonstrated by Hacking et al. a constant-pressure inlet and outlet boundary condition is fundamentally unstable. The other extreme, a constant-flow inlet boundary condition, does admit stable flow configurations. However, unlike constant-pressure, the constant-flow boundary condition causes tight coupling between sibling branches, meaning any change in one branch will generally impact the flow in other branches. This can potentially lead to stability issues. Also, this clearly precludes the ability to draw any additional flow. An intermediate boundary condition involves using pressure boundaries with an internal series resistance. This is the preferred approach, but requires some care to properly match the internal resistance to the specific domain being investigated.

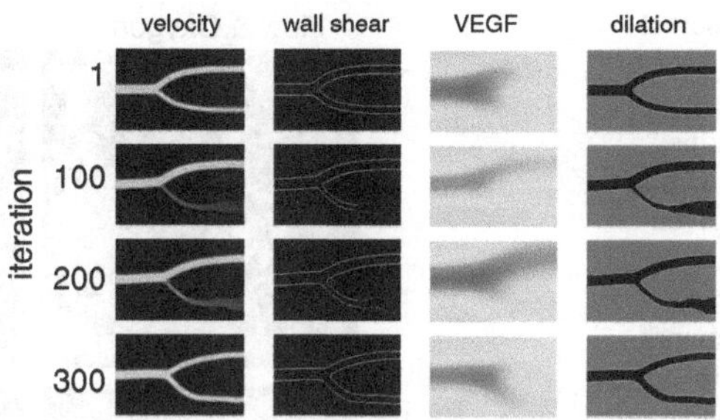

Fig. 3 Maintaining a stable bifurcation. Shown, from left to right, are the fluid velocity, wall shear, VEGF level, and the map of instantaneous wall velocities (which determine dilation or contraction locally). Flow is from left to right. As opposed to models that consider only shear-based remodeling (and consequently regress to a single outlet channel in this configuration), the system recovers from the initial perturbation that causes the drop in flow in the bottom channel (iterations 200–300). Upstream signaling from the low-flow, hypoxic vessel effectively prevents its upstream region from "pinching off"

5.10 Model Performance and Stability

Figure 3 shows remodeling on a bifurcation that initially displays asymmetric flow rates. Without any external metabolic signaling, the lower branch will regress. However, as the VEGF level increases, the lower branch has begun dilating by iteration 100. Continued dilation by iteration 200 increases the flow in the lower branch and begins raising the shear rate along the segment. Finally, by iteration 300, the shear rate in both branches is nominal and the simulation has converged. Interestingly, the final shape after convergence differs from the initial symmetric configuration.

Figure 4 demonstrates the time course of adaptation in an example network. Initially the network is a dense mesh (white outline). By iteration 25, the network has adapted in response to shear forces and local VEGF level. After an initial influx of VEGF driven by hypoxia, the vessels dilate to bring in more oxygen, and in the process, the network is subjected to shear-based remodeling. Note that patches of high VEGF, such as seen at iteration 20, effectively protect vessels in hypoxic regions from being pruned.

Vascular remodeling and normalization are a central feature of tumor networks treated with anti-angiogenic therapy. By including the correct feedback mechanisms, the present model is capable of reproducing the appropriate network dynamics and predict transport of soluble species such as oxygen, growth factors, and drugs. With appropriate calibration of the various thresholds for adaptation for individual tumor types, this model will enable detailed, time-resolved analyses of network changes and drug and nutrient delivery to tissue.

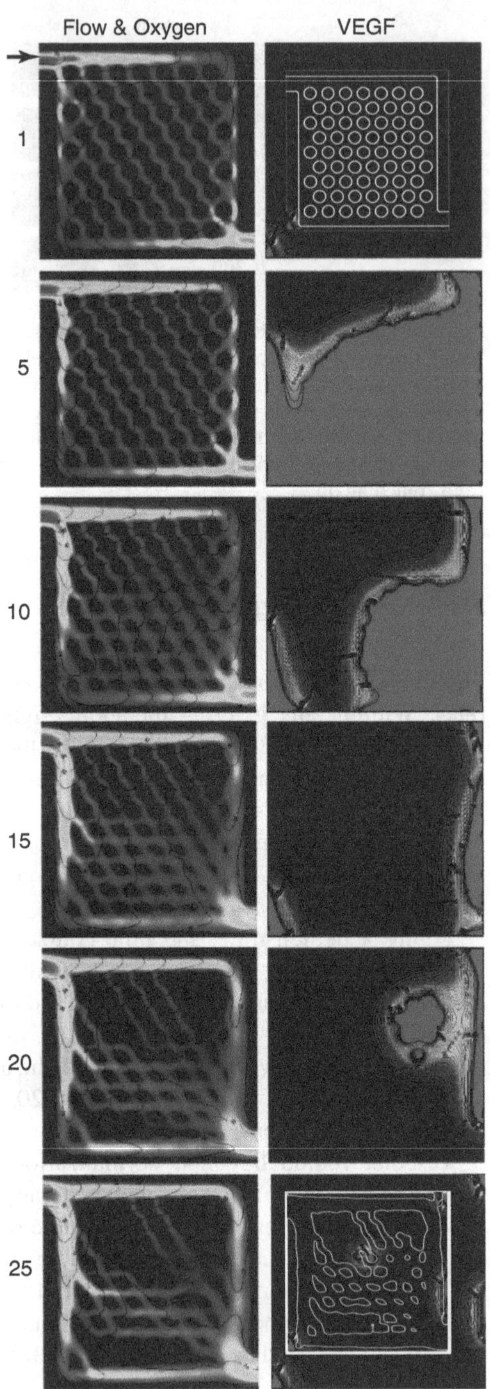

Fig. 4 Simulating vessel network dynamics: Adaptation of a 2D mesh network over 25 iterations (iteration number is shown at left). Flow velocity is given by the color map; the oxygen contour lines are in *black*. The initial and final network boundaries are shown in *white* in the top right and bottom right panels. Blood flows into the top left and exits the bottom right channel. With this parameter set, there is an initial excess of oxygen metabolism that induces VEGF in the beginning. Vessel dilation and remodeling compensate, relieving the initial hypoxia. Local VEGF production protects vessels from pruning, even if shear is below the lower threshold

6 Conclusion and Discussion

It is likely that growth factors such as VEGF enable shear stress-driven adaptive remodeling of immature tumor vessel networks and that there are optimal configurations of network architecture and hydraulic conductivity for delivering drugs to tumors. A better understanding of this complex physiology requires a comprehensive mathematical model such as presented here, which calculates blood flow in complex tumor networks. The model incorporates the necessary and sufficient components for simulating vascular structure and function during anti-VEGF therapy: (1) lattice-Boltzmann calculations of the full flow field within the vasculature and within the tissue, (2) diffusion and convection of soluble species such as oxygen or drugs within vessels and the tissue domain, (3) distinct and spatially resolved vessel hydraulic conductivities and permeabilities for each species, (4) erythrocyte particles advecting in the flow and delivering oxygen with real oxygen release kinetics, (5) hypoxia-driven VEGF production, and (6) shear stress-mediated vascular remodeling. This model, combined with data from intravital imaging of tumor vessel structure and function (Fig. 5), will provide a rational framework for the development of new therapeutic strategies or more effect use of existing drugs.

A Lattice Boltzmann Background

A.1 The Boltzmann Equation

The fundamentals of the lattice-Boltzmann method are the kinetic Boltzmann equation which is well described in textbooks. It is based on the evolution of the single particle distribution function $f(\mathbf{x},\mathbf{v},t)$. This function indicates how many particles are present at a point in space $\mathbf{x}$ with the velocity $\mathbf{v}$ at time t. During the time $\mathrm{d}t$ the particles will now move according to their velocities to the new location $\mathbf{x}^* = \mathbf{x} + \mathbf{v}\mathrm{d}t$. In case of external forces $\mathbf{f}$ the velocity of the particles will change as well to $\mathbf{v}^* = \mathbf{v} + m\mathbf{f}\mathrm{d}t$, with m being the mass of a particle. Without particle–particle collisions they would just move on like

$$\left[f(\mathbf{x}+\mathbf{v}\mathrm{d}t,\mathbf{v}+\frac{1}{m}\mathbf{f}\mathrm{d}t,t+\mathrm{d}t) - f(\mathbf{x},\mathbf{v},t)\right]\mathrm{d}^3x\mathrm{d}^3v = 0$$

but due to intermolecular collisions the particle distribution function f changes. This change can be described by a collision operator Ω that represents the number of particles that enter or leave the given phase space volume, so the Boltzmann equation reads as

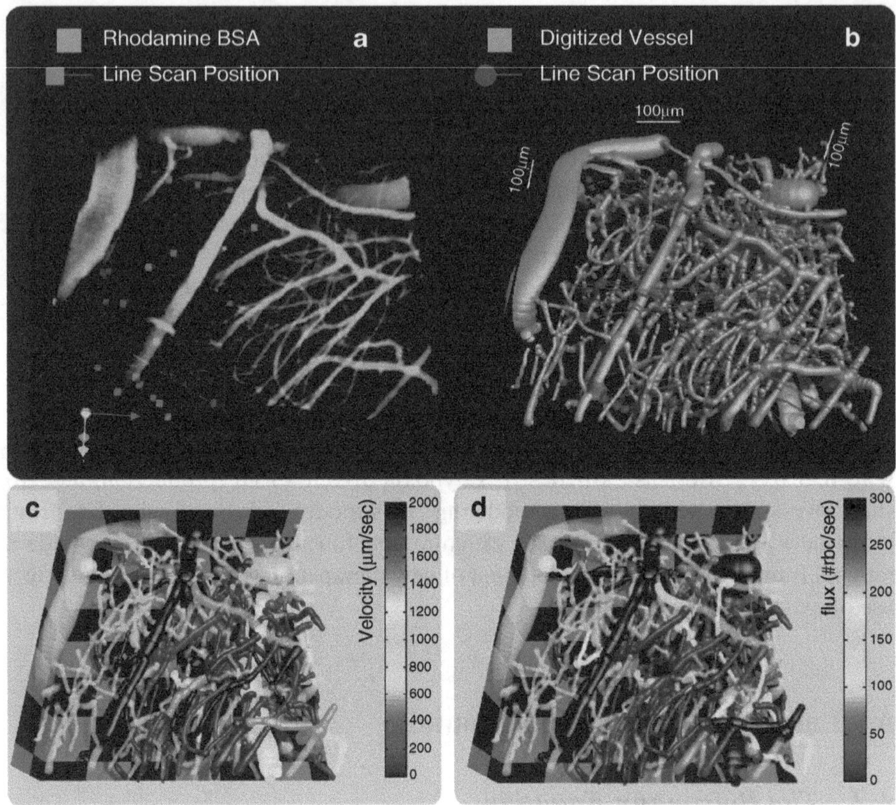

Fig. 5 Measuring blood flow velocities at single vessel resolution over a network [44]. (**a**) To analyze large networks, lines intersect multiple vessels and the velocities and fluxes are deconvolved using the structural information (i.e., angles between the vessel and scan line). (**b**) Network trace of (a), showing the binary representation of the network. (**c, d**) Velocity and flux maps, respectively, superimposed on the traced digital network to visualize the relationships between flow and topology. Image size is 630×630×50 µm

$$\left[f(\mathbf{x}+\mathbf{v}\mathrm{d}t,\mathbf{v}+\frac{1}{m}\mathbf{f}\mathrm{d}t,t+\mathrm{d}t)-f(\mathbf{x},\mathbf{v},t)\right]\mathrm{d}^3x\mathrm{d}^3v=\Omega \tag{14}$$

The problem to solve is how the collision operator Ω has to be constructed. Here, it can be mentioned that Boltzmann himself could not solve this problem. However, he proposed the following thoughts that illustrate the problem.

Ω is the number of particles that enter or leave the phase space volume $\mu = [\mathbf{x},\mathbf{x}+\mathrm{d}\mathbf{x}];[\mathbf{v},\mathbf{v}+\mathrm{d}\mathbf{v}]$. Following assumptions are used: Every collision results in either a particle entering or leaving μ. Further, particles only collide with particles outside μ or particles from outside μ are collide into μ. $W(\mathbf{v},\mathbf{v}_2,\mathbf{v}_3,\mathbf{v}_4)$ gives the probability that two particles with velocity $\mathbf{v}$ and $\mathbf{v}_2$ collide and end up with the velocities $\mathbf{v}_3$ and $\mathbf{v}_4$ while all velocities except $\mathbf{v}$ are outside μ. Therefore, the number of particles lost during a time step is $W(\mathbf{v},\mathbf{v}_2,\mathbf{v}_3,\mathbf{v}_4)$ and due to symmetry the number of particles added to μ is $W(\mathbf{v}_3,\mathbf{v}_4,\mathbf{v}_2,\mathbf{v}) = W(\mathbf{v},\mathbf{v}_2,\mathbf{v}_3,\mathbf{v}_4)$. Taking into account the fact that W is proportional to the number of particles with the relevant velocity $f(\mathbf{v}_i)$ one ends up with

$$\left[\frac{\partial}{\partial t} + v\nabla_x + \frac{1}{m}\mathbf{f}(\mathbf{x})\nabla_v\right] f(\mathbf{x},\mathbf{v},t) = \int \mathrm{d}^3v_2 \int \mathrm{d}^3v_3 \int \mathrm{d}^3v_4 W(\mathbf{v},\mathbf{v}_2,\mathbf{v}_3,\mathbf{v}_4) \times [f(\mathbf{x},\mathbf{v}_3,t)f(\mathbf{x},\mathbf{v}_4,t) - f(\mathbf{x},\mathbf{v}_2,t)f(\mathbf{x},\mathbf{v},t)] \tag{15}$$

which is a nonlinear integrodifferential equation. Since the Boltzmann equation (15) cannot be solved analytically, different approximations have been proposed to find solutions like the one by Bhatnagar, [11]. Their idea was that the system relaxes towards a local equilibrium distribution $f^{\mathrm{eq}}(\mathbf{x},\mathbf{v})$. Further, it can be assumed that in case the system is not far from this equilibrium it will relax with a single relaxation rate $1/\tau$. Thus, the Boltzmann equation with BGK collision operator reads as

$$\left[\frac{\partial}{\partial t} + v\nabla_x + \frac{1}{m}\mathbf{f}(\mathbf{x})\nabla_v\right] f(\mathbf{x},\mathbf{v},t) = \frac{f(\mathbf{x},\mathbf{v},t) - f^{\mathrm{eq}}}{\tau} \tag{16}$$

Here, it should be noted that one can now "choose" the equilibrium function in order to obtain the desired physical properties. However, one needs to fulfill some requirements such as isotropy and conservation of mass and momentum. Usually, the Maxwell distribution or some expansions of it are used. By performing a Chapman Enskog procedure it is possible to show that such a model reproduces the Navier–Stokes equation, if the distribution function is chosen right. It should be noted that this is a two-way process. One can either construct the method in such a way that it simulates a given macroscopic equation or one can show what equation is solved by a given method.

A.2 Concept of Lattice Boltzmann

The core concept of the lattice-Boltzmann method is to discretize the Boltzmann equation in time, velocity and real space [58]. This means instead of having infinitely small phase space elements μ one takes into account the single particle distribution function $f(\mathbf{x}_j,\mathbf{c}_i,t)$ only on a lattice site $\mathbf{x}_j$, with velocities $\mathbf{c}_i$ that point to the nearest and next-nearest neighbors. The lattice-Boltzmann equation [79] then reads as

$$f(\mathbf{x}+\mathbf{c}_i,\mathbf{c}_i,t+\delta t) - f(\mathbf{x},\mathbf{c}_i,t) = \Omega \tag{17}$$

In two-dimensions a D2Q9 model is commonly used. Other lattice types in two and three-dimensions can be used as well but will not be discussed here. The article of Qian et al. provides an overview on different lattice schemes [70]. In his notation that is widely used in the context of lattice Boltzmann the D indicates the dimension and the Q the number of used velocities $\mathbf{c}_i$. The choice of the lattice has an impact on several quantities like the speed of sound c_s. Taking fewer velocities into account decreases the computational costs but computational artifacts due to the lattice nature become more severe, since certain directions might be favored, violating Galilei invariance.

The collision operator Ω represents inter-molecular interactions, i.e., collisions between molecules that lead to a redistribution of $f(\mathbf{x}, \mathbf{c}_i, t)$. This redistribution relaxes the single particle distribution function f towards a local equilibrium f^{eq}, while it conserves mass and momentum. The actual shape of the collision operator Ω can differ in the way how the equilibrium distribution f^{eq} is calculated or how the actual relaxation is realized. For a more detailed overview of the method the reader is referred to the literature [19, 28, 29, 50, 51, 79, 80].

The code as it is used for the simulations in this chapter uses a BGK (or single relaxation time) collision operator, similar to Eq. (16). The equilibrium distribution function is given by

$$f_i^{\mathrm{eq}}(\mathbf{x}, t) = w_i \rho_f \left(1 + 3\mathbf{c}_i \cdot \mathbf{u} + \frac{9}{2}(\mathbf{c}_i \cdot \mathbf{u})^2 - \frac{3}{2}\mathbf{u} \cdot \mathbf{u}\right) \tag{18}$$

which is a polynomial expansion of the Maxwell distribution $c_s = 1/\sqrt{3}$ is the speed of sound for the D2Q9 lattice, $\mathbf{u}(\mathbf{x}, t)$ is the macroscopic velocity of the fluid.

Physical properties of the fluid are carried by the moments of the distribution f. Namely the local density

$$m_0 = \rho = f \sum_i f_i$$

the local momentum

$$m_{1,2,3} = j_{x,y,z} = \mathbf{e}_{x,y,z} \sum_i f_i \mathbf{c}_i$$

and the momentum flux

$$m_{3 \cdots 18} = \Pi = \sum_i f_i \mathbf{c}_i \mathbf{c}_i$$

The equilibrium distribution f^{eq} is constructed such that it conserves local mass and momentum but changes the viscous (or nonequilibrium) part of the momentum flux $\Pi^{\mathrm{neq}} = \Pi - \Pi^{\mathrm{eq}}$. It is realized by

$$\rho^{\mathrm{eq}} = \sum_i f_i \tag{19}$$

$$\mathbf{j}^{\mathrm{eq}} = \sum_i f_i \mathbf{c}_i \tag{20}$$

$$\Pi^{\mathrm{eq}} = \sum_i f_i^{\mathrm{eq}} \mathbf{c}_i \mathbf{c}_i = p\mathbf{1} + \rho \mathbf{u}\mathbf{u} \tag{21}$$

The equation of state is given by $p = \rho c_s^2$, where the speed of sound is given by the advection as $c_s = \sqrt{1/3}$ lattice units per time step. The equilibrium part is unchanged by the collision. Further, in the simulation scheme only the shear modes survive the collision process. Therefore, only the nonequilibrium momentum flux changes. The post collision nonequilibrium momentum flux $\Pi^{\mathrm{neq}*}$ is given by

$$\Pi^{\mathrm{neq}*} = (1+\lambda)\bar{\Pi}^{\mathrm{neq}} + \frac{1}{3}(1+\lambda_v)(\Pi^{\mathrm{neq}} : \mathbf{1})\mathbf{1}, \tag{22}$$

where the over bar indicates the traceless part of the momentum flux. The star indicates that a value is post collision. The parameter λ is an eigenvalue of Λ and controls the rate of relaxation of the stress tensor. By this it as well regulates the shear viscosity

$$\mu = -\frac{1}{6}\rho(2/\lambda + 1). \tag{23}$$

For stability reasons λ has to be in the range $-2 < \lambda < 0$. The kinetic modes can also contribute to the post-collision distribution, but the eigenvalues of these modes were set to be 0, so they have no effect on the post-collision distribution f_i^*. The post-collision distribution is given in terms of the mass-density ρ and the momentum-density $\mathbf{j}$ (or the velocity $\mathbf{u}$ and the updated post-collision momentum flux Π^*)

$$f_i^* = f_i + \Omega_i(f) = w_0\rho + w_1 j_\alpha c_{i\alpha} + w_2 \Pi^*_{\alpha\beta} c_{i\alpha} \bar{c}_{i\beta} + w_3(\Pi^*_{\alpha\alpha} - 3\rho c_s^2). \tag{24}$$

The coefficients w_i have their origin in the different lengths of the vectors $|\mathbf{c}_i|$, i.e., the absolute velocity needed to reach the neighbor cell in one time step, depending whether it is a nearest neighbor or a diagonal neighbor.

B Vessel Tracing

For completeness, this section presents methods to model complex vasculature in 3-D images using *cylindroidal superellipsoids*, along with robust estimation and detection algorithms that are detailed in [90, 91].

B.1 Superellipsoids

Superellipsoids are geometric modeling primitives that have been used in early computer vision literature [6]. A cylindroidal superellipsoid is given in implicit form by

$$S(\mathbf{x};\varepsilon) = \left(|x|^2 + |y|^2\right)^{\frac{1}{\varepsilon}} + |z|^{\frac{2}{\varepsilon}} \tag{25}$$

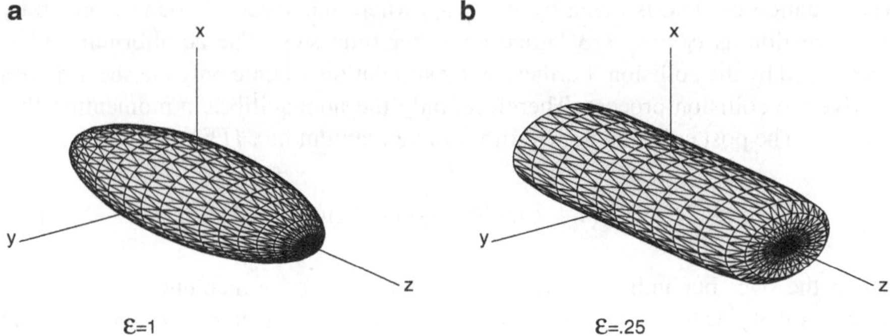

Fig. 6 Illustrating the superellipsoid. Panels (**a**) and (**b**) show the boundary of a superellipsoid, defined as the surface $S(\mathbf{x};\varepsilon) = 1$, while the shape parameter ε is set to 1 and 0.25, respectively

where the parameter ε controls the *shape* of the superellipsoid. One advantage of this shape model is that it allows for the flexibility of probing and traversing tortuous vessels and branches with a *softer* ellipsoidal shape, while enhancing accuracy when vessels are more tubular (Fig. 6).

It is assumed that the image data denoted $I(\mathbf{x})$, with domain $\Omega \subset \mathbb{R}^3$, contain a vessel segment characterized by two homogeneous regions separated by a boundary that is well modeled by the superellipsoid. Hence, a piecewise constant vessel model is defined as

$$V(\mathbf{x};\beta,I_F,I_B) = I_B - (I_B - I_F) \times H(1 - S(\mathbf{x};\beta)) \qquad (26)$$

where I_F and I_B correspond to the foreground and background intensities and $S(\mathbf{x};\beta)$ has been generalized from (25) to express local pose via the parameter set $\beta = (\mu^T,\sigma,\phi,\varepsilon)$, which includes 10 degrees of freedom in 3-D, i.e., 3 rotation (ϕ), 3 scale (σ), 3 shift (μ), and 1 shape parameter (ε). Finally, $H(\cdot)$ is the Heaviside function, which evaluates to unity when the argument is positive and zero otherwise.

B.2 Estimation of the Superellipsoid Vessel Model

A maximum likelihood approach is used to estimate the superellipsoid model parameters. The log-likelihood equation is given by

$$L(\beta,I_F,I_B) = \int_{\mathscr{R}(\beta)} \log f(I(\mathbf{x}) - I_F)\, \mathrm{d}\mathbf{x} + \int_{\Omega\setminus\mathscr{R}(\beta)} \log f(I(\mathbf{x}) - I_B)\, \mathrm{d}\mathbf{x} \qquad (27)$$

where the limits of integration are given by $\mathscr{R}(\beta) := \{x \in \Omega | S(\mathbf{x};\beta) < 1\}$, which includes the points interior to the superellipsoid and its complement with respect to the domain Ω. The marginal distribution f describes the additive noise in the image $I(\mathbf{x})$ and is assumed to be independent and identically distributed.

Maximizing (27) is done by alternately estimating the intensity parameters and then updating the boundary using gradient ascent. Assuming the intensities (I_F, I_B) are known, the gradient of (27) with respect to the parameters β is given by

$$\nabla L(\beta) = \oiint_{C(\beta)} \left[\log \frac{f(I(\mathbf{x}(u)) - I_F)}{f(I(\mathbf{x}(u)) - I_B)}\right] \frac{\partial C(u;\beta)}{\partial \beta}^T \mathbf{n}(u;\beta)\mathrm{d}S \tag{28}$$

where the integral is over the superellipsoid surface $C(\beta) = \partial\mathscr{R}(\beta)$ parameterized by u with surface area element $\mathbf{n}(u;\beta)\mathrm{d}S$, and where $\frac{\partial C(u;\beta)}{\partial \beta}$ is the Jacobian matrix giving the change in the superellipsoid boundary $C(\beta)$ with respect to parameters β. Finally, the gradient step used to update β is given by

$$\beta^{(t+\Delta t)} = \beta^{(t)} + \nabla L(\beta)^T \Delta t \tag{29}$$

where Δt is a constant time step.

Above, it has been demonstrated how to find the optimal superellipsoid assuming the intensity parameters are known. Since they are in fact unknown, they must be estimated. The use of a robust image intensity estimate is important for dealing with various sources of noise. The median is a good choice and can tolerate as much as 50 % outliers before *breaking down*. However, when estimating the image intensities, it's possible that more than 50 % of the voxels will be outliers. This can happen at branch points or when adjacent vessels are very close to each other. For additional robustness, the *skipped median* can be used, which censures very large residuals from the estimate of the median.

Since the maximum likelihood intensity estimator under i.i.d. Laplacian noise gives the population median, a Laplacian noise model is assumed when maximizing (27). The resulting maximum value can be used to determine a robust likelihood ratio test (LRT) statistic for the null hypothesis that no vessel is present. Hence, it becomes possible to measure goodness of fit in a principled manner.

B.3 Traversing Vessels Using the Superellipsoid

This section describes the general procedure for *tracing* vessels using the proposed model. The procedure is outlined in the following pseudo-code:

Seed extraction is done on a regular grid over the 2-D intensity projection image as described in [2]. The model is fit to each candidate seed, and candidates are then prioritized by goodness of fit. For each seed, there are two directions, 180° apart, that are traversed sequentially. The inner traversal loop involves alternately shifting the centerpoint of the model, and then reestimating the parameters. Specifically,

Algorithm 1 Vessel Tracing

Extract a set of candidate seeds S
Estimate and prioritize seeds: $Q \leftarrow S$
for all $s \in Q$ **do**
  Initialize from seed:
  $(\beta, I_F, I_B)_0 \leftarrow (\beta, I_F, I_B)_s$
  for seed direction$\in [0°, 180°]$ **do**
    for $k = 1 : \infty$ **do**
      Shift the model:
      $\tilde{\mu}_k = \mu_{k-1} + \Delta_k$
      Estimate superellipsoid parameters:
      $\beta_k \leftarrow (\beta_{k-1}, \tilde{\mu}_k)$
      Update intensity estimates
      $(I_F, I_B)_k \leftarrow (I_F, I_B)_{k-1}$
      break if LRT $< \tau$ **or** intersection
    end for
  end for
end for

the direction of shift Δ_k is determined by the model orientation. The remaining parameters are initialized directly using the previous estimates. The traversal terminates when the LRT falls below a threshold, or the vessel intersects with the image domain or a previously traced vessel.

C Active Contours

Active contours are deformable shape descriptors that evolve a boundary or *front* dynamically in response to equations of motion derived either by the minimization of an energy functional using variational methods [46] or based on principles of geometric flow [13, 57]. The first type uses an explicit parameterization, e.g., tensor product splines, to represent the boundary of the active contour, and is commonly referred to as *snakes*. The second type is *geodesic active contours*, which incorporates notions of curve shortening geodesics, while using an intrinsic, parameter-free representation of the boundary curve.

C.1 Geodesic Active Contours

The concept of a geodesic curve relates to the minimum distance between points. To see the relation to active contours consider the following functional:

$$L(C) = \oint |C'(u)| \mathrm{d}u \tag{30}$$

which gives the Euclidean length of a curve C. It is well known that the fastest way to reduce $L(C)$ is to evolve C via the flow given by $C_t = \kappa\mathbf{n}$, where κ is the Euclidean curvature and $\mathbf{n}$ is the normal to the curve. This flow is known as the Euclidean curve shortening flow. Importantly, the principle of curve shortening naturally promotes smoothness during the evolution process.

Following on this same idea, a constant speed can be added to the evolution equation, giving a curve that expands or contracts monotonically with speed v relative to the shortening flow, which in this formulation now exists to provide regularization:

$$C_t = (v + \kappa)\mathbf{n} \tag{31}$$

In image processing, this constant speed is known as a *balloon force* [16] and has the tendency to drive the curve outward until the boundary of an object is hit. Therefore, what is required is some criteria for stopping the curve evolution based on the underlying image, giving a final curve evolution of the form:

$$C_t = g(I)(v + \kappa)\mathbf{n}, \tag{32}$$

where $g(I) \to 0$ at the boundary of the region of interest. Typically, a nonlinear response function related to the image gradient is used to denote the edge of an object [13]. What remains to be determined is a flexible and efficient way of representing the boundary curve C and its evolution equations. To this end, the level set method has gained traction in the image processing community, following the application by Osher and Sethian [64] to curvature-dependent boundary evolution. Importantly, the level set technique offers a number of advantages over the classical snakes approach, including the ability to adapt naturally to topology changes (e.g., splitting/merging of regions) as well as handling discontinuities and cusps in the boundary.

C.2 Level Set Method

Given an object $\Omega \subset \mathbb{R}^2$, the level set method represents a curve $C = \partial\Omega$ as the zero level set of a higher-order embedding function denoted ϕ, which satisfies:

$$\text{If} \begin{cases} \phi(\mathbf{x}) = 0, \mathbf{x} \in \partial\Omega \\ \phi(\mathbf{x}) < 0, \mathbf{x} \in \Omega \setminus \partial\Omega \\ \phi(\mathbf{x}) > 0, \mathbf{x} \in \mathbb{R}^2 \setminus \Omega \end{cases} \tag{33}$$

Two important properties of the level set function are that the boundary normal is given by $\mathbf{n} = \nabla\phi/|\nabla\phi|$ and $\kappa = \mathrm{div}(\nabla\phi/|\nabla\phi|)$. Returning to (32), the update equation can now be given in terms of the level set function:

$$\delta\phi = g(I)\,(v + \mathrm{div}(\nabla\phi/|\nabla\phi|))\,|\nabla\phi| * \delta t \tag{34}$$

where ϕ_0 is initialized as the signed distance function of C_0 and (34) is iterated until $\delta\phi < \varepsilon$.

Importantly, in order for the update (34) to work correctly, the level set property of ϕ must be maintained at all times. In practice, this requires that ϕ must be periodically re-initialized. This is costly and has led to the development of efficient *narrow banding* techniques that limit updates to ϕ near the zero level set, while further constraining ϕ to be a distance function in this neighborhood. Techniques have also been developed that optimize and adapt the level set technique specifically for image segmentation tasks. A novel discrete framework was introduced by [75] which achieves unparalleled speedup by avoiding the solution of any costly pdes. Some interesting aspects of the algorithm include the adoption of a level set function with a range limited to the set $(-3,-1,1,3)$. The level set function updates based on neighborhood connectivity over a minimally sparse set of active boundary points, and never needs re-initialization. Smoothing is achieved by periodically convolving the level set function with a Gaussian kernel.

References

1. Al-Kilani, A., Lorthois, S., Nguyen, T-H., Le Noble, F., Cornelissen, A., Unbekandt, M., Boryskina, O., Leroy, L., Fleury, V.: During vertebrate development, arteries exert a morphological control over the venous pattern through physical factors. Phys. Rev. E. **77**, 051912 (2008)
2. Al-Kofahi, K.A., Lasek, S., Szarowski, D., Pace, C., Nagy, G., Turner, J.N., Roysam, B.: Rapid automated three-dimensional tracing of neurons from confocal image stacks. IEEE-TITB. **6**(2), 171–187 (2002)
3. Alemani, D.: A Lattice Boltzmann numerical approach for modelling reaction-diffusion processes in chemically and physically heterogeneous environments. UNIVERSITÉ DE GENÈVE (2007)
4. Ansiaux, R., Baudelet, C., Jordan, B.F., Crokart, N., Martinive, P., DeWever, J., Gregoire, V., Feron, O., Gallez, B.: Mechanism of reoxygenation after antiangiogenic therapy using SU5416 and its importance for guiding combined antitumor therapy. Canc. Res. **66**(19), 9698–9704 (2006)
5. Baish, J.W., Netti, P.A., Jain, R.K.: Transmural coupling of fluid flow in microcirculatory network and interstitium in tumors. Microvasc. Res. **53**, 128–141 (1997)
6. Barr, A.H.: Superquadrics and angle-preserving transformations. IEEE Comput. Graph. Appl. Mag. **1**(1), 11–23 (1981)
7. Batchelor, T.T., Sorensen, A.G., diTomaso, E., Zhang, W.T., Duda, D.G., Cohen, K.S., Kozak, K.R., Cahill, D.P., Chen, P.J., Zhu, M., Ancukiewicz, M., Mrugala, M.M., Plotkin, S., Drappatz, J., Louis, D.N., Ivy, P., Scadden, D.T., Benner, T., Loeffler, J.S., Wen, P.Y., Jain, R.K.: AZD2171, a pan-VEGF receptor tyrosine kinase inhibitor, normalizes tumor vasculature and alleviates edema in glioblastoma patients. Canc. Cell. **11**(1), 83–95 (2007)
8. Baxter, L.T., Jain, R.K.: Vascular permeability and interstitial diffusion in superfused tissues: A two-dimensional model. Microvasc. Res. **36**(1), 108–115 (1988)
9. Baxter, L.T., Jain, R.K.: Vascular and interstitial transport in tumors. In: Vaupel, P., Jain, R.K. (eds) Tumor Blood Supply and Metabolic Microenvironment: Characterization and Therapeutic Implications. Fischer Publications, Stuttgart (1991)

10. Baxter, L.T., Yuan, F., Jain, R.K.: Pharmacokinetic analysis of the perivascular distribution of bifunctional antibodies and haptens: Comparison with experimental data. Canc. Res. **52**(20), 5838–5844 (1992)
11. Bhatnagar, P.L., Gross, E.P., Krook, M.: Model for collision processes in gases. I. Small amplitude processes in charged and neutral one-component systems. Phys. Rev. **94**(3), 511 (1954)
12. Brown, E., McKee, T., diTomaso, E., Pluen, A., Seed, B., Boucher, Y., Jain, R.K.: Dynamic imaging of collagen and its modulation in tumors in vivo using second-harmonic generation . Nat. Med. **9**, 796–800 (2003)
13. Caselles, V., Catte, F., Coll, T., Dibos, F.: A geometric model for active contours in image processing. Numer. Math. **66**(1), 1–31 (1993)
14. Chaplain, M., Anderson, A.: Mathematical modelling of tumour-induced angiogenesis: Network growth and structure. Canc. Treat Res. **117**, 51–75 (2004)
15. Chaplain, M.A., Anderson, A.R.: Mathematical modelling, simulation and prediction of tumour-induced angiogenesis. Invasion Metastasis. **16**(4–5), 222–234 (1996)
16. Cohen, L.D., Cohen, I.L: Finite-element methods for active contour models and balloons for 2-d and 3-d images. IEEE PAMI. **15**(11), 1131–1147 (1993)
17. Dardis, O., McCloskey, J.: Lattice boltzmann scheme with real numbered solid density for the simulation of flow in porous media. Phys. Rev. E. **57**(4), 4834 (1998)
18. d'Onofrio, A., Ledzewicz, U., Maurer, H., Schattler, H.: On optimal delivery of combination therapy for tumors. Math. Biosci. **222**(1), 13–26 (2009)
19. Dünweg, B., Schiller, U.D., Ladd, A.J.C.: Statistical mechanics of the fluctuating lattice Boltzmann equation. Phys. Rev. E, **76**, 36704 (2007)
20. Dupin, M.M., Halliday, I., Care, C.M., Munn, L.L.: Modeling the flow of dense suspensions of deformable particles in three dimensions. Phys. Rev. E. **75**, 066707 (2007)
21. Dupin, M.M., Halliday, I., Care, C.M., Munn, L.L.: Efficiency oriented, hybrid approach for modeling deformable particles in three dimensions. Progr. Comput. Fluid Dynam. **8**, 109–120 (2008)
22. Dupin, M.M., Halliday, I., Care, C.M., Munn, L.L.: Lattice Boltzmann modelling of blood cell dynamics. Int. J. Comput. Fluid Dynam. **22**(7), 481–492 (2008)
23. Eaton, W.A., Henry, E.R., Hofrichter, J., Mozzarelli, A.: Is cooperative oxygen binding by hemoglobin really understood? Nat. Struct. Biol. **6**, 351–358 (1999)
24. Ferrara, N., Hillan, K.J., Gerber, H.P., Novotny, W.: Discovery and development of bevacizumab, an anti-vegf antibody for treating cancer. Nat. Rev. Drug Disc. **3**(5), 391–400 (2004)
25. Ferrara, N., Kerbel, R.S.: Angiogenesis as a therapeutic target. Nature. **438**(7070), 967–974 (2005)
26. Fischer, C., Mazzone, M., Jonckx, B., Carmeliet, P.: Flt1 and its ligands vegfb and plgf: Drug targets for anti-angiogenic therapy? Nat. Rev. Canc. **8**(12), 942–956 (2008)
27. Gillies, R.J., Schornack, P.A., Secomb, T.W., Raghunand, N.: Causes and effects of heterogeneous perfusion in tumors. Neoplasia. **1**(3), 197–207 (1999)
28. Ginzburg, I., Verhaeghe, F., d'Humieres, D.: Study of simple hydrodynamic solutions with the two-relaxation-time lattice Boltzmann scheme. Comm. Comput. Phys. **3**, 519 (2008)
29. Ginzburg, I., Verhaeghe, F., d'Humieres, D.: Two-relaxation-time lattice Boltzmann scheme: About parametrization, veloceity pressure and mixed boundary conditions. Comm. Comput. Phys. **3**, 427 (2008)
30. Gruionu, G., Hoying, J.B., Gruionu, L.G., Laughlin, M.H., Secomb, T.W.: Structural adaptation increases predicted perfusion capacity after vessel obstruction in arteriolar arcade network of pig skeletal muscle. Am. J. Physiol. **288**(6), H2778–84 (2005)
31. Hacking, W.J., VanBavel, E., Spaan, J.A.: Shear stress is not sufficient to control growth of vascular networks: A model study. Am. J. Physiol. **270**(1 Pt 2), H364–75 (1996)
32. Hamzah, J., Jugold, M., Kiessling, F., Rigby, P., Manzur, M., Marti, H.H., Rabie, T., Kaden, S., Grone, H.J., Hammerling, G.J., Arnold, B., Ganss, R.: Vascular normalization in Rgs5-deficient tumours promotes immune destruction. Nature. **453**(7193), 410–414 (2008)

33. Hansen-Smith, F.M., Hudlicka, O., Egginton, S.: In vivo angiogenesis in adult rat skeletal muscle: Early changes in capillary network architecture and ultrastructure. Cell Tissue Res. **286**(1), 123–136 (1996)
34. Huber, P.E., Bischof, M., Jenne, J., Heiland, S., Peschke, P., Saffrich, R., Grone, H.J., Debus, J., Lipson, K.E., Abdollahi, A.: Trimodal cancer treatment: Beneficial effects of combined antiangiogenesis, radiation, and chemotherapy. Canc. Res. **65**(9), 3643–3655 (2005)
35. Hudlicka, O., Brown, M.: Adaptation of skeletal muscle microvasculature to increased or decreased blood flow: Role of shear stress, nitric oxide and vascular endothelial growth factor. J. Vasc. Res. **46**, 504–512 (2009)
36. Ichioka, S., Shibata, M., Kosaki, K., Sato, Y., Harii, K., Kamiya, A.: Effects of shear stress on wound-healing angiogenesis in the rabbit ear chamber. J. Surg. Res. **46**, 29–35 (1997)
37. Jain, R.K.: Normalizing tumor vasculature with anti-angiogenic therapy: A new paradigm for combination therapy. Nat. Med. **7**(9), 987–989 (2001)
38. Jain, R.K.: Normalization of tumor vasculature: An emerging concept in antiangiogenic therapy. Science. **307**(5706), 58–62 (2005)
39. Jain, R.K.: Lessons from multidisciplinary translational trials on anti-angiogenic therapy of cancer. Nat. Rev. Canc. **8**(4), 309–316 (2008)
40. Jain, R.K., Tong, R.T., Munn, L.L.: Effect of vascular normalization by antiangiogenic therapy on interstitial hypertension, peritumor edema, and lymphatic metastasis: insights from a mathematical model. Canc. Res. **67**(6), 2729–2735 (2007)
41. Jones, P.F., Sleeman, B.D.: Angiogenesis - understanding the mathematical challenge. Angiogenesis. **9**(3), 127–138 (2006)
42. Kamoun, W.S., Ley, C.D., Farrar, C.T., Duyverman, A.M., Lahdenranta, J., Lacorre, D.A., Batchelor, T.T., di Tomaso, E., Duda, D.G., Munn, L.L., Fukumura, D., Sorensen, A.G., Jain, R.K.: Edema control by cediranib, a vascular endothelial growth factor receptor-targeted kinase inhibitor, prolongs survival despite persistent brain tumor growth in mice. J. Clin. Oncol. **27**(15), 2542–2552 (2009)
43. Kamoun, W.S., Ley, C.D., Farrar, C.T., Duyverman, A.M., Lahdenranta, J., Lacorre, D.A., Batchelor, T.T., di Tomaso, E., Duda, D.G., Munn, L.L., Fukumura, D., Sorensen, A.G., Jain, R.K.: Edema Control by Cediranib, a Vascular Endothelial Growth Factor Receptor-Targeted Kinase Inhibitor, Prolongs Survival Despite Persistent Brain Tumor Growth in Mice. J. Clin. Oncol. **27**(15), 2542–2552 (2009)
44. Kamoun, W.S., Chae, S.S., Lacorre, D.A., Tyrrell, J.A., Mitre, M., Gillissen, M.A., Fukumura, D., Jain R.K., Munn, L.L.: Simultaneous measurement of rbc velocity, flux, hematocrit and shear rate in vascular networks. Nat. Meth. **7**(8), 655–660 (2010)
45. Kang, Y., Choi, M., Lee, J., Koh, G.Y., Kwon, K., Choi, C.: Quantitative analysis of peripheral tissue perfusion using spatiotemporal molecular dynamics. PLoS ONE. **4**(1), e4275 (2009)
46. Kass, M., Witkin, A., Terzoopoulos, D.: Snakes: Active contour models. Int. J. Comput. Vis. **1**(4), 321–331 (1987)
47. Keller, K.H.: Effect of fluid shear on mass transport in flowing blood. Fed. Proc. **30**(5), 1591–1599 (1971)
48. Kerbel, R.S.: Issues regarding improving the impact of antiangiogenic drugs for the treatment of breast cancer. Breast **18**, S41–S47 (2009)
49. Ko, A.H., Dito, E., Schillinger, B., Venook, A.P., Xu, Z., Bergsland, E.K., Wong, D., Scott, J., Hwang, J., Tempero, M.A.: A phase II study evaluating bevacizumab in combination with fixed-dose rate gemcitabine and low-dose cisplatin for metastatic pancreatic cancer: Is an anti-VEGF strategy still applicable? Investig. New Drugs. **26**(5), 463–471 (2008)
50. Ladd, A.J.C.: Numerical simulations of particulate suspensions via a discretized Boltzmann equation. part 1. theoretical foundation. J. Fluid Mech. **271**, 285 (1994)
51. Ladd, A.J.C., Verberg, R.: Lattice-Boltzmann simulations of particle-fluid suspensions. J. Stat. Phys. **104**, 1191 (2001)
52. Lagerlund, T.D., Low, P.A.: Mathematical modeling of time-dependent oxygen transport in rat peripheral nerve. Comput. Biol. Med. **23**(1), 29–47 (1993)

53. Ledzewicz, U., Marriott, J., Maurer, H., Schattler, H.: Realizable protocols for optimal administration of drugs in mathematical models for anti-angiogenic treatment. Math. Med. Biol. **27**(2), 157–179 (2010)
54. Ledzewicz, U., Naghnaeian, M., Schattler, H.: Optimal response to chemotherapy for a mathematical model of tumor-immune dynamics. J. Math. Biol. **64**(3), 557–77 (2011)
55. Lee, D.S., Rieger, H., Bartha, K.: Flow correlated percolation during vascular remodeling in growing tumors. Phys. Rev. Lett. **96**(5), 58104 (2006)
56. Macklin, P., McDougall, S., Anderson, A.R., Chaplain, M.A., Cristini, V., Lowengrub, J.: Multiscale modelling and nonlinear simulation of vascular tumour growth. J. Math. Biol. **58**(4–5), 765–798 (2009)
57. Malladi, R., Sethian, J.A., Vemuri, B.C.: Shape modeling with front propagation: A level set approach. IEEE PAMI. **17**(2), 158–175 (1995)
58. McNamara, G.R., Zanetti, G.: Use of the Boltzmann equation to simulate lattice-gas automata. Phys. Rev. Lett. **61**(20), 2332 (1988).
59. Migliorini, C., Qian, Y.H., Chen, H., Brown, E., Jain, R.K., Munn, L.L.:. Red blood cells augment leukocyte rolling in a virtual blood vessel. Biophys. J. **83**, 1834–1841 (2002)
60. Munn, L.L.: Aberrant vascular architecture in tumors and its importance in drug-based therapies. Drug Discov. Today. **8**, 396–403 (2003)
61. Munn, L.L., Dupin, M.M.: Blood cell interactions and segregation in flow. Ann. Biomed. Eng. **36**(4), 534–544 (2008)
62. Nakahara, T., Norberg, S.M., Shalinsky, D.R., Hu-Lowe, D.D., McDonald, D.M.: Effect of inhibition of vascular endothelial growth factor signaling on distribution of extravasated antibodies in tumors. Canc. Res. **66**(3), 1434–1445 (2006)
63. Netti, P.A., Roberge, S., Boucher, Y., Baxter, L.T., Jain, R.K.: Effect of transvascular fluid exchange on arterio-venous pressure relationship: Implication for temporal and spatial heterogeneities in tumor blood flow. Microvasc. Res. **52**, 27–46 (1996)
64. Osher, S., Sethian, J.A.: Fronts propagating with curvature-dependent speed: Algorithms based on hamilton-jacobi formulations. J. Comput. Phys. **79**(1), 12–49 (1988)
65. Pries, A.R., Reglin, B., Secomb, T.W.: Remodeling of blood vessels: Responses of diameter and wall thickness to hemodynamic and metabolic stimuli. Hypertension. **46**(4), 725–731 (2005)
66. Pries, A.R., Secomb, T.W.: Modeling structural adaptation of microcirculation. Microcirculation. **15**(8), 753–764 (2008)
67. Pries, A.R., Secomb, T.W.: Origins of heterogeneity in tissue perfusion and metabolism. Cardiovasc. Res. **81**(2), 328–335, (2009)
68. Pries, A.R., Secomb, T.W., Gaehtgens, P.: Structural adaptation and stability of microvascular networks: theory and simulations. Am. J. Physiol. **275**(2 Pt 2), H349–60 (1998)
69. Pries, A.R., Secomb, T.W., Gaehtgens, P.: Design principles of vascular beds. Circ. Res. **77**(5), 1017–1023 (1995)
70. Qian, Y.H., d'Humieres, D., Lallemand, P.: Lattice BGK models for Navier-Stokes equation. Europhys. Lett. **17**, 479 (1992)
71. Reglin, B., Secomb, T.W., Pries, A.R.: Structural adaptation of microvessel diameters in response to metabolic stimuli: Where are the oxygen sensors? Am. J. Physiol. Heart. Circ. Physiol. **297**(6), H2206 (2009)
72. Schattler, H., Ledzewicz, U., Cardwell, B.: Robustness of optimal controls for a class of mathematical models for tumor anti-angiogenesis. Math. Biosci. Eng. **8**(2), 355–369 (2011)
73. Schugart, R.C., Friedman, A., Zhao, R., Sen, C.K.: Wound angiogenesis as a function of tissue oxygen tension: A mathematical model. Proc. Natl. Acad. Sci. USA. **105**(7), 2628–2633 (2008)
74. Sherwood, L.M., Parris, E.E., Folkman, J.: Tumor angiogenesis: Therapeutic implications. New Engl. J. Med. **285**(21), 1182–1186 (1971)
75. Shi, Y., Karl, W.C.: A fast level set method without solving pdes. In: IEEE International Conference on Acoustics, Speech, and Signal Processing, pages –, Mark (2005)

76. Song, J.W., Munn, L.L.: Fluid forces control endothelial sprouting. Proc. Natl. Acad. Sci. **108**(37), 15342–15347 (2011)
77. Staub, N.C.: Alveolar-arterial oxygen tension gradient due to diffusion . J. Appl. Physiol. 18, 673–680 (1963)
78. Stockmann, C., Doedens, A., Weidemann, A., Zhang, N., Takeda, N., Greenberg, J.I., Cheresh, D.A., Johnson, R.S.: Deletion of vascular endothelial growth factor in myeloid cells accelerates tumorigenesis. Nature. **456**(7223), 814–818 (2008)
79. Succi, S.: The Lattice Boltzmann Equation for Fluid Dynamics and Beyond . Oxford University Press, Oxford (2001)
80. Sukop, M.C., Thorne, D.T.: Lattice Boltzmann Modeling: An Introduction for Geoscientists and Engineers. Springer, New York (2007)
81. Sun, C., Jain, R.K., Munn, L.L.: Non-uniform plasma leakage affects local hematocrit and blood flow: implications for inflammation and tumor perfusion. Ann. Biomed. Eng. **35**(12), 2121–2129 (2007)
82. Sun, C., Munn, L.L.: Lattice Boltzmann simulation of blood flow in digitized vessel networks. Comput. Math. Appl. **55**(7), 1594–1600 (2008)
83. Sun, C.H., Hsu, A.T.: Three-dimensional lattice Boltzmann model for compressible flows. Phys. Rev. E. **68**(1), 016303–1–14 (2003)
84. Sun, C.H.,Munn, L.L.: Particulate nature of blood determines macroscopic rheology: A 2-D lattice Boltzmann analysis. Biophys. J. **88**(3), 1635–1645 (2005)
85. Sun, C.H., Munn, L.L.: Influence of erythrocyte aggregation on leukocyte margination in postcapillary expansions: A lattice-Boltzmann analysis. Physica A. **362**, 191–196 (2006)
86. Sun, S., Wheeler, M.F., Obeyesekere, M., Patrick, C.W.: A deterministic model of growth factor-induced angiogenesis. Bull Math. Biol. **67**(2), 313–337 (2005)
87. Szczerba, D., Szekely, G.: Computational model of flow-tissue interactions in intussusceptive angiogenesis. J. Theor. Biol. **234**(1), 87–97 (2005)
88. Tong, R.T., Boucher, Y., Kozin, S.V., Winkler, F., Hicklin, D.J., Jain, R.K.: Vascular normalization by vascular endothelial growth factor receptor 2 blockade induces a pressure gradient across the vasculature and improves drug penetration in tumors. Canc. Res. **64**(11), 3731–3736 (2004)
89. Tyrrell, J.A., Mahadevan, V., Tong, R., Brown, E., Roysam, B., Jain, R.K.: Modeling and analysis of multiphoton tumor microvasculature data using super-gaussians. Microvasc. Res. **70**(3), 149–162 (2005)
90. Tyrrell, J.A., di Tomaso, E., Fuja, D., Tong, R., Kozak, K., Jain, R.K., Roysam, B.: Robust 3-d modeling of vasculature imagery using superellipsoids. IEEE. Trans. Med. Imag. **26**(2), 223–237 (2007)
91. Tyrrell, J.A., Roysam, B., di Tomaso, E., Tong, R., Brown, E.B., Jain, R.K.: Robust 3-d modeling of tumor microvasculature using superellipsoids. In: ISBI, pp. 185–188. IEEE (2006)
92. Vosseler, S., Mirancea, N., Bohlen, P., Mueller, M.M., Fusenig, N.E.: Angiogenesis inhibition by vascular endothelial growth factor receptor-2 blockade reduces stromal matrix metalloproteinase expression, normalizes stromal tissue, and reverts epithelial tumor phenotype in surface heterotransplants. Canc. Res. **65**(4), 1294–1305 (2005)
93. Waite, L., Fine, J.: Applied BioFluid Mechanics. McGraw-Hill, New York (2007)
94. Walsh, Stuart D.C., Burwinklea, H., O Saar, M.: A new partial-bounceback lattice-Boltzmann method for fluid flow through heterogeneous media . Comput. Geosci. **35**(6), 1186–1193 (2008)
95. Welter, M., Bartha, K., Rieger, H.: Vascular remodelling of an arterio-venous blood vessel network during solid tumour growth. J. Theor. Biol. **259**(3), 405–422 (2009)
96. Willett, C.G., Boucher, Y., di Tomaso, E., Duda, D.G., Munn, L.L., Tong, R.T., Chung, D.C., Sahani, D.V., Kalva, S.P., Kozin, S.V., Mino, M., Cohen, K.S., Scadden, D.T., Hartford, A.C., Fischman, A.J., Clark, J.W., Ryan, D.P., Zhu, A.X., Blaszkowsky, L.S., Chen, H.X., Shellito, P.C., Lauwers, G.Y., Jain, R.K.: Direct evidence that the VEGF-specific antibody

bevacizumab has antivascular effects in human rectal cancer.[see comment][erratum appears in Nat. Med. 2004;10(6), 649]. Nat. Med. **10**(2), 145–147 (2004)
97. Winkler, F., Kozin, S.V., Tong, R.T., Chae, S.S., Booth, M.F., Garkavtsev, I., Xu, L., Hicklin, D.J., Fukumura, D., di Tomaso, E., Munn, L.L., Jain, R.K.: Kinetics of vascular normalization by VEGFR2 blockade governs brain tumor response to radiation: Role of oxygenation, angiopoietin-1, and matrix metalloproteinases. Canc. Cell. **6**(6), 553–563 (2004)
98. Wu, J., Long, Q., Xu, S., Padhani, A.R.: Study of tumor blood perfusion and its variation due to vascular normalization by anti-angiogenic therapy based on 3D angiogenic microvasculature. J. Biomech. **42**(6), 712–721 (2009)
99. Zakrzewicz, A., Secomb, T.W., Pries, A.R.: Angioadaptation: Keeping the vascular system in shape. News Physiol. Sci. **17**, 197–201 (2002)
100. Zou, Q., He, X.: On pressure and velocity boundary conditions for the lattice Boltzmann BGK model. Phys. Fluid. **9**(6), 1493–1858 (1997)

Influence of Blood Rheology and Outflow Boundary Conditions in Numerical Simulations of Cerebral Aneurysms

Susana Ramalho, Alexandra B. Moura, Alberto M. Gambaruto, and Adélia Sequeira

1 Introduction

Disease in human physiology is often related to cardiovascular mechanics. Impressively, strokes are one of the leading causes of death in developed countries, and they might occur as a result of an aneurysm rupture, which is a sudden event in the majority of cases. On the basis of several autopsy and angiography series, it is estimated that 0.4–6 % of the general population harbors one or more intracranial aneurysms, and on average the incidence of an aneurysmal rupture is of 10 per 100,000 population per year, with tendency to increase in patients with multiple aneurysms [14, 20].

An aneurysm is a localized pathological dilation of the wall of a blood vessel, due to the congenital or acquired structural weakening of the wall media, and potentially results in severe complications, or even sudden death, through pressing on adjacent structures, or rupturing causing massive hemorrhage [10]. They are primarily located in different segments of the aorta and in the intracranial arteries supplying the brain. Moreover, intracranial aneurysms are most likely to be encountered on or close to the circle of Willis, particularly in apices of the bifurcation of first- and second-order arteries, and in curved arterial segments [28]. The natural history of this pathology is far from being fully understood, which can be related to the paucity of temporal investigations, since aneurysms are rarely detected before rupture. It is believed that the formation, growth, and rupture of intracranial aneurysms are associated with local hemodynamics, other than lumen structural mechanics and biomedical responses.

S. Ramalho • A.B. Moura • A.M. Gambaruto • A. Sequeira (✉)
Department of Mathematics and CEMAT, Instituto Superior Técnico, Technical University of Lisbon, Av. Rovisco Pais, 1, 1049-001 Lisboa, Portugal
e-mail: susana.ramalho@ist.utl.pt; almoura@math.ist.utl.pt; agambar@math.ist.utl.pt; adelia.sequeira@math.ist.utl.pt

U. Ledzewicz et al., *Mathematical Methods and Models in Biomedicine*, Lecture Notes on Mathematical Modelling in the Life Sciences, DOI 10.1007/978-1-4614-4178-6_6,

Blood is a concentrated suspension of formed cellular elements that includes red blood cells (RBC or erythrocytes), white blood cells (or leukocytes), and platelets (or thrombocytes), suspended in an aqueous polymer solution, the plasma. RBC have been shown to exert the most significant influence on the mechanical properties of blood, mainly due to their high concentration (hematocrit $H_t \approx$ 40–45%). Consequently, the rheology of blood is largely affected by the behavior of the RBC, which can range from 3D microstructures to dispersed individual cells, depending predominantly on the shear rates [24]. Hemodynamics is not only related to the fluid properties but also to other mechanical factors, including the forces exerted on the fluid, the fluid motion, and the vessel geometry.

According to the circulatory region of interest and the desired level of accuracy, blood flow may be modeled as steady or pulsatile, Newtonian or non-Newtonian, and laminar or turbulent. In medium to large vessels, blood flow has pulsatile behavior, due to the repeated, rhythmic mechanical pumping of the heart [15]. However, in small arteries sufficiently distant from the heart the flow is predominantly steady. In this work, the importance of including the pulsatility of blood is studied, and both steady and unsteady simulations are considered.

As mentioned above, the RBC play an important role in the blood rheology. While plasma exhibits a nearly Newtonian behavior, whole blood has non-Newtonian characteristics [22]. This is mainly due to the RBC's tendency to form 3D microstructures at low shear rates and to their deformability and alignment with the flow field at high shear rates. Experimental studies suggest that in most part of the arterial system the viscosity of blood can be considered as a constant, and blood can be modeled as a Newtonian fluid. However, the complex processes related to the formation and breakup of the 3D microstructures, as well as the elongation and recovery of individual RBC, contribute in particular to blood shear-thinning viscosity, corresponding to a decrease in the apparent viscosity with increasing shear rate. It has also been observed that blood can present viscoelastic behavior [1, 22]. The variability of the blood viscosity leads to differences in perceived shear stress along the arterial wall. Indeed, in large arteries the instantaneous shear rate over a cardiac cycle has drastic variations, up to two orders of magnitude [25]. Despite these findings, as referred, it is often reasonable to simulate blood flow as a Newtonian fluid, since in sufficiently large nonpathological arteries it experiences high shear rates, over $100\,\mathrm{s}^{-1}$. Many authors adopt this argument however this assumption is not valid when the shear rate is lower than $100\,\mathrm{s}^{-1}$, which is the case of small arteries, veins, capillaries, and aneurysms or in recirculation regions downstream of a stenosis [27]. In these cases the flow is slower and the non-Newtonian models are better suited. Nevertheless, hemodynamics in intracranial aneurysms has been argued to be accurately modeled using the Newtonian assumption [4]. Here, both Newtonian and non-Newtonian fluid mathematical models will be adopted and compared.

Variations in the mathematical modeling of blood rheology lead to modeling uncertainties, which might compromise the reproducibility of the clinical data. The present work also focuses on the uncertainties that arise from considering different boundary conditions at the outflow sections of the computational domain, as well as from the inclusion or exclusion of the main side-branches in the geometry. The geometries employed in this work consist of a patient-specific aneurysm, obtained from medical imaging, and an idealization, for comparison purposes.

The outline of this chapter is as follows. Section 2 is dedicated to the geometry reconstruction of the patient-specific medical image data. The idealization of the anatomically realistic geometry will be also discussed. Section 3 is devoted to the detailed description of the mathematical models. It includes the description of the three-dimensional (3D) fluid model, as well as the reduced one-dimensional (1D), and zero-dimensional (0D) models. The couplings of the reduced models with the 3D one, that serve here as proper outflow boundary conditions, are also discussed. The numerical methods, geometry specifications, and inflow boundary conditions are introduced in Sect. 4. In Sect. 5 the numerical results are presented and discussed. Finally, in Sect. 6 conclusions are drawn.

2 Geometries Definition

The numerical simulations of hemodynamics are performed on both idealized geometries and an anatomically realistic geometry of a patient-specific aneurysm. The patient-specific geometry is reconstructed from medical images obtained in vivo from rotational computerized tomography angiography (CTA), with resulting voxel resolution of 0.4 mm on a 512^3 grid. This volumetric data is segmented using a constant threshold value. The surface triangulation of the vessel wall is extracted using a marching tetrahedra algorithm and hence a linear interpolation. This approach is computationally inexpensive but assumes that the image intensity of the desired object is sufficiently different from the background to permit a constant gray scale threshold choice. It furthermore requires that the medical image resolution is fine enough and isotropic to perform marching tetrahedra directly, instead of performing an interpolation as presented in [8] and references therein.

Several other segmentation methods exist for image data of cerebral aneurysms, such as deformable models and region growing [2, 4, 23]; however these tend to be sensitive to user defined parameter settings. Each segmentation approach will yield a different geometry definition that depends on user-defined coefficients or assumptions made in the approach [26]. Ultimately there is an inherent uncertainty in the model definition limited by the acquisition modality, resolution, contrast, and noise.

The resulting virtual model of the vasculature is then prepared for the numerical simulations by identifying the regions of interest and removing secondary branches. Successively surface smoothing is performed due to medical imaging noise and limited resolution, taking care not to alter the object beyond the pixel size, since this represents the inherent uncertainty size. Smoothing is performed using the bi-Laplacian method, with a final inflation along the local normal by a constant distance in order to minimize the volume alteration and surface distortion [8].

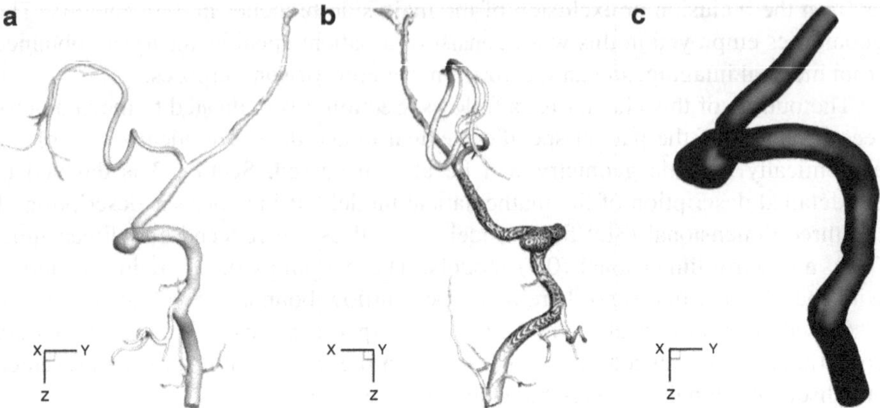

Fig. 1 Cerebral arterial system showing a saccular aneurysm located on the outer bend of the posterior inferior cerebral artery (PICA) in (**a**) coronal view, (**b**) sagittal view with superposition of the region of interest,and (**c**) detail of the region of interest in coronal view

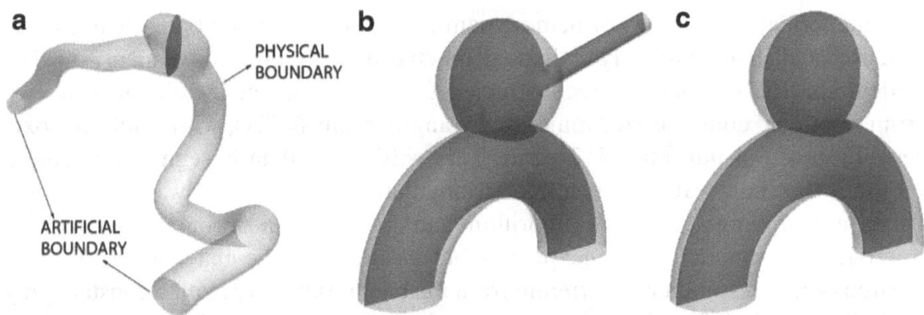

Fig. 2 The geometries considered, including the chosen cross-sections: (**a**) region of interest of the anatomically realistic geometry with side-branches excluded; (**b**) idealized geometry with side-branch in the aneurysm; (**c**) idealized geometry with hole (clipped side-branch) in the aneurysm

The anatomically realistic geometry of the aneurysm and the identification of the region of interest for the computational domain are depicted in Fig. 1. The idealized geometry considered is inspired from [11]. It has a reduced surface definition complexity, introducing however a side-branch in the aneurysm. The aim is to provide a clearer understanding of the sensitivity to the choice of fluid boundary conditions in a similar flow field to that of the anatomically realistic geometry. Nonetheless, the idealization reduces the presence of complex flow structures that arise in the patient-specific case, due to the non-planarity of the main vessel and the small-scale detail in the surface definition. The idealized geometry consists of a main vessel with constant diameter and radius of curvature, a spherical saccular aneurysm, and a side-branch that is represented by either a straight tube or a hole, resulting in a total of two idealized geometries. Figure 2 shows the geometries studied: anatomically realistic, idealized with tube side-branch, and idealized with hole side-branch. For abbreviation these geometries will be referred to as "real," "idealized with branch," and "idealized with hole," respectively.

3 The Mathematical Models

Hemodynamics in the cardiovascular system is modeled through the time-dependent equations for incompressible fluids, derived from the conservation of momentum and mass. They describe a homogeneous fluid in terms of the velocity and the pressure fields. Considering an open and bounded domain $\Omega \subset \mathbb{R}^3$, the system of equations representing such fluid is given by

$$\begin{cases} \dfrac{\partial \mathbf{u}}{\partial t} + (\mathbf{u}\cdot\nabla)\,\mathbf{u} - \dfrac{1}{\rho}\,\mathbf{div}\ \boldsymbol{\sigma}(p,\mathbf{u}) = \mathbf{f}, \\ \qquad\qquad\qquad\qquad\qquad\qquad\qquad\qquad \text{in} \quad \Omega, \forall t > 0, \\ \qquad\qquad\qquad\qquad\quad \operatorname{div} \mathbf{u} = 0, \end{cases} \tag{1}$$

where $\mathbf{f}$ represents the body forces (that will be neglected, $\mathbf{f} = \mathbf{0}$, for the case study at hand), ρ is the fluid constant density, and the Cauchy stress tensor $\boldsymbol{\sigma}(p,\mathbf{u})$ depends on the unknown fluid pressure, p, and velocity, $\mathbf{u}$, and may be generally represented as the sum of the so-called spherical, $p\mathbf{I}$, and deviatoric, $\boldsymbol{\tau}(\mathbf{D}(\mathbf{u}))$, parts [21]

$$\boldsymbol{\sigma}(p,\mathbf{u}) = -p\mathbf{I} + \boldsymbol{\tau}(\mathbf{D}(\mathbf{u})). \tag{2}$$

In the spherical part, p is the Lagrange multiplier associated to the incompressibility constraint $\mathbf{div}(\mathbf{u})$, which defines the mechanical pressure for incompressible fluids, $p = p(\mathbf{x},t)$, and $\mathbf{I}$ is the unitary tensor. Concerning the deviatoric tensor, $\boldsymbol{\tau}$, it depends on the strain rate tensor, $\mathbf{D}(\mathbf{u})$, which is the symmetric part of the velocity gradient

$$\mathbf{D}(\mathbf{u}) = \frac{1}{2}\left(\nabla\mathbf{u} + (\nabla\mathbf{u})^{\mathrm{T}}\right).$$

3.1 Newtonian Fluids

The definition of a constitutive relation for $\boldsymbol{\tau}(\mathbf{D}(\mathbf{u}))$ is related to the rheological properties of the fluid. Under the assumption of incompressible Newtonian fluids, the Cauchy stress tensor is a linear isotropic function of the components of the velocity gradient, and it is given by

$$\boldsymbol{\sigma}(\mathbf{u},p) = -p\mathbf{I} + 2\mu\mathbf{D}(\mathbf{u}), \tag{3}$$

where $\mu > 0$ is the fluid constant Newtonian viscosity and $\boldsymbol{\tau}(\mathbf{u}) = 2\mu\mathbf{D}(\mathbf{u})$.

Thus, applying the constitutive relation (3) to Eq. (1), the Navier–Stokes equations for incompressible Newtonian fluids are obtained:

$$\begin{cases} \rho \dfrac{\partial \mathbf{u}}{\partial t} + \rho (\mathbf{u} \cdot \nabla) \mathbf{u} + \nabla p - \operatorname{div}(2\mu \mathbf{D}(\mathbf{u})) = \mathbf{0}, & \\ & \text{in} \quad \Omega, \forall t > 0, \\ \operatorname{div} \mathbf{u} = 0. & \end{cases} \tag{4}$$

3.2 *Generalized Newtonian Fluids*

The most general form of Eq. (2), for isotropic symmetric tensor functions, under frame invariance requirements [21], is given by

$$\boldsymbol{\sigma} = \phi_0 \mathbf{I} + \phi_1 \mathbf{D} + \phi_2 \mathbf{D}^2, \tag{5}$$

with ϕ_0, ϕ_1, and ϕ_2 dependent on the density ρ and on the three principal invariants of $\mathbf{D}$, $\mathbf{I}_D = tr(\mathbf{D})$, $\mathbf{II}_D = \frac{1}{2}\left((tr(\mathbf{D}))^2 - tr(\mathbf{D}^2)\right)$, and $\mathbf{III}_D = det(\mathbf{D})$, where $tr(\mathbf{D})$ and $det(\mathbf{D})$ denote the trace and the determinant of tensor $\mathbf{D}$, respectively. By setting $\phi_2 = 0$, and ϕ_1 constant, we obtain the relation for a Newtonian fluid, governed by the Navier-Stokes equations (4). Considering $\phi_2 \neq 0$ does not correspond to any existent fluid under simple shear, so that the constitutive relation (5) is often used in the reduced general form, with $\phi_2 = 0$ [21]: $\boldsymbol{\sigma} = \phi_0 \mathbf{I} + \phi_1 \mathbf{D}$. Moreover, respecting the frame invariance requirements and the behavior of real fluids, ϕ_1 becomes the viscosity function [21], and the following general constitutive relation is obtained:

$$\boldsymbol{\sigma} = -p\mathbf{I} + 2\mu(\mathbf{II}_D, \mathbf{III}_D)\mathbf{D}, \tag{6}$$

where the viscosity function μ might depend on the second and third invariants of $\mathbf{D}$.

Since $\mathbf{III}_D = 0$ in simple shear, as well as in other viscometric flows, it is reasonable to neglect the dependence of μ on $\mathbf{III}_D$. Furthermore, $\mathbf{II}_D$ is negative for isochoric motions, where $tr(\mathbf{D}) = 0$, so the positive metrics of the rate of deformation

$$\dot{\gamma} \equiv \sqrt{-4\mathbf{II}_D} = \sqrt{2tr(\mathbf{D}^2)},$$

also known as the shear rate, may be defined. Using the definition of the shear rate as a function of the second invariant of $\mathbf{D}$, relation (6) can be rewritten as follows:

$$\boldsymbol{\sigma} = -p\mathbf{I} + 2\mu(\dot{\gamma})\mathbf{D}. \tag{7}$$

This equation defines the constitutive equation for the generalized Newtonian fluids, such that the equations of motion for these fluids are of the form

Table 1 Some generalized Newtonian models for blood viscosity and corresponding constants

Model	Viscosity model	Model constants for blood
Carreau	$F(\dot{\gamma}) = (1+(\lambda\dot{\gamma})^2)^{(n-1)/2}$	$\mu_0 = 0.456, \mu_\infty = 0.032$ $\lambda = 10.03\,s, n = 0.344$
Cross	$F(\dot{\gamma}) = (1+(\lambda\dot{\gamma})^m)^{-1}$	$\mu_0 = 0.618, \mu_\infty = 0.034$ $\lambda = 7.683\,s, m = 0.810$
Yeleswarapu	$F(\dot{\gamma}) = \dfrac{1+\log(1+\lambda\dot{\gamma})}{1+\lambda\dot{\gamma}}$	$\mu_0 = 1.10, \mu_\infty = 0.035$ $\lambda = 45.23\,s$
Oldroyd	$\mu(\dot{\gamma}) = \mu_0 \dfrac{1+(\lambda_1\dot{\gamma})^2}{1+(\lambda_2\dot{\gamma})^2}$	$\mu_0 = 0.426, \mu_\infty = \mu_0\lambda_1^2\lambda_2^{-2}$ $\lambda_1 = 1.09\,s, \lambda_2 = 3.349\,s$

$$\begin{cases} \rho\dfrac{\partial \mathbf{u}}{\partial t} + \rho\,(\mathbf{u}\cdot\nabla)\,\mathbf{u} + \nabla p - \operatorname{div}(2\mu(\dot{\gamma})\mathbf{D}(\mathbf{u})) = \mathbf{0}, & \\ & \text{in} \quad \Omega, \forall t > 0, \\ \operatorname{div}\,\mathbf{u} = 0. & \end{cases} \tag{8}$$

A variety of non-Newtonian viscosity functions $\mu(\dot{\gamma})$ can be used, only differing on the functional dependence of the viscosity μ on the shear rate $\dot{\gamma}$. To model blood flow, the focus is put on bounded viscosity functions of the form

$$\mu(\dot{\gamma}) = \mu_\infty + (\mu_0 - \mu_\infty)\mathrm{F}(\dot{\gamma}), \tag{9}$$

where the constants μ_0 and μ_∞ are the asymptotic viscosities at zero, $\mu_0 = \lim_{\dot{\gamma}\to 0}\mu(\dot{\gamma})$, and infinity, $\mu_\infty = \lim_{\dot{\gamma}\to\infty}\mu(\dot{\gamma})$, shear rate. $\mathrm{F}(\dot{\gamma})$ is a continuous and monotonic function such that

$$\lim_{\dot{\gamma}\to 0}\mathrm{F}(\dot{\gamma}) = 0, \quad \lim_{\dot{\gamma}\to\infty}\mathrm{F}(\dot{\gamma}) = 1. \tag{10}$$

The definition of function $\mathrm{F}(\dot{\gamma})$ characterizes the generalized Newtonian model. Table 1 was taken from [9] and shows several possible viscosity functions.

The values of the parameters there displayed, corresponding to an hematocrit $H_t = 40\,\%$ and temperature $T = 37\,^\circ$C, were obtained from in vitro blood experimental data, as described in [9]. To set the parameters values, a nonlinear least squares fitting was applied [9, 13]. Notice that, with such parameters, all the viscosity functions in Table 1 correspond to shear-thinning models.

Other generalized Newtonian models for blood viscosity, like the power-law and the Carreau–Yasuda model, have been frequently used to describe blood flow (for further details on these models, see [22]). In this work, following [9, 13], the Carreau viscosity function is used, with the parameters provided in Table 1.

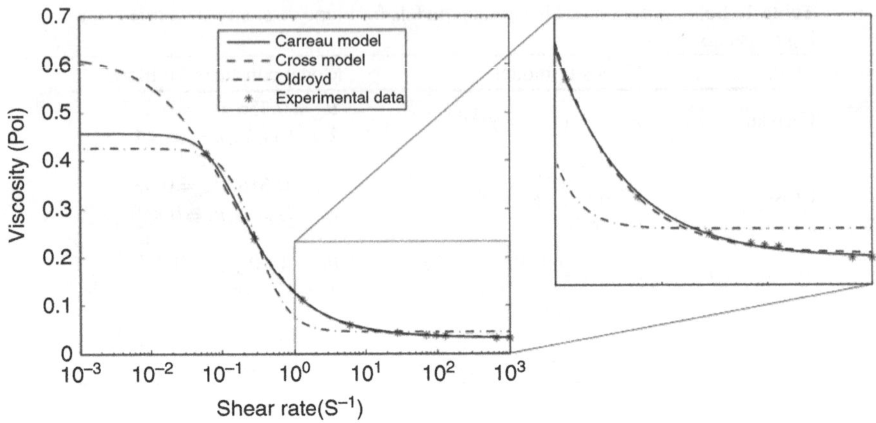

Fig. 3 Apparent viscosity as a function of shear rate for whole blood at $H_t = 40\%$, T = 37°C, taken from [9]

Plots of some non-Newtonian models and the experimental data are shown in Fig. 3. Experimental data for low shear rates is difficult to obtain, resulting in very different behavior as the shear rate approaches zero.

3.3 *Outflow Boundary Conditions*

Equations (4) or (6) have to be provided with initial and boundary conditions, in order to be mathematically well defined and prepared to be solved by numerical methods. The prescription of proper initial and boundary conditions is a crucial step in the numerical procedure to obtain accurate and meaningful computed solutions.

After defining the initial condition, $\mathbf{u} = \mathbf{u}_0$, for $t = 0$ in Ω, an appropriate set of conditions must be imposed on the boundary of the domain Ω. In particular, for the problem of blood flow in arteries, the computational domain is bounded by a physical boundary that is the arterial wall, and by artificial boundaries on the fluid domain due to truncation of the artery, detailed in Fig. 2.

On the physical boundary corresponding to the vascular wall a no-slip condition is imposed, describing the complete adherence of the fluid to the wall. In this study, the compliance of the artery wall will be neglected, that is, a fixed geometry is considered, so that the velocity at the wall is zero. Thus, an homogeneous Dirichlet boundary condition, $\mathbf{u} = \mathbf{0}$, $\forall t > 0$, is imposed at the physical wall of the fluid.

The boundary conditions at the artificial sections cannot be obtained from physical arguments and can be a significant source of numerical inaccuracies in resolving the problem [3]. At these interfaces the remaining parts of the arterial system need to be accounted for and modeled. Typically, it is very difficult to obtain appropriate patient data for the flow boundary conditions.

In this work a traction free boundary condition $\sigma(\mathbf{u},p)\cdot\mathbf{n}=0$ will always be considered at the main vessel outflow (see Fig. 2b). Concerning the outflow section of the side-branch, four types of outflow boundary conditions are explored: zero velocity, $\mathbf{u}=\mathbf{0}$, meaning that the side-branch is neglected and modeled as a no-slip wall [19]; zero normal stress, $\sigma(\mathbf{u},p)\cdot\mathbf{n}=0$ [4]; coupling with a 0D model corresponding to a simple resistance [13]; and coupling with a one-dimensional (1D) model equivalent to the three-dimensional (3D) side-branch [6]. Thus, the first two approaches neglect the effects of the remaining parts of the cardiovascular system, as opposed to the last two which resort to the Geometrical Multiscale Approach [6] to account for the global circulation on the localized numerical simulation.

3.4 The 1D Model

The 1D simplified model is formulated assuming that an artery is a cylindrical compliant tube, with axial symmetry and fixed cylinder axis. The velocity components orthogonal to the vessel axis are neglected and the wall displacements are only accounted for in the radial direction. Moreover, no body forces are considered and the pressure, $P(t,z)$, is assumed constant on each axial section, varying only coaxially. The area of each cross-section S is given by $A(t,z)=\int_S \mathrm{d}\sigma$, and the mean velocity is defined as $\bar{u}=A^{-1}\int_S u_z \mathrm{d}\sigma$, where u_z is the axial velocity. The area, A, the averaged pressure, P, and the mean flux, $Q=A\bar{u}$, are the unknown variables to be determined. The average pressure and flow rate are related to the 3D pressure and velocity, respectively, while the area is related to the 3D wall displacement. Thus, the 1D model provides a fluid–structure interaction (FSI) description of blood flow in arteries, accounting for the wall compliance due to the blood load. For that reason, the 1D model captures very well the wave propagation nature of blood flow in arteries.

Integrating the Navier–Stokes equations on a generic cross-section S of the cylindrical vessel, and after the above mentioned simplifications, explored in [6], the reduced 1D form of the continuity and momentum equations for the flow of blood in arteries is given, for all $t>0$, by

$$\begin{cases} \dfrac{\partial A}{\partial t}+\dfrac{\partial Q}{\partial z}=0, \\ \dfrac{\partial Q}{\partial t}+\alpha\dfrac{\partial}{\partial z}\left(\dfrac{Q^2}{A}\right)+\dfrac{A}{\rho}\dfrac{\partial P}{\partial z}+K_r\left(\dfrac{Q}{A}\right)=0, \end{cases} \quad z\in(a,b), \tag{11}$$

where z is the axial direction, $L=b-a$ denotes the vessel length, K_r is the friction parameter, α is the momentum flux correction coefficient, also known as Coriolis coefficient, defined by $\alpha = \frac{\int_S u_z^2 d\sigma}{A\bar{u}^2}$, and ρ is the fluid mass density. For a parabolic profile, the friction parameter is defined as $K_r = 8\pi\mu$ [6], which is the value generally used in practice. The Coriolis coefficient is set to $\alpha = 1$, corresponding to a flat profile [6], in order to simplify the analysis. The density ρ and the fluid dynamic viscosity μ are considered constant. Hence, the 1D model does not account for the non-Newtonian behavior of blood.

The previous system of two equations for the three unknown variables A, Q, and P needs to be closed. In order to do that, a structural model for the vessel wall movements, relating pressure and area, must be given. Here, the simplest pressure-area algebraic relation [6, 7] is used

$$P(t,z) - P_{\mathrm{ext}} = \beta \frac{\sqrt{A} - \sqrt{A_0}}{A_0}, \tag{12}$$

where A_0 is the initial area and β is a single parameter that describes the mechanical and physical properties of the vessel wall

$$\beta = \frac{\sqrt{\pi} hE}{1 - \xi^2}, \tag{13}$$

where h the wall thickness, E the vessel wall Young, or elasticity, modulus, and ξ the vessel wall Poisson ratio. β is constant along z only when E, h, or A_0 are constant, since they may be functions of z. In this work, the wall parameters are assumed constant along z, and the external pressure is neglected: $P_{\mathrm{ext}} = 0$.

Numerical Discretization of the 1D Model

The 1D model is numerically discretized in time and space by means of a second-order Taylor-Galerkin scheme [6]. It consists in using the Lax-Wendroff scheme to discretize in time and the finite element method to obtain the space approximation. This discretization can be considered as a finite element counterpart of the Lax-Wendroff scheme, which has a very good dispersion error characteristic and can be easily implemented [6].

A uniform mesh is used, meaning that the elements size is constant and equal to h. Moreover, linear ($P1$) finite elements are considered. The Lax–Wendroff scheme is obtained using a Taylor series of the solution $\mathbf{U} = [Q\ A]^{\mathrm{T}}$ truncated to the second order, resulting in an explicit scheme. Being an explicit time advancing method, the Lax–Wendroff scheme requires the verification of a condition bounding the time step [6]

$$\Delta t \leq \frac{\sqrt{3}}{3} \frac{h}{\max(c + |\bar{u}|)}, \tag{14}$$

where h is the size of the spatial mesh and $c = c(A;A_0;\beta) = \sqrt{\frac{\beta}{2\rho A_0}} A^{\frac{1}{4}}$ is the speed of the wave propagation along the vessel. Condition (14) corresponds to a CFL number of $\frac{\sqrt{3}}{3}$.

The final finite element solution for the area and flow rate is obtained by finding, at each time step t^{n+1}, $\mathbf{U}_h^{n+1}(z) = \sum_{i=0}^{N+1} \mathbf{U}_i^{n+1} \psi_i(z) \in V_h(a,b)$, where $V_h(a,b)$ is the space of $P1$ finite elements in 1D for the uniform mesh associated to h spacing, and $\{\psi_i\}_{i=1}^{N}$ its basis, satisfying the following expression for the interior nodes:

$$(\mathbf{U}_h^{n+1}, \psi_j) = (\mathbf{U}_h^n, \psi_j) + \Delta t \left(\mathbf{F}^n - \frac{\Delta t}{2} \mathbf{H}^n \mathbf{B}^n, \frac{\partial \psi_j}{\partial z} \right) - \Delta t \left(\mathbf{B}^n - \frac{\Delta t}{2} \mathbf{B}_{\mathbf{U}}^n \mathbf{B}^n, \psi_j \right)$$
$$- \frac{\Delta t^2}{2} \left(\mathbf{H}^n \frac{\partial \mathbf{F}^n}{\partial z}, \frac{\partial \psi_j}{\partial z} \right) + \frac{\Delta t^2}{2} \left(\mathbf{B}_{\mathbf{U}}^n \frac{\partial \mathbf{F}^n}{\partial z}, \psi_j \right), \; j = 1, \cdots, N, n = 0, \cdots, M-1. \tag{15}$$

Here $\mathbf{U}_h^0$ is a suitable approximation of the initial data, $(\mathbf{u}, \mathbf{v}) := \int_a^b \mathbf{u} \cdot \mathbf{v} dz$ represents the inner product in $V_h(a,b)$, $\{\psi_j\}_{j=1}^{N}$ are the basis functions of $V_h(a,b)$, $\Delta t = t^{n+1} - t^n$, and

$$\mathbf{H} = \begin{bmatrix} 0 & 1 \\ -\alpha \frac{Q^2}{A^2} + \frac{\beta}{2\rho A_0} A^{\frac{1}{2}} & 2\alpha \frac{Q}{A} \end{bmatrix}, \quad \mathbf{B}_{\mathbf{U}} = \begin{bmatrix} 0 & 0 \\ K_r \frac{Q}{A^2} & -K_r \frac{1}{A} \end{bmatrix},$$

$$\mathbf{F} = \begin{bmatrix} Q \\ \alpha \frac{Q^2}{A} + \frac{\beta}{3\rho A_0} A^{\frac{3}{2}} \end{bmatrix}, \qquad \mathbf{B} = \begin{bmatrix} 0 \\ -K_r \frac{Q}{A} \end{bmatrix}.$$

System (15) must be supplemented with proper initial, $\mathbf{U}_h^0$, and boundary conditions for the solution $\mathbf{U}_h^{n+1}$, at the left and right boundary points, $z = a$ and $z = b$, respectively. In the present work, the initial conditions were taken to be $A^0(z) = A_0$ and $Q^0(z) = 0$.

Compatibility Conditions for the 1D Model

By choosing relation (12), the pressure may be eliminated from the momentum equation, and system (11) becomes hyperbolic, with two distinct eigenvalues (see [6, 17] for the characteristic analysis of system (11))

$$\lambda_{1,2} = \bar{u} \pm c, \tag{16}$$

where c is the speed of the propagation of waves along the artery, defined above. The eigenfunctions, or characteristic variables, corresponding to the eigenvalues $\lambda_{1,2}$, are defined by

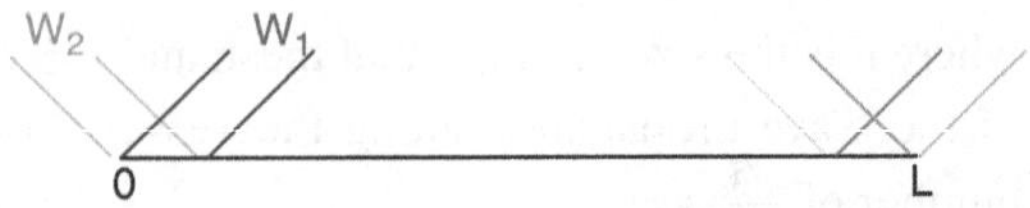

Fig. 4 The characteristic lines

$$W_{1,2} = \bar{u} \pm \int_{A_0}^{A} \frac{c(\tau)}{\tau} \mathrm{d}\tau = \bar{u} \pm 4\sqrt{\frac{\beta}{2\rho A_0}} \left(A^{\frac{1}{4}} - A_0^{\frac{1}{4}} \right). \tag{17}$$

Under physiological conditions, typical values of the flow velocity and mechanical characteristics of the vessel wall are such that $c \gg \bar{u}$, and consequently we have that $\lambda_1 > 0$ and $\lambda_2 < 0$, everywhere. This means that the flow is subcritical, such that the characteristic variable W_1 associated to the first eigenvalue, λ_1, travels forward, while the characteristic variable W_2, associated to the second eigenvalue, λ_2, travels backward (see Fig. 4). Hence, W_1 is the incoming characteristic, and W_2 is the outgoing characteristic, at the upstream left point ($z = a$), and vice versa at the downstream right point ($z = b$), as illustrated in Fig. 4.

Because of this, exactly one boundary condition must be imposed at each extremity of the vessel [18]. However, the discretized model requires two conditions at each boundary node in order to solve the system, corresponding to Q_h^{n+1} and A_h^{n+1}, both at $z = a$ and $z = b$. Thus two additional conditions, which have to be compatible with the problem, are needed at the numerical level. These compatibility conditions can be obtained by means of the outgoing characteristic at each boundary [18], through projecting the equations along the characteristic lines exiting the domain [16]. This results in computing the following additional relations at the boundaries:

$$W_2(Q_h^{n+1}(a), A_h^{n+1}(a)) = W_2(Q_h^n(z_a), A_h^n(z_a)) - \Delta t K_r \frac{Q_h^n(z_a)}{(A_h^n(z_a))^2}, \quad \text{at } z = a, \tag{18}$$

and

$$W_1(Q_h^{n+1}(b), A_h^{n+1}(b)) = W_1(Q_h^n(z_b), A_h^n(z_b)) - \Delta t K_r \frac{Q_h^n(z_b)}{(A_h^n(z_b))^2}, \quad \text{at } z = b, \tag{19}$$

where z_a and z_b are the corresponding foot of the outgoing characteristic lines which, using a first-order approximation [6], are given by

$$z_a = a - \Delta t \lambda_2(Q_h^n(a), A_h^n(a)) = a - \Delta t \left(\frac{Q_h^n(a)}{A_h^n(a)} + \sqrt{\frac{\beta}{2\rho A_0}} (A_h^n(a))^{\frac{1}{4}} \right), \tag{20}$$

$$z_b = b - \Delta t \lambda_1(Q_h^n(b), A_h^n(b)) = b - \Delta t \left(\frac{Q_h^n(b)}{A_h^n(b)} - \sqrt{\frac{\beta}{2\rho A_0}} (A_h^n(b))^{\frac{1}{4}} \right). \tag{21}$$

The solution for the boundary nodes of the domain may then be achieved through resolving a 2×2 nonlinear system given by Eqs. (18) and (19) (for more details see [17]).

Typically, the inflow condition is a flux or a total pressure, while the outflow condition is given by $W_2 = 0$, such that there is no incoming characteristic at $z = b$, corresponding to a completely absorbing boundary condition at the outflow point.

3D–1D Coupling

To couple the artificial boundary, denoted Γ_{art} from here on, of the 3D fluid Eq. (1) with the 1D interface point $z = a$ of the hyperbolic model (11), the continuity of the flow rate and the mean pressure are imposed, for all $t > 0$ (see for instance [7])

$$\int_{\Gamma_{\text{art}}} \mathbf{u} \cdot \mathbf{n} d\gamma = Q^{1D}(a,t), \tag{22}$$

$$\frac{1}{|\Gamma_{\text{art}}|} \int_{\Gamma_{\text{art}}} p d\gamma = P^{1D}(a,t). \tag{23}$$

Here $\mathbf{u}$ and p denote the 3D velocity vector and pressure, respectively, and Q^{1D} and P^{1D} are the 1D flow rate and mean pressure, respectively. The solution of the coupled problem is approximated in an iterative way, by resorting to a splitting strategy. This means that each model is solved separately and yields the resultant information to the other. Thus, at each time step the 3D model returns pointwise data, which is integrated to obtain the averaged quantities to be provided to the 1D model as a boundary condition at $z = a$. On the other hand, the 1D model provides the boundary conditions at the coupling sections of the 3D in terms of average data. The average data is defective for the 3D problem, since it requires pointwise data at the coupling interface. Thus, appropriate techniques must be used in order to prescribe the 1D integrated data onto the 3D model as boundary condition. Precisely, in this work the coupling is performed by passing the flow rate from the 3D to the 1D model, imposing Eq. (22) at the coupling point of the 1D model, $z = a$, and by imposing the mean pressure, computed by the 1D model, to the 3D problem, by means of the condition (23) at the 3D artificial coupling boundary, Γ_{art}. To prescribe the defective mean pressure on the 3D coupling section, Γ_{art}, the approach introduced in [12] is followed, so that the mean pressure is imposed through a Neumann boundary condition

$$\sigma(\mathbf{u}, p) \cdot \mathbf{n} = P_{1D}\mathbf{n}, \quad \text{on } \Gamma_{\text{art}}, \quad \forall t > 0. \tag{24}$$

The 3D–1D iterative coupling algorithm is carried out explicitly in this work. At each time step t^n, the 3D model provides the flow rate computed at the previous time step to the 1D model and receives the mean pressure computed from the 1D model. This is followed by advancing to the next time step (see Fig. 5).

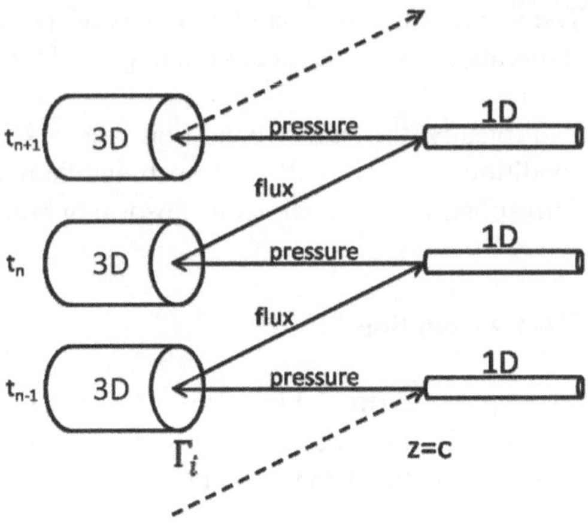

Fig. 5 Scheme of the 3D–1D explicit coupling

3.5 The 0D Model

Lumped parameters models are derived from the 1D ones by further averaging spatially in the coaxial direction [6], thus losing dependence from the spatial coordinates. Because of this, they are also called 0D models. They are represented by a system of ordinary differential equations (ODEs) in time and are analogous to electric circuits, where the flow rate can be identified with the current, the mean pressure with the voltage, and the 3D physical parameters, such as blood viscosity, blood inertia and wall compliance, with the lumped parameters resistance, inductance, and capacitance, respectively [6]. The 0D models are able to represent the circulation in large compartments of the cardiovascular system, such as the venous bed, the pulmonary circulation, or the heart [6].

In the present study a simple 0D model is also used and coupled to the 3D model. It consists of a single resistance, resulting in an algebraic relation between flux and mean pressure, through the resistance parameter: $P = RQ$. This model is constructed using the linear counterpart of the absorbing boundary condition for the 1D model [13]. Precisely, given the expression (17) of W_2, the condition $W_2 = 0$ is equivalent to

$$f(P) = \sqrt{\frac{8\beta}{\rho A_0}\left(P\frac{A_0}{\beta} + \sqrt{A_0}\right)^2\left(\sqrt{P\frac{A_0}{\beta} + \sqrt{A_0}} - A_0^{\frac{1}{4}}\right)} = Q. \quad (25)$$

A linearization of expression (25) is obtained resorting to the first approximation around zero of $f(P) : Q = f(P) = f(0) + f'(0)P$ (see [13]). The pressure is then given by

$$P = \frac{\sqrt{\rho\beta}}{\sqrt{2}A_0^{5/4}}Q. \quad (26)$$

3D–0D Coupling

The coupling of the 0D model (26) with the 3D fluid equations corresponds to imposing the linear counterpart of the absorbing boundary condition for the 1D model, $W_2 = 0$, directly on the 3D artificial section. The coupling is achieved by forcing the pressure given by the resistance of the 0D model at the 3D interface section Γ_{art}, similarly to the 3D–1D coupling. An explicit coupling is applied, meaning that the mean pressure at the current time step, P^{n+1}, is computed by means of expression (26) using the flow rate on the artificial section at the previous time step, Q^n, and it is prescribed at the artificial section at the current time step. Thus, as in [13], the defective averaged data condition

$$P^{(n+1)} = \frac{\sqrt{\rho\beta}}{\sqrt{2}A_0^{5/4}}Q^{(n)}, \quad \text{on } \Gamma_{\text{art}}, \tag{27}$$

is prescribed by means of a Neumann boundary condition (24) on the 3D artificial section.

4 Numerical Simulation Setup

In this work, the hemodynamics inside an intracranial saccular aneurysm is analyzed in an anatomically realistic geometry, as well as in idealized geometries. The idealized geometries serve as test cases with reduced complexity of the flow field, allowing for a better understanding of the effects of changing the fluid models, the boundary conditions, and in evaluating steady and unsteady simulations. Moreover, the numerical simulations in the idealized geometries have lower computational costs than in the realistic ones, allowing to conduct a comprehensive series of tests. While clinical decisions should be based on numerical simulations using anatomically realistic patient-specific geometries, idealized models provide insight into the hemodynamics with respect to choices in modeling and numerical setup.

In both steady and unsteady cases, the fluid is initially at rest and then the inflow flow rate is linearly increased with a parabolic profile to a final steady-state flux $Q = 2.67\,\text{cm}^3\,\text{s}^{-1}$, such that

$$Q_{\text{in}}^{\text{ramp}}(t) = \frac{tQ}{t_{\text{ramp}}}, \quad \text{for } t < t_{\text{ramp}}, \tag{28}$$

$$Q_{\text{in}} = Q, \quad \text{for } t > t_{\text{ramp}}, \tag{29}$$

where t_{ramp} is the time length of the linear ramp, and it is set to $t_{\text{ramp}} = 1$ s in all test cases. The reference value for the inflow condition, $Q = 2.67\,\text{cm}^3\,\text{s}^{-1}$, is obtained through the relationship between flow rate and vessel areas, derived from measurements in internal carotid and vertebral arteries [5].

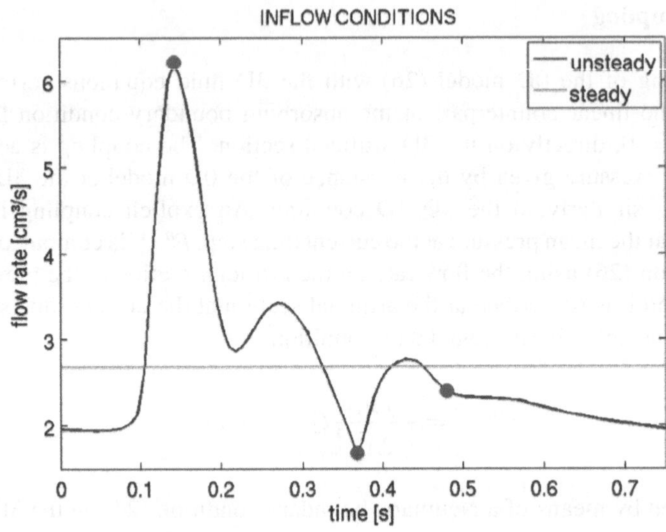

Fig. 6 Steady and unsteady inflow flux profiles versus time. The *points* indicate locations of "peak systole," "minimum diastole" and "mean diastole" used in the discussion section

In the case of pulsatile flow simulations for the idealized geometry, a periodic wave inflow boundary condition is imposed, representing a realistic heart beat waveform, in the carotid, with a mean flux equal to the steady-state flux value. The steady and unsteady inflow flux profiles with respect to time are illustrated in Fig. 6.

Convergent steady-state and pulsatile solutions were identified by checking that the difference between two consecutive time steps (steady case) or two consecutive cycles (unsteady case) was negligible. In the case of the steady state solutions this convergence is of the order of 10^{-7}, while for the unsteady case all the results presented correspond to the 12th cycle where the convergence is of the order of 10^{-6}.

The steady-state simulations were carried out using a time step of 0.01 s, while the pulsatile used a time step of 0.0075 s, corresponding to a hundredth of the heart beating period. Moreover, a time step of 0.5×10^{-4} s was taken when the coupling with the 1D hyperbolic model is used as outflow boundary condition. For both 1D and 0D models, the β parameters used were determined through expression $\beta = \frac{\sqrt{\pi} h_0 E}{1-\xi^2}$, where the thickness of the wall h_0 was set to 10% of the vessel radius, the Young modulus was set to $E = 10^5$, and the Poisson ratio was set to $\xi = 0.5$, assuming the artery wall is incompressible.

A volumetric mesh of about 0.85 M tetrahedra was created for the anatomically realistic geometry, corresponding to a graded mesh with element size of 0.016 cm within the aneurysm, and maximum size of 0.04 cm in the upstream and downstream sections. The idealized geometries are planar with the parent vessel radius of

0.25 cm, the side-branch radius of 0.075 cm, and the aneurysm radius of 0.4 cm. The side-branch length is 1.2 cm. The idealized geometry volumetric mesh is composed of approximately 0.5 *M* tetrahedral elements, with elements of size 0.02 cm.

5 Discussion

5.1 Idealized Geometry

Hemodynamics inside the idealized aneurysm was studied using the Newtonian and Carreau fluid models, both in steady and unsteady inflow regimes, including and excluding a side-branch within the aneurysm, and prescribing four different types of outflow boundary conditions on the side-branch: traction-free (TF), no-slip (NS), 3D–1D coupling (1D), and 3D–0D coupling (0D). At the outflow section of the main vessel a traction-free boundary condition was always prescribed.

The differences between the Newtonian and Carreau solutions, for both steady and unsteady regimes, are depicted in Fig. 7 (velocity) and Fig. 8 (WSS). The geometry considered for these results is the idealized with hole (clipped side-branch), and the traction-free condition at this outflow boundary. The maximum

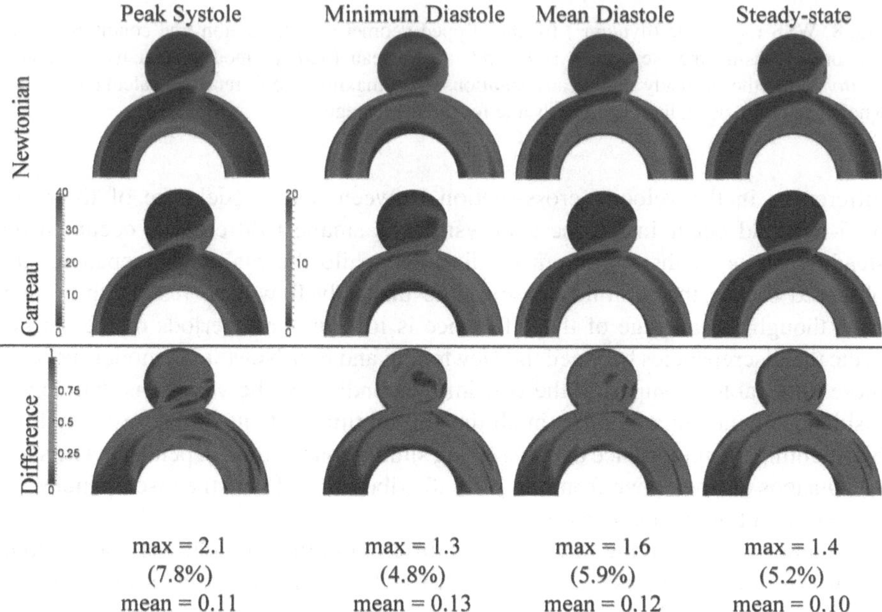

Fig. 7 Velocity magnitude (cm/s) for the clipped geometry with traction-free conditions at the side-branch outflow, using the Newtonian (*top*) and the Carreau (*middle*) models, and its differences (*bottom*), for the unsteady and steady solutions. The maximum difference is calculated for the cross-section, using the maximum value for the percentage

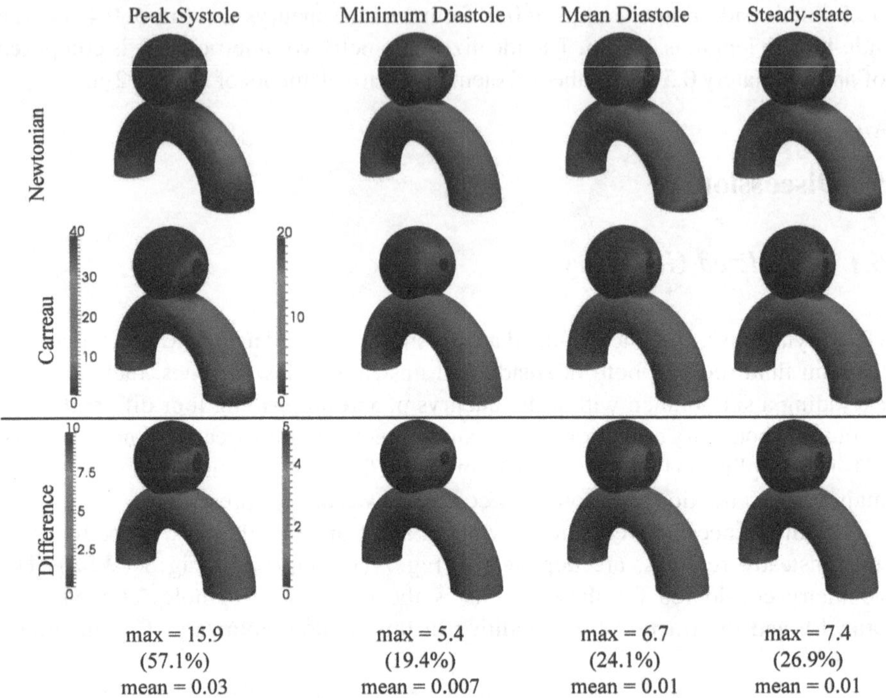

Fig. 8 WSS magnitude (dyn/cm^2) for the clipped geometry with traction-free conditions at the side-branch, using the Newtonian (*top*) and the Carreau (*middle*) models, and its differences (*bottom*), for the unsteady and steady solutions. The maximum difference is calculated for the whole geometry, using the maximum value for the percentage

differences in the velocity cross-section between both models are of the order of 5–8 % and occur inside the aneurysm. The smallest differences occur on the steady case or at the minimum of diastole, while the higher discrepancies are observed during the systolic phase of the unsteady flow. The results show that, even though the average of the difference is low, in some periods of the cardiac cycle the discrepancies between the Newtonian and non-Newtonian models become more noticeable. Comparing the two inflow conditions, the variations that appear inside the aneurysm are higher in all the chosen time instants of the unsteady flow, highlighting the importance of considering simulations as time dependent. The same conclusions can be drawn from the WSS distribution, yet here the discrepancies are more relevant in the main vessel.

In order to analyze the effects of the different outflow conditions, the configuration with the side-branch and a traction-free boundary condition at its outflow is compared with the clipped geometry using all the considered boundary conditions, see Fig. 9 and Table 2. In this particular geometry the side-branch has a substantial influence on the solution, not only due to its location, inside the aneurysm, but also due to the large percentage of flow that enters the branch. The traction-free condition

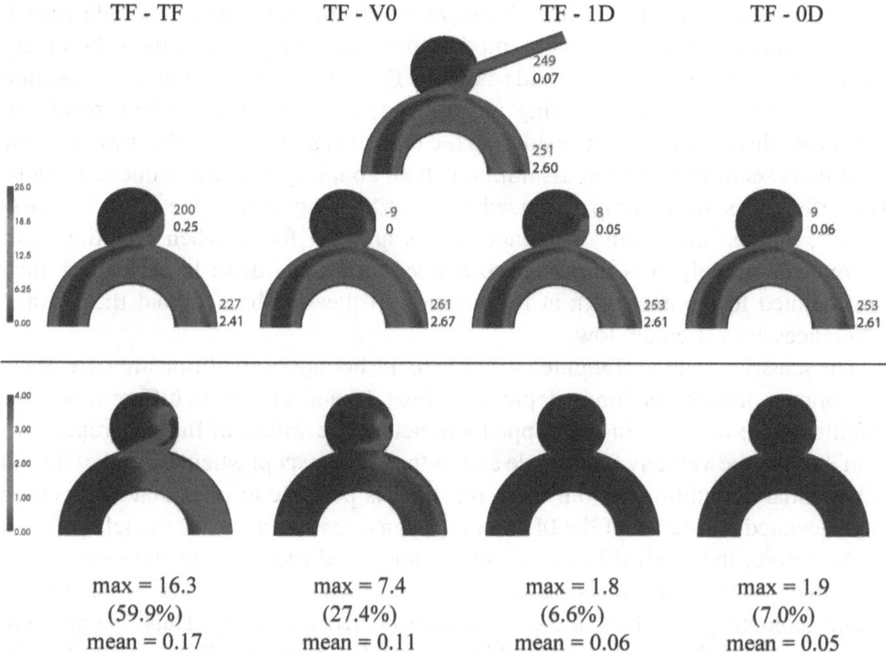

Fig. 9 Velocity magnitude (cm/s) for stead-state simulations of the geometry with side-branch and traction-free boundary condition and the clipped geometry with different boundary conditions and its differences. The maximum difference is calculated on the cross-section, using the maximum value for the percentage. The values in the figure are the pressure drop with the inflow (*red*) and flow rate (*blue*)

Table 2 WSS magnitude differences (dyn/cm^2) for the geometry with side-branch and traction-free boundary condition and the clipped geometry with different boundary conditions

TF - TF	TF - V0	TF - 0D	TF - 1D
Max = 23.5	Max = 13.6	Max = 15.0	Max = 15.0
(81%)	(46.8%)	(51,7%)	(51.7%)
Mean = 4.3e−4	Mean = 8.2e−4	Mean = 4.4e−4	Mean = 4.8e−4

The maximum difference is calculated on the whole geometry, using the maximum value for the percentage

was chosen to be imposed at the end of the side-branch for comparison purposes, here considering the fluid to be fully developed at the branch outflow. From the velocity results of Fig. 9 it is possible to infer that the two reduced models are good approximations of the side-branch, since the differences between imposing these reduced models directly in the clipped configuration and accounting for the side-branch are very small. The disparity between the pressure drops obtained using the reduced 1D and 0D models and the side-branch, are related to the pressure drop across the side-branch. In fact, the pressure drop across the side-branch is 258 dyn/cm^2, which is approximately that found for the reduced models.

These pressure drops, together with the values of the flow rate at the side-branch outflow, indicate that the reduced models provide appropriate outflow boundary conditions, accounting for the side-branch. Prescribing a traction-free condition on the hole section, or neglecting its existence by a $\mathbf{u} = \mathbf{0}$ condition, results in significant discrepancies by considering the branched geometry. Thus, these outflow conditions seem to be worse assumptions than coupling with the reduced models. The differences are more pronounced in the WSS map than in the velocity cross-section, but the minimum difference values are still found when coupling with the reduced models. It is important to notice that despite these larger values, they are confined to the aneurysm at the location of the side-branch, and the average differences are extremely low.

The sensitivity of the computed solution to the boundary condition imposed at the side-branch outflow section is depicted in Figs. 10 and 11, where different outflow conditions are imposed in the clipped geometry. The values of the differences are high, both in the velocity magnitude and in the WSS, except when using the 1D and 0D boundary conditions. From these results it is possible to infer that in this case the calculated resistance of the 0D model is consistent with the 1D model.

As before, the WSS differences are mainly localized close to the side-branch base. In this region the values are very high, yet when considering the average in the whole geometry, the values of the differences decrease abruptly. Thus, as expected, the influence of the side-branch and its outflow boundary condition is particularly important when the side-branch is located very close or within the aneurysm.

Figure 12 displays the differences that exist between the steady-state solution and the time average of the unsteady solution, both for the velocity cross-section and the WSS distribution, in the case of the hole geometry coupled with the 1D model. It is possible to observe that the differences are very small, especially when considering the average. At first sight this could indicate a great resemblance between the steady and unsteady solutions. However, comparing the unsteady solutions of the clipped geometry coupled with the 1D model and the branched geometry with traction-free boundary condition, the differences are magnified at several instants of the cardiac cycle (see Fig. 13). The comparison of these two cases reveals minimal differences for the steady-state inflow conditions, as shown in Fig. 9. Nevertheless, these differences are again significantly higher at different instants of the cardiac cycle, as plotted in Fig. 13. Exhaustive conclusions cannot be drawn from steady-state solutions, since even the average difference of the unsteady solutions of the clipped geometry with the 1D boundary condition and the branched geometry with the traction-free boundary condition is greater than the one for the steady-state. This demonstrates, once again, the relevance of considering unsteady simulations, especially when studying the influence of boundary conditions on the numerical solutions.

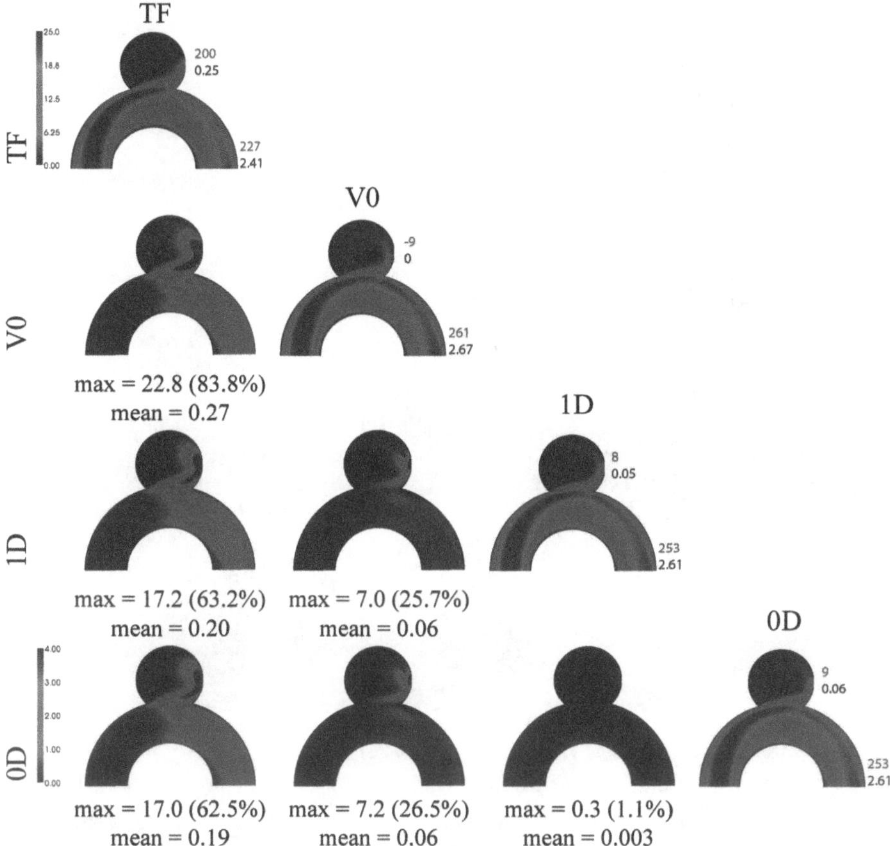

Fig. 10 Velocity magnitude (cm/s) for steady-state simulations of the clipped geometry using different boundary conditions and its differences. The maximum difference is calculated for the cross-section, using the maximum value for the percentage. The values in the figure are the pressure gradient (up) and flow rate (down)

5.2 *Anatomically Realistic Geometry*

The anatomically realistic patient-specific geometry of a cerebral aneurysm (Figs. 1 and 2) was used to study the impact of changing the fluid rheological model. The simulations in this case were performed under a steady inflow regime. As expected also from the results in the idealized geometry, variations in the fluid model influence the computational solution. In Fig. 14 the results for the velocity magnitude, the WSS and WSSG (spatial WSS gradient) for both Newtonian and Carreau solutions are shown, as well as their differences. The discrepancies between the two models reach 11 % in the velocity magnitude inside the aneurysm. The WSS and WSSG differences are even more significant, 19 % and 25 %, respectively, located at the neck of the aneurysm. These results indicate that the use of a constant

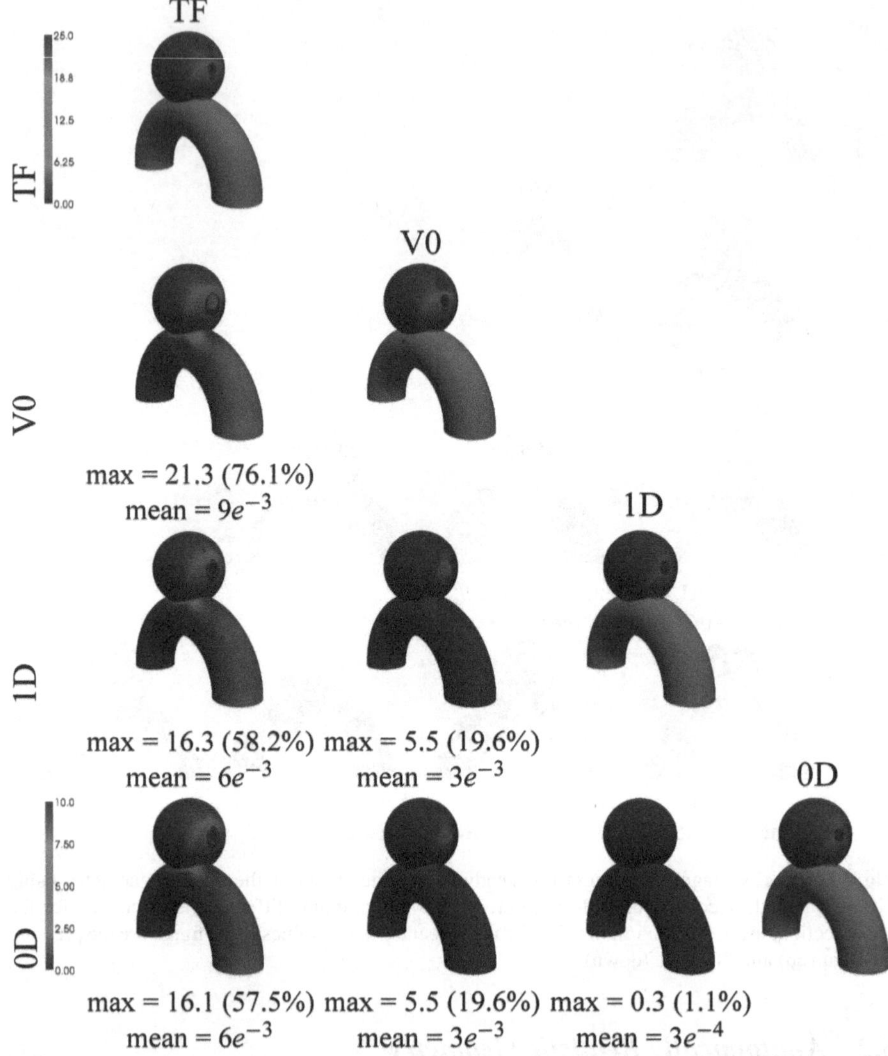

Fig. 11 WSS magnitude (dyn/cm^2) for steady-state simulations of the clipped geometry using different boundary conditions and its differences. The maximum difference is calculated for the whole geometry, using the maximum value for the percentage

viscosity results in overestimated values for the hemodynamic indicators under analysis. Given the special physiological relevance and correlation of low WSS to disease in arteries, the choice of a non-Newtonian model could yield different clinical evaluations. The particle tracing depicted in Fig. 15, where the seeding locations were maintained, shows that the flow structures inside the aneurysm are similar for the Newtonian and non-Newtonian cases; however different locations of

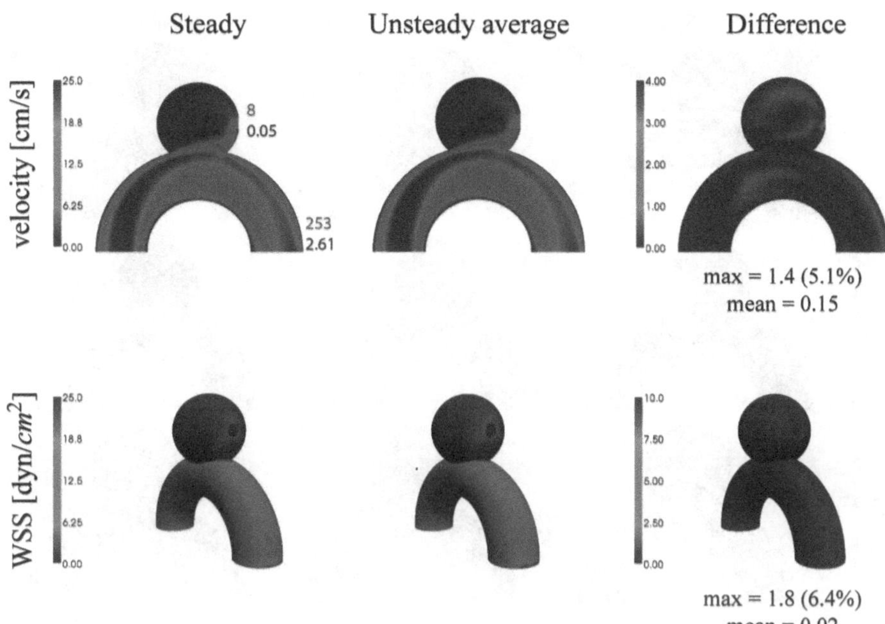

Fig. 12 Velocity (cm/s) and WSS (dyn/cm^2) magnitude for the clipped geometry coupled with the 1D model with time averaged unsteady and steady flow regimes, and corresponding differences

jet impingement and size of the swirling motion are apparent. This indicates that the differences due to the rheological model choice do not only affect the near-wall region, but in cases of complex recirculating flow the free-stream field may also be effectively altered.

6 Conclusions

Two types of geometries were considered: idealized configurations of a curved vessel with an aneurysm, where a side-branch in the aneurysm was included as a tube or a hole, and an anatomically realistic geometry of a cerebral aneurysm with side-branches removed.

Regarding the idealized geometries, both steady and unsteady inflow regimes were considered. Several boundary conditions were prescribed at the outflow section of the side-branch in the aneurysm. Results indicate a large influence of the outflow conditions on the entire domain, but more pronounced near the side-branch base. The reduced 1D and 0D models seem to be fair approaches to take into account the presence of the side-branches, providing appropriate pressure drops. The importance of considering the side-branch increases when located close or in the aneurysm, as it was the case of the idealized geometry.

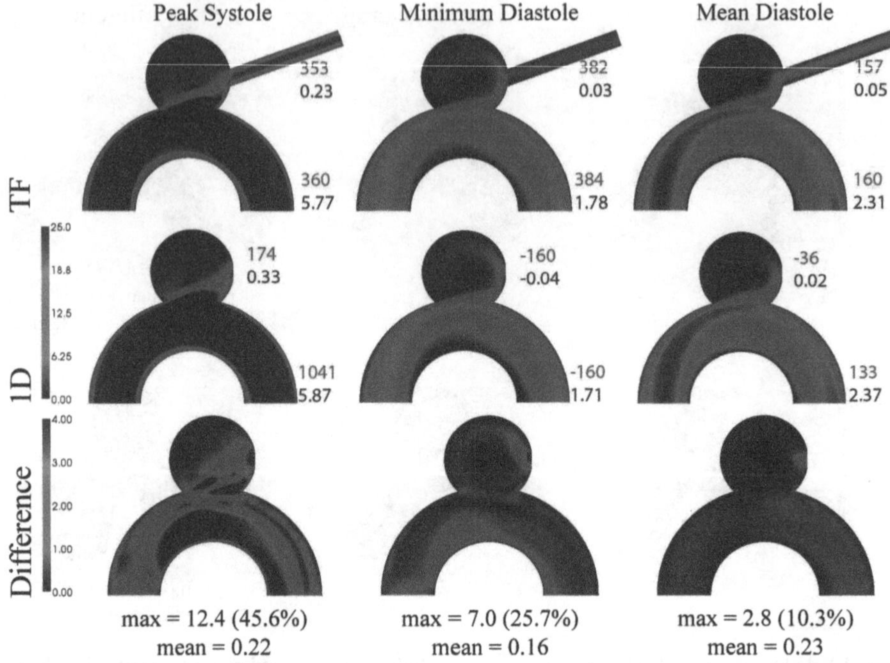

Fig. 13 Velocity magnitude (cm/s) for the geometry with side-branch and traction-free boundary condition and the clipped geometry coupled with the 1D model and its differences, in unsteady flow regime

These conclusions might not be straightforwardly extended to the anatomically realistic geometries, since in such cases the side-branches are not straight tubes. Work is ongoing in applying the approaches here presented to a significant number of patient-specific geometries. The traction-free outflow condition on the clipped geometry compared poorly to the solution of the tube side-branch with a fully developed flow. The differences between steady and unsteady inflow conditions are small and localized when the time averaged solution is compared. However, at specific time instants of the cardiac cycle those differences are much more significant, specially during systole. The Newtonian and Carreau shear-thinning fluid models were used in both realistic and idealized geometries. In both cases differences between the two rheological models are apparent, but less emphatic than the influence of the boundary conditions. Also, the results of the WSS and the WSSG show higher discrepancies between the two blood flow models. The results here presented are preliminary in the sense that they should be complemented with extensive studies in patient-specific geometries in order to obtain conclusions in more general scenarios.

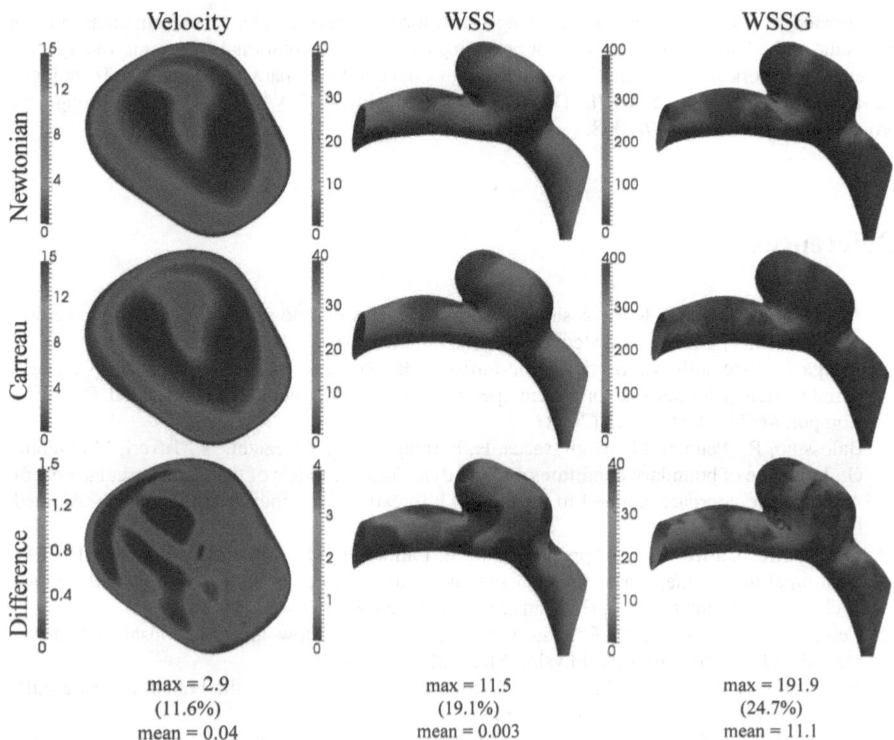

Fig. 14 Results in the realistic geometry for the Newtonian (*top*) and Carreau (*middle*) solutions and their differences (*bottom*). Velocity magnitude (cm/s) in the cross-section depicted in Fig. 2 (*left*), WSS (dyn/cm^2) (*middle*), and WSSG (dyn/cm^2) (*right*). The differences are given in the cross-section for the velocity and over the entire surface for WSS and WSSG. The percentage is calculated using values of the inflow section

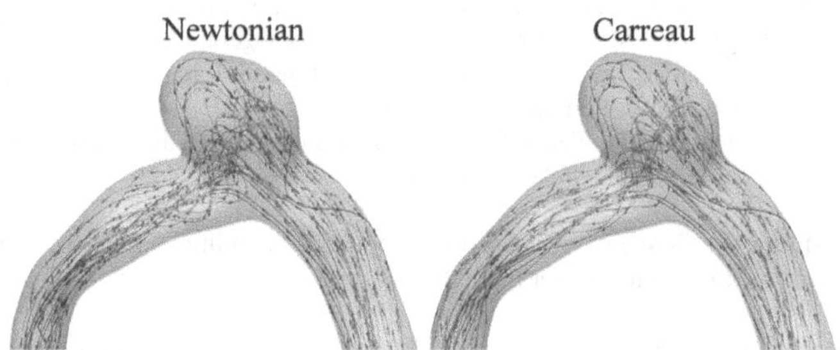

Fig. 15 Particle trace of the Newtonian (*left*) and Carreau (*right*) solutions. Particles are selected at the same location

Acknowledgments We greatly acknowledge Prof. Jorge Campos and his team from the Faculty of Medicine of the University of Lisbon, for providing us the in vivo rotational CTA scans of a specific patient. This work has been partially funded by FCT (Fundação para a Ciência e a Tecnologia, Portugal) through grants SFRH/BPD/34273/2006 and SFRH/BPD/44478/2008 and through the project UT Austin/CA/0047/2008.

References

1. Anand, M., Rajagopal, K.R.: A shear-thinning viscoelastic fluid model for describing the flow of blood. Int. J. Cardiovasc. Med. Sci. **4**(2), 59–68 (2004)
2. Antiga, L., Piccinelli, M., Botti, L., Ene-Iordache, B., Remuzzi, A., Steinman, D.A.: An image-based modeling framework for patient-specific computational hemodynamics. Med. Biol. Eng. Comput. **46**(11), 1097–1112 (2008)
3. Balossino, R., Pennati, G., Migliavacca, F., Formaggia, L., Veneziani, A., Tuveri, M., Dubini, G.: Influence of boundary conditions on fluid dynamics in models of the cardiovascular system: A multiscale approach applied to the carotid bifurcation. Comput. Meth. Biomech. Biomed. Eng. **12**(1) (2009)
4. Cebral, J.R., Castro, M.A., Appanaboyina, S., Putman, C.M., Millan, D., Frangi, A.F.: Efficient pipeline for image-based patient-specific analysis of cerebral aneurysm hemodynamics: Technique and sensitivity. IEEE Trans.Med. Imag. **24**(4), 457–467 (2005)
5. Cebral, J.R., Castro, M.A., Putman, C.M., Alperin, N.: Flow–area relationship in internal carotid and vertebral arteries. Physiol. Meas. **29**, 585 (2008)
6. Formaggia, L., Veneziani, A.: Reduced and multiscale models for the human cardiovascular system. Lecture Notes VKI Lecture Series **7** (2003)
7. Formaggia, L., Moura, A., Nobile, F.: On the stability of the coupling of 3D and 1D fluid-structure interaction models for blood flow simulations. Math. Model. Numer. Anal. **41**(4), 743–769 (2007)
8. Gambaruto, A.M., Peiró, J., Doorly, D.J., Radaelli, A.G.: Reconstruction of shape and its effect on flow in arterial conduits. Int. J. Numer. Meth. Fluid. **57**(5), 495–517 (2008)
9. Gambaruto, A.M., Janela, J., Moura, A., Sequeira, A.: Sensitivity of hemodynamics in patient specific cerebral aneurysms to vascular geometry and blood rheology. Math. Biosci. Eng. **8**(2), 409–423 (2011)
10. Goljan, E.F.: Rapid Review Pathology. Mosby/Elsevier, Philadelphia (2010)
11. Hassan, T., Timofeev, E.V., Saito, T., Shimizu, H., Ezura, M., Matsumoto, Y., Takayama, K., Tominaga, T., Takahashi, A.: A proposed parent vessel geometry based categorization of saccular intracranial aneurysms: Computational flow dynamics analysis of the risk factors for lesion rupture. J. Neurosurg. Pediatr. **103**(4) (2005)
12. Heywood, J.G., Rannacher, R., Turek, S.: Artificial boundaries and flux and pressure conditions for the incompressible Navier-Stokes equations. Int. J. Numer. Meth. Fluid. **22**(5), 325–352 (1996)
13. Janela, J., Moura, A., Sequeira, A.: Absorbing boundary conditions for a 3D non-Newtonian fluid–structure interaction model for blood flow in arteries. Int. J. Eng. Sci. **48**(11), 1332–1349 (2010)
14. Krex, D., Schackert, H.K., Schackert, G.: Genesis of cerebral aneurysms–an update. Acta Neurochir. **143**(5), 429–449 (2001)
15. Ku, D.N.: Blood flow in arteries. Annu. Rev. Fluid Mech. **29**(1), 399–434 (1997)
16. Moura, A.: The geometrical multiscale modelling of the cardiovascular system: Coupling 3D and 1D FSI models. PhD thesis, Politecnico di Milano (2007)
17. Nunes, D., Ramalho, S.: 1D hyperbolic models for blood flow in arteries. Internal Report, CEMAT (2009)

18. Quarteroni, A.M., Valli, A.: Numerical Approximation of Partial Differential Equations. Springer, Berlin (2008)
19. Ramalho, S., Moura, A., Gambaruto, A.M., Sequeira, A.: Sensitivity to outflow boundary conditions and level of geometry description for a cerebral aneurysm. Int. J. Numer. Meth. Biomed. Eng. **28**(6-7), 697–713 (2012)
20. Rinkel, G.J.E., Djibuti, M., Algra, A., Van Gijn, J.: Prevalence and risk of rupture of intracranial aneurysms: A systematic review. Stroke **29**(1), 251 (1998)
21. Robertson, A.M.: Review of Relevant Continuum Mechanics. In: Galdi, G.P., Rannacher, R., Robertson, A.M., Turek, S. (eds.) Hemodynamical Flows: Modeling, Analysis and Simulation, pp. 1–62. Birkhäuser, Boston (2008)
22. Robertson, A.M., Sequeira, A., Kameneva, M.V.: Hemorheology. In Galdi, G.P., Rannacher, R., Robertson, A.M., Turek, S., editors, Hemodynamical Flows: Modeling, Analysis and Simulation, pages 63–120. Birkhäuser, Boston (2008)
23. Sazonov, I., Yeo, S.Y., Bevan, R.L.T., Xie, X., van Loon, R., Nithiarasu, P.: Modelling pipeline for subject-specific arterial blood flow - a review. Int. J. Numer. Meth. Biomed. Eng. (2011) DOI: 10.1002/cnm.1446.
24. Sequeira, A., Janela, J.: An overview of some mathematical models of blood rheology. In: Pereira, M.S. (ed.) A Portrait of Research at the Technical University of Lisbon, pp. 65–87. Springer, Berlin (2007)
25. Sequeira, A., Moura, A., Janela, J.: Towards a geometrical multiscale approach to non-Newtonian blood flow simulations. In: Sequeira, A., Rannacher, R. (eds.) Advances in Mathematical Fluid Mechanics - dedicated to G.P. Galdi on his 60th birthday, chapter 295–309. Springer, Berlin (2009)
26. Sezgin, M., Sankur, B.: Survey over image thresholding techniques and quantitative performance evaluation. J. Electron. Imag. **13**(1), 146–165 (2004)
27. Tu, C., Deville, M.: Pulsatile flow of non-Newtonian fluids through arterial stenosis. J. Biomech. **29**(7), 899–908 (1996)
28. Wulandana, R., Robertson, A.M.: An inelastic multi-mechanism constitutive equation for cerebral arterial tissue. Biomech. Model. Mechanobiol. **4**(4), 235–248 (2005)

Part III
Cancer Modeling

The Steady State of Multicellular Tumour Spheroids: A Modelling Challenge

Antonio Fasano and Alberto Gandolfi

1 Introduction

Cells from different tumour cell lines can be grown in vitro to form spheroidal masses, called multicellular tumour spheroids, currently considered valuable experimental models of avascular tumours [35, 37, 50, 51, 58]. Multicellular tumour spheroids have been extensively investigated in that they provide a useful model to assess the effects of oxygenation and nutrition on growth, as well as the effects of treatments with drugs and radiation.

During the spheroid growth the fraction of proliferating cells decreases, and when cells in the inner region become deprived of oxygen, glucose and other nutrients, and/or metabolic waste accumulates, cell death occurs. Thus, in a late stage of growth, the spheroids consist of an outer viable rim (whose thickness takes values from about 100 μm to 250 μm) surrounding a central necrotic region. The spheroid growth is initially exponential and then it tends to saturate. Examples of reaching the stationary state (with final diameter of 1–3 mm) have been reported [30, 32].

Many mathematical models have been proposed to describe the spheroid evolution, from simple growth models such as Gompertzian models [46], to models that take into account the internal spheroid structure using either continuum or discrete approaches (see [6, 43] for extensive reviews). In almost all these models,

A. Fasano (✉)
Dipartimento di Matematica "U. Dini", Universita' di Firenze,
Viale Morgagni 67/A, 50134 Firenze, Italy

Istituto di Analisi dei Sistemi ed Informatica "A. Ruberti" - CNR,
Viale Manzoni 30, 00185 Roma, Italy
e-mail: fasano@math.unifi.it

A. Gandolfi
Istituto di Analisi dei Sistemi ed Informatica "A. Ruberti" - CNR,
Viale Manzoni 30, 00185 Roma, Italy
e-mail: alberto.gandolfi@iasi.cnr.it

U. Ledzewicz et al., *Mathematical Methods and Models in Biomedicine*, Lecture Notes on Mathematical Modelling in the Life Sciences, DOI 10.1007/978-1-4614-4178-6_7,

cell proliferation and death are assumed to depend on the concentration of a single critical chemical (generally oxygen), diffusing from the external medium into the spheroid mass. According to this view, the boundary between the viable rim and the necrotic core is often defined as the level set of the oxygen concentration corresponding to a given threshold. However, the formation mechanism of the central necrotic region in multicellular spheroids is a much debated and a not yet well-understood process. The diffusion of both glucose and oxygen has been included in the spheroid models proposed in [21, 39, 55]. More recently, the cell energy metabolism, i.e. the intracellular ATP production involving glucose, oxygen and lactate, has been incorporated in models of spheroids [10, 11, 61], as well as in various models of tumour growth [7, 8, 33, 57]. In [10, 11], the formation of the spheroid necrotic region was described by assuming that cell death occurs when the ATP production rate falls below a critical value. The possible role of acidity in determining the onset of the central necrosis in tumours was investigated in [13].

With only a few exceptions [1, 2, 48], the final attainment of a steady state during the sheroid growth has been associated, in the modelling literature, with a loss of volume from the necrotic core that balances the new cellular volume created in the viable rim by cell proliferation. The experimental evidence of this mechanism, however, is indirect and relies on the observation of active cell proliferation even when the growth rate of the spheroid is very small or vanishes [27]. From a biological point of view the way the necrotic core is modelled may look to be a minor question. Nevertheless, the structure attributed to the necrotic zone has a crucial influence on the general mechanical behaviour of the entire spheroid and hence on its evolution.

In the present chapter, we first give a brief survey (Sect. 2) of the modelling options proposed in the literature for describing the necrotic core and for explaining the balance between live and dead cells in the steady state of tumour spheroids. Next we review our recent work on this topic (Sects. 3–5). We adopted the two-fluid scheme, in which the extracellular fluid and cells are schematized as incompressible fluids (inviscid and viscous, respectively), and we introduced several free boundaries, having the role of sharp transition interfaces marking a change in the state of cells. In particular, the necrotic region was subdivided in a shell of dead cells surrounding a purely liquid core. The advantage of sharp interfaces is the possibility of writing the velocity field of each fluid in an explicit way. The equilibrium size of the spheroid (if it exists) can then be obtained through the analysis of the stress accompanying the flow of the two fluids, generated by proliferation. Although the two-fluid approach is certainly naïve, it allowed to reach meaningful quantitative results, with no need of postulating any mechanism for the removal of necrotic material. Some concluding remarks are given in Sect. 6.

2 A Brief History of the Necrotic Core Modelling

In the influential paper by Greenspan [34], the necrotic core, composed of "dead cells and cellular material in various stages of disintegration" is viewed as a "jelly-like" material "capable of supporting the pressure exerted on it by the outer viable layers". This "solid" debris, while keeping a constant density, continually dissolves into "simpler permeable compounds" capable of moving easily through the outer region of the spheroid. Because of this motion, a volume loss occurs from the region of necrosis. Although it does not play a direct role in the model, a surface tension is postulated to maintain the compactness of the aggregate. The degradation of the necrotic material occurs according to a first-order kinetics, with uniform degradation rate constant through the whole core. In Greenspan's model, however, the volume loss is not the only mechanism allowing the attainment of a steady state: another important role is played by a mitosis inhibitor which is supposed to be produced at a constant rate in the necrotic core or as a waste from living cells (for comments about this conjecture, see [3]). Cell death occurs when the oxygen concentration decreases to some critical threshold (a feature incorporated in many subsequent models) and mitosis stops when the inhibitor concentration raises above the inhibition threshold. According to this picture, the model contains two free boundaries. A substantial gap of the model is the absence of any mechanical explanation of how the postulated material loss from the necrotic core can take place. In other words, a study of the flow of the various components based on the general principles of mechanics is missing. This kind of analysis came much later in cancer modelling.

The Greespan's viewpoint was largely adopted in the following years (see Deakin [26], MacElwain and Ponzo [47], Maggelakis and Adam [44], Adam and Maggelakis [3], Byrne and Chaplain [17], Cui and Friedman [23], Bertuzzi et al. [12]), and in the next section we will illustrate a simple model based on it. Also the model studied by Cui and Friedman in [24] describes the central zone of a spherical tumour (although without a sharp interface) as essentially full of dead cells subjected to degradation according to a uniform rate constant.

A different mechanism for the attainment of a stationary state during the spheroid growth was proposed by McElwain and Morris [48]. These authors, following Burton [16], assumed the necrotic material immune from degradation (at least in the time horizon of interest) and supposed that the relevant volume loss happens in the inner viable rim via cell apoptosis and phagocytosis of the resulting apoptotic bodies by the viable neighbouring cells. This mechanism then accounts for some experimental observations of stationary spheroids without central necrosis [59]. Volume loss was totally absent in the models by Adam [1] and Adam and Maggelakis [2], where instead a diffusing endogenous mitotic inhibitor, possibly produced inside the necrotic region [2], eventually blocks the proliferation of all the cells. Some different views of the necrotic region derived from the explicit modelling of the multi-phase nature of the cell aggregates. Ward and King [62] distinguished in the spheroid the viable cells (that can occupy a varying local volume fraction) and a diffusible "cellular material" originated by the immediate degradation of cells upon death (a death rate is introduced depending on the concentration of a critical chemical). This material may be reused to sustain the cell proliferation, so that growth saturation can be achieved. The necrotic core is then a zone deprived of cells but occupied by the cellular material, and the volume loss from the necrotic core is given by the diffusing flow of such a material towards the outer region. This view was reconsidered in [60], where, however, the lack of reutilization of the material coming from cell disintegration prevents the saturation of growth.

Following some ideas of the model in [54], Landman and Please [41] described the spheroid as a liquid–cells mixture whose mechanics is borrowed from a model for suspensions [42]. The force balance equation is explicitly included together with the mass balance, and not only the liquid, but also the cell component has isotropic stress tensor. Thus, stresses are expressed by two pressures: the liquid pressure and the intercellular pressure. The net proliferation and death rate are expressed as a function of oxygen concentration, switching sign across a critical threshold. Cells immediately degrade into liquid after death (then all cells are living cells), whereas maintaining a constant local volume fraction (and a compact arrangement) until the cellular pressure is greater than the liquid pressure. Complete mass exchange between liquid and cellular phase occurs during cell proliferation and at cell death. An interesting feature introduced in [41] is claiming that when the cellular pressure tends to drop below the liquid pressure, cells detach from the compact arrangement and "float" in the liquid. Thus the necrotic core is essentially described as a liquid with a small fraction of viable cells committed to death. This fraction vanishes at the steady state, when the necrotic core is purely liquid. However, the existence of the steady state is related to the presence of a suitable surface tension: if the surface tension is insufficient, the spheroid eventually will grow linearly.

A two-phase model based on a more complex mechanics was proposed by Byrne and Preziosi [18] (see also [4, 15, 19]). In this "two-fluid" model, cells are represented by a viscous fluid whose pressure contains an extra-term depending on the cell volume fraction and describing the cell-to-cell interaction, whereas the extracellular liquid is represented by an inviscid fluid. Again, in this model cells degrade instantaneously into liquid after death, and complete mass exchange between liquid and cellular phase occurs at cell proliferation and cell death, which are under the control of a critical nutrient. At the steady state, the local volume fraction of (living) cells continuously decreases towards the centre of the spheroid, at which it does not vanish, so that the necrotic core is mimicked by a region in which the density of living cells is reduced. A similar view is also present in the model by Ambrosi and Preziosi [5], in which the cell component is represented by a visco-elasto-plastic fluid, and in the model proposed by Cristini et al. [25], which was focussed on the derivation of the interaction potential.

In two recent papers [28, 29], we have proposed that at the steady state the necrotic core (N) may be partitioned into two zones: a "solid" domain NS where dead cells are supposed to keep the mechanical properties they had before death and the same volume fraction and an inner core NL simply liquid. This partition follows from the assumption that cell membrane degradation occurs after a fixed time from cell death, and that this degradation marks the transition from "solid" to "liquid".

Some support to the NS/NL partition comes from nuclear magnetic resonance (NMR) measurements of the self-diffusion of water in EMT-6 spheroids [52]. These measurements have shown that whereas in the viable rim water appears confined into two compartments with different diffusion coefficients (intracellular and extracellular water), the central part of the necrotic core looks like a single

Table 1 Main features of the description of the necrotic core (N) in some literature models

Model	Dead cell degradation	Necrotic core structure	Reaching steady state	Mechanical aspects
[34]	First-order kinetics, mitotic inhibitor production	Solid	Volume loss from N[a] + mitotic inhibition	Mono-phase model, kinematic approach
[48]	Absent	Solid	Phagocytosis of apoptotic cells	Mono-phase model, kinematic approach
[2]	Absent, mitotic inhibitor production	Solid	Proliferation blocked by mitotic inhibitors diffusing from N	Mono-phase model, kinematic approach
[62]	Immediate	Reservoir of diffusible "cellular material"	Volume loss from N[b]	Two-phase model (one phase diffusible)
[41]	Immediate	Liquid at the steady state	Volume loss from N[c]	Two-fluid model, momentum balance, surface tension (needed for equilibrium)
[18]	Immediate	Reduced density of living cells	Volume loss from N[c]	Two-fluid model, momentum balance, surface tension may be included
[12]	After a Gamma-distributed time	Solid	Volume loss from N[a]	Multi-phase model, kinematic approach
[29]	After a fixed time	Includes a liquid core within a solid shell	Volume loss from N[c]	Two-fluid model, kinematic approach + stress analysis, surface tension (needed for equilibrium)

Volume loss: [a]Postulated [b]By diffusion [c]Supported by a fluid dynamical theory

compartment characterized by a single diffusion coefficient. Moreover, NMR imaging evidenced an intermediate zone between the viable rim and the centre of the necrotic region, which still appears to have two diffusion compartments, although the fraction of volume of the diffusion-restricted compartment was found lower than the corresponding fraction in the viable rim.

The presence of an inner liquid core precludes, however, the possibility of studying the spheroid on a pure kinematic basis, and requires the introduction of a mechanical scheme. In the above cited papers, the two-fluid model was revisited to investigate the spheroid steady state .

The main features of the necrotic core description in some literature models are summarized in Table 1.

3 The Internal Structure of a Multicellular Spheroid

We assume the spheroid be a mixture of two components, cells, and extracellular liquid, whose velocities are denoted respectively by $\mathbf{u}$ and $\mathbf{v}$. The local volume fractions of living cells, dead cells and extracellular liquid are denoted by ν_C, ν_N and ν_E respectively. Assuming no voids, we have $\nu_C + \nu_N + \nu_E = 1$. We will consider oxygen as the limiting nutrient, so oversimplifying the description of metabolism in order to concentrate on the mechanical aspects. In spherical symmetry, we denote by $\sigma(r,t)$ the oxygen concentration, r being the radial distance from the spheroid centre and t the time. Let $R(t)$ be the outer radius of the spheroid. To gain some conceptual simplification, it is convenient to divide the spheroid into spherically symmetric domains, separated by sharp interfaces. The partition of the spheroid is obtained by introducing thresholds for the oxygen concentration. More precisely, we introduce a proliferation threshold σ_{P} and a necrosis threshold $\sigma_{\mathrm{N}} < \sigma_{\mathrm{P}}$ assuming that all cells die when the oxygen concentration reaches σ_{N}. So, all cells in the region $P = \{r : \sigma(r,t) > \sigma_{\mathrm{P}}\}$ are proliferating, while the cells in the region $Q = \{r : \sigma_{\mathrm{N}} < \sigma(r,t) < \sigma_{\mathrm{P}}\}$ are quiescent. The necrotic region is given by $N = \{r : \sigma(r,t) = \sigma_{\mathrm{N}}\}$. We will assume that in P and in Q it is $\nu_C = \nu = constant$. For simplicity, we take that all cells in P consume oxygen at the same rate and proliferate with a common constant proliferation rate χ. The above scheme of a spheroid includes two interfaces:

- $r = \rho_P$, the $P - Q$ interface
- $r = \rho_N$, the $Q - N$ interface

The determination of ρ_P, ρ_N requires the solution of the following *oxygen diffusion–consumption problem*: given the radius R of the spheroid, find a piecewise twice continuously differentiable function $\sigma(r)$, and ρ_P, ρ_N, such that

$$D_{O_2}\Delta\sigma(r) = f(\sigma(r))\nu, \qquad \text{in } P, \tag{1}$$

$$D_{O_2}\Delta\sigma(r) = \frac{1}{m}f(\sigma(r))\nu, \qquad \text{in } Q, \tag{2}$$

$$\sigma(R) = \sigma^*, \tag{3}$$

$$\sigma(\rho_P) = \sigma_{\mathrm{P}}, \tag{4}$$

$$\sigma(\rho_N) = \sigma_{\mathrm{N}}, \tag{5}$$

$$\sigma_r(\rho_N) = 0. \tag{6}$$

In the above equations, D_{O_2} is the oxygen diffusivity in the spheroid, $\Delta = \frac{1}{r^2}\frac{\mathrm{d}}{\mathrm{d}r}\left(r^2\frac{\mathrm{d}}{\mathrm{d}r}\right)$ is the Laplacian operator, $f(\sigma(r))$ is the consumption rate per unit cell volume in P, reduced by the factor $1/m < 1$ in Q, and σ^* is the given oxygen concentration at the external boundary ($\sigma^* > \sigma_{\mathrm{P}}$). This problem is not trivial, but it can be proved (with the techniques of [9]) that:

- For any given R sufficiently large there exists one and only one solution $\rho_P = \hat{\rho}_P(R)$, $\rho_N = \hat{\rho}_N(R)$ (otherwise at least one of the interface does not exist).

- The differences $R-\rho_P$, $\rho_P-\rho_N$ tend to stabilize, as R increases, to values depending on σ^* and obtainable by solving the much simpler system in plane geometry.

3.1 The "Solid" Necrotic Core Model

We give here a short description of a simple model based on a two-phase approach, assuming, as in Greenspan [34], that the necrotic core is "solid". The necrotic core will be then filled by dead cells whose local volume fraction ν_N is constant while they are dissolving into liquid with a rate constant μ_N. Supposing that all the components of the mixture have equal mass density and that $\nu_N=\nu$, the mass balance yields the following equations for $\mathbf{u}$ and $\mathbf{v}$:

$$\nabla\cdot\mathbf{u} = \chi, \qquad \text{in } P, \tag{7}$$

$$\nabla\cdot\mathbf{u} = 0, \qquad \text{in } Q, \tag{8}$$

$$\nabla\cdot\mathbf{u} = -\mu_N, \qquad \text{in } N, \tag{9}$$

$$\nabla\cdot\mathbf{v} = -\chi\frac{\nu}{1-\nu}, \qquad \text{in } P, \tag{10}$$

$$\nabla\cdot\mathbf{v} = 0, \qquad \text{in } Q, \tag{11}$$

$$\nabla\cdot\mathbf{v} = \mu_N\frac{\nu}{1-\nu}, \qquad \text{in } N. \tag{12}$$

By multiplying Eqs. (7)–(9) by ν and Eqs. (10)–(12) by $1-\nu$, and summing up, we obtain

$$\nabla\cdot(\nu\mathbf{u}+(1-\nu)\mathbf{v})=0, \tag{13}$$

both in $P\cup Q$ and in N.

In spherical symmetry the velocities are expressed by the scalars $u(r,t)$ and $v(r.t)$, and symmetry imposes

$$u(0,t)=0, \qquad v(0,t)=0. \tag{14}$$

Therefore, taking into account the continuity of the velocities at $r=\rho_N(t)$ and at $r=\rho_P(t)$, from Eqs. (7) to (9) and (14), the following expression for u can be obtained:

$$u(r,t)=\begin{cases} -\frac{\mu_N}{3}r, & \text{in N},\\ -\frac{\mu_N}{3}\frac{\rho_N^3(t)}{r^2}, & \text{in Q},\\ \frac{\chi}{3}\left(r-\frac{\rho_P^3(t)}{r^2}\right)-\frac{\mu_N}{3}\frac{\rho_N^3(t)}{r^2}, & \text{in P}.\end{cases} \tag{15}$$

From Eqs. (13) and (14) it follows that

$$v(r,t) = -\frac{\nu}{1-\nu} u(r,t).$$

The evolution of the outer radius is determined by the equation

$$\dot{R}(t) = u(R(t),t),$$

and a steady state exists for the R value such that

$$\chi\left(R^3 - \hat{\rho}_P^3(R)\right) = \mu_N \hat{\rho}_N^3(R).$$

Note that at the interface $r = \rho_N$ the velocity u is negative and v is positive: so, there is continual *loss of liquid* from the necrotic core induced by the constraint $\nu_N = \nu = constant$. It can be easily verified that this volumetric loss $4\pi\rho_N^2 v(\rho_N,t)(1-\nu)$ is equal to $\frac{4}{3}\pi\rho_N^3 \nu \mu_N$. At the steady state, it is $u(R) = v(R) = 0$, so that all the liquid mass necessary for cell proliferation comes from the necrotic core. Assuming a first-order kinetics for the degradation of dead cells corresponds to supposing that degradation occurs randomly according to the Poisson distribution, i.e. that the time interval from cell death to cell dissolution is exponentially distributed with mean value equal to $1/\mu_N$. In [12], a Gamma distribution for the degradation time was considered, and the distributed delay of cell dissolution was modelled by the passage of dead cells through a chain of n equal stages with Poisson exit, the last one marking the actual transition to the liquid waste. According to that model, Eq. (9) is changed into the following set of equations:

$$\frac{\partial \nu_{N_1}}{\partial t} + \nabla \cdot (\mathbf{u}\nu_{N_1}) = -\mu_N' \nu_{N_1},$$

$$\frac{\partial \nu_{N_2}}{\partial t} + \nabla \cdot (\mathbf{u}\nu_{N_2}) = \mu_N' \nu_{N_1} - \mu_N' \nu_{N_2},$$

$$\vdots$$

$$\frac{\partial \nu_{N_n}}{\partial t} + \nabla \cdot (\mathbf{u}\nu_{N_n}) = \mu_N' \nu_{N_{n-1}} - \mu_N' \nu_{N_n},$$

where $\nu_{N_i}(r,t)$ is the local volume fraction of dead cells in the i-th subcompartment, $i = 1,\ldots,n$, and μ_N' is the exit rate constant from each subcompartment. For the volume fractions of cells in different stages of death, the constraint $\sum \nu_{N_i} = \nu = constant$ is assumed, so that the velocity u can still be determined by $u(0,t) = 0$.

3.2 The NS/NL Partition of the Necrotic Core

In [28, 29], as previously mentioned, we have assumed that the necrotic core at the steady state is partitioned into two zones: a "solid" domain *NS* where cells are supposed to keep the mechanical properties they had before death and the same volume fraction ν and an inner liquid core NL. As we said in Sect. 2, the actual presence of the latter structure appears to have some experimental support. As a matter of fact, by "liquid" we mean a mixture that may contain solid fragments and macromolecules. The important feature from the mechanical point of view is that the stress, in static condition, is isotropic. In the biological literature we may find evidences of more complex states, like, e.g. coagulative necrosis [45],which would require, however, much more complicated constitutive equations.

Dead cells are supposed to degrade into liquid after a fixed time, θ_D, upon death. Such an assumption makes a new interface appear, $r = \rho_D$, dividing NS from NL.

In order to find ρ_D at the stationary state, it is necessary to calculate the velocity field $\mathbf{u}$ of the cells in $P \cup Q \cup NS$. From the mass balance, we have the system

$$\nabla \cdot \mathbf{u} = \chi, \qquad \text{in } P, \tag{16}$$

$$\nabla \cdot \mathbf{u} = 0, \qquad \text{in } Q \cup NS, \tag{17}$$

$$\nabla \cdot \mathbf{v} = -\chi \frac{\nu}{1-\nu}, \qquad \text{in } P, \tag{18}$$

$$\nabla \cdot \mathbf{v} = 0, \qquad \text{in } Q \cup N, \tag{19}$$

which keeps into account the incompressibility of the mixture, i.e $\nabla \cdot [\nu \mathbf{u} + (1-\nu)\mathbf{v}] = 0$. Note that $\mathbf{u}(\rho_D, t)$ is unknown, and this fact makes it *impossible* to determine the evolution of the spheroid and the stationary radius by means of a purely kinematic approach. This was instead possible in the case of a "solid" necrotic core because in such a case we could impose $u(0,t) = 0$.

By imposing the global flux continuity at $r = \rho_D$, namely

$$\mathbf{v}(\rho_D^-) = \nu \mathbf{u}(\rho_D^+) + (1-\nu)\mathbf{v}(\rho_D^+),$$

since $\mathbf{v}(0) = 0$, which holds by symmetry, and Eq. (19) implies $\mathbf{v}(\rho_D^-) = 0$, we get

$$\nu \mathbf{u}(\rho_D^+) + (1-\nu)\mathbf{v}(\rho_D^+) = 0.$$

Thus, for any $r \in (\rho_D, R)$ we have

$$\nu \mathbf{u} + (1-\nu)\mathbf{v} = 0, \tag{20}$$

i.e. a global no flux condition holds. Therefore, at the steady state both $\mathbf{u}$ and $\mathbf{v}$ vanish at $r = R$. Note that having taken the same density for the cells and for the liquid, proliferation and degradation do not imply volume changes.

Since at the steady state $\mathbf{u}$ is zero on $r = R$, the radial component $u(r)$ of the cell velocity, for a given R, can be easily computed:

$$u(r) = -\frac{\chi}{3r^2}(R^3 - r^3), \qquad \text{for } \rho_P < r < R, \tag{21}$$

$$u(r) = -\frac{\chi}{3r^2}(R^3 - \rho_P^3), \quad \text{for } \rho_D < r < \rho_P. \tag{22}$$

The latter formula emphasizes the occurrence of a singularity if ρ_D is allowed to vanish. Following the motion along the velocity field Eq. (22), we can deduce the value of ρ_D imposing that

$$\theta_D = -\int_{\rho_D}^{\rho_N} \frac{\mathrm{d}r}{u(r)},$$

so that ρ_D is given by

$$\rho_D^3 = \rho_N^3 - \chi\theta_D(R^3 - \rho_P^3). \tag{23}$$

Equation (23) represents a constraint on the system, meaning that R has to be sufficiently large to allow Eq. (23) to have a positive solution. Through Eqs. (22) and (23) we recognize indeed that a transition from the "solid" to the "liquid" phase that occurs with a fixed delay from death is not compatible (at the steady state) with $\rho_D = 0$, i.e. with a necrotic core fully "solid".

At this point it is clear that the internal structure of the stationary spheroid can be found once R is known. To proceed further for determining R we must address the mechanical description of the spheroid.

4 A Mechanical Scheme Based on the Two-Fluid Model

Two-fluid models adopt the point of view that a spheroid is a two-component mixture consisting of an inviscid fluid (the extracellular fluid) and another fluid (representing cells) for which an appropriate rheological model has to be chosen. In some papers (cf. [28, 29]) a simple Newtonian scheme is assumed in which the effect of cell-cell interactions is somehow translated into a viscosity. In other models cells are treated as an inviscid fluid too [41] or according to some nonlinear constitutive law. In this section we present the implications of identifying cells with a Newtonian fluid. The limitations which are intrinsic to this approach will be discussed in the next section.

In the Newtonian framework, the Cauchy stress tensors for the two components are written in the form

$$\mathbf{T}_C = \nu\left[-p_C\mathbf{I} + 2\eta_C\mathbf{D}_C - \frac{2}{3}\eta_C\nabla\cdot\mathbf{u}\mathbf{I}\right], \tag{24}$$

$$\mathbf{T}_E = (1-\nu)\big[-p_E\mathbf{I}\big], \tag{25}$$

where $\mathbf{D}_C = \frac{1}{2}[\nabla\mathbf{u} + (\nabla\mathbf{u})^T]$ is the cell strain rate tensor, and η_C is the cell viscosity. In Eq. (24) the so-called Stokes' assumption has been used. The pressures p_C, p_E have to stay distinct. The reason for that will become apparent when we consider for instance the conditions at the boundary $r = R$. Each component is incompressible, but the two velocity fields $\mathbf{u}$, $\mathbf{v}$ are not divergence free in the proliferation region, as we have seen in the previous section. This leads to the definition of the discontinuous function $\hat{\chi}(r)$, equal to χ in the region P and vanishing elsewhere. It is useful to recall that in spherical coordinates and for a radial flow the tensor $\mathbf{D}_C$ has the diagonal structure Diag$(u',\ u/r,\ u/r)$.

Neglecting body forces, and denoting by $\frac{d}{dt}$ the material derivative, we write down the momentum balance equations for the two components (supposing they have the same mass density δ):

$$\delta v \frac{d\mathbf{u}}{dt} = \nabla \cdot \mathbf{T}_C + \mathbf{m}_C, \tag{26}$$

$$\delta(1-v)\frac{d\mathbf{v}}{dt} = \nabla \cdot \mathbf{T}_E + \mathbf{m}_E, \tag{27}$$

in which we define the interaction forces $\mathbf{m}_C$, $\mathbf{m}_E$ to be

$$\mathbf{m}_C = \lambda_C(\mathbf{v}-\mathbf{u}), \tag{28}$$

$$\mathbf{m}_E = \lambda_E(\mathbf{u}-\mathbf{v}). \tag{29}$$

The coefficients λ_C, λ_E can be found by imposing two conditions:

1. The global balance of momentum exchange rate

$$0 = \mathbf{m}_C + \hat{\chi}\delta v\mathbf{u} + \mathbf{m}_E - \hat{\chi}\delta v\mathbf{v} = \frac{\lambda_E - \lambda_C}{1-v}\mathbf{u} + \frac{\hat{\chi}\delta v}{1-v}\mathbf{u}. \tag{30}$$

2. The Darcy's law for the flow of the extracellular liquid relative to cells

$$\mathbf{v} - \mathbf{u} = -K\nabla p_E, \tag{31}$$

where $K(1-v)$ plays the role of hydraulic conductivity.

The final result deduced from Eqs. (28) to(31) is that the interaction forces have the expressions:

$$\mathbf{m}_C = -\left(\frac{1}{K} + \hat{\chi}\delta\frac{v}{1-v}\right)\mathbf{u}, \tag{32}$$

$$\mathbf{m}_E = \frac{\mathbf{u}}{K}. \tag{33}$$

In practice Eq. (32) reduces to $\mathbf{m}_C = -\mathbf{u}/K$ with very good approximation. Coming back to Eqs. (26) and (27), we note that the inertia terms can be neglected. It is not difficult to show that those equations provide the governing differential system for the two pressures $p_C(r)$, $p_E(r)$, namely

$$p_C' = -\frac{u}{Kv} + \frac{4}{3}\eta_C\chi\delta(r-\rho_P), \tag{34}$$

$$p_E' = \frac{u}{K(1-v)}, \tag{35}$$

where $\delta(\cdot)$ denotes the Dirac function.

5 Looking for Steady States

As we said at the end of Sect. 3, if we are given the spheroid radius R, we can find all other unknowns. Thus we need just one more equation to find R. The most natural way of proceeding is to impose that the normal component of the total stress is continuous across the critical interface, i.e. $r = \rho_D$. This is the technique we used in [29].

However, before we come to that, we want to discuss the possibility that the missing equation could be derived from considerations based on power dissipation. Looking at the problem from the point of view of energy is advantageous because it highlights the relative contribution of cell–cell friction and of liquid–cell friction, ultimately related with the coefficients η_C and K, respectively.

5.1 The Energy Based Approach

Coming back to our initial goal of obtaining one more equation from energy considerations, we may think of different options. First of all, we must look at the spheroid in its equilibrium configuration as an "engine", in which mechanical power is produced by proliferating cells and then dissipated by the internal motion so generated. At the stationary state, indeed, cells move inwards until they reach the interface $r = \rho_D$ whereas extracellular liquid moves in the opposite way, the liquid in NL staying at rest (see Sect. 3.2). A very tempting criterion is to say that the radius at equilibrium corresponds to the minimal energy produced and dissipated. In [28], instead, we started from the general principle that the equilibrium size must guarantee the balance of the power dissipated and the one produced by proliferating cells. Note that power balance for each species can be derived from the momentum balance equations (26) and (27). Indeed we can write the equations

$$\delta\nu \frac{\mathrm{d}}{\mathrm{d}t}\left(\frac{1}{2}\mathbf{u}\cdot\mathbf{u}\right) = \nabla\cdot(\mathbf{T}_C\cdot\mathbf{u}) - \mathbf{T}_C : \mathbf{D}_C + \mathbf{m}_C\cdot\mathbf{u}, \tag{36}$$

$$\delta(1-\nu) \frac{\mathrm{d}}{\mathrm{d}t}\left(\frac{1}{2}\mathbf{v}\cdot\mathbf{v}\right) = \nabla\cdot(\mathbf{T}_E\cdot\mathbf{v}) - \mathbf{T}_E : \mathbf{D}_E + \mathbf{m}_E\cdot\mathbf{v}, \tag{37}$$

which take the explicit form

$$\begin{aligned}\delta\nu \frac{\mathrm{d}}{\mathrm{d}t}\left(\frac{1}{2}\mathbf{u}\cdot\mathbf{u}\right) &= -\nu\mathbf{u}\cdot\nabla p_C - \frac{2}{3}\nu\eta_C\nabla\chi_P\cdot\mathbf{u} \\ &\quad +2\nu\eta_C\nabla\cdot(\mathbf{D}_C\cdot\mathbf{u}) - 2\nu\eta_C\mathbf{D}_C : \mathbf{D}_C \\ &\quad -\left(\frac{1}{K} + \hat{\chi}\delta\frac{\nu}{1-\nu}\right)\mathbf{u}\cdot\mathbf{u},\end{aligned} \tag{38}$$

$$\delta(1-\nu)\frac{\mathrm{d}}{\mathrm{d}t}\left(\frac{1}{2}\mathbf{v}\cdot\mathbf{v}\right) = -(1-\nu)\mathbf{v}\cdot\nabla p_E + \frac{\mathbf{u}}{K}\cdot\mathbf{v}$$

$$= -(1-\nu)\mathbf{v}\cdot\nabla p_E - \frac{\mathbf{u}\cdot\mathbf{u}}{K}\frac{\nu}{1-\nu}, \tag{39}$$

where again the left-hand sides can be neglected. Note the presence of the term $-\frac{2}{3}\nu\eta_C\nabla\hat{\chi}\cdot\mathbf{u}$, producing a Dirac distribution centred at the interface $r=\rho_P$. Such a singularity is a consequence of the extreme schematization of the transition $P\to Q$ linked to a threshold of oxygen concentration. In a model with a gradual transition the jump of $\hat{\chi}$ would be replaced by some steep variation, corresponding to a peak in the derivative, but with no substantial change in the qualitative behaviour. A very similar remark can be made for the Laplacian of $\mathbf{u}$.

Since Eqs. (38) and (39) derive from the momentum balance, in view of our purpose we do not have a new piece of information. In [28] the equations above have been used to recognize the dissipation terms, namely $w_C(r) = 2\nu\eta_C\mathbf{D}_C:\mathbf{D}_C$, representing the power dissipated per unit volume because of cell-cell friction, and $w_E(r) = \mathbf{u}\cdot\mathbf{u}/[K(1-\nu)]$, due to liquid–cell friction (the other terms are either negligible or represent power production or transmission). Dissipation due to the conversion of liquid into cells in the proliferating zone can be checked to be absolutely negligible compared to w_C and to w_E. Thus it is possible to calculate the power globally dissipated, W_{diss}, by summing the two contributions integrated over the spheroid.

The explicit expression of the cell–cell friction dissipation term is

$$w_C(r) = \begin{cases} 2\nu\eta_C\left[\left(\chi-2\frac{u}{r}\right)^2+2\frac{u^2}{r^2}\right] = 2\nu\eta_C\left[\chi^2-4\frac{u}{r}\chi+6\frac{u^2}{r^2}\right], & \text{in } P, \\ 12\nu\eta_C\frac{u^2}{r^2}, & \text{in } Q\cup NS. \end{cases} \tag{40}$$

Integrating over the spheroid, the global cell-cell friction dissipation power is obtained as

$$W_C = \frac{16}{9}\pi\nu\eta_C\chi^2R^3\left\{y^3-\frac{1}{2}\left(1+\frac{1}{y^3}\right)+\left(1-\frac{1}{y^3}\right)^2\left[\left(\frac{\rho_P}{\rho_D}\right)^3-1\right]y^3\right\}, \tag{41}$$

where $y = R/\rho_P > 1$. The global contribution of liquid-cell friction is

$$W_E = \frac{4\pi\chi^2}{9K(1-\nu)}R^5\left[y-1-\left(1-\frac{1}{y^2}\right)+\frac{1}{5}\left(1-\frac{1}{y^5}\right)\right.$$
$$\left.+\left(1-\frac{1}{y^3}\right)^2\left(\frac{\rho_P}{\rho_D}-1\right)y\right]. \tag{42}$$

The dissipated power $W_{\mathrm{diss}} = W_C + W_E$ results therefore a function of R.

The idea proposed in [28] was to define the *average power production* of a proliferating cells, w_{cell}, as an independent cell *parameter*, thus computing independently the global power production, W_{prod}, as a quantity simply proportional to the volume of the proliferating region (in turn a function of R). The new equation, from which the stationary radius can be determined, was then obtained by equating $W_{\text{diss}}(R)$ to $W_{\text{prod}}(R)$, i.e. to w_{cell} multiplied by the number of proliferating cells.

In [28] we considered, as reference, a spheroid having at the steady state $R = 1\,\text{mm}$ when the outer oxygen concentration is $\sigma^* = 0.28\,\text{mM}$. The selected parameter values were the following: $\nu = 0.6$, $\chi = \log 2/48\,\text{h}^{-1}$, $\theta_{\text{D}} = 48\,\text{h}$, $D_{O_2} = 1.82 \cdot 10^{-5}\,\text{cm}^2/\text{s}$ [49], $\sigma_{\text{P}} = 0.05\,\text{mM}$, $\sigma_{\text{N}} = 0.01\,\text{mM}$, $m = 2$ [14]. The right-hand side of Eq. (1) was written as $nQ\sigma(r)/(H + \sigma(r))$, where Q is the maximum oxygen consumption rate per cell and n is the cell concentration, and we assumed $Q = 8.3 \cdot 10^{-17}\,\text{mol/(cell s)}$ [31], $n = 5 \cdot 10^8\,\text{cell/cm}^3$ [31]and $H = 4.64 \cdot 10^{-3}\,\text{mM}$ [20]. In models in which the volume loss of dead cells occurred after a chain of successive stages, the mean time for this process was estimated greater than $\sim 80\,\text{h}$ by fitting growth curves of treated sheroids [12] or xenografts [56]. We chose for θ_{D} a shorter value, taking into account that the above estimates reflect the full process of volume loss and account also for the dynamics of the effect of treatment [56].

Concerning the choice of the parameters η_{C} and K, we managed to obtain values for the two quantities W_C, W_E of the same order. This was achieved by assuming $\eta_{\text{C}} = 10^4\,\text{Poise}$ according to [38] (compared to $10^{-2}\,\text{Poise}$ for water at room temperature) and $K = 10^{-7}\,\text{cm}^3\,\text{s/g}$ (i.e. a permeability of $10^{-9}\,\text{cm}^2$, typical of a moderately permeable material). Tumours in vivo have a much lower permeability (two orders of magnitude less), as healthy tissues do [53], but that is due to a considerable compactness provided by a substantial extracellular matrix. In spheroids extracellular matrix is a much lighter structure [35], and we have even neglected its volume fraction. This justifies the assumption of a relatively large value for K. The value taken for viscosity may also look quite large (in the viscosity range of a paste). In the Newtonian scheme, however, viscosity mimics not just pure friction in the relative motion of cells, but also the influence of the forming and breaking of intercellular links (which suggest that a Bingham scheme would be more appropriate). Here we are in a domain of large uncertainty and the choices made in [28], especially for the K value, are certainly questionable. Their main motivation was to keep both kinds of dissipation in the game, waiting for the acquisition of more precise information.

Remark 5.1. The dimensionless ratio of the factors multiplying the brackets in the expressions of W_C, W_E is $4\nu(1-\nu)\eta_C K/R^2$. A typical value for ν in spheroids is 0.6, thus when $R = 1$ mm we obtain a value close to 0.1. If the product $\eta_C K$ is reduced (which can easily be the case), then dissipation is dominated by W_E.

The value of w_{cell} might be estimated from the knowledge of the steady-state radius at a given oxygen concentration (provided the values of the other parameters are known). In [28], for instance, for a spheroid having radius $R = 1\,\text{mm}$ at

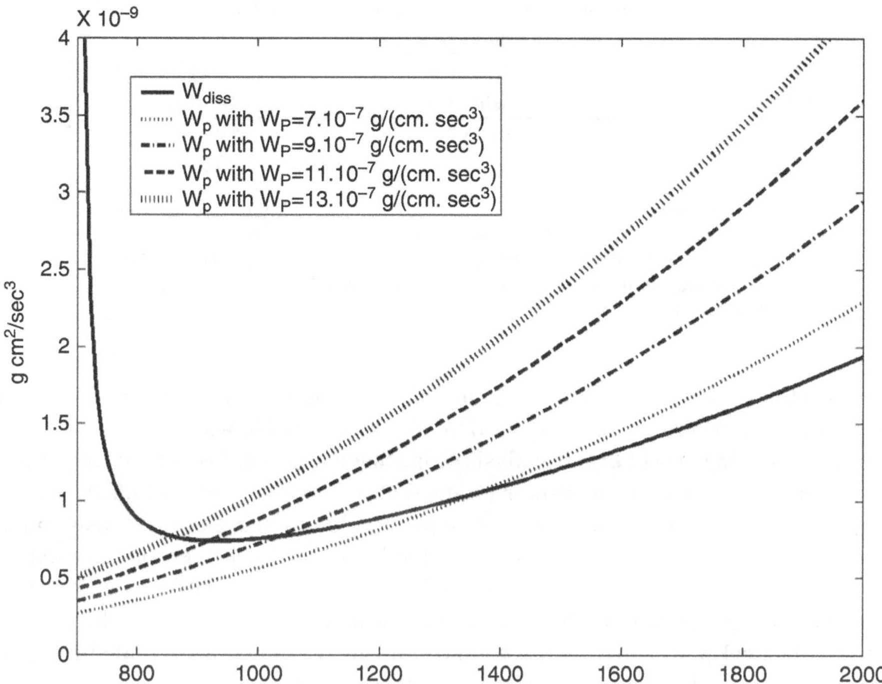

Fig. 1 W_{diss} and W_{prod} for different values of w_{P}

$\sigma^* = 0.28\,\text{mM}$, we calculated $W_{\text{diss}} = 6.2 \cdot 10^{-10}\,\text{erg/s}$. On the basis of the following formula derived by equating dissipated power and produced power:

$$w_{\text{cell}} = W_{\text{diss}} \left(\frac{\nu V_{\text{P}}}{v_{\text{cell}}} \right)^{-1}, \tag{43}$$

where v_{cell} is the cell volume and $V_{\text{P}} = \frac{4\pi}{3}\left(R^3 - \rho_{\text{P}}^3\right)$ is the volume of the proliferating rim, we estimated for the quantity $w_{\text{P}} = w_{\text{cell}}/v_{\text{cell}}$, namely the power supplied by proliferating cells per unit cell volume, the value $w_{\text{P}} = 9.05 \cdot 10^{-7}\,\text{g/(cm s}^3)$. If we now apply our conjecture that this value represents a characteristic of the proliferating cells of that cell line, and so is irrespective of the spheroid size, then we can use Eq. (43) (with the estimated w_{cell}, and $W_{\text{diss}}(R) = W_C(R) + W_E(R)$) as the equation determining the spheroid radius R for values of the outer oxygen concentration σ^* different from 0.28 mM. As a sensitivity test, in our simulation we checked that increasing the estimated w_{P} value by 20 % makes R decrease by 10 %.

In Fig. 1 we plot W_{diss} and W_{prod} as functions of R, for different values of w_{P}. Table 2 shows the values of R for different values of σ^*, deduced by means of the criterion illustrated above and assuming $w_{\text{P}} = 9 \cdot 10^{-7}\,\text{g/(cm s}^3))$, compared to the experimental values in [32].

Table 2 Radius of the stationary spheroid predicted by assuming $w_P = 9 \cdot 10^{-7}$ g/(cm s^3), for different values of σ^*

σ^* (mM)	Predicted radius (μm)	Experimental values (μm)
0.28	1,033	1,380
0.14	598	526
0.07	329	265

The reported experimental values are the saturation values obtained by a Gompertzian fitting of the measured growth curves [32] The glucose concentration in the medium was 5.5 mM (0.8 mM in the case of $\sigma^* = 0.07$ mM)

As Table 2 shows, despite the many approximations, this method is able to reproduce the experimentally observed trend in a reasonable way.

An interesting aspect of the analysis of mechanical power dissipation comes out by comparing w_{cell} with the power production of a tumour cell calculated on the basis of glucose consumption rate. A typical value of the glucose consumption rate in a tumour cell is $21.5 \cdot 10^{-17}$ mol/(cell s) [31]. Since the consumption of 1 g of glucose corresponds in average to the production of 4 Kcal, we conclude that the average power production per cell is about $6.5 \cdot 10^{-3}$ erg/(cell s). From the mechanical power dissipation we have obtained a much lower value ($w_{\text{cell}} = 3.6 \cdot 10^{-15}$ erg/(cell s), for a cell of radius 10 μm), suggesting that mechanical energy is just a very small fraction of the energy required to sustain the cell life.

5.2 *Imposing Normal Stress Continuity*

As we have seen, the analysis of mechanical power dissipation has some biological relevance, but it relies on the simplifying assumption that we can regard the quantity w_P as a parameter characterizing the cells of a given cell line, irrespective of their location inside the spheroid, of the spheroid size, and of the growth conditions. Such an assumption, actually, has no biophysical justification. A different approach, having a rigorous basis in the selected mechanical framework, consists in imposing the *continuity of normal stress* throughout the spheroid. As we shall see, this approach, that has been pursued in [29], will eventually lead to rejecting (in a strict sense) the conjecture adopted in the previous section, although in some cases it may result in an acceptable approximation. In any case, the approach based on the requirement of normal stress continuity has the considerable advantage of being fully consistent with the selected mechanical model (no matter how reliable the latter is claimed to be).

In Sect. 4 we derived the differential equations (34) and (35) for the pressure profile in each component. Now we perform the integration (using the expressions (21) and (22) for the cell velocity field), obtaining

$$p_E(r) = p_{\text{ext}} + \frac{\chi}{3K(1-\nu)}\left(\frac{R^3}{r} + \frac{r^2}{2} - \frac{3}{2}R^2\right), \quad \text{for } \rho_P \le r \le R, \tag{44}$$

$$p_E(r) = p_E(\rho_P) + \frac{\chi(R^3-\rho_P^3)}{3K(1-\nu)}\left(\frac{1}{r} - \frac{1}{\rho_P}\right), \quad \text{for } \rho_D \le r < \rho_P, \tag{45}$$

$$p_C(r) = \hat{p} + \frac{2\gamma}{R} - \frac{\chi}{3K\nu}\left(\frac{R^3}{r} + \frac{r^2}{2} - \frac{3}{2}R^2\right), \quad \text{for } \rho_P < r < R, \tag{46}$$

$$p_C(r) = p_C(\rho_P^-) - \frac{\chi(R^3-\rho_P^3)}{3K\nu}\left(\frac{1}{r} - \frac{1}{\rho_P}\right), \quad \text{for } \rho_D < r < \rho_P. \tag{47}$$

The jump relation

$$p_C(\rho_P^+) - p_C(\rho_P^-) = \frac{4}{3}\eta_C\chi \tag{48}$$

is generated by the Dirac distribution in Eq. (34), that we have already commented. In Eq. (44) p_{ext} is the value taken by p_E at $r = R$, supposed to coincide with the pressure of the water component in the outer medium (which can be just water or a gel, containing in any case a preponderant water fraction). In Eq. (46) γ denotes the surface tension, and a new quantity appears, namely $\hat{p} = p_C(R) - \frac{2\gamma}{R}$, that needs some explanation. Its value has to be found by imposing the balance of normal stress when passing from the spheroid to the external medium. This operation may not be trivial.

If the spheroid is grown in water, then the external normal stress reduces to the pressure p_{ext}. If the outer medium is a gel, then on the cells there will be an extra action due to the deformation of the polymer network making the skeleton of the gel. Spheroids which are subjected to such an extra compression have been reported to exhibit a reduced growth [36]. This question would deserve a deeper investigation, since it can be related to inhibition of proliferation (which would make the proliferation χ depend on pressure [22]), but can also have an independent mechanical origin, as we shall see. It has to be emphasized that the extra compression, coming from the solid component of gel, acts only on the solid component of the spheroid (namely the cells, despite their schematization as a fluid). Similarly, surface tension acts exclusively on the cells.
Thus, while we just have pressure continuity for the liquid component, the boundary condition for p_C can be stated by imposing the following jump condition to the total stress $\mathbf{T}$, relative to radial direction (see, e.g. [40]):

$$\left[\mathbf{Te}\cdot\mathbf{e}\right]_{r=R} = -\nu\frac{2\gamma}{R}, \tag{49}$$

where $\mathbf{e}$ is the radial unit vector. Equation (49) is equivalent to

$$-\nu\left(p_C(R) + \frac{2}{3}\eta_C\chi\right) + 2\nu\eta_C u'(R) - (1-\nu)p_{\text{ext}} = -\nu p_{\text{gel}} - \nu\frac{2\gamma}{R} - p_{\text{ext}}, \tag{50}$$

where we have denoted by p_{gel} the pressure selectively exerted on cells by the gel polymeric network. Such a quantity should be represented by a monotone function of R, stabilizing to an asymptotic finite value, since the network deformation can only have a somehow localized influence. Here p_{ext} is atmospheric pressure. We recall that $u'(R) = \chi$. Hence Eq. (50) implies

$$\hat{p} = p_{\text{gel}} + p_{\text{ext}} + \frac{4}{3}\eta_C \chi. \tag{51}$$

Next we turn our attention to the normal stress continuity at $r = \rho_D$, which takes the form

$$\nu p_C(\rho_D^+) - 2\nu\eta_C u'(\rho_D^+) + (1-\nu) p_E(\rho_D) = p_E(\rho_D), \tag{52}$$

i.e.

$$p_C(\rho_D^+) = p_E(\rho_D) + 2\eta_C u'(\rho_D^+), \tag{53}$$

finally leading, for $p_{\text{gel}} = 0$, to

$$2\gamma = \frac{1}{\nu(1-\nu)} \frac{\chi R^3}{3K} \left\{ \frac{R}{\rho_D} \left[1 - \left(\frac{\rho_P}{R} \right)^3 \right] - \frac{3}{2} \left[1 - \left(\frac{\rho_P}{R} \right)^2 \right] \right\}$$
$$+ \frac{4}{3}\eta_C \chi R \left(\frac{R}{\rho_D} \right)^3 \left[1 - \left(\frac{\rho_P}{R} \right)^3 \right]. \tag{54}$$

The study of Eq. (54) has been performed in [29] (actually a slightly different expression was used there, corresponding to a simplified definition of $\hat{p}$). The right-hand side of Eq. (54) is a function of R tending to infinity both for $R \to \infty$ and in correspondence of the critical value of R for which $\rho_D \to 0$ and below which the interface $r = \rho_D$ is not defined. There is only one minimum, which we may call $2\gamma^*$, which defines a critical value of the surface tension, discriminating between existence and non-existence of a steady state. Thus the problem of finding R is solvable if and only if $\gamma > \gamma^*$. In [29] we found that a value slightly greater than 0.01 dyne/cm for γ is compatible with our reference situation with $R = 1$ mm. The value of γ increases to about 0.05 dyne/cm, if K is reduced to $10^{-8}\,\text{cm}^3\,\text{s/g}$. Clearly when $\gamma > \gamma^*$ Eq. (54) has two solutions. Since the spheroid grows to a steady state from a small initial size, we can say that the physical solution is the smaller.

Remark 5.2. In the liquid-dominated case (see Remark 5.1) the term in Eq. (54) containing viscosity can be neglected. In such a case the solution is going to depend just on the product $K\gamma$.

Remark 5.3. Equation (54) shows that, keeping γ fixed, equilibrium becomes impossible when η_C is raised above some threshold. This fact has a physical interpretation. Indeed a steady-state configuration requires that all cells possess a radial inward directed velocity. If viscosity is too large, the inward motion is hindered and the spheroid tends to grow indefinitely to the exterior.

Remark 5.4. By recalculating the values of w_P corresponding to the solutions predicted by Eq. (54) for different values of the oxygen concentration σ^*, w_P is actually found to vary. So the conjecture of the energy-based approach that the power w_P can be defined as a characteristic cell parameter is disconfirmed. Nevertheless, in the simulation of [29] the variation of w_P was only of the order of 15 %.

As a final comment, we want to point out some internal contradictions of the two-fluid model that we summarize in the following remarks.

Remark 5.5. The sheer fact that we use Darcy's law to describe the motion of the extracellular fluid relative to cells implies that a fluid–cell friction does exist, though we have supposed that the fluid is inviscid (which would make it flow among cells with no resistance). But certainly its viscosity is many orders of magnitude smaller than η_C, thus the above compromise is reasonable.

Remark 5.6. As we have seen, in many models (including the model in [29]) the action of "surface tension" is necessary to reach equilibrium. However, we must not identify the concept of surface tension in a spheroid with the one arising in a liquid drop. Indeed cells mutually interact through macromolecular bridges which can provide some limited tensile stress and evolve according to the dynamical state of the spheroid. At the outer surface of a growing spheroid such stresses can produce an effect similar to surface tension, but only if the number of cells in the spheroid is large enough. Thus it would be wrong to use the classical Laplace formula in the early stage of the spheroid growth, introducing in the model abnormally high pressures which simply are not there. The correct way of using surface tension in spheroids should be instead to let it come into play in a gradual way as the size of the spheroid grows.

Remark 5.7. The action of intercellular links cannot be fully taken into account in the framework of the two-fluid model if the "cell fluid" is Newtonian. Thus, in view of the previous remark, including surface tension is in fact an internal contradiction. For this reason it may be of interest to study model extensions in which the cell component is represented by a Bingham fluid (such an extension was preliminarily considered in [29]).

However, any model in which mechanics is excessively simplified cannot provide an accurate description, since the adopted governing laws are in fact trying to provide a simple representation of phenomena whose nature can be more complex. At the same time, going deeper in investigating the mechanical structure clashes inevitably with the practical impossibility of getting experimental information on the many parameters involved. Therefore, if on one side we must be conscious of the great limitations of a mechanically naïve approach, on the other we must recognize that simplicity is the basic component of a practical strategy.

6 Conclusion and Discussion

It is well known that multicellular spheroids in an advanced stage of their evolution contain a considerable fraction of dead cells and debris, mostly concentrated in the central region. It is therefore quite natural to speak of a necrotic core. Describing the structure of that region, as it results from cells degradation, turns out to be a crucial step in modelling the whole system, since, irrespective of the constitutive equations selected for the various components, the growth of the spheroid will depend on how the viable region interacts mechanically with the necrotic core. Various schemes have been proposed in the literature since the early paper by Greenspan [34], ranging between two extremes: from a completely solid to a completely liquid core. The necrotic zone is frequently described as a region bounded by a sharp interface. Clearly, the presence of interfaces within a spheroid separating cells in different states is an extrapolation which is frequently adopted (the spherical symmetry itself is an idealization). In our opinion the sharp boundary approach is quite sensible, since it simplifies the conceptual geometrical scheme without deeply altering the actual cell distribution.

In this chapter we paid special attention to the modelling of the necrotic zone, confining ourselves to the analysis of the steady state (when it exists). We started with a short summary of the relevant literature, trying to point out which specific assumption in each of the considered models (most of the times related with the necrotic region) guarantees the existence of a steady state. After having illustrated the implications of assuming a completely solid necrotic core, we review some theories developed in the papers [28, 29] in which the necrotic region consists of a solid shell encasing a liquid nucleus. We took this opportunity for carrying out a critical analysis not only of the approach of those papers, in which we have adopted a two-fluid scheme, but also of the general conceptual difficulties accompanying such a representation of the spheroids mechanics. However, we emphasize that there are good motivations for selecting a relatively simple model, since not only the number of parameters to be determined increases with the complexity of the mathematical scheme, but also the uncertainty of their identification becomes more and more serious. On the other hand, there is a price to pay for simplicity, since one cannot expect that models with few parameters can give a particularly accurate description of intrinsically complex systems. Thus, as a general rule in mathematical biology, the focal point in modelling is to find a reasonable compromise. Of course it is only the comparison with experimental data which can say to what extent the compromise is acceptable.

In this respect we found that the approach followed in [28], based on energy balance, and the one used in [29], imposing the continuity of normal stress throughout the system, perform in a comparable way, despite the fact that—strictly speaking—they are not mutually compatible. Indeed, while the principle of normal stress continuity always applies, the energy balance proposed in [28] involved the conjecture that all proliferating cells deliver the same amount w_P of mechanical power. Such a conjecture is not confirmed by the results of [29]. This is an indication that defining a quantity like w_P may not be correct. As usual, one has to be cautious in stating what is or is not correct, since it has to be considered that all these results have been obtained in a framework of a model that we already know to be largely approximated. In the same spirit we cannot even refuse the third (quite appealing) option that equilibrium is characterized as the configuration minimizing the dissipation of mechanical power, which does not appear to be met by any of the previous models.

That said, once it is made clear that we cannot ask too much to a mechanically simple model, the studies performed both in [28] and in [29] lead to results that are quantitatively acceptable and qualitatively interesting, since they permit to ascertain the influence of the basic parameters on the possible attainment of equilibrium, and on the size eventually reached by the spheroid under specified environmental conditions. A quantity which may play a critical role is the so called tumour surface tension, which in some of the models reviewed is necessary for equilibrium to be attained. This is indeed the case of the model in [29].

A step further, already envisaged in [29] (see also [5]), consists in modelling the cell component as a Bingham fluid, thus possessing a yield stress acting as a threshold to allow deformations (and hence the radial flow typical of cells in a spheroid). The reason to shift to a Bingham flow is to better represent the effect of intercellular bonds that can bear some limit tension and have to be broken to allow deformations, a responsibility that is totally assigned to viscosity in the Newtonian framework. We find the Bingham approach particularly stimulating and we plan to study the evolution of "Bingham spheroids" in a future paper. We may anticipate that the selection of a Bingham-like constitutive law is a delicate issue, owing to the peculiar feature of the "fluid" considered, which is actually incompressible, but in which volume is not preserved due to proliferation.

As a general conclusion we may say that the scheme in which the necrotic core has an interface separating a liquid nucleus from a solid shell allows to describe the complex radial flow within a spheroid at equilibrium, either in a fully Newtonian framework, or adopting a Bingham scheme for the fluid representing cells. Despite all the limitations accompanying the two-fluid models, which we have carefully pointed out, the results obtained are meaningful and they do not require the arbitrary definition of any special mass removing mechanism from the necrotic core. The predictions about the influence of the model mechanical parameters on the equilibrium size could be in principle verified on the basis of ad hoc designed experiments. Based on the positive outcome of the approach of [29], we believe that the whole analysis carried out for the steady state can be extended to describe the spheroid evolution from an early fully proliferative state to its asymptotic configuration. We plan to do it in a forthcoming paper.

Possible extensions of the model may include other aspects that are biologically important, among them the cell inhibition by contact. In the review section of this chapter we have mentioned models which include inhibitors of proliferation generated by dead cell disaggregation. A possibly more important cause of inhibition is related with crowding. Since the cell volume fraction does not vary much within the spheroid, crowding can be sensed via the stress. For instance in the Newtonian scheme cell proliferation can be assumed to stop when p_C exceeds some threshold.

Acknowledgments The present work was partially supported by the PRIN (2008): "Modelli matematici per sistemi a molte componenti nelle scienze mediche ed ambientali" (MIUR).

References

1. Adam, J.A.: A mathematical model of tumour growth. iii. Comparison with experiment. Math. Biosci. **86**, 213–227 (1987)
2. Adam, J.A., Maggelakis, S.A.: Mathematical models of tumour growth. iv. Effects of a necrotic core. Math. Biosci. **97**, 121–136 (1989)
3. Adam, J.A., Maggelakis, S.A.: Diffusion regulated growth characteristics of a spherical prevascular carcinoma. Bull. Math. Biol. **52**, 549–582 (1990)
4. Ambrosi, D., Preziosi, L.: On the closure of mass balance for tumour growth. Math. Mod. Meth. Appl. Sci. **12**, 737–754 (2002)
5. Ambrosi, D., Preziosi, L.: Cell adhesion mechanisms and stress relaxation in the mechanics of tumours. Biomech. Model. MechanoBiol. **8**, 397–413 (2009)
6. Araujo, R.P., McElwain, D.L.S.: A history of the study of solid tumour growth: The contribution of mathematical modelling, Bull. Math. Biol. **66**, 1039–1091 (2004)
7. Astanin, S., Preziosi, L.: Mathematical modelling of the Warburg effect in tumour cords, J. Theor. Biol. **258**, 578–590 (2009).
8. Astanin, S., Tosin, A.: Mathematical model of tumour cord growth along the source of nutrient, Math. Model. Nat. Phenom. **2**, 153–177 (2007)
9. Bertuzzi, A., Fasano, A., Gandolfi, A.: A free boundary problem with unilateral constraints describing the evolution of a tumour cord under the influence of cell killing agents. SIAM J. Math. Analysis **36**, 882–915 (2004).
10. Bertuzzi, A., Fasano, A., Gandolfi, A., Sinisgalli, C.: ATP production and necrosis formation in a tumour spheroid model. Math. Model. Nat. Phenom. **2**, 30–46 (2007)
11. Bertuzzi, A., Fasano, A., Gandolfi, A., Sinisgalli, C.: Necrotic core in EMT6/Ro tumor spheroids: Is it caused by an ATP deficit?. J. Theor. Biol. **262**, 142–150 (2010)
12. Bertuzzi, A., Bruni, C., Fasano, A., Gandolfi, A., Papa, F., Sinisgalli, C.: Response of tumor spheroids to radiation: Modeling and parameter identification. Bull. Math. Biol. **72**, 1069–1091 (2010)
13. Bianchini, L., Fasano, A.: A model combining acid-mediated tumour invasion and nutrient dynamics. Nonlin. Anal. Real World Appl. **10**, 1955–1975 (2009)
14. Bredel-Geissler, A., Karbach, U., Walenta, S., Vollrath, L., Mueller-Klieser, W.: Proliferation-associated oxygen consumption and morphology of tumour cells in monolayer and spheroid culture. J. Cell. Physiol. **153**, 44–52 (1992)
15. Breward, C.J., Byrne, H.M., Lewis, C.E.: The role of cell-cell interactions in a two-phase model for avascular tumour growth. J. Math. Biol. **45**, 125–152 (2002)
16. Burton, A.C.: Rate of growth of solid tumours as a problem of diffusion. Growth **30**, 157–176 (1966)
17. Byrne, H.M., Chaplain, M.A.J.: Growth of necrotic tumors in the presence and absence of inhibitors. Math. Biosci. **135**, 187–216 (1996)
18. Byrne, H., Preziosi, L.: Modeling solid tumor growth using the theory of mixtures. Math. Med. Biol. **20**, 341–366 (2003)
19. Byrne, H.M., King, J.R., McElwain, D.L.S., Preziosi, L.: A two-phase model of solid tumour growth. Appl. Math. Lett. **16** 567–573 (2003)
20. Casciari, J.J., Sotirchos, S.V., Sutherland, R.M.: Variation in tumor cell growth rates and metabolism with oxygen concentration, glucose concentration, and extracellular pH. J. Cell. Physiol. **151**, 386–394 (1992)
21. Casciari, J.J., Sotirchos, S.V., Sutherland, R.M.: Mathematical modelling of microenvironment and growth in EMT6/Ro multicellular tumour spheroids. Cell Prolif. **25**, 1–22 (1992)

22. Chaplain, M., Graziano, L., Preziosi, L.: Mathematical modelling of the loss of tissue compression responsiveness and its role in solid tumour development. Math. Med. Biol. **23**, 197–229 (2006)
23. Cui, S., Friedman, A.: Analysis of a mathematical model of the growth of necrotic tumors. J. Math. Anal. Appl. **255**, 636–677 (2001)
24. Cui, S., Friedman, A.: A hyperbolic free boundary problem modeling tumor growth. Interf. Free Bound. **5**, 159–182 (2003)
25. Cristini, V., Li, X., Lowengrub, J.S., Wise, S.M.: Nonlinear simulations of solid tumor growth using a mixture model: Invasion and branching. J. Math. Biol. **58**, 723–763 (2009)
26. Deakin, A.S.: Model for the growth of a solid in vitro tumour. Growth **39**, 159–165 (1975)
27. Durand, R.E.: Cell cycle kinetics in an in vitro tumor model. Cell Tissue Kinet. **9**, 403–412 (1976)
28. Fasano, A., Gandolfi, A., Gabrielli, M.: The energy balance in stationary multicellular spheroids. Far East J. Math. Sci. **39**, 105–128 (2010)
29. Fasano, A., Gabrielli, M., Gandolfi, A.: Investigating the steady state of multicellular spheroids by revisiting the two-fluid model. Math. Biosci. Eng. **8**, 239–252 (2011)
30. Folkman, J., Hochberg, M.: Self-regulation of growth in three dimensions. J. Exp. Med. **138** 745–753 (1973)
31. Freyer, J.P., Sutherland, R.M.: A reduction in the in situ rates of oxygen and glucose consumption of cells in EMT6/Ro spheroids during growth. J. Cell. Physiol. **124**, 516–524 (1985)
32. Freyer, J.P., Sutherland, R.M.: Regulation of growth saturation and development of necrosis in EMT6/Ro multicellular spheroids by the glucose and oxygen supply. Cancer Res. **46**, 3504–3512 (1986)
33. Gerlee, P., Anderson, A.R.: A hybrid cellular automaton model of clonal evolution in cancer: The emergence of the glycolytic phenotype. J. Theor. Biol. **250**, 705–722 (2008)
34. Greenspan, P.: Models for the growth of a solid tumour by diffusion. Stud. Appl. Math. **51**, 317–340 (1972)
35. Hamilton, G.: Multicellular spheroids as an in vitro tumor model. Cancer Lett. **131**, 29–34 (1998)
36. Helmlingen, G., Netti, P.A., Lichtembeld, H.C., Melder, R.J., Jain, R.K.: Solid stress inhibits the growth of multicellular tumor spheroids. Nature Biotech. **15**, 778–783 (1997)
37. Hirschhaeuser, F., Menne, H., Dittfeld, C., West, J., Mueller-Klieser, W., Kunz-Schugart, L.A.: Multicellular tumor spheroids: An underestimated tool is catching up again. J. Biotech. **148**, 3–15 (2010)
38. Iordan, A., Duperray, A., Verdier, C.: A fractal approach to the rheology of concentrated cell suspensions. Phis. Rev. E **77**, 011911 (2008)
39. Jiang, Y., Pjesivac-Grbovic, J., Cantrell, C., Freyer, J.P.: A multiscale model for avascular tumor growth. Biophys. J. **89**, 3884–3894 (2005)
40. Kang, M., Fedkiw, R.P., Liu, X.: A boundary condition capturing method for multiphase incompressible flow. J. Scient. Comput. **15**, 323–360 (2000)
41. Landman, K.A., Please, C.P.: Tumour dynamics and necrosis: Surface tension and stability. IMA J. Math. Appl. Med. Biol. **18**, 131–158 (2001)
42. Landman, K.A., White L.R.: Solid/liquid separation of flocculated suspension. Adv. Colloids Interface Sci. **51**, 175–246 (1994)
43. Lowengrub, J.S, Frieboes, H.B., Jin, F., Chuang, Y-L., Li, X., Macklin, P., Wise, S.M., Cristini, V.: Nonlinear modelling of cancer: Bridging the gap between cells and tumours. Nonlinearity **23**, 1–91 (2010)
44. Maggelakis, S.A., Adam, J.A.: Mathematical model of prevascular growth of a spherical carcinoma. Math. Comput. Modelling **13**, 23–38 (1990)
45. Majno, G., La Gattuta, M., Thompson, T.E.: Cellular death and necrosis: Chemical, physical and morphologic chenges in rat liver. Virchows Arch. path. Anat. **333**, 421–465 (1960)
46. Marusic, M., Bajzer, Z., Freyer, J.P., Vuk-Pavlovic, S.: Analysis of growth of multicellular tumor spheroids by mathematical models. Cell. Prolif. **27**, 73–94 (1994)

47. McElwain, D.L.S., Ponzo, P.J.: A model for the growth of a solid tumor with non-uniform oxygen consumption. Math. Biosci. **35**, 267–279 (1977)
48. McElwain, D.L.S., Morris, L.E.: Apoptosis as a volume loss mechanism in mathematical models of solid tumor growth. Math. Biosci. **39**, 147–157 (1978)
49. Mueller-Klieser, W.: Method for the determination of oxygen consumption rates and diffusion coefficients in multicellular spheroids. Biophys. J. **46**, 343–348 (1984)
50. Mueller-Klieser, W.: Three dimensional cell cultures: From molecular mechanisms to clinical applications. Am. J. Physiol. **273**, C1109–1123 (1997)
51. Mueller-Klieser, W.: Tumor biology and experimental therapeutics. Crit. Rev. Hematol. Oncol. **36**, 123–139 (2000)
52. Neeman, M., Jarrett, K.A., Sillerud, L.O., Freyer, J.P.: Self-diffusion of water in multicellular spheroids measured by magnetic resonance microimaging. Cancer Res. **51**, 4072–4079 (1991)
53. Netti, P.A., Jain, R.K.: Interstitial transport in solid tumours. In: Preziosi, L. (ed.) Cancer Modelling and Simulation. Chapman&Hall/CRC, Boca Raton (2003)
54. Please, C.P., Pettet, C.J., McElwain, D.L.S.: A new approach to modelling the formation of necrotic regions in tumours. Appl. Math. Lett. **11**, 89–94 (1998)
55. Schaller, G., Meyer-Hermann, M.: Continuum versus discrete model: A comparison for multicellular tumour spheroids. Phil. Trans. R. Soc. A **364**, 1443–1464 (2006)
56. Simeoni, M., Magni, P., Cammia, C., De Nicolao, G., Croci, V., Pesenti, E., Germani, M., Poggesi, I., Rocchetti, M.: Predictive pharmacokinetic-pharmacodynamic modeling of tumor growth kinetics in xenograft models after administration of anticancer agents. Cancer Res. **64**, 1094–1101 (2004)
57. Smallbone, K., Gatenby, R.A., Gillies, R.J., Maini, P.K., Gavaghan, D.J.: Metabolic changes during carcinogenesis: Potential impact on invasiveness. J. Theor. Biol. **244**, 703–713 (2007)
58. Sutherland, R.M.: Cell and environment interactions in tumor microregions: the multicell spheroid model. Science **240**, 177–184 (1988)
59. Sutherland, R.M., Durand, R.E.: Hypoxic cells in an in vitro tumour model. Int. J. Radiat. Biol. **23**, 235–246 (1973)
60. Tindall, M.J., Please, C.P., Peddle, M.J.: Modelling the formation of necrotic regions in avascular tumours. Math. Biosci. **211**, 34–55 (2008)
61. Venkatasubramanian, R., Henson, M.A., Forbes, N.S.: Incorporating energy metabolism into a growth model of multicellular tumor spheroids. J. Theor. Biol. **242**, 440–453 (2006)
62. Ward, J.P., King, J.R.: Mathematical modelling of avascular tumor growth II. Modelling growth saturation. IMA J. Math. Appl. Med. Biol. **16**, 171–211 (1999)

Deciphering Fate Decision in Normal and Cancer Stem Cells: Mathematical Models and Their Experimental Verification

Gili Hochman and Zvia Agur

1 Introduction

All tissues in the body are derived from stem cells (SCs). SCs are undifferentiated cells with two essential properties: unlimited replication capacity and the ability to differentiate into one or more specialized cell types. Embryonic SCs are pluripotent, meaning that they can give rise to nearly all cell types. Non-embryonic, adult SCs are found in various tissues and are capable of generating a limited set of tissue-specific cell types. The first discovered and most extensively studied type of adult SC is the hematopoietic SC, found in the bone marrow, which can give rise to all lineages of mature blood cells [12, 84]. Organ-specific SCs have been identified in many other tissues, including the liver, skin, brain, and mammary gland (see [19] for review).

Adult SCs are responsible for tissue maintenance and renewal throughout the life of an organism. They replenish cell populations after normal cell death and following more extensive tissue damage caused by disease or injury. This regenerative ability has made SCs a key focus of scientific research, much of which is aimed at developing treatment for a broad variety of diseases [86, 87]. For many years, hematopoietic SCs have been successfully used to treat leukemia and other hematological disorders, through bone marrow transplantation [32]. Recently, clinical trials have been conducted to evaluate SC-based treatment for cardiovascular diseases [20], neurological diseases [43], spinal cord injuries [93], and diabetes [63]. Researchers have also attempted to exploit SCs in tissue engineering, aspiring to replace damaged tissues or cells by transplanting SCs that have been induced in vitro to differentiate into specific phenotypes [37].

G. Hochman (✉) • Z. Agur
Institute for Medical BioMathematics, 10 Hate'ena St., Bene Ataroth, Israel
e-mail: gili@imbm.org; aguri@imbm.org

U. Ledzewicz et al., *Mathematical Methods and Models in Biomedicine*, Lecture Notes on Mathematical Modelling in the Life Sciences, DOI 10.1007/978-1-4614-4178-6_8,

SCs do not proliferate or differentiate at a constant rate. Rather, their behavior is highly complex and closely regulated, attuned to the exact needs of the tissue at any given time. For example, under normal conditions SCs might produce only a few differentiated tissue cells (DCs) at a continuous rate, but if a tissue is injured, the SCs may suddenly be required to produce larger quantities of DCs to repair it. It is crucial that SC proliferation and differentiation correspond precisely to the requirements of the tissue. Insufficiently rapid proliferation and differentiation may impair tissue function, whereas overproliferation may result in uncontrolled growth and increase the occurrence of mutations, which might be cancerous [7]. The need to maintain the delicate balance between proliferation and differentiation implies the existence of a dynamic regulatory mechanism that, at each point in time, determines the *fate* of each SC in the tissue: according to the requirements of the tissue, the SC either proliferates, differentiates, or is quiescent.

The SC fate decision mechanism is a key component of *homeostasis*, or the maintenance of a stable internal environment, which is a fundamental condition for life. The fate decision mechanism is responsible, for example, for ensuring that the blood continuously contains enough red blood cells to carry oxygen to remote corners of the body, while at the same time triggering immune responses to unexpected, immediate threats. An understanding of SC fate decision can shed light on the very essence of homeostasis. Correspondingly, if we examine what happens when the fate decision mechanism malfunctions, we might be able to understand what happens in diseases in which homeostasis is interrupted—such as cancer.

One approach to investigating the role of SC fate decision in cancer relates to the theory of cancer stem cells. This theory suggests that, like healthy tissues, cancers are characterized by a hierarchical structure, in which a small minority of cancer cells (called cancer stem cells, or CSCs) have stem cell-like properties [6, 18, 75]. CSCs can proliferate indefinitely and are responsible for tumor growth, whereas the majority of (differentiated) cancer cells have only a limited ability to proliferate [57]. Even a few CSCs can regenerate a depleted tumor following treatment, and therefore, according to the CSC theory, the only way of effectively curing disease is to eliminate the CSC population [39]. Therapeutic approaches that target CSCs may entail simply killing these cells (elimination therapy) or, alternatively, inhibiting their proliferation (inhibition therapy), or driving them to differentiation (differentiation therapy), which eliminates their unlimited replication capacity [78]. The latter two kinds of therapy involve interfering with CSC fate decision mechanisms. A deeper understanding of SC and CSC fate decision could be instrumental in the development of such treatments.

Herein we review a series of mathematical models formulated by Agur and colleagues, aimed at elucidating fate decision mechanisms in SC and CSC populations. These models are, then, used to gain insight into cancer therapy.

The first SC model by Agur et al. is aimed to decipher homeostasis in developing systems, using as few assumptions as possible [4]. This model is a cellular automaton, general enough to represent any normally functioning tissue. The model assumes that SC fate decision is determined by negative feedback, depending on local cell–cell interactions between the SCs. Specifically, Agur and colleagues assume that cells are able to "count" the numbers of cells in their area and make decisions accordingly. This counting ability is known to exist in bacteria and is referred to as quorum sensing (QS). Analysis of this model [4, 45] shows that QS is sufficient for maintaining the homeostatic properties of a tissue. Moreover, this is the simplest model capable of retrieving homeostasis.

This model was followed by an effort to study the derangement of homeostasis, i.e., to learn what causes a normal, homeostatic tissue to become cancerous. To this end, Agur et al.'s original model was refined to incorporate a specific three-dimensional structure of the tissue and varying intensities of intracellular signaling (i.e., variation of the distance at which cells can detect the presence of other cells) [3]. Results confirm that excessive SC proliferation may be triggered by change in the intensity of intercellular communication.

In a subsequent study, the model was adjusted in order to explore the behavior of a cancerous tissue containing CSCs [90]. Exploring the system behavior under various parameter values enabled the authors to identify general therapeutic approaches that are likely to be effective in targeting CSC populations.

A separate model aimed to identify the molecular mechanism underlying fate decision control in a single SC, by incorporating intracellular molecular signaling pathways that are sensitive to microenvironmental signals [5, 44]. This intracellular model was integrated within the previously studied tissue model, to create a multiscale model, which, if verified experimentally, could also serve as a useful tool for distinguishing specific possible therapeutic targets for eliminating CSCs [5]. Mathematical analysis [44] and simulations [5] of this model show that one of the key factors for fate decision regulation is the Dickkopf1 (Dkk1) ligand, which is secreted by SCs into the microenvironment, and may serve as a potential modulator of the negative feedback (QS) mechanism.

The rest of this chapter is organized as follows. Sections 2 and 3 provide background about the SC fate decision mechanism and about the theory of CSCs. Section 4 discusses mathematical modeling of SC fate decision. Section 5 discusses the tissue models, and Sect. 6 discusses the molecular mechanism model. Section 7 discusses the results of the analysis of these models, the implications of considering the concept of feedback regulation through SC-to-SC interactions, and possible future applications for these models in CSC research.

2 Fate Decision in Stem Cells: Managing the Replication–Differentiation Balance

Tissues containing SCs are organized as cellular hierarchies, in which SCs make up a small fraction of the cell population [34]. SCs can divide either symmetrically or asymmetrically. In symmetric division, two similar SCs are produced, i.e., the SC proliferates. Asymmetric division, in contrast, yields one SC and one daughter

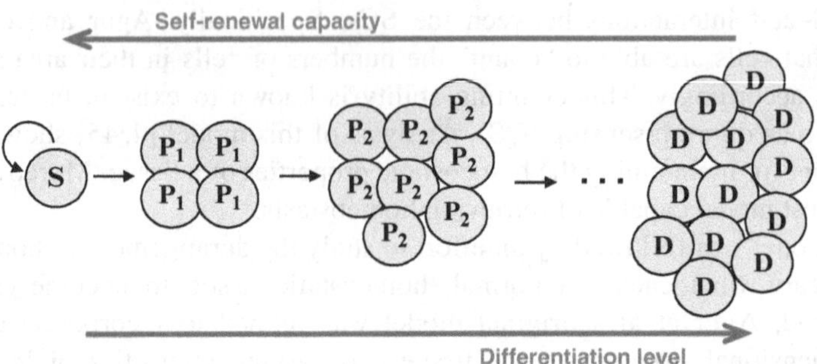

Fig. 1 Schematic description of the cell hierarchy in a tissue. A stem (*S*) cell can replicate indefinitely, while producing early progenitor cells (P_1), which in turn produce a larger population of more differentiated progenitors (P_2). The differentiation process is naturally continuous and can go on through several lineages of PCs, eventually resulting in fully differentiated (*D*) cells

cell that is more differentiated, termed a progenitor cell (PC). The PC transiently amplifies, meaning that it replicates for a limited time. The PC produces either additional PCs that are at an even more advanced stage of differentiation or terminally differentiated cells (DCs), which cannot replicate (Fig. 1). DCs fulfill the tissue's functionality (e.g., blood cells, skin cells).

As noted above, the SC proliferation and differentiation rates must conform to the tissue's development and changing needs. The SCs must constantly supply the required quantities of DCs under various constraints, for example, in growing tissues or following disease or injury. At the same time, the size of the SC population must be restricted in order to prevent uncontrolled growth and crowding out of the DC population and in order to decrease the risk of cancerous mutations [7].

Control over an SC's fate is exerted through the cell's microenvironment. The SC receives signals from its environment and, according to these signals, "decides" whether to replicate, differentiate, die (apoptosis), or remain quiescent. The signals regulating SC decisions might come from any number of sources: they may be determined by biochemical and mechanical characteristics of the environment, such as cytokine concentrations, cell-to-cell signals, extracellular matrix properties, and possibly somatic properties of the SC itself [65, 70, 89]. Some theories suggest that an external, physical tissue structure transmits the various signals that regulate SC fate [14]. Other theories propose that SCs are capable of sending signals to one another without relying on additional structures. The QS theory, which forms the basis of the work by Agur et al., stems from the latter approach.

The SC fate decision mechanism controls the cell-production rate, and this control is key to tissue homeostasis. Derangement of this mechanism might lead to the development of cancer. The theory of CSCs, elaborated in the following section, creates an opportunity to further explore this notion.

3 Cancer Stem Cell Theory

The CSC theory asserts that some elements of the normal cellular hierarchy exist also in cancer. The theory states that in cancerous tissue, as in normal tissue, a small percentage of cells possess the ability of unlimited self-renewal [6, 18, 75]. These cells, called CSCs, drive the growth and spread of the disease, whereas their more differentiated progeny are destined to die, as they have limited or no ability to undergo further mitotic divisions [57]. It was originally postulated that CSCs arose from normal SCs that escaped the bounds of self-renewal [29, 52]. However, it is also possible that these cells are the result of mutations that caused a progenitor cell to reacquire the ability of self-renewal [18].

In the 1990s, studies in patients with chronic myelogenous leukemia (CML) and acute myelogenous leukemia (AML) provided compelling evidence for the existence of CSCs [11, 29, 88]. Since then, cells with SC characteristics have been identified in solid cancer diseases, such as brain cancer and breast cancer. Putative SC populations have also been observed in cancer types such as colon, pancreas, prostate, and melanoma (see review by Lobo et al. [57]). However, there is still controversy about the generality of the CSC theory [1, 42].

CSCs seem to be relatively resistant to conventional therapy. In several *in vitro* experiments, putative SCs in different cancer types, for example multiple myeloma and breast cancer, did not respond to conventional chemotherapeutic agents [56, 62]. Radioresistance was also shown for ex vivo Glioma stem cells [9]. This may be because CSCs have a slow proliferation rate, in comparison to differentiated transiently amplifying tumor cells, while chemotherapy and radiotherapy generally target rapidly proliferating cells [92]. Moreover, owing to their limitless replication capacity, CSCs that have survived treatment are capable of replenishing a depleted tumor. This may explain the high occurrence of cancer relapse after seemingly successful therapy with strong clinical response [66]. According to this hypothesis, effective tumor eradication must include agents that target CSCs [23]. Recently, outcomes of clinical trials in both myeloma [40] and breast cancer [21] patients have supported this theory by showing correlation between CSC quantities and patient survival after treatment.

Agents that efficaciously attack CSCs and cause their death (elimination therapy) are scarce, owing to these cells' resistance to drugs. Alternative therapy modalities that target CSCs include inhibiting CSC proliferation (inhibition therapy), or driving them to differentiate into transiently amplifying tumor cells (differentiation therapy), which leads to their terminal differentiation and eventual death and facilitates their elimination through conventional therapy [78].

CSC theory suggests that cancerous tissues might have some kind of homeostatic regulation analogous to that in normal tissues. Thus, an understanding of fate decision mechanisms can shed light on CSC population sizes and dynamics, just as it can for SCs in normal tissue. Some of the main signaling pathways that participate in the regulation of SC fate decision in developmental processes have been found to be mutated in cancer [57, 83]. Researchers have begun to seek ways of targeting CSCs by blocking or modifying these pathways, with the aim of allowing specific CSC therapy without affecting normal SCs [77].

4 Mathematical Modeling of Stem Cell Fate Decision

Understanding the mechanisms regulating SC fate decision is fundamental to understanding homeostasis—a basic condition for life. Specifically, deciphering fate decision in CSCs may be key to controlling and eliminating tumor growth. Although more and more biological data have become available regarding multiple factors in the microenvironment that affect SC fate decision [57], it is still not fully understood what controls an SC's decision to replicate or to differentiate into self-amplifying progenitors.

Over the last few years, mathematical models based on biological data have been proposed to describe SC fate decision processes at the cellular and intracellular levels. Some models have described the kinetics of molecular dynamical mechanisms, such as signaling pathways (e.g., [2, 44]). Systems biology approaches have been employed to investigate intracellular signaling pathways and transcription factor networks that play a role in determining SC fate (for a review see [70]).

In order to understand the dynamics of normal and cancerous tissues, which might enable researchers to identify drug targets for controlling tumor cell populations, it is not sufficient to investigate intracellular molecular processes. Rather, it is necessary to examine the tissue as a whole. Several mathematical models have been proposed to describe the role of SCs and CSCs in tissue balance. Many of these models used continuous ordinary differential equations (ODE) systems to describe the dynamics of different cell subpopulations (e.g., SCs and DCs) [22–24, 30, 51, 60, 61, 67, 71, 73, 80, 85, 96]. Others are discrete cellular automata models, where the behavior of individual cells is followed [3–5, 8, 28, 59, 64, 91]. Most of these studies did not focus on the regulation of fate decision and did not examine the validity of the methods used to model this decision. SC control was either considered stochastic, with fixed probabilities of differentiation and replication (e.g., [85]), or described by generic feedback from a homogeneous environment, with no specified underlying mechanisms [22–24, 30, 61, 67, 73, 80]. Some of the models [51, 71, 96] introduce regulation by specific environmental signals (e.g., NF-κB, GDF11 or EGFR), but they did not consider cell-to-cell interactions. Many of the models apply to specific systems and cannot be generalized [8, 59, 60, 64, 91].

In what follows we describe a series of models by Agur and colleagues, which focus both on tissue-level cell-population dynamics and on intracellular molecular signaling in order to describe SC and CSC behavior. The models rely on a minimum of assumptions, all of which concern the SC fate decision mechanism. This minimalism enables the models to provide generalizable conclusions and concrete therapeutic recommendations that are not restricted to specific tissue or disease types.

5 General Description of Stem Cell Dynamics in Tissue: A Discrete Model

5.1 A General Cellular Automaton Tissue Model

The first model by Agur et al. was a general model describing tissues with hierarchical (SC-based) structures [4]. This model formed the basis for all SC models that followed, and its aim was to describe the simplest possible system capturing the essential properties of developing tissues which is capable of retrieving homeostasis in living systems.

The model is a simple, discrete dynamical system that can represent any tissue containing SCs. As the replication–differentiation balance in SCs is essential for maintenance of tissue homeostasis, the model assumes that replication and differentiation decisions are regulated by feedback regarding the condition of the tissue as a whole. Specifically, an SC's fate is assumed to be determined by feedback it receives from neighboring cell populations (referred to as *quorum sensing*, QS). The SC "reads" and responds to signals from other SCs in its local microenvironment. Thus, QS is the fate decision mechanism controlling the SC replication–differentiation balance. The QS mechanism exists among Gram-negative bacteria, e.g., *Vibrio harveyi* and *Vibrio cholera* [10, 54]. In these bacteria, gene expression is regulated through the monitoring of population density, using diffusible molecules for communication.

To be able to take cell–cell feedback interactions into account, without assuming spatial homogeneity of the environmental signals, Agur et al. [4] used a *cellular automata* (CA) model, in which the behavior of each individual cell is tracked. In CA models, cells are discrete sites on a lattice. Time is also discretized, and at every time step, the state of each cell is defined by fixed rules. The rules can be deterministic or include stochasticity and probability distributions, but they must be determined by local conditions at the site of the specific cell.

The basic conceptual model includes the minimum of details necessary to represent a normally functioning tissue, as can be seen in the scheme in Fig. 2. Tissue cells are represented by three types of automata cells: stem (S), differentiated (D), and null (N) cells, the latter representing vacant space in the tissue. An SC can either replicate, generating new SCs, or differentiate and become a DC. A DC is assumed to live in the system for a certain maturation time, and then die or migrate from the tissue, leaving an unoccupied space (N cell). This N cell may eventually become occupied by a new SC, created via a proliferation process (i.e., when a neighboring SC replicates). A DC in the model represents an entire cell line of progenitors and differentiated cells before they die or migrate from the tissue, generalized in the model through the DC life span. An SC's "decision" to differentiate or proliferate depends on the number of SCs and N cells in its neighborhood, respectively. This dependency represents the effects of a variety of secreted cytokines in the cell's microenvironment, enabling the cell to sense which types of cells are in its proximity.

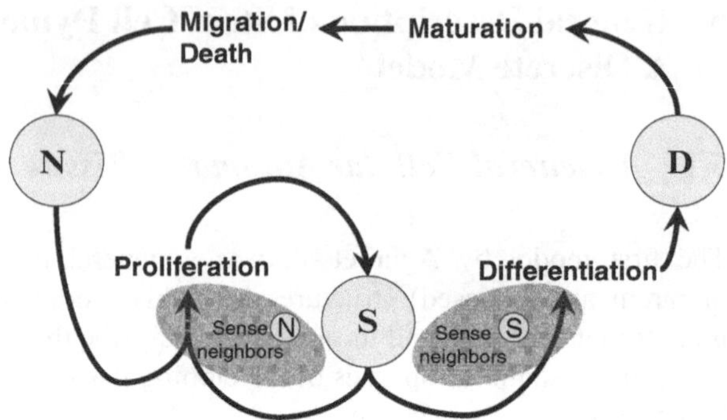

Fig. 2 Schematic description of the general tissue model. Three cell types—stem (S), differentiated (D), and null (N) cells—are represented. The colored areas show QS regulation on the SC fate decision

Mathematically, this system is represented by dynamics on a connected undirected graph $G = (V,E)$, where V and E are sets of vertices and edges, respectively. Each vertex is a cell, and the edges connect each vertex with its closest neighbors. The distance between each two vertices joined by an edge is defined as 1. Each vertex is equipped with an internal counter τ, measuring the cell's progress towards replication or differentiation, if it is an SC, or progress of maturation in the case of DCs. Note that the connected graph formulation compels no restrictions on the geometrical structure or dimensionality of the cellular automaton.

The state x of a vertex v at any time t (denoted $x^t(v)$) is a two-component variable, the first dimension denoting the cell's "type" (either S, D, or N), while the second is a nonnegative integer that denotes its internal counter status. Agur et al. assumed that at each time step, the cell state can be changed due to differentiation (from S to D), proliferation (from N to S), or cell death (from D to N). These changes happen according to the following rules, depending on three nonnegative integer parameters, namely Φ, Ψ, and Θ:

A DC increases its lifetime counter at each time step from τ to $\tau+1$, until when $\tau = \Phi$ it dies, and its state becomes $(N,0)$. Φ represents DC maturation time.

An SC increases its internal counter in the same way, until $\tau = \Psi$, where Ψ represents the duration of SC differentiation time. Then, if all of the SC's closest neighbors are SCs, the cell differentiates (its state becoming $(D,0)$). However, if an SC has a non-stem neighbor when $\tau = \Psi$, it does not differentiate but remains in the same state. This stipulation corresponds to the QS hypothesis of an SC receiving negative feedback signals from the other SCs in its microenvironment.

An N cell does not change its state, unless it has a stem neighbor, which provides the N cell with the potential to become occupied by the SC's daughter cell following the SC's replication. If the N cell has a stem neighbor, it increases its internal counter over time, until $\tau = \Theta$, where Θ represents the cell-cycle time period for SC proliferation. Then the N cell is replaced with a new SC (i.e., its state becomes $(S,0)$).

These rules are described by an iterative operator, which defines what happens to a single vertex during the transition between time t and time $t+1$. This operator is applied simultaneously at each time step on all vertices in V to define the state of the system at any time t. The operator definition is as follows:

$$x^t(v) = (D,\tau) \longrightarrow x^{t+1}(v) = \begin{cases} (N,0) & \text{if } \tau = \Phi, \\ (D,\tau+1) & \text{otherwise;} \end{cases} \tag{1}$$

$$x^t(v) = (S,\tau) \longrightarrow x^{t+1}(v) = \begin{cases} (D,0) & \text{if } \tau = \Psi \text{ and each } v\text{'s neighbor is a stem cell,} \\ (S,\tau) & \text{if } \tau = \Psi \text{ and } v \text{ has a non-stem neighbor} \\ (S,\tau+1) & \text{otherwise;} \end{cases} \tag{2}$$

$$x^t(v) = (N,\tau) \longrightarrow x^{t+1}(v) = \begin{cases} (N,0) & \text{if } v \text{ has no stem neighbor,} \\ (S,0) & \text{if } v \text{ has a stem neighbor and } \tau = \Theta, \\ (N,\tau+1) & \text{otherwise;} \end{cases} \tag{3}$$

where a vertex is defined as a neighbor of v if the distance between the two vertices in the shortest-path metric induced by G is equal to 1.

5.2 *Tissue Homeostasis*

In order to prove that this simple description of fate-decision regulation is sufficient to reproduce tissue homeostasis, Agur and colleagues conducted a mathematical analysis of the model [4, 45]. This resulted in a set of propositions, analytically proven, that together show that the model retains the basic properties essential for maintaining tissue homeostasis, reaching stable SC and DC populations. These theorems are nonquantitative and are robust for any potential refinements involving more elaborate rules. In other words, the model represents a family of cellular automata, and it can be modified to describe more specifically the cell-population control of specific cell types in different tissues. For example, imposing limitations on the kinetic parameters Φ, Ψ, and Θ or imposing a certain geometrical structure will not affect the system's homeostatic properties, since the theorems that follow directly from the basic model assumptions will stay valid.

It was proven that, after some limited initial number of time steps, the tissue model sustains a minimal density of SCs at any time point. A constant supply of mature cells is also assured, owing to the existence of a lower bound for the rate of production of DCs. (The proofs are detailed in [4].) The authors also analyzed the dynamics leading to a state in which the system dies out, i.e., when all vertices are in the state of N. They proved that the system never dies out, regardless of the initial SC population size, except under specific extreme conditions. This feature of the model reflects the tissue's ability to recover after SC depletion.

As will be shown later, the homeostatic balance reproduced by the model depends primarily on the minimal fraction of SCs in the particular SC's immediate neighborhood that would lead to initiating its differentiation. For simplicity, in the first, general model this parameter (referred to as the QS parameter) was set to 1. The second condition guaranteeing homeostasis is a strictly positive time-delay between a cell's "birth" and its differentiation (Ψ). Since the latter condition exists for all biological cells, it will not be discussed any further. The other parameters of the model determine factors such as speed of cell production but do not influence the ability of tissue cell populations to reach homeostasis. This demonstrates the importance of the negative feedback, depicted in the model by rule (2), in which an SC does not differentiate unless its immediate microenvironment is saturated with SCs. This regulatory feedback has a crucial role in the homeostatic characteristics described above.

Moreover, further analysis of the model shows that under certain assumptions, the model guarantees stability in the proportion of SCs in the population [45]. Minimalistic and biologically plausible limitations on the cells' kinetic parameters, and some constraints on the symmetry of the initial SC subset, enable derivation of an expression for the fraction of SCs (and of DCs) in the population, averaged over a period of $\Psi + \Theta + \Phi + 3$ time steps. During this time period, which is the minimal time for an automaton cell to go through all states (proliferation, differentiation, and death), the SC population size fluctuates. However, for a special case of tube-like tissues, the size of the SC population is bounded from above and from below. When cylindrical symmetry is imposed on the graph, by constructing it as $h+1$ similar-sized layers, the numbers of all SCs and DCs at each time step do not differ from the average value by more than $\gamma\%$, where

$$\gamma = \frac{400(\Psi + \Theta + \Phi + 3)}{h+1} < \frac{1{,}600(\Phi + 1)}{h} \tag{4}$$

(proof in [45]). Importantly, given such a cylindrical structure, it is possible to calculate how many initial SCs are needed in the system in order to generate a stable cell population. This is of interest for tissue engineering, where tube-like tissues are constructed using SCs implemented in an artificial scaffold [81].

What can go wrong in tissue homeostasis? To examine the effect of deranged intercellular communication in the microenvironment, Agur and colleagues modified the model slightly [3]. They allowed the QS parameter to be less than unity, now denoting it K_i, representing the intensity of a signal that reaches an SC from another SC located at a distance of i on the connected graph. Rule (2) of the CA iterative operator was generalized, such that an SC differentiates only if the overall signal intensity it is exposed to (from all SCs in its proximity) is above a certain threshold. The model was modified to have a cubic geometrical structure in order to simplify quantification of this demand (see Fig. 1 in [3]).

Numerical simulations of this model were performed under various values of K_i and with $\sim 10^4$ possible triplets of values for cell kinetic parameters Φ, Ψ, Θ, and different randomly chosen initial states. Most of the simulations resulted in one of two states: (i) system death, i.e., when all SCs differentiate and eventually die, or (ii) uncontrolled proliferation, i.e., when most of the SCs keep proliferating and do not differentiate throughout the duration of the simulation. In the latter case, when the modeled tissue becomes saturated with cells, the system achieves a quasi-steady-state, where a small stable fraction of the cell population is DCs, and a much greater part of the CA is occupied by SCs. Statistical segmentation of all simulation results showed that the magnitude of intercellular communication, represented by the QS parameter, dominantly affects the probability of uncontrolled proliferation and the probability of system death. The conclusion is that tissue homeostatic balance is highly dependent on signal intensity, which implies that QS is a crucial mechanism in fate decision.

Analysis and simulations, examining the effect of relations between the kinetic parameters, show that shortening DC life span can increase the proliferation of SCs. Analysis also shows that proliferation may become unlimited when the initial SC population is large. A possible implication for SC therapy would be a necessity to limit the initial number of implanted SCs. Regarding cancer, these results are consistent with the CSC theory rationalization that conventional therapy fails because it mainly eliminates non-CSC tumor cells (as represented in the simulation of shortening DC life span). Moreover, these results imply that such therapy may intensify CSC proliferation. Implications of the conceptual QS model for a cancerous tissue will be discussed in detail in the following section.

5.3 *Model of Cancerous Tissue*

The existence of the QS mechanism implies that the trigger for cancer may lie in the SC's ability to sense its microenvironment. The results of the model analysis described above suggest that excessive cell proliferation may result from changes in the kinetic parameters of the SCs changing their inherent ability to receive signals, or from changes in the microenvironment, affecting the magnitude of the signals transduced to SCs. Hence, cancer initiation may be stimulated by factors that cause microenvironmental changes (e.g., inflammation) rather than by increased mutagenesis, as suggested elsewhere [58]. On the other hand, a natural outcome of excessive proliferation is an increase in the expected number of random mutations, including irreversible oncogenic mutations. If this explanation for carcinogenesis is valid, it means that in the first stage of cancer development, namely, during extensive proliferation of normal SCs, carcinogenesis can be reversed by inducing environmental changes that modify cell signaling intensity.

This also means that the SCs' microenvironment is where we should look for keys to possibly control, prevent, or reverse the direction of tumor growth. If we adopt the theory that CSCs are largely responsible for tumor growth, then controlling the dynamics of cancer progression might become possible through imposing changes in the environment of these SC-like cells. Drugs affecting local signals in the interactions between CSCs can be used for manipulating their differentiation and proliferation rates. Yet any attempt to eliminate CSCs must take into consideration the feedback of the CSC population on itself. For example, elimination of DCs may accelerate the CSC replication rate, owing to the negative feedback that CSCs receive from the population. Hence, cancer therapy based on targeting only DCs (or progenitor tumor cells) may be counterproductive, as it may stimulate CSC proliferation.

To analyze the dynamics of cancer cell populations containing CSCs, Vainstein et al. [90] adapted the SC model by Agur and colleagues, under the CSC theory assumption that hierarchical dynamics in cancer resemble those of normal tissues. Several changes were made in an attempt to increase the model's realism. In Vainstein et al.'s model, a CSC can be in a non-cycling (quiescent) state, or in a cycling state, in which a proliferation process takes place. Furthermore, whereas the original model described proliferation as a "decision" of an empty space to become occupied by an SC, in this model proliferation is initiated by the proliferating cell (i.e., the internal counter for proliferation belongs to the dividing cell and not to the vacant space). Finally, the model is probabilistic, where QS control is achieved by setting the probability of differentiation and of entering proliferation cycle as a function of the number of stem and vacant neighbor cells, respectively.

The model is implemented in a honeycomb-shaped CA grid, where each automata cell has six neighbors. The probability p_d of a non-cycling CSC A to differentiate is

$$p_d = p_{\max} - \frac{a^m (p_{\max} - p_{\min})}{a^m + \left(den(A)\right)^m} \tag{5}$$

where $\mathrm{den}(A) = N_1 + \frac{N_2}{2k}$ is the density of SCs in the proximity of A, N_i being the number of CSCs at a distance i from A, and k is the damping coefficient reflecting a reduction in signal intensity as the distance from the neighbor grows. $1/a$ represents the sensitivity to this microenvironmental signal, and m, $p_{\max}$ and $p_{\min}$ are parameters for steepness and maximal and minimal borders of the function, respectively.

The probability p_c of a non-cycling CSC A to enter the proliferation cell cycle is

$$p_c = 1 - (1 - p_0)^n, \tag{6}$$

where n is the number of vacant automata cells in the proximity of A, calculated in the same way as $den(A)$, and p_0 is a parameter representing the proliferation probability when one neighboring vacant cell is available. When a CSC enters the cell cycle, an adjacent empty cell is randomly chosen, and after a certain proliferation time Θ this site becomes occupied by a new CSC. As in the previous models [3, 4], DCs possess an internal counter as well, to force their death after an estimated life span Φ.

The model was simulated under many different combinations of model parameter values, in biologically plausible ranges based on published information (see [90] for details). These model parameters include parameters determining a CSC's level

Table 1 Summary of effects of varying model parameters on all population sizes and on the total tumor size

	Cycling CSC density	Non-cycling CSC density	DC density	Total tumor cell population
Increasing Differentiation rate	–	↓	–	↓
Shortening DC life span	↑	–	↓	↓
Decreasing Proliferation rate	↓	–	↓	↓

Up arrow means increasing effect of the indicated change in parameters on the specified cell density, down arrow means decreasing effect, and "–" means no effect. A change of no single parameter reduced both cycling and non-cycling CSC densities [90]

of sensitivity to microenvironmental signaling (a, k) and other parameters that influence proliferation and differentiation rates ($p_{\max}$, p_0, Φ), as well as initial size and distribution of the cell-population subsets (i.e., CSCs and DCs) in the CA.

Numerical simulations of the model, under almost all conditions tested, reproduced the dynamics of tumor growth in three phases: initial slow growth in the cell population size, accelerated growth, and decelerated growth until a state of saturation (due to the space limitations of the CA model). This saturation constitutes a "quasi-steady-state" of cell-population size with small fluctuations, which demonstrates the homeostatic tissue balance induced by the QS control mechanism, similar to the quasi-stability observed in simulations of the previous model [3]. This is also similar to the QS-controlled SC–DC balance that was observed in the analysis of the first general model [4] described in Sect. 5.1. Multiple simulations of the CSC model showed that in the quasi-steady-state, cell densities and spatial distributions of the cells were robust to stochastic effects, as well as to changes in the initial conditions and CA size.

The model can be used to examine possible methods of controlling tumor progression, by trying to pinpoint critical parameters that can be targeted in order to eliminate the CSC population. For this purpose, we can look at the simulation results (summarized in Table 1) to observe what happens to the various cell populations when each model parameter is manipulated in various ways. Stimulating differentiation by increasing $p_{\max}$ or decreasing a or k (see Eq. (5)) reduced the density of non-cycling CSCs but did not affect cycling-CSC and DC cell populations. Shortening DC life span Φ, which is expected to indirectly cause acceleration of CSC differentiation, resulted in a decrease in the size of the total tumor cell population; however, the non-cycling CSC density was not reduced, and the cycling-CSC density increased. On the other hand, decreasing the proliferation rate (p_0) caused a reduction in cycling-CSC density, but the non-cycling CSC population was not affected. The effect of changing each of the parameters was found to be independent of other changes.

These results indicate that there is no single parameter that can be manipulated in order to decrease densities of all cell types. Rather, only therapy that both inhibits proliferation and promotes differentiation can be effective. Simulation results of this combinational therapy showed that it can indeed successfully eradicate tumor cells of all cell types.

5.4 Model Prediction of Power Law Tumor Growth Rate and Supporting Experimental Results

Examining the simulated macroscopic dynamics of tumor growth reveals an interesting result regarding tumor growth rate. Model dynamics in the intermediate stage of accelerated growth support previous results, suggesting power law tumor growth [26, 31, 36], as opposed to the widely accepted assumption that the tumor growth rate is exponential or Gompertzian (e.g., [49]). In the simulation results for the two-dimensional CA, the total number of cells is well approximated by a parabola, i.e., it is proportional to the square of time [90]. Similar model simulations of a one-dimensional automaton show that growth of the total number of cells is linear [48]. Therefore, the model suggests that a tumor radius should grow linearly with time. This is corroborated by experimental findings in breast cancer [36] and malignant glioma [82].

To test this, *in vitro* experiments [48] have been conducted in a breast cancer MCF-7 cell line. Small colonies of these cells were seeded in a thin channel or a Petri dish, and their growth was monitored for several days. One-dimensional growth of cells in channels showed that the progression rate of the cell-colony front line was linear (Fig. 3). The two-dimensional area growth of cell colonies showed good fit with the model's prediction of quadratic growth (Fig. 4). Measurements of 3D tumor growth, done in a mouse xenograft model of human breast cancer cells, also support this hypothesis of linear growth of tumor radius [48]. Analysis of these results and of the possible implications of power law tumor growth rate on clinical therapy is to be published in [48].

5.5 Experimental Results Supporting the Quorum Sensing Concept

In vitro experiments [3] with CSCs or "stem-like cells" from the breast cancer MCF-7 cell line were conducted in order to test the theory of the QS control mechanism underlying the model. CSCs, or "stem-like cells" positive for the CD44 marker, were isolated from the breast cancer cell line and plated at different proportions with remaining cell populations. The proportion of CSCs was evaluated several times, until the culture was confluent, and cell populations' proportions reached equilibrium.

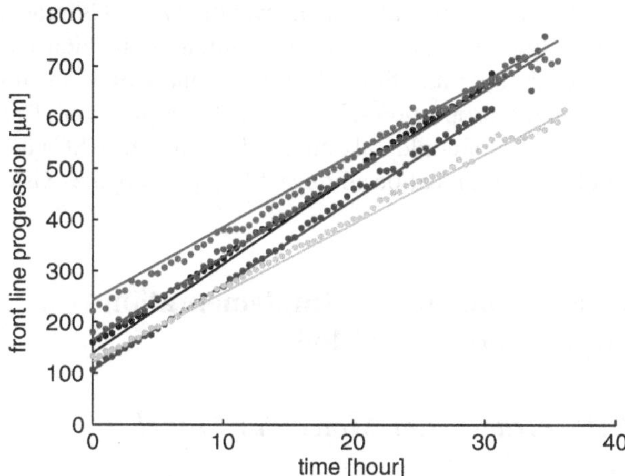

Fig. 3 Experiments prospectively confirming model's predictions in one dimension. One-dimensional front line progression of five MCF-7 cell colonies (each denoted by a different color) shows a linear growth pattern. The colony growth in a thin channel was experimentally measured in 0.5-hr intervals (*dots*). Comparing the slopes of the linear fit (*lines*) of all the different independent replicates shows that the growth rate is similar in the different replicates (In collaboration with Björn Boysen, Andreas Lankenau, Claus Duschl, Fraunhofer Institute for Biomedical Engineering; IBMT)

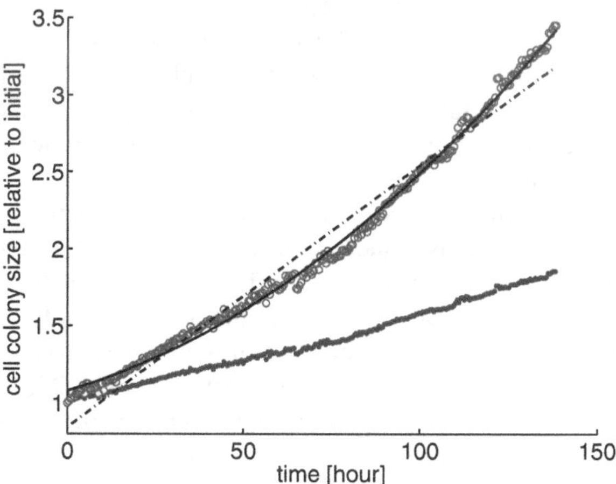

Fig. 4 Experiments prospectively confirming model's predictions in two dimensions. Two-dimensional growth of an MCF-7 cell colony shows a quadratic growth pattern. The colony area growth in a Petri dish was experimentally measured in 0.5-hr intervals (*hollow circles*). Quadratic fit (*solid line*) is presented, in comparison to linear fit (*dashed–dotted line*). Growth of the cell colony radius (*dots*), calculated from area measurements assuming circular structure, demonstrates the linearity of radial growth rate (In collaboration with Björn Boysen, Andreas Lankenau, Claus Duschl, Fraunhofer Institute for Biomedical Engineering; IBMT)

The experimental results revealed that, eventually, CSCs reached a constant proportion in the population, regardless of their initial plating density. These results imply the existence of some additional factor, beyond CSCs' intrinsic replication rate, that determines the proportion of CSCs in the population. The experimental setup dictates that this factor could only have come from the CSCs themselves (e.g., through intercellular communication among CSCs), proving the existence of QS [3].

6 A Molecular Model and Its Implementation in the Large-Scale Tissue Model

6.1 Stem Cell Intracellular Molecular Model

The models described in Sect. 5 show that the balance between SCs and DCs in a tissue (normal or cancerous) is controlled by QS and specifically by SCs' sensitivity to microenvironmental signals. If one wishes to control the balance of different cell populations in a tissue, it is necessary to understand the molecular mechanisms that enable SCs to monitor their environment and, thus, to modulate tissue homeostasis. Understanding this molecular mechanism could enable prediction of the consequences of specific environmental changes, and this knowledge may be used to find ways to externally influence SC fate.

Several intracellular signaling pathways are known to be important for SC fate decision. These include the Wnt canonical pathway and the Notch and Shh pathways. These pathways take part in the fate decision process in embryonic SCs and are also suspected of being active in CSCs. Mutations in these pathways have been found in different cancer types [57, 83, 94].

Agur et al. [5] and Kirnasovsky et al. [44] formulated a new model, describing the pathways in a single breast cancer stem cell (BCSC). They implemented this intracellular network within the tissue model. The model's objective was to evaluate the fate decision process in BCSCs. Hence, the Wnt and Notch pathways were selected to be modeled in this work, because of their central role in the mammary tissue homeostasis, and in transformation to breast cancer [15, 25, 35, 74].

The Wnt canonical pathway is activated by binding of the Wnt ligands to Frizzled/LRP membrane receptors, causing accumulation of β-catenin [72, 76]. High β-catenin levels in the nucleus induce transcription of target genes, which leads to cell proliferation [13]. β-catenin is also involved in regulation of the adhesion molecule E-cadherin, which mediates SC contacts with neighboring cells [17]. The bound E-cadherins affect the efficiency of gene transcription induced by β-catenin [38]. The Notch pathway also plays an important role in SC self-renewal [35, 95]. The binding of the membrane-bound Notch receptors to neighbor cell transmembrane ligands, Delta, Serrate, Lag-2 (DSL), activates transcription of genes such as Hes, which suppress differentiation. A kinetic model of the intracellular steps of the Wnt pathway, down to the level of β-catenin regulation, was first introduced by Lee and Heinrich [53] and further extended and analyzed by others (see review by [46]). The Notch pathway was also mathematically modeled [2]. However, to our knowledge, the model presented in [5, 44] is the first model that specifically merges these pathways together. The approach used in this model is supported by recent information about crosstalk between the pathways [83].

The model of a single BCSC [5,44] was built on the basis of the above biological information. It comprises descriptions of the Wnt, Notch, and E-cadherin pathways, including feedback loops and crosstalk between the pathways. This intracellular network was implemented [5] within a tissue model, where SCs and non-SCs are interconnected through signals in the microenvironment. The CA tissue model is similar to that described in the previous section (Sect. 5.3), except that the SC decision to differentiate and its decision to enter the proliferation cycle are not simply a function of numbers of neighbor cells. Rather, these decisions are dictated deterministically by accumulation of proliferation factors (PF) and differentiation factors (DF) above certain thresholds (C_P and C_M, respectively). These factors are quantitatively estimated for each SC, taking into account the specific inter-cellular signal intensities, as illustrated in the scheme shown in Fig. 5 [5].

In [5, 44], the intracellular processes in a BCSC are modeled according to the following assumptions: activated LEF/TCF transcription factors (denoted in the equations as L) encourage proliferation by increasing PF levels (denoted as P). The activation of LEF/TCF is positively controlled by the Wnt signal intensity (denoted as S) and negatively controlled by the E-cadherins, which are bound to E-cadherins in neighboring cells. (The levels of total and bound E-cadherins are denoted as E and E_b, respectively.) E-cadherin synthesis is negatively regulated by Wnt signal intensity. The Wnt pathway is assumed to be activated by the Wnt ligand (W) in the close environment of the cell, while Dkk1 proteins (D) form a negative feedback loop on the pathway, since their secretion is enhanced as a function of the signal intensity, and they in turn inhibit the Wnt signal [16]. The Notch pathway is activated by Notch receptors (N) binding to DSL proteins in neighboring cells, which are assumed to be expressed by every cell in the model at a constant level. An activated Notch receptor stimulates a sequence of molecular events that increases Hes protein (H) synthesis, which inhibits cell differentiation by reducing DF (M). The Notch pathway is also regulated by a positive feedback loop, as LEF/TCF inhibits the degradation of the Notch receptor [41].

On the basis of these assumptions, a hybrid model was constructed, where cells in the tissue were represented as automata cells, while the intracellular realm of each SC in the tissue model was described by an ODE model of the protein–protein interactions in the Wnt and Notch signaling pathways. The ODE system is as follows:

$$S(t) = f_F^{\uparrow}\left(W_t(t)\right) \cdot f_S^{\downarrow}\left(D_t(t)\right) \tag{7}$$

$$\dot{D} = f_D^{\uparrow}(L) - \mu_D \cdot D \tag{8}$$

$$\dot{L} = S \cdot f_L^{\downarrow}(E_b) - \mu_L \cdot L \tag{9}$$

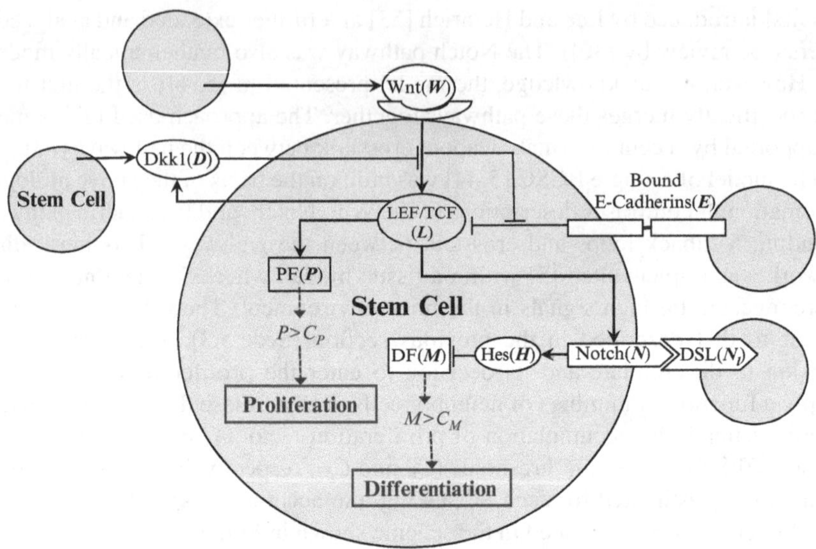

Fig. 5 Schematic representation of the mathematical model of an SC fate decision, regulated by signals in the cell's microenvironment. In this model, the SC's decision to proliferate and to differentiate is caused by the accumulation of proliferation and differentiation factors (PF and DF), respectively, above certain thresholds (C_P and C_M, respectively). The regulation of these factors by Wnt, Notch, and E-cadherin signaling pathways is represented, including feedback loops and crosstalk between the pathways. The role of the proteins Wnt, Dickkopf1 (Dkk1), LEF/TCF and Hes and of the cell-surface receptors Notch, Delta, Serrate, Lag-2 (DSL) and E-cadherin in these pathways is demonstrated, using pointed arrows ($\rightarrow$) to represent activation and blunt arrows ($\dashv$) to represent inhibition. Neighboring cells increase the levels of Wnt, E-cadherins, and Notch–DSL bindings. A stem neighbor may also increase Dkk1 levels. The threshold-dependent effects of PF and DF, respectively, on the SC fate decision are also shown (*dashed arrows*). Notation for the level of every factor/protein, as used in the equations, is written in parentheses [5]

$$\dot{E} = f_E^{\downarrow}(S) - \mu_E \cdot E \tag{10}$$

$$E_{b,i}(t) = \max\left(E_{b,i}(0), \max_{\tau \leq t}(k_b \cdot E(t) \cdot E_i(t))\right) \tag{11}$$

$$\dot{P} = f_P^{\uparrow}(L) - \mu_P \cdot P \tag{12}$$

$$\dot{N} = p_N - f_N^{\downarrow}(L) \cdot N \tag{13}$$

$$\dot{H} = f_H^{\uparrow}(N_r) - \mu_H \cdot H \tag{14}$$

$$\dot{M} = f_M^{\downarrow}(H) - \mu_M \cdot M, \tag{15}$$

where the dependence of each protein on another protein is described using a sigmoid-shaped Hill function of the form: $f(x) = \dfrac{u \cdot a^m + v \cdot x^m}{a^m + x^m}$. Different functions (with different parameters u, v, a, m for each, which determine the exact shape and limits of the sigmoid) are denoted by subscripts (f_F, f_S, etc.). Increasing and decreasing functions are denoted by $f^{\uparrow}$ and $f^{\downarrow}$, respectively.

In Eq. (7), W_t is the total expression level of Wnt proteins in the environment of the considered cell, calculated as $W_t = \frac{1}{2} \cdot (W + \frac{W_n}{6})$, where W is Wnt produced by the particular cell, and W_n is the sum of Wnt produced by all of the cells in the adjacent environment of that cell (maximum of six neighbors, due to the CA grid structure). D_t, representing the total expression level of Dkk1 in the environment of the cell, is calculated in a similar way, while D is the Dkk1 produced by the cell itself.

The parameters μ are the constant degradation rates for each different protein, respectively, denoted by subscript (μ_D, μ_L, etc.).

The number of bound E-cadherins in the cell, E_b (Eq. (9)), is the sum of the number of E-cadherins bound to any adjacent cell: $E_b(t) = \sum_{i=1}^{6} E_{b,i}(t)$, where $E_{b,i}(t)$ is the number of E-cadherins bound to the neighboring cell in the ith direction. $E_{b,i}(t)$ is dependent (Eq. (11)) on the level of E-cadherins in the considered cell E, in the considered neighboring cell E_i, and on the E-cadherin binding coefficient k_b.

p_N (Eq. (13)) is the Notch receptor synthesis rate.

N_r (Eq. (14)) is the level of Notch receptor ready to be activated, which is dependent on the number of Notch receptors in the cell (N) and also on the level of DSL in the microenvironment, in the following way: $N_r = \min(N, N_l)$, $N_l = \frac{1}{6} \cdot \sum_{i=1}^{6} N_{l,i}$, where $N_{l,i}$ is the DSL level of the neighboring cell being in ith direction from the considered cell and N_l is the total level of DSL directed to the cell.

6.2 Hybrid Cellular Automata Multi-scale Model

The new tissue model [5], formed by implementation of this ODE model into the CA model described above, is considered a hybrid cellular automata (HCA) model since it contains both continuous protein activities and discrete cellular developmental and spatial states. This multi-scale model can be used to study consequences of specific intracellular changes on the structure of the tissue.

In order to analyze the molecular proliferation–differentiation regulation mechanism, the ODE system describing the signaling pathways was slightly simplified, and stability analysis was conducted, concluding that the system has a unique equilibrium point (i.e., stable values for all variables, in particular P and M), which is locally asymptotically stable [44]. Simulations showed that the system converges to an equilibrium point under a wide range of biologically relevant values of parameters and initial conditions.

The authors studied how the steady-state values for P and M, which reflect the SC's tendency to proliferate or differentiate, are dependent on the microenvironmental conditions [44]. This was performed by examining the system's response to changes in the various external signals, e.g., Dkk1 level, Wnt level, and the level of DSL receptors on the neighboring cells. Results of this analysis showed that the steady-state values for P and M depend on the level of local cell density. Under high local density, the high E-cadherin signal is dominant and causes differentiation. Under lower cell density, Wnt and Dkk1 signal intensities are dominant, and SCs proliferate at a rate that is dependent on the Dkk1 signal intensity. As will be explained later, the Dkk1 signal reflects the feedback regulation of the SC proportion in the population. Low cell density is generally characterized by a high proliferation rate; however, under extremely low cell density, low Notch signal leads to SC differentiation.

In addition, numerical simulations of the HCA model dynamics have been carried out. The CA honeycomb grid was initially seeded with randomly placed cells. To provide a stable model, parameter values for normal SCs were chosen in a range that promises tissue survival to confluence. These parameters were estimated to fit real characteristics of mammary SCs, based on relevant literature (for details see [5]). Then, for every time step, intracellular dynamics for all the cells were simulated by calculation of per-cell expression levels of all modeled proteins. Accordingly, cell fate was determined for each of the automata cells. This way, the effects of changes in specific protein concentrations, e.g., Dkk1, on the tissue dynamics, could be explored.

Simulations were also used to examine possible effects of defects in signaling pathways on SC proliferation-differentiation balance. Parameters were changed such that Notch receptor synthesis, Wnt ligand expression, or E-cadherin concentration required for LEF/TCF activity inhibition would increase by 5–20 %. The model was re-simulated, and results were compared to the control simulation result with "normal" parameter values.

6.3 *Dkk1 as a Key Regulating Factor for Fate Decision Regulation*

Mathematical analysis of the model [44] showed that Dkk1 is a key, biologically plausible factor in fate decision regulation. The protein Dkk1 is secreted by SCs into the microenvironment and hence may serve as a potential QS modulator, as it can indicate the number of SCs in the close neighborhood. The model predicts that above a specific level, Dkk1 reduces proliferation, thus increasing differentiation.

The numerical simulation results suggest that the Dkk1 effect is biphasic. Below a critical concentration Dkk1 will not affect, and may even somewhat increase, the proportion of SCs in the population. Above this threshold, increasing Dkk1 concentration leads to a significant decrease in the numbers of both proliferating and quiescent SCs, as a result of differentiation.

Simulating dose effects of Dkk1 with changed model parameters, representing increasingly activated pathways due to mutations, did not change the qualitative dependence of SC proportion on Dkk1. However, the critical Dkk1 concentration under which proliferating SCs switch to differentiation depends on the pathway activity, as affected by the specific mutation. This implies that application of

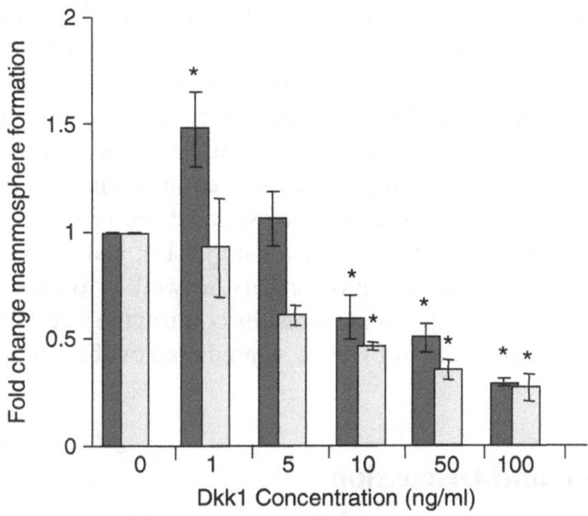

Fig. 6 Effect of Dkk1 treatment on mammosphere formation. Dkk1 effect was measured by MCF-7 cells (*dark grey bars*) and by primary human invasive breast cancer cells (*bright grey bars*). MCF-7 cells were pre-incubated with graded concentrations of Dkk1 in serum-free medium for 24 h and then plated for mammosphere-forming assays for 7 days. Primary breast cancer cells were plated in the presence of Dkk1 and cultured for 7 days. Data for each concentration of Dkk1 are expressed as the fold change in mammosphere formation compared to untreated controls (0 ng/mL). Asterisks mark statistically significant differences [5]

exogenous Dkk1 can be used to control the number of SCs transitioning from proliferation to differentiation and thus to maintain tissue homeostasis, even in situations of derangement of the intracellular mechanism controlling SC fate decision.

6.4 Experimental Validation in Breast Cancer Stem Cells

To test model predictions, *in vitro* experiments on breast cancer MCF-7 cell line, and on primary cells from breast cancer patients, were conducted [5]. Agur and colleagues treated these cell colonies with graded doses of Dkk1 and measured each dose's effect on the proportion of breast cancer cells characterized by the SC phenotype $CD44^{+}CD24^{-/\text{low}}$ [6, 33] and on mammosphere formation [33]. For both the cell line and primary breast cancer cells, the results validated the model prediction that high Dkk1 levels would reduce the number of CSCs. This is demonstrated in Fig. 6, where mammosphere formation under high Dkk1 levels is shown to decrease in a dose-dependent manner. At low levels of Dkk1, there was no increase in CSC numbers or mammosphere formation in primary breast cancer cells, but there was a significant increase in the number of CSCs observed for MCF-7 cells. Overall, the results were variable and highly dependent on the particular experimental protocol, i.e., duration of treatment with Dkk1 before and during the mammosphere-forming assay (see Fig. 4 in [5]) or before the flow cytometry of $CD44^{+}CD24^{-/\text{low}}$ cells. The latter result lends support to the prediction that the CSC proliferation rate under low Dkk1 levels may vary in different tissues, depending on parameters such as pathway activity levels and DCs' mortality.

In addition, the effect of Notch pathway activation and inhibition was examined. For this purpose, Agur and colleagues investigated the response of MCF-7 cells to a recombinant human Notch-receptor ligand DLL4, and to exposure to DAPT, an inhibitor that blocks Notch receptor activity, as well as to knocking out Notch4 expression by siRNA. The experimental results confirm the role of Notch activation in increasing proliferation rate in BCSCs, as predicted by the model.

7 Conclusion and Discussion

7.1 Establishment of the Quorum Sensing Theory in Healthy Stem Cells and in Cancer

The series of mathematical models reviewed in this chapter was aimed at revealing what determines homeostasis in developing tissues. The fundamental question of homeostasis of tissue composition can, actually, be narrowed down to the question of how SC fate is decided, between continued replication and commitment to maturation. The understanding of this important control function might also illuminate the possible derangements of SC fate decision in cancer, thus leading to improved ways of controlling cancer progression.

The first SC model formulated by Agur and colleagues aimed to decipher the basic regulation of SC fate decision that yields homeostasis in developing tissues. Using a simple mathematical model, Agur et al. [4] showed that an extrinsic control—QS, or negative feedback on SC replication—is sufficient for maintaining homeostasis in a developing tissue, given the existence of an intrinsic control—a cell-cycle clock.

The developed CA model was general, describing only basic universal properties of SCs. No specific assumptions were made about tissue spatial structure, growth rate, duration of cell cycle, or DC population characteristics, such as their life span. This underlines the generality of the model's conclusion, regarding the significance of the feedback of cell densities in the SC's environment on its fate decision. In other words, the SC's ability to "count" its stem neighbors lies at the core of homeostasis. The QS concept in the context of oncogenesis was established experimentally in BCSCs as described above [3].

A more realistic model [90], implemented as a probabilistic CA, shed new light on the role of QS-regulated fate decision in CSC-based tumor progression. Simulations showed that the model yielded a quasi-steady-state of the proportion of CSCs in the tumor cell population, which is comparable to the homeostatic resilient state of a normal tissue as described by the general model. Examination of how various changes in model parameters affect cell-population size resulted in significant conclusions: First, accelerated death of DCs weakened the negative feedback that these cells posed on CSC proliferation, which, rather counter-intuitively, increased the number of cycling CSCs. This observation is in line with the CSC hypothesis that to achieve tumor elimination CSCs must be targeted instead or in addition to the transient amplifying tumor cells [78]. Second, simulation results suggested that neither inhibition of proliferation alone nor stimulation of differentiation alone was sufficient to reduce both cycling and non-cycling CSC populations. Moreover, the model enabled analysis of the tumor growth dynamics, and the results implied that the tumor radius grows linearly with time.

Attempting to decipher the mechanism that enables QS, Agur et al. [5] introduced a new hybrid CA model, which described processes at the molecular level in addition to dynamics at the tissue level. The model included a detailed description of the intracellular system of signaling pathways, triggered by microenvironmental signals received from neighboring cells, which were found to balance SC replication and differentiation in developing tissues and in particular in the mammary tissue and in breast cancer. Analyzing this model enabled the authors to explore the means by which tissue balance can be controlled. In the case of cancer, this would mean controlling tumor growth. Analysis of this model [44] pinpointed the Dkk1 protein as a key factor in breast cancer SC fate decision regulation, as it increases the probability of SC differentiation, in addition to reducing the probability of its proliferation. Numerical simulations of the model [5], corroborated by experiments, suggested the existence of a critical Dkk1 concentration, below which SC replication remains largely unaffected. Above this threshold, SC replication is significantly suppressed.

Overall, these models present a new concept, in which QS is viewed as the basic regulatory mechanism driving SC and CSC fate decision. This mechanism is the foundation for the maintenance of healthy tissue homeostasis [4, 45], and its disruption is at the source of cancer initiation [3]. Deciphering the explicit molecular mechanism that enables SCs to monitor their environment and, thus, to modulate tissue homeostasis could pave the way to controlling fate decision. In the case of CSCs, this could lead to identifying new therapeutic agents to be used for controlling tumor progression, as demonstrated by a model of the network of intracellular signaling pathways controlling fate decision in breast cancer SCs [5, 44].

Recently, there is growing interest in the theory of the *stem cell niche*, suggesting that SCs reside in a supporting physiological microenvironment of a defined structure within the tissue [55]. The existence of such a niche for CSCs has been proposed, and experimental evidence for this structure has been found at least in colon cancer, where SCs seem to be localized in a narrow ring near the base of the crypts, and in breast cancer (for reviews see [14, 50, 79]). The normal or cancer SC niche is usually described as a physiological microenvironment, consisting of specialized cells that provide the necessary conditions for SCs to remain undifferentiated and proliferate. These supporting niche-cells are thought to participate in the regulation of SC fate decision and control their range of function [14]. However, the QS model presented here shows that the creation of an external, well-defined niche structure is not necessary for controlling the replication–differentiation balance. The niche could be formed spontaneously, with required conditions for SC proliferation and differentiation being supplied by the SC population itself. This proposition is supported by mathematical analysis [44] of the model for intra- and intercellular feedback mechanisms.

7.2 Implications of Model Analysis for Cancer Therapy

Theoretical analysis and simulations of these mathematical models have already yielded some conclusions that may help open new directions for cancer therapy. First, the intensity of SC-to-SC signaling was found to be a critical factor in the maintenance of tissue balance. Insufficient signal intensity, either due to environmental factors, or due to insufficient signal receipt, as a result of mutations inherent in the SCs themselves, was shown to lead to excessive SC proliferation until they completely deplete the DC population from the tissue [3]. This implies that cancer initiation may be stimulated by changes in the microenvironment, affecting the magnitude of the signals transduced to SCs, and that this process can be reversible under environmental changes that modify the signal intensity. If cancer initiation is caused by increased mutagenesis [58], no epigenetic process can prevent it.

Exploring the system behavior under various possible manipulations, changing factors that influence proliferation and differentiation rates, suggested that only combinational therapy that targets both CSC proliferation and differentiation can be effective [90]. This is in line with clinical experience, since drugs targeting CSCs were found to be clinically more effective when combined with each other, or with conventional therapy that mainly targets DCs [27, 68].

The implementation of a molecular model of processes on the intracellular scale pointed to Dkk1 as a key factor in SC fate decision regulation [5]. Dkk1 can be used for differentiation therapy and is expected to be more effective than other agents stimulating SC differentiation, since it also reduces proliferation. According to the model, Dkk1 therapy challenges the QS-regulated fate decision, which is a general cellular homeostasis mechanism; hence, it should be more robust than other methods. However, the generality of the model does not allow parameter estimation that would be accurate enough to estimate the optimal Dkk1 dosage. Optimizing the dosage of Dkk1 administration is crucial to effectiveness of therapy, since both simulations and experimental results showed that too low administration of Dkk1 may stimulate CSC proliferation. This may not only be ineffective in eliminating the tumor, but also lead to the opposite result.

The simulation results also provide new insights into the tumor growth rate. Simulation results imply that tumor radius grows linearly with time, i.e., the growth of tumor volume is cubic in time [90]. This finding is corroborated by experimental results [48]. Yet only a few previous studies have related to the possibility of a power-law tumor growth rate [26, 31, 36], whereas it is widely accepted to model tumor growth as exponential or Gompertzian (e.g., [49]). This could have therapeutic implications, for example with regard to the design of schedules for radiotherapy [48], which are usually optimized assuming exponential tumor growth during the intervals between irradiation [69].

7.3 *A Robust Tool for Exploring and Manipulating Stem Cells Behavior*

The generality of the basic CA model makes it relevant for the research of adult SCs of any kind. Refinement of the basic model by implementation of various explicit limitations, describing specific tissue-dependent characteristics, could enable researchers to model SC behavior in any tissue of interest, including solid and non-solid tumors. The model can be adjusted to describe, for example, the bone marrow with the migration of mature cells to the peripheral blood or colon cancer with the specific spatial structure of the crypt.

The CA form of the model allows for consideration of the influence of neighboring cells on fate decision in the dynamical process of tumor growth. This is not possible in continuous CSC dynamical models, which describe the macroscopic behavior of CSCs and rely on assumptions about tumor growth rate or spatial homogeneity of environmental signals (cf. [23, 30]). Agur et al. [4] were the first to use a CA formulation to create a general model of SC behavior; other CA models, in contrast, were built to model SC spatial behavior in the specific tissue structure of the colon [59, 64, 91] or breast [8]. Enderling et al. [28] also used a CA model to describe tumor growth dynamics, but they did not try to simulate homeostatic properties in the tissue and had no constraints on the tissue's resilience, considering no feedback of the SC population on SC differentiation.

The multi-scale model [5], which includes modeling of intracellular-level dynamics in conjunction with the dynamics on the tissue level, is used for distinguishing possible therapeutic targets for eliminating CSCs. Notwithstanding, the model still captures the principal mechanism of SC fate decision regulation, i.e., the QS mechanism. Analysis of the model could point out the most effective therapeutic agents, those that attack the main control of CSCs' self-maintenance. The intracellular part was modeled in view of the biological data for BCSCs; however, a different approach could be adopted in order to gain insights for other specific cancer types and therapies.

Currently, the intracellular SC model is being expanded to combine more of the main relevant signaling pathways, including detailed molecular models for these pathways. For example, a detailed model of the Wnt pathway has been built, and its parameters were fitted and validated using experimental data [47]. Implementation of this detailed molecular model in the multi-scale tissue model will have the advantage of parameter availability. Thus, the resultant multi-scale model will be able to make quantitative predictions of the effects of different therapeutic agents. Such a model could be useful for the research of all cancer types, unlike the multi-scale model of van Leeuwen et al. [91], which is specific for colorectal cancer SCs.

Beyond the interest for cancer, manipulating SC fate decision can help in controlling *in vitro* developed tissues, engineered for the purpose of transplantation or designed for experimental research. A model describing the regulation of fate decision in a tissue *in vitro* could contribute to optimization of tissue engineering. For example, analysis of the basic tissue model presented here [45] suggested the possibility of evaluating the minimal number of SCs necessary for replenishing an empty scaffold. Furthermore, the ability to control SC proliferation and differentiation *in vitro* might help to increase the availability of adult SCs for transplantations. In addition, the model can be used for exploring other diseases caused by SC malfunctions.

In conclusion, the presented mathematical models suggest that QS is the key to SC fate decision regulation, and they also begin to decipher the molecular mechanisms underlying it. These efforts bring us closer to the goal of controlling fate decision in real tumors, using mathematical models as tools for quantitative predictions of the efficacy of concrete therapeutic agents for specific cancer types.

Acknowledgments We thank Yuri Kogan and Karin Halevi-Tobias for helpful discussions and support, Karen Marron for helpful advice and careful editing of the manuscript, and the Chai Foundation for supporting the study.

References

1. Adams, J.M., Strasser, A.: Is tumor growth sustained by rare cancer stem cells or dominant clones? Cancer Res. **68**, 4018–4021 (2008)
2. Agrawal, S., Archer, C., Schaffer, D.V.: Computational models of the Notch network elucidate mechanisms of context-dependent signaling. PLoS Comput. Biol. **5**, e1000390 (2009)
3. Agur, Z., Kogan, Y., Levi, L., Harrison, H., Lamb, R., Kirnasovsky, O.U., Clarke, R.B.: Disruption of a quorum sensing mechanism triggers tumorigenesis: A simple discrete model corroborated by experiments in mammary cancer stem cells. Biol Direct. **5**, 20 (2010)
4. Agur, Z., Daniel, Y., Ginosar, Y.: The universal properties of stem cells as pinpointed by a simple discrete model. J. Math. Biol. **44**, 79–86 (2002)
5. Agur, Z., Kirnasovsky, O.U., Vasserman, G., Tencer-Hershkowicz, L., Kogan, Y., Harrison, H., Lamb, R., Clarke, R.B.: Dickkopf1 regulates fate decision and drives breast cancer stem cells to differentiation: An experimentally supported mathematical model. PLoS One **6**, e24225 (2011)
6. Al-Hajj, M., Wicha, M.S., Benito-Hernandez, A., Morrison, S.J., Clarke, M.F.: Prospective identification of tumorigenic breast cancer cells. Proc Natl Acad Sci USA **100**, 3983–3988 (2003)
7. Alon Eron, S.: Maximizing lifespan in view of constant hazards during stem cells proliferation: A mathematical model. MA Thesis, Tel-Aviv University, Tel Aviv (2006)
8. Bankhead, A., Magnuson, N.S., Heckendorn, R.B.: Cellular automaton simulation examining progenitor hierarchy structure effects on mammary ductal carcinoma in situ. J. Theor. Biol. **246**, 491–498 (2007)
9. Bao, S., Wu, Q., McLendon, R.E., Hao, Y., Shi, Q., Hjelmeland, A.B., Dewhirst, M.W., Bigner, D.D., Rich. J.N.: Glioma stem cells promote radioresistance by preferential activation of the DNA damage response. Nature, **444**, 756–760 (2006)

10. Bassler, B.L.: How bacteria talk to each other: Regulation of gene expression by quorum sensing. Curr Opin Microbiol **2**, 582–587 (1999)
11. Baum, C.M., Weissman, I.L., Tsukamoto, A.S., Buckle, A.M., Peault, B.: Isolation of a candidate human hematopoietic stem-cell population. Proc Natl Acad Sci USA **89**, 2804–2808 (1992)
12. Becker, A.J., McCulloch, E.A., Till, J.E.: Cytological demonstration of the clonal nature of spleen colonies derived from transplanted mouse marrow cells. Nature **197**, 452–454 (1963)
13. Behrens, J., von Kries, J.P., Kuhl, M., Bruhn, L., Wedlich, D. et al.: Functional interaction of beta-catenin with the transcription factor LEF-1. Nature **382**, 638–642 (1996)
14. Borovski, T., De Sousa E Melo, F., Vermeulen, L., Medema, J.P.: Cancer stem cell niche: The place to be. Cancer Res. **71**, 634–639 (2011)
15. Brennan, K.R., Brown, A.M.: Wnt proteins in mammary development and cancer. J Mammary Gland Biol Neoplasia **9**, 119–131 (2004)
16. Chamorro, M.N., Schwartz, D.R., Vonica, A., Brivanlou, A.H., Cho, K.R. et al.: FGF-20 and DKK1 are transcriptional targets of beta-catenin and FGF-20 is implicated in cancer and development. EMBO J. **24**, 73–84 (2005)
17. Chen, H., Paradies, N.E., Fedor-Chaiken, M., Brackenbury, R. E-cadherin mediates adhesion and suppresses cell motility via distinct mechanisms. J. Cell. Sci. **110**, 345–356 (1997)
18. Clarke, M.F., Fuller, M.: Stem cells and cancer: Two faces of eve. Cell **124**, 1111–1115 (2006)
19. Cogle, C.R., Guthrie, S.M., Sanders, R.C., Allen, W.L., Scott, E.W., Petersen, B.E.: An overview of stem cell research and regulatory issues. Mayo Clin. Proc. **78**, 993–1003 (2003)
20. Cook, M.M., Kollar, K., Brooke, G.P., Atkinson, K.: Cellular therapy for repair of cardiac damage after acute myocardial infarction. Int J Cell Biol. **2009**, 906507 (2009)
21. Creighton, C.J., Li, X., Landis, M., Dixon, J.M., Neumeister, V.M., Sjolund, A., Rimm, D.L., Wong, H., Rodriguez, A., Herschkowitz, J.I. et al.: Residual breast cancers after conventional therapy display mesenchymal as well as tumor-initiating features. Proc. Natl. Acad. Sci. USA **106**, 13820–13825 (2009)
22. Deasy, M., Jankowski, R.J., Payne, T.R., Cao, B., Goff, G.P., Greenberger, J.S., Huard, J.: Modeling stem cell population growth: Incorporating terms for proliferative heterogeneity. Stem Cells **21**, 536–545 (2003)
23. Dingli, D., Michor, F.: Successful therapy must eradicate cancer stem cells. Stem Cells **24**, 2603–2610 (2006)
24. Dingli, D., Traulsen, A., Michor, F.: (A)symmetric stem cell replication and cancer. PLoS Comput. Biol. **3**, e53 (2007)
25. Dontu, G., Jackson, K.W., McNicholas, E., Kawamura, M.J., Abdallah, W.M. et al.: Role of Notch signaling in cell-fate determination of human mammary stem/progenitor cells. Breast Cancer Res. **6**, R605–R615 (2004)
26. Drasdo, D., Hoehme, S.: A single-cell based model to tumor growth in-vitro: Monolayers and spheroids. Phys. Biol. **2**, 133–147 (2005)
27. Dubrovska, A., Elliott, J., Salamone, R.J., Kim, S., Aimone, L.J., Walker, J.R., Watson, J., Sauveur-Michel, M., Garcia-Echeverria, C., Cho, C.Y., Reddy, V.A., Schultz, P.G.: Combination therapy targeting both tumor-initiating and differentiated cell populations in prostate carcinoma. Clin. Cancer Res. **16**, 5692–5702 (2010)
28. Enderling, H., Anderson, A.R., Chaplain, M.A., Beheshti, A., Hlatky, L., Hahnfeldt, P.: Paradoxical dependencies of tumor dormancy and progression on basic cell kinetics. Cancer Res. **69**, 8814–8821 (2009)
29. Fialkow, P.J.: Stem cell origin of human myeloid blood cell neoplasms. Verh. Dtsch. Ges. Pathol. **74**, 43–47 (1990)
30. Ganguly, R., Puri, I.K., Mathematical model for the cancer stem cell hypothesis. Cell Prolif. **39**, 3–14 (2006)
31. Gatenby, R.A., Frieden, B.R.: Information dynamics in carcinogenesis and tumor growth. Mutat. Res. **568**, 259–273 (2004)
32. Gratwohl, A., Baldomero, H., Aljurf, M. et al.: Hematopoietic stem cell transplantation: A global perspective. J. Am. Med. Assoc. **303**, 1617–1624 (2010)

33. Gregory, C.A., Singh, H., Perry, A.S., Prockop, D.J.: The Wnt signaling inhibitor dickkopf-1 is required for reentry into the cell cycle of human adult stem cells from bone marrow. J. Biol. Chem. **278**, 28067–28078 (2003)
34. Harrison, D.E., Stone, M., Astle, C.M.: Effects of transplantation on the primitive immunohematopoietic stem cell. J. Exp. Med. **172**, 431–437 (1990)
35. Harrison, H., Farnie, G., Howell, S.J., Rock, R.E., Stylianou, S., Brennan, K.R., Bundred, N.J., Clarke, R.B.: Regulation of breast cancer stem cell activity by signaling through the Notch4 receptor. Cancer Res. **70**, 709–718 (2010)
36. Hart, D., Shochat, E., Agur, Z.: The growth law of primary breast cancer as inferred from mammography screening trials data. Br J Cancer **78**, 382–387 (1998)
37. Hodgkinson, T., Yuan, X.F., Bayat, A.: Adult stem cells in tissue engineering. Expert Rev. Med. Dev. **6**, 621–640 (2009)
38. Huber, A.H., Weis, W.I.: The structure of the beta-catenin/E-cadherin complex and the molecular basis of diverse ligand recognition by beta-catenin. Cell **105**, 391–402 (2001)
39. Huff, C.A., Matsui, W., Smith, B.D., Jones, R.J.: The paradox of response and survival in cancer therapeutics. Blood **107**, 431–434 (2006)
40. Huff, C.A., Wang, Q., Rogers, K., Jung, M., Borrello, I.M., Jones, R.J., Matsui, W.: Correlation of clonogenic cancer stem cell growth with clinical outcomes in multiple myeloma (MM) patients undergoing treatment with high dose cyclophosphamide and rituximab. Proc AACR Late Breaking Abstract, LB**87** (2008)
41. Katoh, M., Katoh, M.: Notch ligand, JAG1, is evolutionarily conserved target of canonical WNT signaling pathway in progenitor cells. Int. J. Mol. Med. **17**, 681–685 (2006)
42. Kelly, P.N., Dakic, A., Adams, J.M., Nutt, S.L., Strasser, A.: Tumor growth need not be driven by rare cancer stem cells. Science **317**, 337 (2007)
43. Kim, S.U., de Vellis, J.: Stem cell-based cell therapy in neurological diseases: A review. J. Neurosci. Res. **87**, 2183–2200 (2009)
44. Kirnasovsky, O.U., Kogan, Y., Agur, Z.: Analysis of a mathematical model for the molecular mechanism of mammary stem cell fate decision. Math. Model. Nat. Phenom. **3**, 78–89 (2008)
45. Kirnasovsky, O.U., Kogan, Y., Agur, Z.: Resilience in stem cell renewal: Development of the Agur-Daniel-Ginossar model. Disc. Cont. Dyn. Systems **10**, 129–148 (2008)
46. Kofahl, B., Wolf, J.: Mathematical modelling of Wnt/beta-catenin signalling. Biochem. Soc. Trans. **38**, 1281–1285 (2010)
47. Kogan, Y., Halevi-Tobias, K.E., Hochman, G., Baczmanska, A.K., Leyns, L., Agur, Z.: A new validated mathematical model of the Wnt signaling pathway predicts effective combinational therapy by sFRP and Dkk. Biochem. J. **444**, 115–125 (2012)
48. Kogan, Y., Hochman, G., Vainstein V., Shukron O., Lankenau, A., Boysen, B., Lamb, R., Berkman, T., Clarke, R.B., Duschl, C., Agur, Z.: Evidence for power law tumor growth and implications for cancer radiotherapy. Submitted
49. Kozusko, F., Bajzer, Z.: Combining Gompertzian growth and cell population dynamics. Math. Biosci. **185**, 153–167 (2003)
50. LaBarge, M.A.: The difficulty of targeting cancer stem cell niches. Clin. Cancer **16**, 3121–3129 (2010)
51. Lander, A.D., Gokoffski, K.K., Wan, F.Y.M., Nie, Q., Calof, A.L.: Cell lineages and the logic of proliferative control. PLoS Biol. **7**, e1000015 (2009)
52. Lapidot, T., Sirard, C., Vormoor, J., Murdoch, B., Hoang, T., et al.: A cell initiating human acute myeloid leukaemia after transplantation into SCID mice. Nature **17**, 645–648 (1994)
53. Lee, E., Salic, A., Kruger, R., Heinrich, R., Kirschner, M.W.: The roles of APC and Axin derived from experimental and theoretical analysis of the Wnt pathway. PLoS Biol. **1**, e10 (2003)
54. Lenz, D., Mok, K., Lilley, B., Kulkarni, R., Wingreen, N., Bassler, B.: The small RNA chaperone Hfq and multiple small RNAs control quorum sensing in Vibrio harveyi and Vibrio cholerae. Cell **118**, 69–82 (2004)
55. Li, L., Xie, T.: Stem cell niche: Structure and function. Annu. Rev. Cell Dev. Biol. **21**, 605–631 (2005)

56. Li, X., Lewis, M.T., Huang, J., Gutierrez, C., Osborne, C.K., Wu, M.F., Hilsenbeck, S.G., Pavlick, A., Zhang, X., Chamness, G.C. et al.: Intrinsic resistance of tumorigenic breast cancer cells to chemotherapy. J. Natl. Cancer Inst. **100**, 672–679 (2008)
57. Lobo, N.A., Shimono, Y., Qian, D., Clarke, M.F.: The biology of cancer stem cells. Annu. Rev. Cell Dev. Biol. **23**, 675–699 (2007)
58. Loeb, L.A.: A mutator phenotype in cancer. Cancer Res. **61**, 3230–3239 (2001)
59. Loeffler, M., Stein, R., Wichmann, H.E., Potten, C.S., Kaur, P., Chwalinski, S.: Intestinal cell proliferation. I. A comprehensive model of steady-state proliferation in the crypt. Cell Tissue Kinet. **19**, 627–645 (1986)
60. Loeffler, M., Wichmann, H.E.: A comprehensive mathematical model of stem cell proliferation which reproduces most of the published experimental results. Cell Tissue Kinet. **13**, 543–561 (1980)
61. Marciniak-Czochra, A., Stiehl, T., Ho, A., Jaeger, W., Wagner, W.: Modeling of asymmetric cell division in hematopoietic stem cells-regulation of self-renewal is essential for efficient repopulation. Stem Cells Dev. **18**, 377–386 (2009)
62. Matsui, W., Wang, Q., Barber, J.P., Brennan, S., Smith, B.D., Borrello, I., McNiece, I., Lin, L., Ambinder, R.F., Peacock, C. et al.: Clonogenic multiple myeloma progenitors, stem cell properties, and drug resistance. Cancer Res. **68**, 190–197 (2008)
63. Matsumoto, S.: Islet cell transplantation for Type 1 diabetes. J. Diabetes **2**, 16–22 (2010)
64. Meineke, F.A., Potten, C.S., Loeffler, M.: Cell migration and organization in the intestinal crypt using a lattice-free model. Cell Prolif., **34**, 253–266 (2001)
65. Metallo, C.M., Mohr, J.C., Detzel, C.J., De Pablo, J.J., Van Wie, B.J., Palecek S.P.: Engineering the stem cell microenvironment. Biotechnol. Prog. **23**, 18–23 (2007)
66. Michor, F., Hughes, T.P., Iwasa, Y., Branford, S., Shah, N.P. et al.: Dynamics of chronic myeloid leukaemia. Nature, **435**, 1267–1270 (2005)
67. Michor, F., Nowak, M.A., Frank, S.A., Iwasa, Y.: Stochastic elimination of cancer cells. Proc. Biol. Sci. **270**, 2017–2024 (2003)
68. Mueller, M.T., Hermann, P.C., Witthauer, J., Rubio-Viqueira, B., Leicht, S.F., Huber, S., Ellwart, J.W., Mustafa, M., Bartenstein, P., D'Haese, J.G., Schoenberg, M.H., Berger, F., Jauch, K.W., Hidalgo, M., Heeschen, C.: Combined targeted treatment to eliminate tumorigenic cancer stem cells in human pancreatic cancer. Gastroenterology **137**, 1102–1113 (2009)
69. O'Rourke, S.F., McAneney, H., Hillen, T.: Linear quadratic and tumour control probability modelling in external beam radiotherapy. J. Math. Biol. **58**, 799–817 (2009)
70. Peltier, J., Schaffer, D.V.: Systems biology approaches to understanding stem cell fate choice. IET Syst. Biol. **4**, 1–11 (2010)
71. Piotrowska, M.J., Widera, D., Kaltschmidt, B., an der Heiden, U., Kaltschmidt, C.: Mathematical model for NF-kappaB-driven proliferation of adult neural stem cells. Cell Prolif. **39**, 441–455 (2006)
72. Polakis, P.: Wnt signaling and cancer. Genes Dev. **14**, 1837–1851 (2000)
73. Prudhomme, W.A., Duggar, K.H., Lauffenburger, D.A.: Cell population dynamics model for deconvolution of murine embryonic stem cell self-renewal and differentiation responses to cytokines and extracellular matrix. Biotechnol. Bioeng. **88**, 264–272 (2004)
74. Reya, T., Clevers, H.: Wnt signalling in stem cells and cancer. Nature **434**, 843–850 (2005)
75. Reya, T., Morrison, S.J., Clarke, M.F., Weissman, I.L.: Stem cells, cancer, and cancer stem cells. Nature **414**, 105–111 (2001)
76. Rubinfeld, B., Albert, I., Porfiri, E., Fiol, C., Munemitsu, S. et al.: Binding of GSK3beta to the APC-beta-catenin complex and regulation of complex assembly. Science **272**, 1023–1026 (1996)
77. Sell, S.: Cancer and stem cell signaling: A guide to preventive and therapeutic strategies for cancer. Stem Cell Rev. **3**, 1–6 (2007)
78. Sell, S.: Stem cell origin of cancer and differentiation therapy. Crit. Rev. Oncol. Hematol. **51**, 1–28 (2004)
79. Sneddon, J.B., Werb, Z.: Location, location, location: The cancer stem cell niche. Cell Stem Cell **1**, 607–611 (2007)

80. Stiehl, T., Marciniak-Czochra, A.: Characterization of stem cells using mathematical models of multistage cell lineages. Math. Comp. Model. **53**, 1505–1517 (2011)
81. Stosich, M.S., Mao, J.J.: Adipose tissue engineering from human adult stem cells: Clinical implications in plastic and reconstructive surgery. Plast. Econstr. Surg. **119**, 71–83 (2007)
82. Swanson, K.R., Harpold, H.L., Peacock, D.L., Rockne, R., Pennington, C., Kilbride, L., Grant, R., Wardlaw, J.M., Alvord, E.C. Jr.: Velocity of radial expansion of contrast-enhancing gliomas and the effectiveness of radiotherapy in individual patients: A proof of principle. Clin Oncol (R Coll Radiol) **4**, 301–308 (2008)
83. Takebe, N., Ivy, S.P.: Controversies in cancer stem cells: Targeting embryonic signaling pathways. Clin. Cancer Res. **16**, 3106–3112 (2010)
84. Till, J.E., McCulloch, E.A.: A direct measurement of the radiation sensitivity of normal mouse bone marrow cells. Radiat. Res. **14**, 213–222 (1961)
85. Tomlinson, I.P.M., Bodmer, W.F.: Failure of programmed cell death and differentiation as causes of tumors: Some simple mathematical models. Proc. Natl. Acad. Sci. USA **92**, 1130–1134 (1995)
86. Trounson, A., Thakar, R.G., Lomax, G., Gibbons, D.: Clinical trials for stem cell therapies. BMC Med. **9**, 52 (2011)
87. Tuch, B.E.: Stem cells—a clinical update. Aust. Fam. Physician. **35**, 719–721 (2006)
88. Uchida, N., Sutton, R.E., Friera, A.M., He, D., Reitsma, M.J., Chang, W,C., Veres, G., Scollay, R., Weissman, I.L.: HIV, but not murine leukemia virus, vectors mediate high efficiency gene transfer into freshly isolated G0/G1 human hematopoietic stem cells. Proc. Natl. Acad. Sci. USA **95**, 11939–11944 (1998)
89. Underhill, G.H., Bhatia, S.N.: High-throughput analysis of signals regulating stem cell fate and function. Curr. Opin. Chem. Biol. **11**, 357–366 (2007)
90. Vainstein, V., Kirnasovsky, O.U., Kogan, Y., Agur, Z.: Strategies for cancer stem cell elimination: Insights from mathematical modeling. J. Theor. Biol. **298**, 32–41 (2012)
91. van Leeuwen, I.M., Mirams, G.R., Walter, A., Fletcher, A., Murray, P., Osborne, J., Varma, S., Young, S.J., Cooper, J., Doyle, B. et al.: An integrative computational model for intestinal tissue renewal. Cell Prolif. **42**, 617–636 (2009)
92. Wicha, M.S., Liu, S., Dontu, G.: Cancer stem cells: An old idea—a paradigm shift. Cancer Res. **66**, 1883–1890 (2006)
93. Wright, K.T., Masri, W.E., Osman, A., Chowdhury, J., Johnson, W.E.: Concise review: Bone marrow for the treatment of spinal cord injury: Mechanisms and clinical applications. Stem Cells **29**, 169–178 (2011)
94. Zardawi, S.J., O'Toole, S.A., Sutherland, R.L., Musgrove, E.A.: Dysregulation of Hedgehog, Wnt and Notch signalling pathways in breast cancer. Histol. Histopathol. **24**, 385–398 (2009)
95. Zhang, X.P., Zheng, G., Zou, L., Liu, H.L., Hou, L.H., Zhou, P., Yin, D.D., Zheng, Q.J., Liang, L., Zhang, S.Z., Feng, L., Yao, L.B., Yang, A.G., Han, H., Chen, J.Y.: Notch activation promotes cell proliferation and the formation of neural stem cell-like colonies in human glioma cells. Mol. Cell. Biochem. **307**, 101–108 (2008)
96. Zhu, X., Zhou, X., Lewis, M.T., Xia, L., Wong, S.: Cancer stem cell, niche and EGFR decide tumor development and treatment response: A bio-computational simulation study. J. Theor. Biol. **269**, 138–149 (2011)

Data Assimilation in Brain Tumor Models

Joshua McDaniel, Eric Kostelich, Yang Kuang, John Nagy, Mark C. Preul, Nina Z. Moore, and Nikolay L. Matirosyan

1 Introduction

A typical problem in applied mathematics and science is to estimate the future state of a dynamical system given its current state. One approach aimed at understanding one or more aspects determining the behavior of the system is mathematical modeling. This method frequently entails formulation of a set of equations, usually a system of partial or ordinary differential equations. Model parameters are then measured from experimental data or estimated from computer simulation or other methods, for example chi-squared parameter optimization as done in [26] or genetic algorithms which are frequently used in neuroscience [33]. Solutions to the model are then studied through mathematical analysis and numerical simulation usually for qualitative fit to the dynamical system of interest and any relative time-series data that is available. While mathematical modeling can provide meaningful insight, it may have limited predictive value due to idealized assumptions underlying the model, measurement error in experimental data and parameters, and chaotic behavior in the system. In this chapter we explore a different approach focused on optimal state estimation given a model and observational data of a biological process, while accounting for the relative uncertainty in both. The case explored

J. McDaniel (✉) • E. Kostelich • Y. Kuang
School of Mathematical and Statistical Sciences, Arizona State University, Tempe, AZ 85287, USA
e-mail: Joshua.McDaniel@asu.edu; kostelich@asu.edu; kuang@asu.edu

J. Nagy
Department of Biology, Scottsdale Community College, Scottsdale, AZ 85256-2626, USA
e-mail: john.nagy@sccmail.maricopa.edu

M.C. Preul • N.Z. Moore • N.L. Matirosyan
Department of Neurosurgery Research, Barrow Neurological Institute, St. Josephs Hospital and Medical Center, Phoenix, AZ 85013, USA
e-mail: Mark.Preul@chw.edu; nina.z.moore@gmail.com; nikolay.martirosyan@chw.edu

U. Ledzewicz et al., *Mathematical Methods and Models in Biomedicine*, Lecture Notes on Mathematical Modelling in the Life Sciences, DOI 10.1007/978-1-4614-4178-6_9,

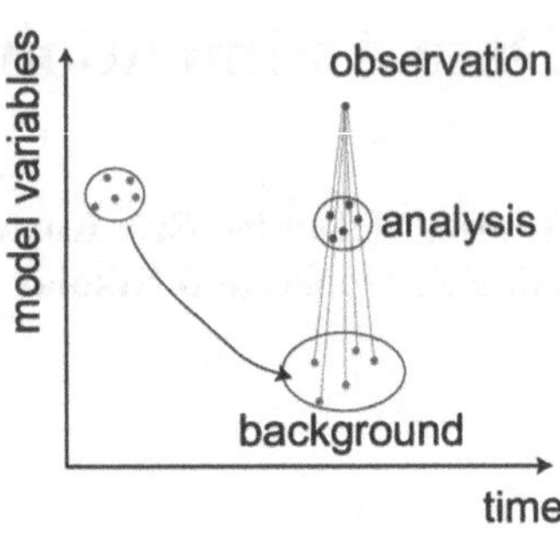

Fig. 1 Schematic illustration of the data assimilation procedure

here is the growth and spread of glioblastoma multiforme (GBM), a very aggressive form of glioma brain tumor which remains extremely difficult to manage clinically. The method employed is different from other approaches used in biology in that it is independent of the mathematical model and seeks an optimal initial condition. This is in contrast to other techniques such as those discussed in [21], which are model dependent and seek to find an optimal model parameterization given the observations and uncertainties in the system of interest.

The method discussed and implemented in this chapter is derived from numerical weather prediction in which initial conditions to atmospheric weather models must be updated frequently from noisy measurements. Chaos exhibited in the underlying weather models leads to the propagation of uncertainty in the initial conditions over short time periods. Thus accurate forecasting requires update of forecast model initial conditions frequently (typically 6 h in global models) based on current observational data. The algorithm for computing the update is known as *data assimilation*. Figure 1 summarizes the main elements of this procedure. One begins with a model trajectory that represents the best estimate of the true state of the system. This is advanced under the model, until a new observation becomes available, producing the *background*. An update called the *analysis* is computed and the process repeats. Under assumptions regarding the probability distribution of the background and observation error, the analysis is a maximum-likelihood estimate of the true system state.

In this chapter we demonstrate how the data assimilation approach might be adapted to help anticipate the progression of clinical patient cases of GBM. GBM is chosen because the dynamics of tumor growth are dependent on the location and density of the underlying tumor cell population; hence, its dynamics are described by a spatiotemporal model. Also noisy observations in the form of magnetic resonance (MR) images are available which should provide insight into the current state of the tumor (meaning the cancer cell density at every point on the brain geometry). The overall approach is independent of the model; hence, this method may be used on a broad range of spatiotemporally complex biological systems.

There are two principal sources of uncertainty in estimating the future growth of GBM. First unlike the motions of the atmosphere, which can be described by well-established physical models, many details of the growth of glioblastomas

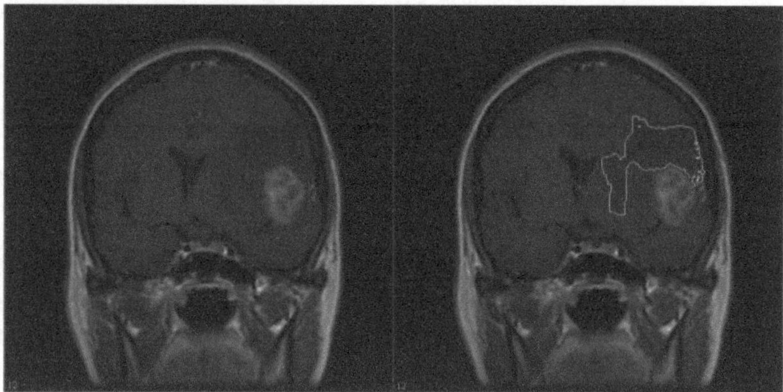

Fig. 2 A representative contrast enhanced magnetic resonance image of a GBM patient at initial diagnosis. The *right panel* shows the edema outlined in a closed curve and the enhancement region below the edema. See text for additional details

tumor cells are poorly understood and stochastic in nature. Typically tumors exhibit genetic abnormalities leading to heterogeneous compositions of cells, metabolism, and vasculature. Brain anatomy varies on an individual basis, and tumors deform the brain geometry depending on their size and location (known as the mass effect).

The second problem arises from interpretation of MR imaging studies. Due to the risks associated with surgical treatment and the problem of accessing the tumor through the skull, noninvasive MR scans are the primary method for assessing the progression of gliomas, including GBM. MR scans are usually taken at intervals from weeks to months depending on the treatment regime. Patients are administered a contrast agent that highlights visibility of the tumor vasculature from the surrounding brain tissue. GBM tumors are highly vascular; thus, strength of contrast is typically interpreted as a measure of glioma cell density. The region of highest intensity is called the enhancing region and corresponds to area of highest blood vessel density, and most likely blood plasma that has leaked into surrounding brain tissue. This area along with the nearby necrotic regions of less contrast is sometimes called the tumor core. The region of lower contrast, usually near the enhancing region, is called the *edema*, which is interpreted to be swelling of noncancerous brain tissue due to leaky and abnormal tumor vasculature and infiltration of "healthy" tissue by the tumor [10]. Figure 2 shows a contrast enhanced MR image of a patient tumor. The red and yellow boundaries mark the enhancing and edema, respectively. The image also shows the mass effect of the tumor in the right hemisphere.

It is difficult to infer quantitative information about the underlying cell density from MRI data for many reasons. First the relationship (if any exists) between a pixel intensity at a given a point and the cell density there is unknown. Quantitative comparison of patient MRI data to a simulated tumor requires a common geometry. Thus a second problem, known as registration error, arises from inaccuracies in mapping the patient MR scan onto the associated brain atlas (model geometry) of the forecast model. Some sources of registration error are individual variances in patient brain geometry and the mass effect. Radiation necrosis resulting from treatment, which may appear similar to active tumor on the image, can further complicate MR image interpretation [25].

The goal of this chapter is to demonstrate the use of ensemble forecasting and data assimilation to make improved short-term (30-day) estimates of the growth and spread of a simulated brain tumor. The remainder of the chapter is organized as follows. Section 2 provides a brief overview of two models of glioblastoma employed to simulate virtual brain tumors. Section 3 introduces the concept of ensemble forecasting and the history and rationale underlying this method. Section 4 introduces the mathematical aspects of data assimilation and presents a detailed derivation of both the basic Kalman filter for linear dynamical systems and a state-of-the-art extension to nonlinear problems known as the local ensemble transform Kalman filter (LETKF). Section 5 describes numerical experiments that demonstrate implementation of the LETKF on a brain tumor models where synthetic observations of a simulated tumor are generated. The results illustrate the potential feasibility of this approach to better forecast the evolution of individual patient lesions.

2 Models of Glioblastoma

GBM is the most common primary malignant brain tumor, and its prognosis is very poor; patient survival is less than 15 months, on average, from initial diagnosis [23]. GBM tumors are highly aggressive, typically develop resistance to chemotherapy and radiotherapy [1], and can quickly invade large and sensitive regions of the brain, making complete surgical resection of the tumor impossible and postsurgical recurrence inevitable [6]. Because little clinical progress has been made against GBM, its biology remains the subject of intense study.

In this chapter we use two mathematical models to simulate tumor growth and expansion. These models are chosen because they each represent syntheses of some previous GBM modeling efforts and reflect increasing levels of complexity. Both models are simulated on a static but realistic brain geometry where diffusion rates are differentiated between white matter, gray matter, and other intracranial tissues. The first model, initially proposed by Swanson et. al. [27, 28], considers a single class of glioma cells exhibiting exponential growth with cell motility governed by Fick's law. These assumptions yield the equation

$$\frac{\partial g}{\partial t} = \nabla \cdot \left(D(\mathbf{x})\nabla g\right) + \alpha g. \tag{1}$$

The diffusion rate of GBM cells is assumed to be faster in white matter than in gray. Each tissue type is further assumed to be homogeneous with respect to its

Table 1 Representative parameters for the logistic Swanson model, Eq. (2), in two dimensions

Parameter	Meaning	Value			
α	Maximum glioma growth rate	0.2 day^{-1}			
T_{max}	Glioma carrying capacity	10,000 cells mm^{-2}			
Parameter	Meaning	White matter	Corpus callosum	Gray matter	CSF
$D(\mathbf{x})$	Diffusion rate (mm^2 day^{-1})	0.0065	0.001	0.0013	0.001

diffusivity. Therefore, D is taken to be piecewise constant within each tissue type, but may vary among types. Parameters in Eq. (1) can be estimated from in vitro studies, sequential patient MR studies, or the Einstein–Stokes relation [27].

Since the exponential growth term in Eq. (1) leads to unbounded tumor cell densities, a more realistic approach assumes that cells grow logistically with some carrying capacity, T_{max}, at any given point in the model's domain. This modification gives [29]

$$\frac{\partial g}{\partial t} = \nabla \cdot (D(\mathbf{x})\nabla g) + \alpha g \left(1 - \frac{g}{T_{max}}\right). \tag{2}$$

We refer to Eq. (2) as the *Logistic Swanson model.* Baseline parameter values for this model are reported in Table 1.

A more sophisticated modeling approach by Eikenberry et al. [7] considers two distinct phenotypic classes of tumor cells: proliferating and migrating. The proliferating cells grow logistically and produce a generalized chemorepellent which is assumed to induce transition to the migrating cell class at sufficiently high concentrations. Migrating cells degrade the extracellular matrix (ECM) and migrate away from the main tumor mass along the ECM gradient via haptotaxis. The transition between the two cell classes is stochastic. The four dependent model variables are

$$\begin{aligned} g(\mathbf{x},t) &= \text{proliferating cell density} \\ m(\mathbf{x},t) &= \text{migrating cell density} \\ c(\mathbf{x},t) &= \text{chemorepellent density} \\ w(\mathbf{x},t) &= \text{ECM density} \end{aligned}$$

and the Eikenberry model is expressed as a coupled system of four stochastic partial differential equations given by

$$\frac{\partial g}{\partial t} = \underbrace{\nabla \cdot (D_G(\mathbf{x})\nabla g)}_{\text{diffusion}} - \underbrace{\nabla \cdot (\chi(\mathbf{x})m\nabla w)}_{\text{haptotaxis}} + \underbrace{\alpha g\left(1 - \frac{g+m}{T_{\max}}\right)}_{\text{logistic growth}} - \underbrace{\Phi}_{g\text{–}to\text{–}m} + \underbrace{\Psi}_{m\text{–}to\text{–}g}, \quad (3)$$

$$\frac{\partial m}{\partial t} = \underbrace{\nabla \cdot (D_M(\mathbf{x})\nabla m)}_{\text{diffusion}} + \underbrace{\nabla \cdot (\chi(\mathbf{x})m\nabla w)}_{\text{haptotaxis}} + \underbrace{\Phi}_{g\text{–}to\text{–}m} - \underbrace{\Psi}_{m\text{–}to\text{–}g} - \underbrace{\sigma(g,m)}_{\text{death}}, \quad (4)$$

$$\frac{\partial c}{\partial t} = \underbrace{D_C\nabla^2 c}_{\text{diffusion}} + \underbrace{\gamma g}_{\text{production}} - \underbrace{\beta_C c}_{\text{degradation}}, \quad (5)$$

$$\frac{\partial w}{\partial t} = \underbrace{-\rho w\left(\frac{g+m}{\theta_W + T}\right)}_{\text{degradation}} + \underbrace{\alpha_W w(1-w)}_{\text{repair}}, \quad (6)$$

where Φ is the rate at which proliferating cells become migratory and Ψ is the rate at which migratory cells become proliferating. Eikenberry et al. [7] assumed that the transition rate to the migratory phenotype is greatest when the cell density is low (such as at the edges of the tumor):

$$\Phi = \Phi(g,m,c) = \tau g\left(\frac{\theta_G^2}{(g+m)^2 + \theta_G^2}\right)\left(\frac{c}{\phi_M + c}\right). \quad (7)$$

The stochasticity arises from how Ψ is implemented over the discretization of the model. The probability of transition from a migrating to a proliferating phenotype is assumed to be an exponential random variable with rate depending on the chemorepellent concentration

$$\lambda = \lambda_0\left(\frac{\theta_G}{\theta_G + c}\right). \quad (8)$$

Once λ is determined, Ψ is computed by finding the number of migrating cells that transition at each grid point and converting this to a cell density. At a given grid point the cell density must constitute at least one cell for any such transition to occur.

Cell death is assumed to result from crowding and occurs only when the cell density reaches a critical threshold, M_D:

$$\sigma(g,m) = \begin{cases} \left(\frac{g+m}{T_{\max}}\right)^2, & g+m > M_D, \\ 0, & \text{otherwise.} \end{cases} \quad (9)$$

Note the haptotaxis term in Eq. (3) is a modification of the original model and is included to preserve conservation of mass.

Table 2 Representative parameters for the Eikenberry model, Eqs. (3)–(6), in 2 dimensions

Parameter	Meaning	Value		
α	Maximum glioma growth rate	$0.15\ \text{day}^{-1}$		
$T_{\max}$	Glioma carrying capacity	10000 cells mm^{-2}		
α_{W}	Maximum ECM recovery rate	$0.01\ \text{day}^{-1}$		
β_{C}	Chemorepellent degradation rate	$0.25\ \text{day}^{-1}$		
D_{C}	Chemorepellent diffusion rate	$2.0\ \text{mm}^2\ \text{day}^{-1}$		
γ	Chemorepellent production rate	$2.5 \times 10^{-5}\ \text{day}^{-1}$		
λ_0	Maximum migrating to grow rate	$2.5\ \text{day}^{-1}$		
M_{D}	Crowding death threshold	$5{,}000$ cells mm^{-2}		
ϕ_{M}	Half-maximum grow to migrating coefficient	$0.5\ \text{mm}^{-2}$		
ρ	Maximum ECM remodeling rate	$0.02\ \text{day}^{-1}$		
τ	Maximum grow to migrating transition rate	$0.1\ \text{day}^{-1}$		
θ_{G}	Half-maximum grow to migrating cell density	100 cells mm^{-2}		
θ_{W}	Cell density at half-maximum ECM degradation	100 cells mm^{-2}		
Parameter	Meaning	White matter	Gray matter	CSF
$D_{\text{G}}(\mathbf{x})$	Growing cell diffusion rate ($\text{mm}^2\,\text{day}^{-1}$)	0.02	0.004	0.0001
$D_{\text{M}}(\mathbf{x})$	Migrating cell diffusion rate ($\text{mm}^2\,\text{day}^{-1}$)	0.1	0.02	0.0001
$\chi(\mathbf{x})$	Haptotaxis coefficient ($\text{mm}^4\,\text{day}^{-1}$)	1.0	0.2	0

Table 2 displays the nominal parameter values for the Eikenberry model. The values used here differ slightly from those in [7] and were chosen so that the total tumor cell populations from both the Logistic Swanson model, Eq. (2), and the Eikenberry model grow at approximately the same rate. The diffusion rates, D_{G} and D_{M}, in the Eikenberry model are smaller in regions corresponding to cerebrospinal fluid (CSF) than those in the Swanson model. See [7] for further details.

Both sets of equations are integrated using a brain geometry from the BrainWeb database, developed by the McConnel Brain Imaging Center of the Montreal Neurological Institute at McGill University [2]. We use the discrete anatomical model of a normal brain generated by McGill's MR simulator, which consists of a $181 \times 217 \times 181$ isotropic grid of $1\ \text{mm}^3$ voxels in Talairach space [31]. Each voxel is classified as background, CSF, gray or white matter, fat, muscle/skin, skin, skull, or glial matter. The classification determines the boundary of the model domain and the diffusion and haptotaxis coefficients at each grid point. To reduce the computational expense, the equations are integrated over a representative coronal slice at the center of the three-dimensional domain, from which voxels representing the skull and similar tissue have been removed. The resulting two-dimensional domain is a fixed 145×143 grid (the mass effect is not modeled). For simulation purposes, glial matter is treated as white matter, and the domain has been edited by hand to enumerate voxels comprising the corpus callosum (which is a favored migratory pathway for GBM cells [1]).

The spatial derivatives are approximated by finite differences, and the resulting set of ordinary differential equations is integrated using the second-order (in time) Heun's method with a fixed time step 0.5 day^{-1} in the case of Logistic Swanson and a fourth-order Runge–Kutta method for the Eikenberry with fixed timestep of 0.1 day^{-1}. The numerical methods chosen here are for computational efficiency and haven't exhibited instabilities with the chosen time steps. Given the discrete nature of the brain geometry, location-dependent parameters (such as the diffusion constants) are taken to be piecewise constant.

These models are chosen for this initial investigation because they are inexpensive to integrate, particularly in two dimensions, and are adequate for illustrating the basic ideas behind data assimilation, which is our principal focus here. Other efforts have modeled important aspects of GBM growth such as various forms of treatment [32], the mass effect of the tumor, and directed diffusion of GBM cells along white matter tracts [3]. Such refinements would undoubtedly be part of any data assimilation system intended for clinical application.

3 Ensemble Forecasting

Ensemble forecasting, a technique used to assess and quantify the effect of uncertainty in a mathematical model of a dynamical system, was developed from early studies of chaotic behavior. A classic example, formulated by Edward Lorenz in 1963 [19], consisted of a system of three coupled ordinary differential equations modeling fluid flow. The system exhibited sensitive dependence on initial conditions. That is, small errors in non-fixed-point initial conditions quickly propagated in time, leading to large differences in solutions. Even though trajectories had similar limit sets, they became uncorrelated over time even when the initial conditions were very similar. In the case of weather forecasting, like many other systems, one cannot sample the atmosphere at every point, observations are corrupted by noise, and any given weather model is imperfect. This, coupled with chaotic behavior, leads to models that offer no predictive advantage over climatological averages. Even on timescales of a few days or less, uncertainties in the initial state of the atmosphere may lead to substantial forecast errors.

To account for uncertainties in a forecast model, Lorenz suggested in 1965 [20] that, instead of simulating a single initial condition under the model from a best estimate of the state of the atmosphere, one should evolve a set or *ensemble* of initial conditions, each from a statistically equivalent estimate of the true initial state. Then the ensemble gives a Monte Carlo estimate of the uncertainty in a given weather model. Under assumptions discussed in the next section, the ensemble mean constitutes an empirical maximum-likelihood estimate of the true state of the atmosphere. Ensemble forecasting became part of the routine operations at the U.S. and European weather centers in 1992 [16].

Figure 3 shows representative ensemble forecasts of geopotential height contours at 500 hPa. Each curve shows the result, from one initial condition on Oct. 12, 2010,

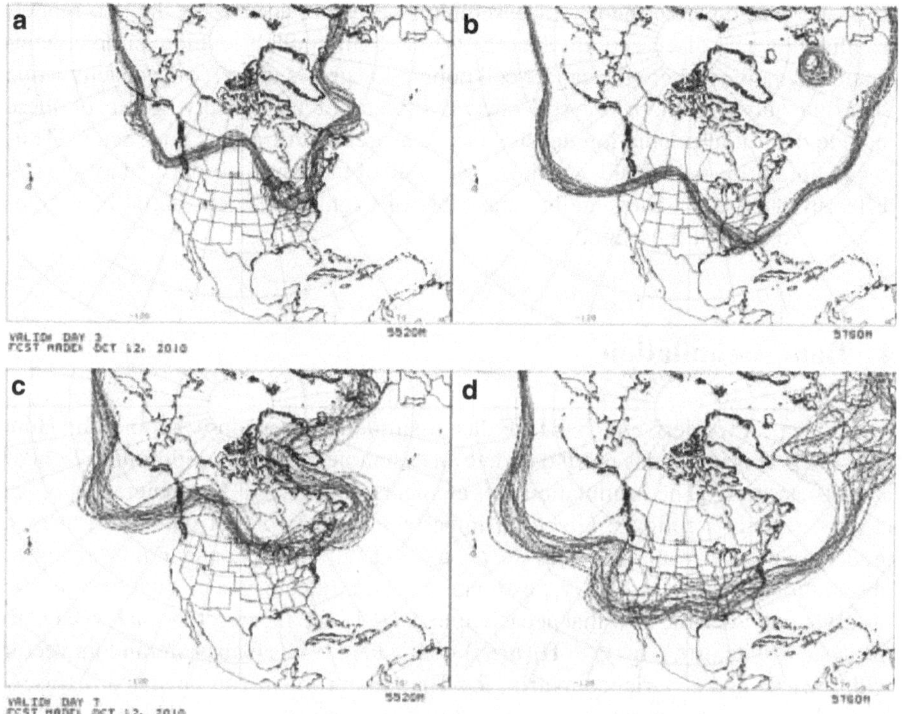

Fig. 3 Ensemble forecasts. Shown are contours of the 500 hPa geopotential height over North America for forecasts started on Oct. 12, 2010. The *top panels* show the predicted values after 3 days and the *bottom panels*, after 7 days

of a forecast obtained by running the weather model for 3 days (top panels) and 7 days (bottom panels). Roughly speaking, the maps show the predicted locations where half the atmosphere's mass is above 5,520 m (left panels) and 5,760 m (right panels).[1]

Of greatest interest here is how the maps illustrate the magnitude of forecast uncertainty, which varies considerably in space as well as time. Because of the chaotic dynamics, the forecast uncertainty generally is larger at 7 days than at 3 days. The 5,760-m contours (right panels) show considerable spread over the North Atlantic Ocean at 7 days, corresponding to especially large uncertainties in the forecast locations of the 500-hPa geopotential height.

[1] The geopotential, $\Phi(z)$, is the work needed to raise a unit mass a vertical distance z from mean sea level and accounts for the variation of the earth's gravitational field with latitude and elevation. The geopotential height is $\Phi(z)/g_0$, where $g_0 = 9.80665\,\mathrm{m\,s^{-2}}$ is the global average of gravitational acceleration at mean sea level. For more details, see Chap. 1 of [12].

In Sect. 5 we illustrate the use of ensemble forecasting adapted to the two models of glioblastoma discussed in Sect. 2. While both models exhibit simple, non-chaotic dynamics where the cancer cell population grows to carrying capacity while diffusing outward, there are still several sources of uncertainty. Some of these include the initial population density and measurement of model parameters. Thus in our simulation we consider both an ensemble of models distinguished by slightly different parameter values and an ensemble of slightly different initial conditions assigned to each of these models.

4 Data Assimilation

In this section we derive the LETKF data assimilation procedure authored by Hunt et al. [13]. This method is used to update an ensemble of initial conditions in light of new observations. The formulation uses elements from [9,13]. The general approach may be stated as follows: Given an imperfect forecast model which advances a model trajectory from time t_{n-1} to t_n, $\mathbf{u}_{t_n} = \mathbf{F}(\mathbf{u}_{t_{n-1}}, t_{n-1})$, and noise-corrupted observational data, $\mathbf{y}_1^o, \mathbf{y}_2^o, \ldots, \mathbf{y}_n^o$, estimate solution trajectory, $\mathbf{u}(t)$, that best fits the observations. Assume now that at each time t_i, $i \in 1 \ldots n$, the observation is related to the system state, $\mathbf{u}(t_i)$, by $\mathbf{y}_i = \mathbf{H}_i(\mathbf{u}(t_i)) + \varepsilon_i$, where ε_i is a Gaussian random vector with mean $\mathbf{0}$ and covariance matrix, $\mathbf{R}_i$. Then the problem can be stated precisely by seeking the maximum-likelihood estimator of the trajectory that best fits the observational data. In other words, we wish to maximize the likelihood function

$$L[\mathbf{u}(t)] = \prod_{j=1}^{n} \exp\Big(-\frac{1}{2} [\mathbf{y}_j^o - \mathbf{H}_j(\mathbf{u}(t_j))]^{\mathrm{T}} \mathbf{R}_j^{-1} [\mathbf{y}_j^o - \mathbf{H}_j(\mathbf{u}(t_j))] \Big). \tag{10}$$

Taking the log of Eq. (10) we see that its maximizer corresponds to minimization of the cost function defined by

$$J[\mathbf{u}(t)] = \sum_{j=1}^{n} [\mathbf{y}_j^o - \mathbf{H}_j(\mathbf{u}(t_j))]^{\mathrm{T}} \mathbf{R}_j^{-1} [\mathbf{y}_j^o - \mathbf{H}_j(\mathbf{u}(t_j))]. \tag{11}$$

When the forecast model and observation operator are nonlinear, cost function J may not have a unique minimizer, and even if it does, finding it can be computationally difficult. The LETKF combats this problem by approximating the minimum in a manner based on the Kalman filter [14, 15]. We proceed to derive the Kalman filter below, which produces the minimizer analytically for the case of a linear model and observation operator under the assumption that the minimizer represents Gaussian distribution.

4.1 Kalman Filter

The Kalman filter is an iterative update scheme that minimizes the cost function J when the system state evolves according to a linear model, $\mathbf{u}_n = \mathbf{M}_n\mathbf{u}_{n-1}$, and the observation operator is also linear

$$\mathbf{y}_i = \mathbf{H}_i\mathbf{u}_i + \varepsilon_i. \tag{12}$$

We begin the derivation with some assumptions. Suppose at time t_{n-1} we have the minimizer, $\bar{\mathbf{u}}_{\mathrm{a}_{n-1}} = \bar{\mathbf{u}}(t_{n-1})$, which we assume represents a Gaussian distribution with an associated covariance matrix $\mathbf{P}_{\mathrm{a}_{n-1}}$. This assumption is motivated by the fact that a Gaussian distribution propagates to a Gaussian distribution under a linear model. In the absence of new observations the most likely estimate of the true system state is the background

$$\bar{\mathbf{u}}_{\mathrm{b}_n} = \mathbf{M}_n\bar{\mathbf{u}}_{\mathrm{a}_{n-1}}. \tag{13}$$

The background covariance matrix is

$$\mathbf{P}_{\mathrm{b}_n} = \mathbf{M}_n\mathbf{P}_{\mathrm{a}_{n-1}}\mathbf{M}_n^{\mathrm{T}} + \mathbf{C}_n. \tag{14}$$

Here $\mathbf{C}_n$ is assumed to be positive definite and represents model error. For simplicity we assume that $\mathbf{C}_n = 0$. If $\mathbf{u}$ is a state vector and c an arbitrary constant, we assume algebraically that the analysis completes the square of the cost function:

$$\sum_{j=1}^{n-1}[\mathbf{y}_j^o - \mathbf{H}_j\mathbf{M}_{n-1,j}\mathbf{u}]^{\mathrm{T}}\mathbf{R}_j^{-1}[\mathbf{y}_j^o - \mathbf{H}_j\mathbf{M}_{n-1,j}\mathbf{u}] = [\mathbf{u} - \bar{\mathbf{u}}_{\mathrm{a}_{n-1}}]^{\mathrm{T}}\mathbf{P}_{\mathrm{a}_{n-1}}^{-1}[\mathbf{u} - \bar{\mathbf{u}}_{\mathrm{a}_{n-1}}] + c. \tag{15}$$

When a new observation vector, $\mathbf{y}_n^o$, becomes available at time t_n, a simple induction argument applied to Eq. (15) at the new time shows how the updated analysis minimizes the cost

$$\begin{aligned} J[\mathbf{u}(t)] &= [\mathbf{u} - \bar{\mathbf{u}}_{\mathrm{b}_n}]^{\mathrm{T}}\mathbf{P}_{\mathrm{b}_n}^{-1}[\mathbf{u} - \bar{\mathbf{u}}_{\mathrm{b}_n}] + [\mathbf{y}_n^o - \mathbf{H}_n\mathbf{u}]^{\mathrm{T}}\mathbf{R}_n^{-1}[\mathbf{y}_n^o - \mathbf{H}_n\mathbf{u}] \\ &= [\mathbf{u} - \bar{\mathbf{u}}_{\mathrm{a}_n}]^{\mathrm{T}}\mathbf{P}_{\mathrm{a}_n}^{-1}[\mathbf{u} - \bar{\mathbf{u}}_{\mathrm{a}_n}] + \tilde{c}. \end{aligned} \tag{16}$$

If one thinks of the covariance matrices in Eq. (16) as numbers, the effect of updating the analysis is intuitive. For example suppose $\mathbf{P}_{\mathrm{b}_n}$ is large compared to $\mathbf{R}_n$. Then the inverse $\mathbf{P}_{\mathrm{b}_n}^{-1}$ will be smaller than $\mathbf{R}_n^{-1}$. Hence, the cost function, J, gives more weight to the observation and the resulting analysis will give more preference to the observations. Figure 1 illustrates the result of this process. The analysiss from the previous step (leftmost circle and set of dots) is evolved under the model resulting in propagation of the uncertainty, represented by the spread of the green dots and size of the blue ovals. The Kalman filter algorithm yields a new analysis that has reduced uncertainty. The lines connecting the dots from the background, analysis, and observation represent the fact that the analysis is computed by finding the optimal linear combination of the discrepancy between the background and observation.

To derive the updated analysis $\bar{\mathbf{u}}_{\mathrm{a}_n}$ we seek a minimum variance, unbiased estimator of the true state, $\mathbf{u}$, that satisfies the recursive update equation

$$\bar{\mathbf{u}}_{\mathrm{a}_n} = \mathbf{K}'_n \bar{\mathbf{u}}_{\mathrm{b}_n} + \mathbf{K}_n \mathbf{y}^o_n. \tag{17}$$

The matrices $\mathbf{K}'$ and $\mathbf{K}$ can be thought of as telling us how much we should trust the background and observations. Below all quantities are at timepoint n so we drop this subscript. We first simplify Eq. (17) by eliminating the matrix $\mathbf{K}'$. To do so consider the estimation error equations

$$\bar{\mathbf{u}}_{\mathrm{a}} = \mathbf{u} + \tilde{\mathbf{u}}_{\mathrm{a}}, \tag{18}$$

$$\bar{\mathbf{u}}_{\mathrm{b}} = \mathbf{u} + \tilde{\mathbf{u}}_{\mathrm{b}}. \tag{19}$$

Substituting Eq. (17) in for $\bar{\mathbf{u}}_{\mathrm{a}}$ and solving for $\tilde{\mathbf{u}}_{\mathrm{a}}$ yield

$$\tilde{\mathbf{u}}_{\mathrm{a}} = \mathbf{K}'\bar{\mathbf{u}}_{\mathrm{b}} + \mathbf{K}\mathbf{y}^o - \mathbf{u}. \tag{20}$$

Now substitute Eq. (19) for $\bar{\mathbf{u}}_{\mathrm{b}}$ and Eq. (12) for $\mathbf{y}^o$ into Eq. (20). This gives, after rearrangement,

$$\tilde{\mathbf{u}}_{\mathrm{a}} = [\mathbf{K}' + \mathbf{K}\mathbf{y}^o - \mathbf{I}]\mathbf{u} + \mathbf{K}'\,\tilde{\mathbf{u}}_{\mathrm{b}} + \mathbf{K}\varepsilon. \tag{21}$$

Since we want the updated analysis to be unbiased, we force $\mathrm{E}[\tilde{\mathbf{u}}_{\mathrm{a}}] = 0$. Taking the expected value of both sides of Eq. (21) we get

$$0 = [\mathbf{K}' + \mathbf{K}\mathbf{H} - \mathbf{I}]\mathbf{u} + \mathbf{K}'\,\mathrm{E}[\tilde{\mathbf{u}}_b] + \mathbf{K}\,\mathrm{E}[\varepsilon] \tag{22}$$

$$= [\mathbf{K}' + \mathbf{K}\mathbf{H} - \mathbf{I}]\mathbf{u}, \tag{23}$$

where we have used the fact that ε is Gaussian with zero mean, and background is unbiased. It follows that

$$\mathbf{K}' = \mathbf{I} - \mathbf{K}\mathbf{H}. \tag{24}$$

Substituting this into Eq. (17) we get the Kalman filter update equation

$$\bar{\mathbf{u}}_{\mathrm{a}} = \bar{\mathbf{u}}_{\mathrm{b}} + \mathbf{K}[\mathbf{y}^o - \mathbf{H}\bar{\mathbf{u}}_{\mathrm{b}}]. \tag{25}$$

In Eq. (25), $\mathbf{K}$ is called the *Kalman Gain* matrix. It represents the proper linear combination of the discrepancy between the background and observations that minimizes the variance of the updated analysis.

To determine $\mathbf{K}$ we first find a convenient form for the analysis covariance matrix $\mathbf{P}_a$, then minimize variance which is defined by the cost function

$$J(\mathbf{K}) = tr(\mathbf{P}_a), \tag{26}$$

where tr stands for the trace of a matrix. Below we assume that the observation and background error are uncorrelated; i.e., $E[\varepsilon\tilde{\mathbf{u}}_b^T] = E[\tilde{\mathbf{u}}_b\varepsilon^T] = 0$. Then since $\tilde{\mathbf{u}}_a$ and $\tilde{\mathbf{u}}_a$ differ only by the state vector, $\bar{\mathbf{u}}$, the analysis covariance matrix is given by

$$\mathbf{P}_a = E[\tilde{\mathbf{u}}_a\,\tilde{\mathbf{u}}_a^T] \tag{27}$$

$$= E\Big\{(\mathbf{I}-\mathbf{KH})\,\tilde{\mathbf{u}}_b\left[\tilde{\mathbf{u}}_b^T(\mathbf{I}-\mathbf{KH})^T + \varepsilon^T\mathbf{K}^T\right] \tag{28}$$

$$+\mathbf{K}\varepsilon\left[\tilde{\mathbf{u}}_b^T(\mathbf{I}-\mathbf{KH})^T\varepsilon^T\mathbf{K}^T\right]\Big\}. \tag{29}$$

Using standard properties of expected value and the assumption of uncorrelated errors we see that Eq. (29) simplifies to

$$\mathbf{P}_a = (\mathbf{I}-\mathbf{KH})\,E[\tilde{\mathbf{u}}_b\,\tilde{\mathbf{u}}_b^T](\mathbf{I}-\mathbf{KH})^T + \mathbf{K}\,E[\varepsilon\varepsilon^T]\,\mathbf{K}^T \tag{30}$$

$$= (\mathbf{I}-\mathbf{KH})\,\mathbf{P}_b\,(\mathbf{I}-\mathbf{KH})^T + \mathbf{KRK}^T. \tag{31}$$

Next let $\mathbf{N} = \mathbf{HP}_b\mathbf{H}^T + \mathbf{R}$. Expanding Eq. (31) we deduce that

$$\mathbf{P}_a = \mathbf{P}_b - \mathbf{KHP}_b + \mathbf{KHP}_b\mathbf{H}^T\mathbf{K}^T - \mathbf{P}_b\mathbf{H}^T\mathbf{K}^T + \mathbf{KRK}^T \tag{32}$$

$$= \mathbf{P}_b - \mathbf{KHP}_b - \mathbf{P}_b\mathbf{H}^T\mathbf{K}^T + \mathbf{K}(\mathbf{HP}_b\mathbf{H}^T + \mathbf{R})\mathbf{K}^T \tag{33}$$

$$= \mathbf{P}_b - \mathbf{KHP}_b - \mathbf{P}_b\mathbf{H}^T\mathbf{K}^T + \mathbf{KNK}^T. \tag{34}$$

To derive the minimizer we differentiate Eq. (26) with respect to $\mathbf{K}$ and apply the identities

$$\frac{\partial}{\partial \mathbf{A}}(tr[\mathbf{ABA}^T]) = 2\mathbf{AB}, \tag{35}$$

$$\frac{\partial}{\partial \mathbf{A}}(tr[\mathbf{AB}]) = \frac{\partial}{\partial \mathbf{A}}(tr[\mathbf{BA}]) = \mathbf{B}^T. \tag{36}$$

This yields

$$\frac{\partial J}{\partial \mathbf{K}} = -2\mathbf{P}_b\mathbf{H}^T + 2\mathbf{KN}. \tag{37}$$

Setting the derivative to zero and solving for $\mathbf{K}$ we find the candidate for $\mathbf{K}$ is

$$\mathbf{K} = \mathbf{P}_b\mathbf{H}^T\mathbf{N}^{-1}. \tag{38}$$

To ensure this minimizes Eq. (26) we show that the Hessian is positive semi-definite. A second calculation using Eqs. (35) and (36) shows that

$$\frac{\partial^2 J}{\partial \mathbf{K}^2} = 2\mathbf{N}^{\mathrm{T}}, \tag{39}$$

which is positive semi-definite because $\mathbf{N}$ is the sum of two covariance matrices which are positive definite by definition. Note there are many equivalent forms for $\mathbf{K}$ and $\mathbf{P}_{\mathrm{a}}$. Here we state the most computationally efficient forms

$$\bar{\mathbf{u}}_{\mathrm{a}} = \bar{\mathbf{u}}_{\mathrm{b}} + \mathbf{P}_{\mathrm{a}}\mathbf{H}^{\mathrm{T}}\mathbf{R}^{-1}[\mathbf{y}^{o} - \mathbf{H}\bar{\mathbf{u}}_{\mathrm{b}}], \tag{40}$$

$$\mathbf{P}_{\mathrm{a}} = (\mathbf{I} + \mathbf{P}_{\mathrm{b}}\mathbf{H}^{\mathrm{T}}\mathbf{R}^{-1}\mathbf{H})^{-1}\mathbf{P}_{\mathrm{b}}, \tag{41}$$

$$\mathbf{K} = \mathbf{P}_{\mathrm{a}}\mathbf{H}^{\mathrm{T}}\mathbf{R}^{-1}. \tag{42}$$

In the next section we derive the LETKF which is an extension of the basic Kalman filter.

4.2 *Local Ensemble Transform Kalman Filter*

Extension of the Kalman filter to the nonlinear scenario entails many difficulties. First, the propagation of the analysis covariance under the model is no longer traceable by Eq. (14). Second, the equations derived for the analysis mean and covariance matrix must be adapted because of the assumed nonlinearity of $\mathbf{H}$ and $\mathbf{F}$. One approach to this problem that has proven useful in operational meterology is ensemble Kalman filtering [8]. The main technique is to select an ensemble of k trajectories whose covariance is used to approximate $\mathbf{P}_{\mathrm{a}_{n-1}}$. Each ensemble is then advanced under the model to time t_n, and the resulting background ensemble sample covariance is used to estimate $\mathbf{P}_{\mathrm{b}_n}$.

Challenges arise in this approach because the size of k is limited (usually less than a few hundred) due to computational resources and is typically smaller than the model resolution, m. If the spread of the ensemble sufficiently approximates $\mathbf{P}_{\mathrm{b}_n}$, one can generate an accurate analysis. On the other hand, if the spread poorly estimates the background covariance, as is the case when the forecast model has more than k Lyapunov exponents, then analysis fails to correct errors in the forecast model.

The LETKF [13] makes use of localization to overcome the challenges related to the size of k. The strategy is to perform the analysis at each point individually by forming a local ensemble over a subset of the model domain. The hope is that the dynamics at a given point are captured over the local region and relatively low-dimensional. If so the local ensemble will sufficiently estimate the background uncertainty and subsequently correct the background forecast at each point, thereby yielding an updated global analysis over the entire model grid. In the case of weather models it has been shown that dynamics in local regions can be regarded as relatively low-dimensional in comparison to the global weather dynamics [24].

The LETKF has several advantages. First it is computationally efficient because the analysis in each local region can be done in parallel where the dimensionality of the matrices in the Kalman filter Eqs. (40)–(42) is reduced. Second the LETKF uses a convenient basis to perform the optimization contributing significantly to the computational performance of the algorithm. Third the analysis is computed without use of the model equations; thus, the LETKF is *model independent*, a key feature with regard to GBM where the biology is poorly understood and modeling efforts are in their infancy. Studies have proven the LETKF to be an accurate and efficient data assimilation approach for weather and ocean models when the analysis is updated over sufficiently small intervals [11, 13, 30].

Notation

We now derive the LETKF data assimilation procedure. At time t_{n-1} we start with an analysis ensemble consisting of m-dimensional model vectors

$$\{\mathbf{u}_{t_{n-1}}^{a(i)} : i = 1 \ldots k\}.$$

For the case of the Eikenberry model each $\mathbf{u}_{t_{n-1}}^{a(i)}$ is comprised of density of the proliferating and migrating cells, chemorepellent, and ECM at every grid point of the model geometry. The mean is regarded as the best estimate of the most likely state of the system. Each ensemble member is advanced under the model until time t_n. This yields the background ensemble

$$\{\mathbf{u}_{t_n}^{b(i)} : i = 1 \ldots k\},$$

where

$$\mathbf{u}_{t_n}^{b(i)} = \mathbf{F}(\mathbf{u}_{t_{n-1}}^{a(i)}, t_{n-1}).$$

For the remainder of this chapter we will simplify the notation by omitting all time subscripts, which are assumed to be t_n. The sample background mean and analysis mean are defined by

$$\bar{\mathbf{u}}_\mathrm{b} = k^{-1} \sum_{i=1}^{k} \mathbf{u}^{b(i)},$$

$$\mathbf{P}_\mathrm{b} = (k-1)^{-1} \sum_{i=1}^{k} (\mathbf{u}^{b(i)} - \bar{\mathbf{u}}_\mathrm{b})(\mathbf{u}^{b(i)} - \bar{\mathbf{u}}_\mathrm{b})^\mathrm{T} \quad (43)$$

$$= (k-1)^{-1} \mathbf{U}^b (\mathbf{U}^b)^\mathrm{T}, \quad (44)$$

where the ith column of the $m \times k$ background ensemble perturbation matrix, $\mathbf{U}^b$, is given by $\mathbf{u}^{b(i)} - \bar{\mathbf{u}}_\mathrm{b}$. The problem now is to determine a suitable analysis ensemble, $\{\mathbf{u}^{a(i)} : i = 1 \ldots k\}$, that has the appropriate mean and covariance matrix

$$\bar{\mathbf{u}}_\mathrm{a} = k^{-1} \sum_{i=1}^{k} \mathbf{u}^{a(i)},$$

$$\mathbf{P}_\mathrm{a} = (k-1)^{-1} \sum_{i=1}^{k} (\mathbf{u}^{a(i)} - \bar{\mathbf{u}}_\mathrm{a})(\mathbf{u}^{a(i)} - \bar{\mathbf{u}}_\mathrm{a})^\mathrm{T} \tag{45}$$

$$= (k-1)^{-1} \mathbf{U}^a (\mathbf{U}^a)^\mathrm{T}. \tag{46}$$

Cost Function

Formally the LETKF computes the analysis by approximately minimizing the analogue of Eq. (16) adapted to the nonlinear observation operator:

$$J(\mathbf{u}) = [\mathbf{u} - \bar{\mathbf{u}}_\mathrm{b}]^\mathrm{T} \mathbf{P}_\mathrm{b}^{-1} [\mathbf{u} - \bar{\mathbf{u}}_\mathrm{b}] + [\mathbf{y}^o - \mathbf{H}(\mathbf{u})]^\mathrm{T} \mathbf{R}^{-1} [\mathbf{y}^o - \mathbf{H}(\mathbf{u})]. \tag{47}$$

Difficulties ensue in adapting the Kalman filter Eqs. (40)–(42) because the columns of $\mathbf{U}^b$ sum to zero, which implies $rank(\mathbf{P}_\mathrm{b}) = rank(\mathbf{U}_\mathrm{b}) < k$. It follows then that $\mathbf{P}_\mathrm{b}^{-1}$ is not generally defined for all model vectors. However, $\mathbf{P}_\mathrm{b}$ is symmetric, thus one-to-one of its column space, S. This space is the same as the column space of $\mathbf{U}_\mathrm{b}$, i.e., the span of the background ensemble perturbations vectors. Now if we let $\tilde{S}$ denote a general k-dimensional space, we can treat $\mathbf{U}_\mathrm{b}$ as a linear transformation from $\tilde{S}$ onto S. Then our strategy is to find the appropriate linear combination of background ensemble perturbations, $\bar{\mathbf{w}}_\mathrm{a} \in \tilde{S}$, so that $\bar{\mathbf{u}}_\mathrm{a} = \bar{\mathbf{u}}_\mathrm{b} + \mathbf{U}_\mathrm{b} \bar{\mathbf{w}}_\mathrm{a}$ minimizes Eq. (47).

To justify this approach rigorously, suppose $\mathbf{w}$ has a Gaussian distribution with mean $\mathbf{0}$ and covariance matrix, $(k-1)^{-1}\mathbf{I}$. From properties of Gaussian random vectors we know $\mathbf{u} = \bar{\mathbf{u}}_\mathrm{b} + \mathbf{U}_\mathrm{b}\mathbf{w}$ is Gaussian with mean $\bar{\mathbf{u}}_\mathrm{b}$ and covariance given by Eq. (44). Then the analogous cost function for $\mathbf{w}$ defined on the space $\tilde{S}$ is

$$\tilde{J}(\mathbf{w}) = (k-1)\mathbf{w}^\mathrm{T}\mathbf{w} + [\mathbf{y}^o - \mathbf{H}(\bar{\mathbf{u}}_\mathrm{b} + \mathbf{U}_\mathrm{b}\mathbf{w})]^\mathrm{T} \mathbf{R}^{-1} [\mathbf{y}^o - \mathbf{H}(\bar{\mathbf{u}}_\mathrm{b} + \mathbf{U}_\mathrm{b}\mathbf{w})]. \tag{48}$$

We prove that if $\bar{\mathbf{w}}_\mathrm{a}$ minimizes $\tilde{J}$, then $\bar{\mathbf{u}}_\mathrm{a} = \bar{\mathbf{u}}_\mathrm{b} + \mathbf{U}_\mathrm{b}\bar{\mathbf{w}}_\mathrm{a}$ minimizes J. To see this let $\mathbf{P}$ be the orthogonal projection matrix from $\tilde{S}$ onto the subspace spanned by the columns of $\mathbf{U}_\mathrm{b}$. Then $\mathbf{P} = \mathbf{U}_\mathrm{b}[\mathbf{U}_\mathrm{b}^\mathrm{T}\mathbf{U}_\mathrm{b}]^{-1}\mathbf{U}_\mathrm{b}^\mathrm{T}$. Next decompose the vector $\mathbf{w}$ as $\mathbf{w} = \mathbf{P}\mathbf{w} + (\mathbf{I} - \mathbf{P})\mathbf{w}$. Substituting this into Eq. (48) for $\mathbf{w}$ only in the first term we get

$$(k-1)\mathbf{w}^\mathrm{T}\mathbf{w} = (k-1)\mathbf{w}^\mathrm{T}[\mathbf{P}\mathbf{w} + (\mathbf{I} - \mathbf{P})\mathbf{w}] \tag{49}$$

$$= (k-1)\mathbf{w}^\mathrm{T}\mathbf{P}\mathbf{w} + (k-1)^{-1}\mathbf{w}^\mathrm{T}(\mathbf{I} - \mathbf{P})\mathbf{w}. \tag{50}$$

Utilizing $\mathbf{w} = \mathbf{U}_\mathrm{b}^{-1}(\mathbf{u} - \bar{\mathbf{u}}_\mathrm{b})$ and the formula for $\mathbf{P}$, we have

$$(k-1)\mathbf{w}^{\mathrm{T}}\mathbf{P}\mathbf{w} = (k-1)(\mathbf{U}_{\mathrm{b}}^{-1}[\mathbf{u}-\bar{\mathbf{u}}_{\mathrm{b}}])^{\mathrm{T}}\mathbf{U}_{\mathrm{b}}[\mathbf{U}_{\mathrm{b}}^{\mathrm{T}}\mathbf{U}_{\mathrm{b}}]^{-1}\mathbf{U}_{\mathrm{b}}^{\mathrm{T}}\mathbf{U}_{\mathrm{b}}^{-1}[\mathbf{u}-\bar{\mathbf{u}}_{\mathrm{b}}] \quad (51)$$

$$= (k-1)(\mathbf{U}_{\mathrm{b}}^{-1}[\mathbf{u}-\bar{\mathbf{u}}_{\mathrm{b}}])^{\mathrm{T}}\mathbf{U}_{\mathrm{b}}^{-1}[\mathbf{u}-\bar{\mathbf{u}}_{\mathrm{b}}] \quad (52)$$

$$= (k-1)[\mathbf{u}-\bar{\mathbf{u}}_{\mathrm{b}}]^{\mathrm{T}}(\mathbf{U}_{\mathrm{b}}^{-1})^{\mathrm{T}}\mathbf{U}_{\mathrm{b}}^{-1}[\mathbf{u}-\bar{\mathbf{u}}_{\mathrm{b}}] \quad (53)$$

$$= (k-1)[\mathbf{u}-\bar{\mathbf{u}}_{\mathrm{b}}]^{\mathrm{T}}([\mathbf{U}_{\mathrm{b}}(\mathbf{U}_{\mathrm{b}})^{\mathrm{T}}]^{-1})[\mathbf{u}-\bar{\mathbf{u}}_{\mathrm{b}}] \quad (54)$$

$$= [\mathbf{u}-\bar{\mathbf{u}}_{\mathrm{b}}]^{\mathrm{T}}(\mathbf{P}_{\mathrm{b}})^{-1}[\mathbf{u}-\bar{\mathbf{u}}_{\mathrm{b}}]. \quad (55)$$

Combining Eqs. (50) and (55) in Eq. (48) we deduce that

$$\tilde{J}(\mathbf{w}) = (k-1)\mathbf{w}^{\mathrm{T}}(\mathbf{I}-\mathbf{P})\mathbf{w} + J(\bar{\mathbf{u}}_{\mathrm{b}} + \mathbf{U}_{\mathrm{b}}\mathbf{w}). \quad (56)$$

The first term on the right is the orthogonal projection of $\mathbf{w}$ onto the null space, N, of $\mathbf{U}_{\mathrm{b}}$; thus, it depends only on the components of $\mathbf{w}$ in the null space. Similarly the second term only depends on the components in the column space, S, of $\mathbf{U}_{\mathrm{b}}$. It follows that $\bar{\mathbf{w}}_{\mathrm{a}}$ minimizes $\tilde{J}$ if and only if it is orthogonal to N and $\bar{\mathbf{u}}_{\mathrm{a}}$ minimizes J.

Analysis Mean and Covariance

We now proceed to derive the updated analysis and covariance matrix by computing an approximate minimizer to $\tilde{J}$ based on the Kalman filter equations. The results are the analogue of Eqs. (40) and (41). To do so we first obtain the linear approximation of the observation operator. This is accomplished by first applying $\mathbf{H}$ to each background trajectory, $\mathbf{u}^{b(i)}$. This produces the ℓ-dimensional ($\ell \le m$) vectors that comprise the background observation ensemble

$$\mathbf{y}^{b(i)} = \mathbf{H}(\mathbf{u}^{b(i)}). \quad (57)$$

Here ℓ denotes the spatial dimension of the OBSERVATIONS. Denote $\bar{\mathbf{y}}_{\mathrm{b}}$ as the mean background observation and $\mathbf{Y}_{\mathrm{b}}$ the $\ell \times k$ background observation ensemble perturbation matrix whose ith column is $\mathbf{y}^{b(i)} - \bar{\mathbf{y}}_{\mathrm{b}}$. We then take the linear approximation

$$\mathbf{H}(\bar{\mathbf{u}}_{\mathrm{b}} + \mathbf{U}_{\mathrm{b}}\mathbf{w}) \approx \bar{\mathbf{y}}_{\mathrm{b}} + \mathbf{Y}_{\mathrm{b}}\mathbf{w}. \quad (58)$$

Replacing $\mathbf{H}$ with the above and using the assumption that the background mean $\bar{\mathbf{w}}_{\mathrm{b}} = 0$, the analysis mean satisfies the analogue of Eq. (25):

$$\bar{\mathbf{w}}_{\mathrm{a}} = \mathbf{K}[\mathbf{y}^{o} - \bar{\mathbf{y}}_{\mathrm{b}} - \mathbf{Y}_{\mathrm{b}}\mathbf{w}]. \quad (59)$$

Applying Eqs. (40) and (41) with $\mathbf{Y}_{\mathrm{b}}$ playing the role of $\mathbf{H}$ produces

$$\bar{\mathbf{w}}_{\mathrm{a}} = \tilde{\mathbf{P}}_{\mathrm{a}}\mathbf{Y}_{\mathrm{b}}^{\mathrm{T}}\mathbf{R}^{-1}(\mathbf{y}^{o} - \bar{\mathbf{y}}_{\mathrm{b}}), \quad (60)$$

$$\tilde{\mathbf{P}}_{\mathrm{a}} = [(k-1)\mathbf{I} + \mathbf{Y}_{\mathrm{b}}^{\mathrm{T}}\mathbf{R}^{-1}\mathbf{Y}_{\mathrm{b}}]^{-1}. \quad (61)$$

In model variables the analysis mean and covariance matrix are determined by

$$\bar{\mathbf{u}}_a = \bar{\mathbf{u}}_b + \mathbf{U}_b \bar{\mathbf{w}}_a, \tag{62}$$

$$\mathbf{P}_a = \mathbf{U}_b \tilde{\mathbf{P}}_a \mathbf{U}_b^T. \tag{63}$$

Analysis Ensemble

To complete the derivation of the LETKF we determine an analysis ensemble which has the above mean and covariance. Our strategy is to choose a matrix whose columns sum to zero and have the desired covariance, then shift the columns by $\bar{\mathbf{u}}_a$ to achieve the mean in Eq. (62). Since the LETKF makes use of localization, we also desire the ensemble to continuously depend on the analysis covariance matrix. This ensures that nearby grid points that have similar matrices, $\tilde{P}_a$, yield similar analysis ensembles. One choice with the desired properties is $\mathbf{U}_a = \mathbf{U}_b \mathbf{W}_a$, where $\mathbf{W}_a$ is the symmetric square root defined by

$$(k-1)\tilde{\mathbf{P}}_a = \mathbf{W}_a \mathbf{W}_a^T. \tag{64}$$

Direct calculation from Eq. (46) shows that

$$\begin{aligned} \mathbf{P}_a &= \mathbf{U}_b \tilde{\mathbf{P}}_a \mathbf{U}_b^T \\ &= \mathbf{U}_b (k-1)^{-1} \mathbf{W}_a \mathbf{W}_a^T \mathbf{U}_b^T \\ &= (k-1)^{-1} \mathbf{U}_a \mathbf{U}_a^T, \end{aligned}$$

which agrees with Eq. (63). The final step is to show the columns of $\mathbf{U}_a$ sum to zero and then adjust them by the appropriate vector to ensure that Eq. (62) holds. This is equivalent to showing $\mathbf{U}_a \mathbf{v} = \mathbf{0}$ where $\mathbf{v}$ is the k by 1 vector defined by $\mathbf{v} = (\mathbf{1}, \mathbf{1}, \ldots, \mathbf{1})^T$. To confirm this note that because the columns of $\mathbf{Y}_b$ sum to zero we have

$$\begin{aligned} \tilde{\mathbf{P}}_a^{-1} \mathbf{v} &= (k-1)\mathbf{v} + \mathbf{Y}_b^{-1} \mathbf{R}^{-1} \mathbf{Y}_b \mathbf{v} \\ &= (k-1)\mathbf{v}. \end{aligned}$$

Thus $\mathbf{v}$ is an eigenvector for $\tilde{\mathbf{P}}_a$ with corresponding eigenvalue $(k-1)^{-1}$. This implies $\mathbf{v}$ is also an eigenvector for $\tilde{\mathbf{P}}_a$ with eigenvalue $(k-1)$. But then from Eq. (64) we have $\mathbf{P}_a \mathbf{v} = (k-1)^{-1} \mathbf{W}_a \mathbf{W}_a^T \mathbf{v}$. It follows then that $\mathbf{v}$ is an eigenvector for $\mathbf{W}_a$ with eigenvalue 1. From this we see that

$$\begin{aligned} \mathbf{U}_a \mathbf{v} &= \mathbf{U}_b \mathbf{W}_a \mathbf{v} \\ &= \mathbf{U}_b \mathbf{v} \\ &= 0, \end{aligned}$$

due to the columns of $\mathbf{U}_b$ summing to 0. To form the analysis we first add $\bar{\mathbf{w}}_a$ to each column vector, $\mathbf{w}^{a(i)}$, of $\mathbf{W}_a$. Denote the resulting vectors $\tilde{\mathbf{w}}^{a(i)}$ and matrix $\widetilde{\mathbf{W}}_a$. Then the ith analysis ensemble member is defined by $\mathbf{u}^{a(i)} = \bar{\mathbf{u}}_b + \mathbf{U}_a \tilde{\mathbf{w}}^{a(i)}$. The updated analysis mean is

$$\begin{aligned}\bar{\mathbf{u}}_a &= k^{-1} \sum_{i=1}^{k} (\bar{\mathbf{u}}_b + \mathbf{U}_a \tilde{\mathbf{w}}^{a(i)}) \\ &= \bar{\mathbf{u}}_b + \mathbf{U}_b \bar{\mathbf{w}}_a + k^{-1} \mathbf{U}_b \sum_{i=1}^{k} \mathbf{w}^{a(i)} \\ &= \bar{\mathbf{u}}_b + \mathbf{U}_b \bar{\mathbf{w}}_a + k^{-1} \mathbf{U}_b \mathbf{W}_a \mathbf{v} \\ &= \bar{\mathbf{u}}_b + \mathbf{U}_b \bar{\mathbf{w}}_a,\end{aligned}$$

as desired.

4.3 Computational Implementation of the LETKF

Computation of the analysis mean, covariance and ensemble as derived above is accomplished through the following steps. The LETKF procedure begins with several preliminary calculations carried out over the entire model grid. First the observation operator is applied to the m-dimensional background ensemble vectors $\mathbf{u}^{b(i)}$ to form the background observation ensemble $\mathbf{y}^{b(i)}$. Next both ensembles are averaged and the vectors $\mathbf{y}^{b(i)} - \bar{\mathbf{y}}_b$ and $\mathbf{u}^{b(i)} - \bar{\mathbf{u}}_b$ are computed. These vectors are then used to form the perturbation matrices $\mathbf{U}_b$ and $\mathbf{Y}_b$. The remaining steps below are performed for each local region.

1. Select the components of $\mathbf{u}^{b(i)}$, $\mathbf{y}^{b(i)}$, $\mathbf{U}_b$, $\mathbf{Y}_b$, and $\mathbf{R}$ that correspond to the local region. We denote the resulting local background ensemble and background ensemble perturbation matrix $\mathbf{x}^{b(i)}$ and $\mathbf{X}_b$, respectively.
2. Compute the $k \times \ell$ matrix $\mathbf{C} = \mathbf{Y}_b^T \mathbf{R}^{-1}$ (If the observations are not independent and $\mathbf{R}$ is not diagonal, it is computationally more efficient to solve the system $\mathbf{R}\mathbf{C}^T = \mathbf{Y}_b$ instead of inverting $\mathbf{R}$.).
3. Compute the $k \times k$ symmetric matrix $\tilde{\mathbf{P}}_a = [(k-1)\mathbf{I}/\rho + \mathbf{C}\mathbf{Y}_b]^{-1}$ (See below for more discussion of ρ.).
4. Compute the $k \times k$ matrix $\mathbf{W}_a = [(k-1)\tilde{\mathbf{P}}_a]^{1/2}$. This choice ensures that W_a depends continuously on the elements of $\tilde{\mathbf{P}}_a$ (Otherwise, small changes in $\tilde{\mathbf{P}}_a$ at neighboring grid points can lead to very different analysis ensembles [13, 34].).
5. Compute the k-vector $\bar{\mathbf{w}}_a = \tilde{\mathbf{P}}_a \mathbf{C}(\mathbf{y}^o - \bar{\mathbf{y}}_b)$ and add it to each column of $\mathbf{W}_a$ to form the $k \times k$ analysis weight matrix $\widetilde{\mathbf{W}}_a$.
6. Compute the analysis perturbation matrix $\mathbf{X}_a = \mathbf{X}_b \widetilde{\mathbf{W}}_a$.
7. The analysis ensemble, $\mathbf{x}^{a(i)}$, is formed by adding $\bar{\mathbf{x}}_b$ to the ith column of $\mathbf{X}_a$, $i = 1, 2, \ldots, k$.

The data assimilation process is completed by forming the global analysis ensemble, $\mathbf{u}^{a(i)}$, which consists of the collection of local analysis ensembles, $\mathbf{x}^{a(i)}$, at the center of each local region. In principle, the only free parameters in the LETKF scheme are the ensemble size, k, and the size of each local region. In practice, however, the model is always an imperfect representation of the underlying dynamics. As a result, ensemble methods tend to underestimate the actual background uncertainty, which causes them to underweight the observations when the new analysis is computed. In severe cases, the filter can diverge. One ad hoc remedy is to "inflate" the background ensemble covariance by a tunable parameter. The procedure described above has the effect of multiplying the background ensemble perturbations by $\sqrt{\rho}$, thereby helping to ensure the analysis gives appropriate weight to the observations.

5 Results

In meteorology, tests of proposed data assimilation systems are called *observing system simulation experiments* (OSSEs). Because the weather is a complex multiscale process, one hopes to separate the effects of observation density and error from model error. In a *perfect model* simulation, one creates a "truth run" from a fixed initial condition with the same model that is used to make the ensemble forecasts. At intervals, synthetic noisy observations are generated from the "truth." The goal of the OSSE is to determine how well a forecast ensemble tracks the truth when the synthetic observations are assimilated using a forecast model that is identical to the model used for the truth run. Such experiments can quantify the effect of noise and observation density and frequency on the accuracy of the analyses, since there is no model error.

Contrary to weather prediction, where the models are well developed, efforts to forecast a true patient GBM are likely to have significant sources of model error because current models, such as the ones used here, represent crude idealizations of the true tumor dynamics. Thus our numerical experiments explore a range of sources of model error which are likely to be found in the clinical setting. While a model with a given set of parameter values may reasonably predict the growth of a tumor, the underlying heterogenous and genetically unstable cell population may cause unaccounted for perturbations in key growth and migration rates. Additionally, GBM patients typically experience a combination of treatments including surgery, chemotherapy, and radiotherapy whose effects on the tumor are not well understood or considered here.

Given the above anticipated sources of error, we establish proof of concept for the data assimilation method in the GBM setting. This is accomplished by adapting the LETKF procedure to the two models of glioblastoma discussed in this chapter while accounting for several sources of uncertainty. We start with a "truth" tumor whose dynamics are assumed to be governed exactly by the Eikenberry model, Eqs. (3)–(6), with parameter values given by Table 2. An ensemble of 25 tumors is

Table 3 Ensemble parameter value ranges

Parameter	Range of values
α	$[0.15, 0.35]$
$T_{\max}$	$[8,000, 12,000]$
D_{w}	$[6.5 \times 10^{-4}, 0.16]$

evolved under the Logistic Swanson model, Eq. (2). Tumor heterogeneity and error due to parameter estimation are approximated by assigning each ensemble tumor a unique logistic growth rate (α), carrying capacity ($T_{\max}$), and diffusion rate in white matter (D_{w}) (the preferred path of migration [27]), from a uniformly distributed random variable in the intervals reported in Table 3. The parameter values are assigned at initialization and remain fixed over the entire simulation (In reality parameters are likely to vary in time.) All other parameter values for the ensemble tumors are identical to those reported in Table 1. The truth and ensemble tumors are initialized by integrating each under their respective models until they reach an approximate diameter between 15 mm and 20 mm, measured over grid points with cell density over 80 % of carrying capacity.

The range of parameter values in Table 3 is chosen for several reasons. First because the models used here exhibit simple dynamics, where the cell population generally grows to carrying capacity and diffuses outward, it is necessary numerically to vary parameters such as the carrying capacity over the ensemble. This prevents the background covariance matrix from becoming ill-conditioned and its inverse in Eq. (47) tending to infinity which can cause the filter to diverge or neglect the observations. Second, as our results will show, we wish to demonstrate the ability of the LETKF to better estimate the true tumor state even in the presence of significant model parameter error. For example, we vary D_{w} in Table 3 by several orders of magnitude. The result is a set of tumors that vary greatly in degree of invasion. This is clinically relevant because diffusion rates may vary greatly on a patient by patient basis resulting in substantial uncertainty in the tumor growth.

Our OSSE then begins with generation of a synthetic observation of the truth and the LETKF data assimilation procedure performed using a local region size of 7 mm by 7 mm. The choice of region size is motivated by the empirical assessment that the areas of greatest forecast uncertainty are along the edges of the tumor, meaning the boundary of the enhancing region. This yields an updated analysis after which the truth and ensemble are integrated for 30 days, and the process is repeated for 6 cycles totaling 180 days worth of simulation. The process is finite due to the fatal nature of GBM tumors.

In our simulations the observation operator, $\mathbf{H}(x)$, represents the MRI that would be observed if the tumor state were $\mathbf{x}$. In our initial experiment we regard the tumor state to be the cell density at each point on the model geometry. As previously discussed in the introduction, many details about the relationship between the tumor state and the contrast enhancement are not well characterized, and there is intrinsic variability arising from uptake of the contrast agent and other aspects of MR image generation such as interrater reliability. For our purposes we assume the enhancement varies linearly with the tumor cell density at each point on the domain, up to a random error.

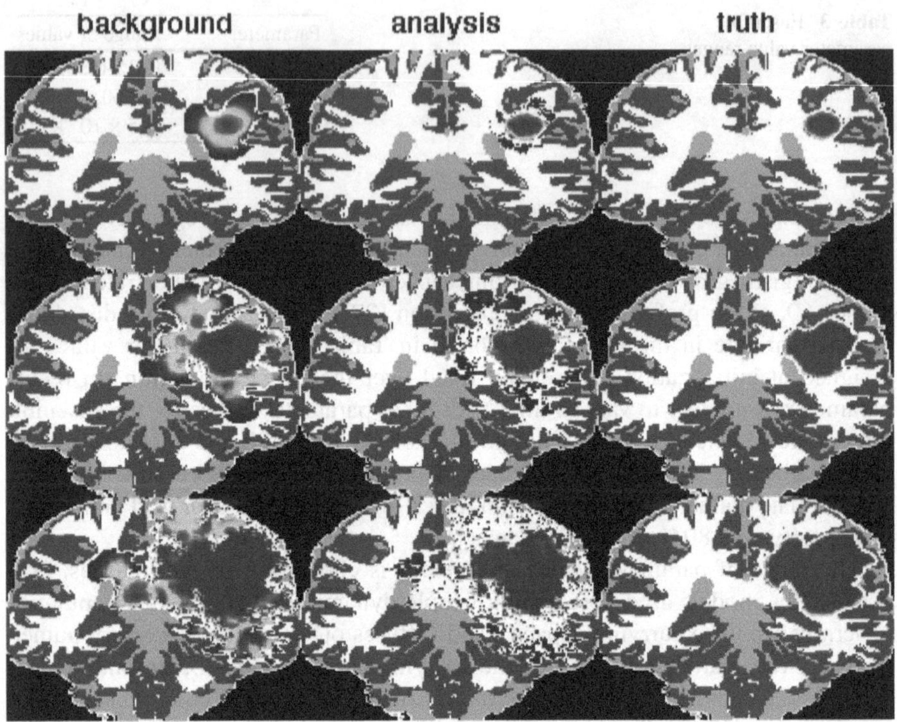

Fig. 4 Cell density plots for our first observing system simulation experiment. The first, second, and third rows correspond to day 1, 90, and 180 of the experiment. See text for additional details

The value of **H** is computed pointwise as follows. Let $u_k(\mathbf{x},t)$ be the tumor cell density for the kth ensemble member at location $\mathbf{x}$ and time t. Then

$$h_k(\mathbf{x}) = \max\left(0, \min\left(1, \frac{u_k(\mathbf{x},t)}{T^k_{\max}} + \eta(\mathbf{x})\right)\right), \tag{65}$$

where $\eta(\mathbf{x})$ is a uniformly distributed random value in the interval $[-0.1, 0.1]$ and $T^k_{\max}$ is the carrying capacity for the kth ensemble solution. The value of h_k (which is confined to the unit interval) is the component of the **H** corresponding to the location $\mathbf{x}$ in the brain domain. Note that the η's are independent. The synthetic observations are computed by applying **H** to the truth with the appropriate carrying capacity.

Figure 4 shows the results of our first experiment. The brain geometry is displayed where the cell density is below 5 cells mm^{-2}. The color coding is done on a 128 color linear scaling analogous to a temperature plot where red corresponds to cell densities at or near the carrying capacity of the truth and blue represents lower cell densities. The first, second, and third rows correspond to days 1, 90, and 180 of the experiment. The left column, labeled "background," shows the background ensemble mean. The middle column, labeled "analysis," shows the analysis mean.

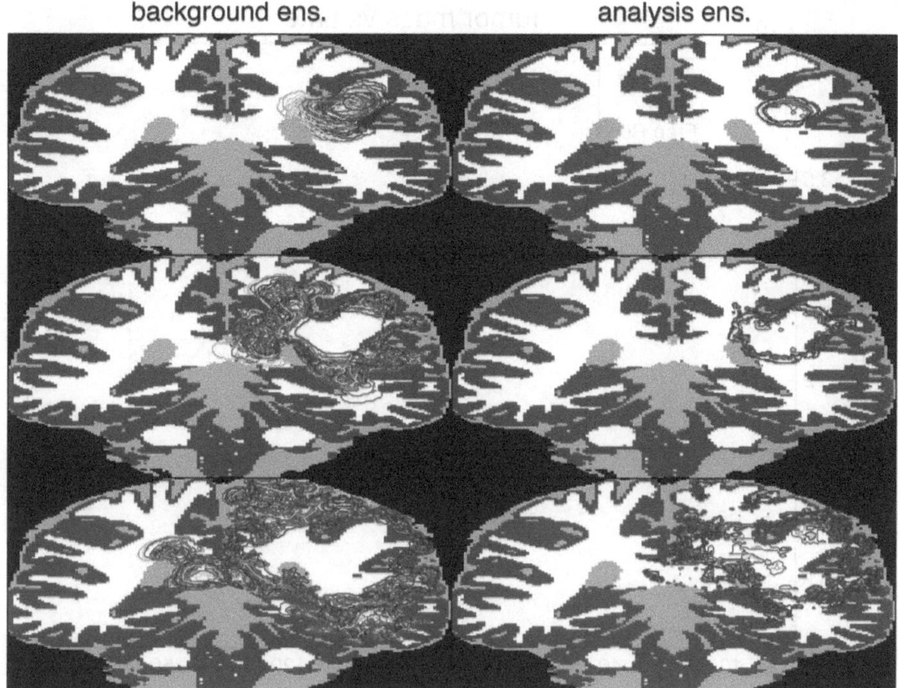

Fig. 5 Contour plot for a subset of the ensemble tumors. See text for additional details

The right column, labeled "truth," displays the truth. We see that each time the data assimilation is performed, the resulting analysis has improved tracking of the truth.

Figure 5 displays the level curves for a five member subset of the ensemble. Two level curves are shown for each ensemble tumor corresponding to 25 % and 80 % of truth carrying capacity. The left and right columns display the background and analysis ensemble, respectively. The decrease in the spread of the level curves following data assimilation illustrates a reduction in uncertainty in estimated size and location of the truth tumor. In clinical practice this would be useful in assessing the location of the tumor boundary and the overall size of the tumor.

For comparison we also conducted a free run of the experiment where data assimilation is not performed and the ensemble is simply integrated for 180 days under their respective models. Figure 6 displays the tumor mass every tenth day of the experiment. The blue curve shows the ensemble mean tumor mass for the simulation where data assimilation is performed. The error bars corresponding to one standard deviation evaluated over the ensemble are also displayed. The red curve displays the tumor mass for the truth. The green curve shows the ensemble mean tumor mass for the free run. Note how data assimilation yields a forecast that more accurately estimates the truth tumor mass. Every 30th day when the LETKF

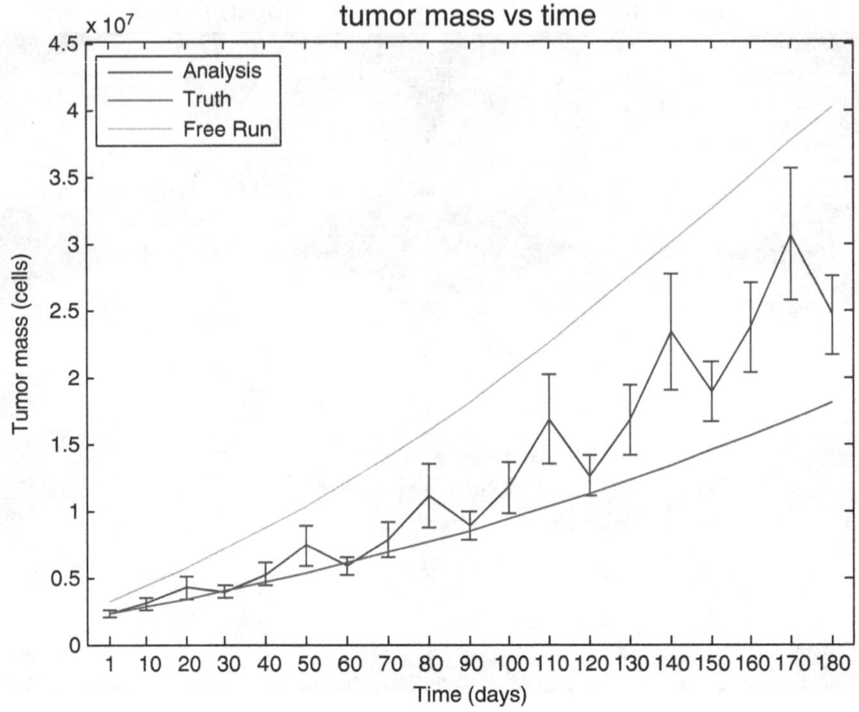

Fig. 6 Plot of tumor mass vs. time for first observing system simulation experiment. See text for additional details

is applied we see a reduction in uncertainty in the estimated tumor mass as well as improved quantitative agreement with the truth.

We perform a second experiment where we assumed the enhancement level is linearly related to the time derivative of the growing cell population (proliferating cells for solutions to Eikenberry model). That is, regions of highest enhancement correspond to the areas with the greatest cell growth rate. The formulation for h_k in this case is derived from the fact that solutions to the logistic equation have maximal derivative at half carrying capacity. Thus to find the upper bound on the time derivative of the growing cells we simply evaluate the right-hand side of the logistic equation, $g\prime = \alpha g(1 - \frac{g}{T_{\max}})$, at $g = \frac{T_{\max}}{2}$. The upper bound on the derivative is $\frac{\alpha T_{\max}}{4}$. Hence the analogue of Eq. (65) is

$$h_k(\mathbf{x}) = \max\left(0, \min\left(1, \frac{4\,\frac{\partial u_k(\mathbf{x},t)}{\partial t}}{\alpha_k\, T_{\max}^k}\right) + \eta(\mathbf{x})\right), \tag{66}$$

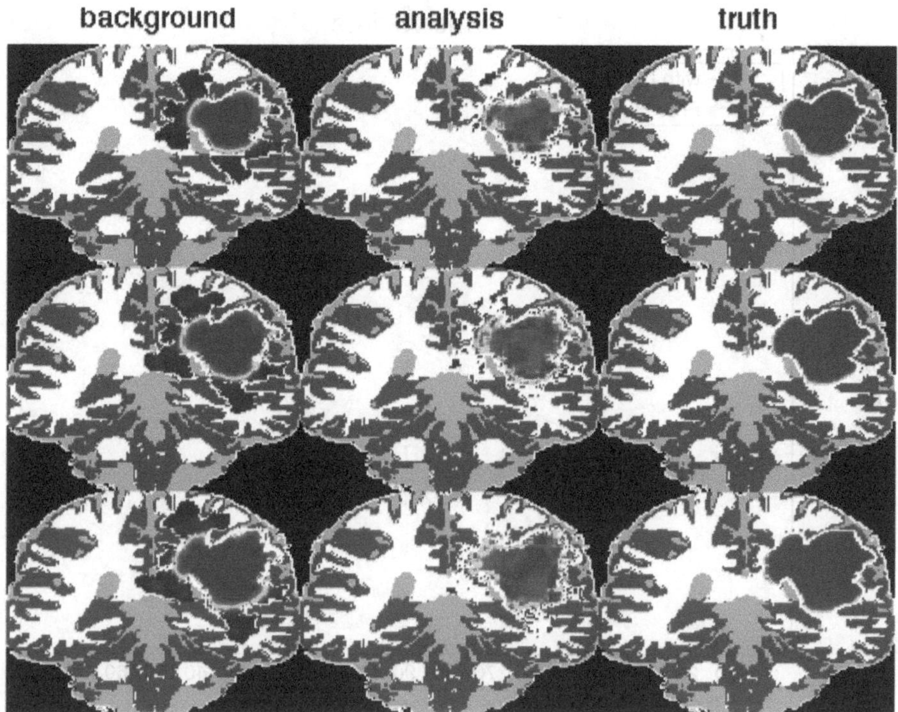

Fig. 7 Cell density plots for the second observing system simulation experiment where Eq. (66) is used to form the components of the observation operator. The first, second, and third rows correspond to days 120, 150, and 180 of the experiment

where α_k is the logistic growth rate of the kth ensemble solution and $\frac{\partial u_k(\mathbf{x},t)}{\partial t}$ is the time derivative of the growing cell density for the kth ensemble member. $\eta(x)$ is as in Eq. (65).

Figure 7 displays results of the experiment with Eq. (66) used to form the components of the observation operator. The rows correspond to days 120, 150, and 180 of the experiment. In this case assimilation of observations yields an analysis that has invaded essentially the same area of the truth, but the tumor core (region composed red grid points) is heterogenous and slightly smaller in overall size. If we compare the analysis on day 120 to the background at day 150, we see that the analysis grows to have about the same tumor core size as the truth between data assimilation steps, but has a greater area of infiltration. Taking into consideration that for the chosen parameter values, the ensemble mean is more invasive than the truth and that the contrast enhancement for this case depends on the time rate of change of the growing cells, we can think of the data assimilation as having a "braking" effect that temporarily slows down the overall growth of the ensemble mean.

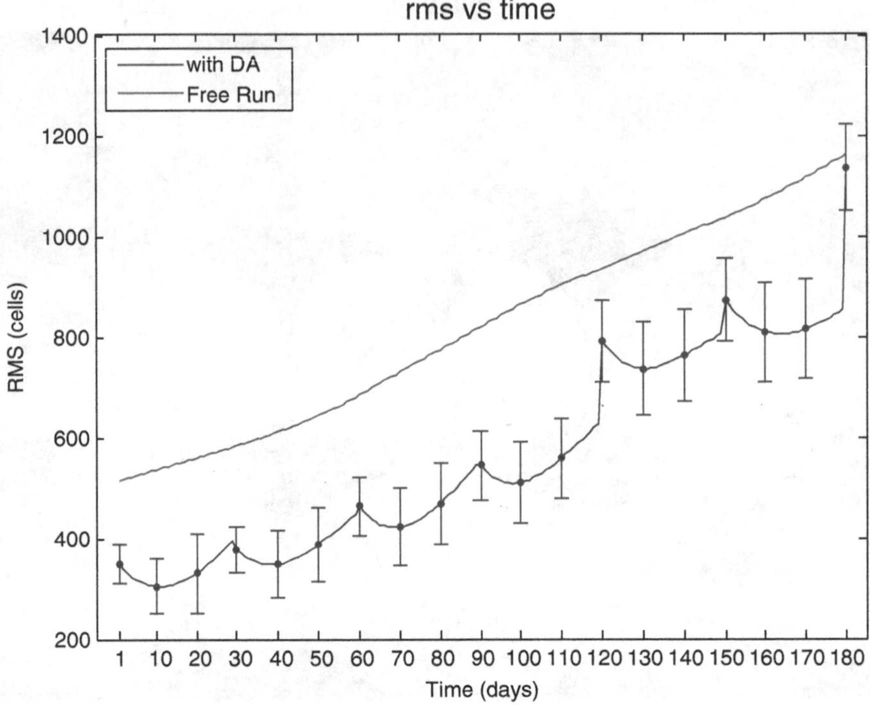

Fig. 8 Plot of the rms vs. time for every tenth day of the second presented experiment. See text for additional details

For this experiment we assess forecast error in terms of the root-mean-square (RMS) difference of the total tumor cell population:

$$\mathrm{rmse}(\overline{T}, t_n) = \left(V^{-1} \sum_{\mathbf{x}} \left[T_{\mathrm{true}}(\mathbf{x}, t_n) - \overline{T}(\mathbf{x}, t_n) \right]^2 \right)^{1/2}, \tag{67}$$

where $\overline{T}$ gives the tumor cell density at each point of the analysis ensemble mean and the sum is over every point of the computational domain (whose two-dimensional volume is V). Figure 8 displays the RMS vs. time plot for this experiment (blue curve) and the free run where no data assimilation is performed (red curve). The error bars on the blue curve display one standard deviation in the RMS evaluated over the ensemble. Observe how application of the LETKF yields an ensemble forecast that has better overall accuracy than the free run of the experiment.

We conducted similar experiments to those presented here where the range of parameter values is changed as well as simulations where the ensemble and truth evolve according to the same model. The overall results were similar in all experiments to what we have reported in this chapter.

Effect of Varying Observation Frequency

We explore the effect of varying the period between assimilation steps. To do so our first experiment is repeated with synthetic observations generated every 1, 7, 30, 60, 90, and 180 day(s). Each simulation is still conducted over a period of 180 days beginning and ending with a data assimilation cycle. Figure 9 shows the final analysis mean cell density. The subplot titled "Daily" corresponds to observations everyday. "Weekly" corresponds to observations every 7 days. The other subplots are titled analogously. In every case but the quarterly run, the final ensemble mean gives a good approximation of the truth tumor core (region in red) with varying degrees of satellitosis. The edema (blue region) surrounding the main tumor mass varies significantly over the experiments and in some cases gives a poor approximation to the truth (see the truth in Fig. 4). This is because when observations are available frequently, for example in the daily run, the errors in the model do not sufficiently propagate between assimilation cycles. The resulting background ensemble has little variance in the edema region causing the updated analysis to essentially ignore the observations locally. Despite varying the frequency of observations, the simulations show that the LETKF data assimilation scheme can still provide an improved approximation to the true tumor state. This is clinically relevant because usually patient images are collected at irregular intervals and typically only at initial diagnosis and before and after surgery.

6 Conclusion and Discussion

The results of this chapter demonstrate the use of ensemble forecasting and data assimilation to make improved estimates of future growth of a simulated glioblastoma given synthetically generated observations of the tumor. The two experiments presented explore different models of the relationship between the tumor state and the contrast enhancement in an MR image. In both cases the ensemble mean had improved tracking of the truth when data assimilation is performed despite the substantial sources of model and observational error. This demonstrates the potential feasibility of this framework for use in human cases of glioblastoma with real patient MR image data.

Several related problems must be considered before the data assimilation approach can be used as a clinical aid. First the effect of surgery, radiotherapy, and chemotherapy on the glioma cell population needs to be incorporated into the forecast model. Similarly any ensemble forecast must account for the uncertainty and variance in key treatment parameters such as the size of the surgical resection cavity, treatment dosage and timing, the application field for radiotherapy or chemotherapy, and development of resistance by glioma cells.

A third problem concerns the use of patient imaging data in an ensemble forecast. Here we have modeled the relationship in an ad hoc way where we assume the pixel intensity directly relates to the tumor state under the statistical assumption of

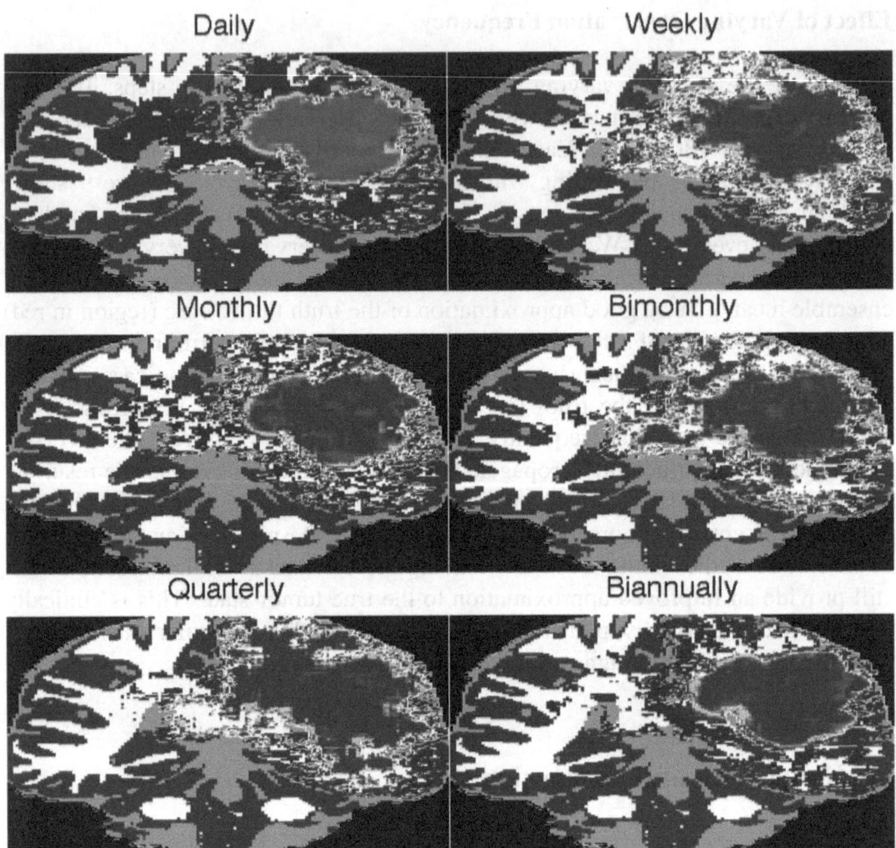

Fig. 9 Final cell density plots for the first experiment repeated where the time period between observations is varied. See text for details

uncorrelated errors that have a Gaussian distribution. In reality it is difficult to model the relationship between cell density and pixel intensity due to the complex means by which tumors evolve. Additionally there are many factors previously discussed that lead to uncertainty in a patient tumor MR image. Also any assumed observation operator will likely have non-Gaussian distributed error with covariance matrix that is difficult to estimate.

Future work will focus on refinement of the forecast model and validation of this approach with real clinical imaging data. Our ultimate goal is to apply data assimilation methods as a treatment aid to help improve management of patient GMB tumors and positively affect their quality of life.

Acknowledgments Portions of this work were funded by the Barrow Neurological Institute Women's Foundation and by funds from the Newsome Family Endowed Chair of Neurosurgery Research held by Dr. Preul. J.M. was supported in part by an Achievement Reward for College Scientists Scholarship. Y.K. was supported by NSF grants DMS-0436341 and DMS-0920744.

References

1. Amberger, V.R., Hensel, T., Ogata, T.N.,and Schwab, M.E.: Spreading and migration of human glioma and rat C6 cells on central nervous system myelin in vitro is correlated with tumor malignancy and involves a metalloproteolytic activity. Cancer Res., **58**, 149–158 (1998)
2. http://www.bic.mni.mcgill.ca/brainweb/.
3. Clatz, O., Sermesant, M., Bondiau, P.-Y., Delingette, H., Warfield, S.K., Malandian, G., and Ayache, N.: Realistic simulation of the 3d growth of brain tumors in MR images coupling diffusion with biomechanical deformation. IEEE Trans. Med. Imaging, **24**, 1334–1346 (2005)
4. Cocosco, C.A., Kollokian, V., Kwan, R.K.-S., and Evans, A.C.: BrainWeb: Online Interface to a 3D MRI Simulated Brain Database NeuroImage, vol.5, no.4, part 2/4, S425, 1997 – Proceedings of 3-rd International Conference on Functional Mapping of the Human Brain, Copenhagen, (1997)
5. Collins, D.L., Zijdenbos, A.P., Kollokian, V., Sled, J.G., Kabani, N.J., Holmes, C.J., and Evans, A.C.: Design and Construction of a Realistic Digital Brain Phantom IEEE Transactions on Medical Imaging, vol.17, No.3, p.463–468, (1998)
6. Demuth, T. and Berens, M.E.: Molecular mechanisms of glioma cell migration and invasion. J. Neurooncol. **70**, 217–228 (2004)
7. Eikenberry, S.E., Sankar, T., Preul, M.C., Kostelich, E.J., Thalhauser, C.J., and Kuang, Y.: Virtual glioblastoma: growth, migration and treatment in a three-dimensional mathematical model. Cell Prolif. **42**, 511–528 (2009)
8. Evensen, G.:Data Assimilation: The Ensemble Kalman Filter, Springer (2006)
9. Gelb A. (ed): Appliede Optimal State Estimation. MIT Press, Cambridge, Ma., (1974)
10. Grossman, A., Helbich, T.H., Kuriyama, N., Ostrowitzki, S., Roberts, T. P., Shames, D.M., van Bruggen, N., Wendland, M.F., Israel, M.A., and Brasch, R.C.: Dynamic contrast-enhanced magnetic resonance imaging as a surrogate marker of tumor response to anti-angiogenic therapy in a xenograft model of glioblastoma multiforme. J. Magn. Reson. Imaging, **15**, 233–240 (2002)
11. Hoffman, R.N, Ponte, R.M., Kostelich, E.J., Blumberg, A., Szunyogh, I., Vinogradov, S.V., and Henderson, J.M.: A simulation study using a local ensemble transform Kalman filter for data assimilation in New York Harbor. J. Atmos. Ocean Tech., **25**, 1638–1656 (2008)
12. Horton, J.R.: An Introduction to dynamic meteorology. 4th ed. Amsterdam: Elsevier Academic Press (2004)
13. Hunt, B.R., Kostelich, E.J., and Szunyogh, I.: Efficient data assimilation for spatiotemporal chaos: A local ensemble transform Kalman filter. Physica D, **230**, 112–126 (2007)
14. Kalman, R.E.: A new approach to linear filtering and prediction problems. Trans. ASME Ser. D: J. Basic Eng., **82**, 35–45 (1960)
15. Kalman, R.E., and Bucy, R.S.: New results in linear filtering and prediction theory. Trans. ASME Ser. D: J. Basic Eng., **83**, 95–108 (1961)
16. Kalnay, E.: Atmospheric modeling, data assimilation, and Predictability. Cambridge University Press (2003)
17. Kwan, R.K.-S., Evans, A.C., and Pike, G.B.: An Extensible MRI Simulator for Post-Processing Evaluation. Visualization in Biomedical Computing (VBC'96). Lecture Notes in Computer Science, vol. 1131. Springer-Verlag, 135–140 (1996)
18. Kwan, R.K.-S., Evans, A.C., and Pike, G.B.: MRI simulation-based evaluation of image-processing and classification methods IEEE Transactions on Medical Imaging. **18(11)**, 1085–97 Nov (1999)
19. Lorenz, E.N.: Deterministic non-periodic flow. J. Atmos. Sci., **20**, 130–141 (1963)
20. Lorenz, E.N.: A study of the predictability of a 28-variable atmospheric model. Tellus, **17**, 321–333 (1965)
21. Marino, S., Hogue, I.B., Ray, C.J., and Kirschner, D.E.: A methodology for performing global uncertainty and sensitivity analysis in systems biology. J. Theor. Biol., **254**, 178-196 (2008)

22. Mohamed A., and Davatzikos, C.: Finite element modeling of brain tumor mass-effect from 3D medical images. In: Duncan J.S., Gerig, G. (eds) 8th International Conference on Medical Image Computing and Computer Assisted Intervention(MICCAI). Springer, Palm Springs CA 400–408 (2005)
23. Norden, A.D., and Wen, P.Y.: Glioma therapy in adults. Neurologist. **12**, 279–292 (2006)
24. Patil, D.J., Hunt, B.R., Kalnay, E., Yorke, J.A., and Ott E.: Local low dimensionality of atmospheric dynamics. Phys. Rev. Lett., **86**, 5878–5881 (2001)
25. Rijpkema, M., Kaanders, J.H., Joosten, F.B., van der Kogel, A.J., and Heerschap, A.: Method for quantitative mapping of dynamic MRI contrast agent uptake in human tumors. J. Magn. Reson. Imaging, **14**, 457–463 (2001)
26. Stein A.M., Demuth T., Mobley D., Berens M., and Sander L.: A mathematical model of glioblastoma tumor spheroid invasion in a 3D *in vitro* experiment. Biophys. J., **92**, 356–365 (2007)
27. Swanson, K.R., Alvord, Jr., E.C., and Murray, J.D.: A quantitative model of differential motility of gliomas in white and grey matter. Cell Prolif., **33**, 317–329 (2000)
28. Swanson, K.R., Bridge C., Murray, J.D., and Alvord, E.C.: Virtual and real brain tumors: using mathematical modeling to quantify glioma growth and invasion. J. Neurol. Sci., **216**, 1–10 (2003)
29. Swanson, K.R., Rostomily, R.C., and Alvord, Jr., E.C.: A mathematical modeling tool for predicting survival of individual patients following resection of glioblastoma: A proof of principle. Brit. J. Cancer, **98**, 113–119 (2008)
30. Szunyogh, I., Kostelich, E.J., Gyarmati, G., Kalnay, E., Hunt, B.R., Ott, E., Satterfield, E., and Yorke, J.A.: A local ensemble Kalman filter data assimilation system for the NCEP global model. Tellus A, **60**, 113–130 (2008)
31. Talairach, J. and Tournoux, P.: Co-Planar Stereotaxic Atlas of the Human Brain: 3-D Proportional System: An Approach to Cerebral Imaging. Thieme Medical Publishers, New York (1988)
32. Tian, J.P., Friedman, A., Wang, J., and Chiocca, E.A.: Modeling the effects of resection, radiation and chemotherapy. J. Neurooncol, **91**, 287–293 (2009)
33. Vose, M.D.: The Simple Genetic Algorithm: Foundations and Theory. The MIT Press, Cambridge, MA (1999)
34. Wang, X., Bishop, C.H., and Julier, S.J.: Which is better, an ensemble of positive negative pairs or a centered spherical simplex ensemble?. Mon. Wea. Rev., **132**, 1590–1605 (2004)

Part IV
Cancer Treatment

Optimisation of Cancer Drug Treatments Using Cell Population Dynamics

Frédérique Billy, Jean Clairambault, and Olivier Fercoq

1 Introduction

Cancer is primarily a disease of the physiological control on cell population proliferation. Tissue proliferation relies on the cell division cycle: one cell becomes two after a sequence of molecular events that are physiologically controlled at each step of the cycle at so-called checkpoints, in particular at transitions between phases of the cycle [105]. Tissue proliferation is the main physiological process occurring in development and later in maintaining the permanence of the organism in adults, at that late stage mainly in fast renewing tissues such as bone marrow, gut and skin.

Proliferation is normally controlled in such a way that tissue homeostasis is preserved. By tissue homeostasis we mean permanence in the mean of tissue in volume, mass and function to ensure satisfaction of the needs of the whole organism. In cancer tissues, this physiological control, which also relies on the so-called checkpoints in the division cycle of individual replicating cells, is disrupted, leading to an overproduction of cells that eventually results in the development of tumours.

Anticancer drugs all attack the cell division cycle, either by slowing it down (possibly until quiescence, i.e., non-proliferation, cells remaining alive), or by blocking it at checkpoints, which in the absence of cell material repair eventually leads to cell death.

Various mathematical models have been proposed to describe the action of anticancer drugs in order to optimise it, that is to minimise the number of cancer cells or a related quantity, as the growth rate of the cancer cell population.

F. Billy (✉) • J. Clairambault
INRIA BANG team, BP 105, F78153 Rocquencourt, and LJLL,
UPMC, 4 Place Jussieu, F75005 Paris, France
e-mail: frederique.billy@inria.fr; jean.clairambault@inria.fr

O. Fercoq
INRIA MAXPLUS team, CMAP, Ecole Polytechnique, F91128 Palaiseau, France
e-mail: olivier.fercoq@inria.fr

U. Ledzewicz et al., *Mathematical Methods and Models in Biomedicine*, Lecture Notes on Mathematical Modelling in the Life Sciences, DOI 10.1007/978-1-4614-4178-6_10,

The constraints at stake, met everyday in the clinic of cancers, are related mainly to resistance to treatment in cancer cell populations and to unwanted toxicity in healthy tissues.

We briefly review some of these models, namely ordinary differential equation (ODE) models, partial differential equation (PDE) models with spatial structure, phase structured cellular automata and physiologically structured PDE models. We do not claim to be exhaustive in a field where so much has been published in the last 50 years. However, we present the main models used in cancer treatment in the last decades, together with the biological phenomena that can be described by each of them.

We then present some techniques used for the identification of the parameters of population dynamic models used in chemotherapy. We also briefly review theoretical therapeutic optimisation methods that can be used in the context of different models of cell population growth, according to the clinical problem at stake, to the available data and to the chosen model, with their advantages and pitfalls.

We then focus on a novel method of optimisation under unwanted toxicity constraints, presented here in the context of cancer chronotherapeutics. This method is based on optimisation of eigenvalues in an age-structured model of cell population dynamics, the parameters of which can be identified in cell cultures by using recent intracellular imagery techniques relying on fluorescence quantification. Thanks to these fine level quantitative cell observations, the structured cell population model, which takes the cell division cycle into account, gives interesting results for the optimisation of the pharmacological treatments of cancer.

Finally, bases for cell population dynamic models, with external control targets, that ought to be used to physiologically represent the effects of different anticancer drugs in use in the clinic, are sketched, as are possible schemes for multitarget multidrug delivery optimisation, designed to meet present clinical challenges in everyday oncology.

2 Drugs Used in Cancer Treatments and Their Targets

Although these are important co-occurring phenomena in cancer and make all its malignancy, tissue invasion and the development of distant metastases will not be mentioned here, for the simple reason that no efficient treatments exist against them specifically so far. For instance, matrix metalloprotease inhibitors (anti-invasive agents) were tested in the past, even until phase II clinical trials, but their development has been arrested in particular due to high toxicity. We will thus stick to drugs that impact on local cancer growth, a process which itself is initiated by growing impairment of the normal physiological control on the cell division cycle in cell populations.

From this point of view, drugs used in cancer treatments may be roughly divided into those that drive cells to their death, which we define here as cytotoxic, and those that just slow down the cell division cycle, letting cells alive, but containing tumour development, which we will define here as cytostatic. Note that "cytotoxic" and "cytostatic" are terms on which consensus is not so widespread, hence this necessary precision for our purposes.

2.1 Fate of Drugs in the Organism: Molecular Pharmacokinetics- Pharmacodynamics

Anticancer drugs are delivered into the general circulation, either directly by intravenous infusion, or indirectly by oral route, intestinal absorption and enterohepatic circulation (*i.e.*, entry in the general blood circulation from the intestine via the portal vein towards the liver, and possibly back from the liver to the intestine via bile ducts). Their fate, from introduction in the circulation until presence of an active metabolite in the neighbourhood of their intracellular targets, can be represented by *pharmacokinetic (PK)* compartmental ODEs for their concentrations. It is also theoretically possible to represent this fate by spatial PDEs with boundary conditions instead of exchange rules between compartments when data on spatial diffusion of the drugs and some geometry of their distribution domain is known—but this is seldom the case.

Then, in the cell medium, either an individual cell, or a mean intracellular medium in a population of cells, *pharmacodynamic (PD)* differential equations must be used to relate local drug concentrations with molecular effects on their targets. At this level of description, it is a priori more relevant to describe by physiologically structured than by spatially structured models the population of cells under pharmacological attack, since anticancer drugs act mainly by blocking the cell division cycle, which does not give rise to a spatially structured cell population (apart from the very early stages of avascular spheroid tumour growth, little geometry is relevant to describe a tumour seen under the microscope).

2.2 Cytotoxics and Cytostatics

Driving cells to their death may be obtained either by damaging the genome, or more indirectly by impairing essential mechanisms of the cell division cycle, such as enzymes thymidylate synthase (an enzyme that plays an essential role in DNA synthesis and is one of the main targets of cytotoxic drug 5-FU) or topoisomerase I (another essential enzyme of DNA synthesis, target of cytotoxic drug irinotecan). The resulting damaged cell, unable to proceed until division into two viable cells, is normally blocked at one or the other checkpoint, mainly G_1/S or G_2/M (recall that the cell division cycle is classically divided into four successive phases, G_1, S for DNA synthesis, G_2, and M for mitosis).

Then, unless it may be repaired by specific enzymes—that are often overexpressed in cancer cells—these impaired cells, blocked at a checkpoint, are subsequently sent to "clean death" by the physiological mechanism of apoptosis (also possibly impaired in cancer, resulting in abnormal cells bypassing these checkpoints). As mentioned above, we define here this class of drugs that have for their ultimate mission to kill cancer cells—even if their primary action is not to directly damage the genome, but rather to damage cell cycle enzymes—as *cytotoxics*.

We reserve the term *cytostatic* to those non-cell-killing drugs that merely slow down proliferation, usually by maintaining cells in G_1 with possible exit to G_0, that is by definition the quiescent phase, i.e., the subpopulation of cells that are not committed in the cell division cycle. Indeed, before the restriction point inside G_1, cells may stop their progression in the cell cycle and go back to quiescence in G_0. This last category comprises all drugs that act as antagonists of growth factors, which may be monoclonal antibodies or tyrosine kinase inhibitors.

2.3 *Drugs That Act on the Peritumoral Environment*

We might also attach to this "cytostatic" category drugs that act on the tumour environment, such as anti-angiogenic drugs (that impoverish it in oxygen and nutrients) or molecules that could be able to modify local pH in a sense unfavourable to cancer cells, assuming that their action is not to entrain cell death, but to indirectly slow down progression in the cell cycle, which is possible mainly, if not only, in the G_1 phase.

Acting on cancer cell populations without killing any of these cells but only by limiting their thriving, even inducing their decay, by unfavourable environmental conditions is certainly an ideal therapy, which avoid cell killing also in healthy cell populations, and it has been achieved in some rare cases in monotherapy, e.g., by Imatinib (initially known as STI571) for chronic myelogenous leukaemia (CML) [53, 100]. Yet most treatments of cancer use combinations of cytotoxics and cytostatics (e.g., irinotecan and cetuximab [44]), for non-cell-killing therapeutics alone are seldom sufficient in advanced stages of cancers.

Another instance of such combination of cell-killing and non-cell-killing therapeutics is the treatment of premyelocytic acute myeloblastic leukaemia (APL or AML3) by combining an anthracyclin and a redifferentiating agent (for AML is characterised by a blockade of differentiation in non-proliferating hematopoietic cells, and redifferentiating agents such as All-TransRetinoic Acid (ATRA) lift such blockade in the case of APL) [66]. Here, maturation, and not proliferation (i.e., not the cell cycle, which is not affected by them), is the target of redifferentiating agents, which may thus by no means be considered as cytostatic.

2.4 *Representation of Drug Targets*

It is appropriate to consider anticancer drugs, cytotoxic or cytostatic, and their targets, through their effects on the cell cycle in cell populations. This of course assumes that a model of the cell cycle in a proliferating cell population is given. Nevertheless we also firstly mention other models, which either do not include the cell cycle or do not describe events at the level of a cell population.

In a review article [75], Kimmel and Świerniak considered two possibilities to represent in a mathematical model the action of cytotoxic drugs on their targets in a proliferating cell population: either by a possible direct effect on cell death, enhancing it, presumably by launching or accelerating the apoptotic cascade in one or more phases of the cell cycle, or by a blockade of one or more transitions between two phases, arresting the cycle at some checkpoint, most often with the involvement of protein p53, and only secondarily launching cell death.

This is indeed a general alternative in the representation of the effects of cytotoxics. If no cell cycle phase structure has been put in the population dynamic model used to represent the evolution of the cell populations at stake, i.e., when no account is taken of cell cycle phases in these populations, then only the first possibility exists: modulation of a death term.

As regards cytostatics (which by definition are not supposed to kill cells, at least not directly), the representation of their action in physiological models with age structure for the cell cycle should be somewhat different. It can be done either by a slowdown of the progression speed in the G_1 phase (or in the proliferating phase in a one-phase model) or by an action on the exchanges between non-proliferating (G_0) and proliferating phases when a G_0 phase is represented in the model.

It is also possible to combine cytostatic and cytotoxic effects in the same model. In [69], for instance, the authors use an age-structured model with a 1-phase proliferative subpopulation exchanging cells with a non-proliferative cell compartment to combine a slowdown effect on proliferation for the cytostatic effect with an increase in the cell death term for the cytotoxic effect—of the same drug, lapatinib, a tyrosine kinase inhibitor, in their case, the variation between these effects depending on the dose. Acting on two different targets in a cell cycle model by two different drugs, a cytotoxic and a cytostatic one, in the same cell population is thus possible, and such models are thus amenable to study and optimise combination therapies, such as cetuximab+irinotecan advocated in [44].

3 Overview of Cell Population Dynamics Designed for Cancer Treatment

In this section, we present an overview of several kinds of cell population models that have been used in modelling anticancer treatments. Some works we refer to do not include cancer treatment optimisation, but it is important to mention them as they could be used as a first step towards cancer therapy optimisation. As so much has been published in the last 50 years, we do not claim to be exhaustive, only recollecting the main models used to describe the fate of cell populations submitted to cancer treatments.

3.1 ODE Models for Growing Cell Populations with Drug Control

The first models of tumour growth were developed to reproduce and explain experimentally observed tumour growth curves. The most common ones are the exponential model ($\frac{dN}{dt} = \lambda N$), the logistic ($\frac{dN}{dt} = \lambda N\left(1 - \frac{N}{K}\right)$, where K is the maximum tumour size, or "carrying capacity" of the environment), and the Gompertz ($\frac{dN}{dt} = \lambda N \ln\left(\frac{K}{N}\right)$, where again K is the carrying capacity). Contrary to the exponential model, the logistic and the Gompertz models take into account the possible limitation of growth due, for instance, to a lack of space or resources, assuming that the instantaneous growth rate $\frac{dN}{Ndt}$ depends on the carrying capacity of the environment. The Gompertz model was initially developed in the context of insurance [62] and was first used in the nineties to fit experimental data of tumour growth [81]. A lot of studies on drug control are based on these models [13, 15, 35, 96–98, 106–108]. For instance, Murray [106, 107] considered a two-population Gompertz growth model with a loss term to model the effect of the cytotoxic drug. He considered both tumour and normal cells in order to take into account possible side-effects of the treatment on the population of normal cells. Murray's purpose was to minimise the size of the population of tumour cells at the end of the treatment while keeping the population of normal cells above a given threshold. In [108], Murray took into account cell resistance to chemotherapy and applied the problem of optimising drug schedules to a two-drug chemotherapy. In [96], Martin developed a model to determine chemotherapy schedules that would minimize the size of the tumour at the end of the treatment, under constraints of maximal drug doses (individual doses and cumulative dose), ensuring that the tumour decrease might be faster than a given threshold. In further works, Martin et al. also introduced tumour cell resistance to chemotherapy [97, 98].

More recently, one of us and his co-workers [15, 35] investigated the effects of oxaliplatin on tumour cells and healthy cells. To model tumour growth, they used a Gompertz model modified by a "therapeutic efficacy term" as a death term depending on the drug concentration.

In [35], the work presented in [15] is extended by coupling this model of tumour growth to a model of healthy cell growth and to a three-compartment model describing the time evolution of the concentration of oxaliplatin in plasma, healthy tissue and tumour tissue. This work was done on the basis of experimental data related to oxaliplatin PK-PD and tumour growth curves with or without drug injection to determine the model parameters and compared time-scheduled infusion schemes with constant ones. In the same work, the drug infusion schedule was optimised by determining drug infusion patterns that should maximise tumour cell death under the constraint of minimising healthy cell death. Barbolosi and Iliadis [13, 70] coupled a Gompertz model of tumour growth, perturbed by a cytotoxic efficacy term, to a two-compartment model of the chemotherapy PK (plasmatic and active drug concentration). They investigated optimal drug delivery schedules under constraints of maximal allowed drug (single doses and cumulative dose) and leukopenia.

In an attempt to design a more realistic model of tumour growth under angiogenic stimulator and inhibitor control, Hahnfeldt et al. [67] proposed a two-variable model derived from the Gompertz model. It is based on observations made on experimental data from lung tumours in mice treated by an antiangiogenic drug. Hahnfeldt et al. introduced as a variable the carrying capacity of the environment, K:

$$\frac{\mathrm{d}N}{\mathrm{d}t} = \lambda N \ln\left(\frac{K}{N}\right), \tag{1}$$

$$\frac{\mathrm{d}K}{\mathrm{d}t} = bN - (\mu + dN^{2/3})K - \eta g(t)K, \tag{2}$$

where b is the rate of the tumour-induced vasculature formation, $\mu + dN^{2/3}$ represents the rate of spontaneous and tumour-induced vasculature loss, $g(t) \geq 0$ represents the antiangiogenic drug concentration.

This model enables to take into account the vasculature that provides nutrients and oxygen to tumour cells, and thus to study the effects of several anti-angiogenic factors on tumour growth. D'Onofrio et al., based themselves on this model, proposed different expressions for K, modelling for instance endothelial cell proliferation or delayed death, to investigate the action on tumour growth of anti-angiogenic therapies [46–48] and of combined therapies [49]. In the same way, Ledzewicz et al. [84], basing themselves on [55], considered the following model

$$\frac{\mathrm{d}N}{\mathrm{d}t} = -\lambda N \ln\left(\frac{N}{K}\right) - \varphi v N, \tag{3}$$

$$\frac{\mathrm{d}K}{\mathrm{d}t} = bK^{2/3} - dK^{4/3} - (\mu + \gamma u - \eta v)K, \tag{4}$$

where u and v represent the doses of an anti-angiogenic drug and of a cytotoxic drug, respectively, and φ, γ, η their effects on tumour cells and on vasculature. The authors introduced an optimisation problem to minimise the tumour cell mass under constraints on the quantity of drug to be delivered.

Most cancer chemotherapies are cell cycle phase-specific, which means that they act only on cells that are in a specific phase of the cell division cycle, for instance in S or in M. To take into account such specificities, models describing the cell cycle have been designed. Without always entering into the details of cell cycle phases, several models distinguish between proliferating and nonproliferating cells [114, 115, 139, 140]. They take into account two kinds of cell populations, proliferative and quiescent, by representing time variations of their densities, and allow exchanges of cells from one population to the other. They are based on the fact that only proliferative cells are sensitive to chemotherapies and they allow to study the effect on tumour growth of several treatment schedules and to determine the optimal ones. For instance, in [114, 115], Panetta et al. proposed the following model

$$\frac{\mathrm{d}}{\mathrm{d}t}\begin{pmatrix} x_1 \\ x_2 \end{pmatrix} = \begin{pmatrix} \alpha-\mu-\eta & \beta \\ \mu & -\beta-\gamma \end{pmatrix}\begin{pmatrix} x_1 \\ x_2 \end{pmatrix}, \tag{5}$$

where x_1 and x_2 represent the cycling and non-cycling tumour cell mass, respectively, α the cycling growth rate, μ the rate at which cycling cells become non-cycling, η the natural decay of cycling cells, β the rate at which non-cycling cells become cycling, and γ the natural decay of non-cycling cells. All these parameters are supposed to be constant and positive. By adding a drug-induced death term in the equation on cycling cells, the authors investigated the effects on tumour growth of two kinds of periodic chemotherapies: a pulsed one and a piecewise continuous one. They also considered the effect on a population of non-tumour cells (or normal cells) in order to determine optimal drug schedules. Some authors later based themselves on this model to determine optimal chemotherapy schedules [56, 83]. Using experimental data, Ribba et al. [124] introduced a third kind of cells, the necrotic cells, and the carrying capacity in order to investigate the effect of an antiangiogenic treatment on tumour growth dynamics and on hypoxic and necrotic tissues within the tumour.

Kozusko et al. [80] deepened the work of Panetta et al. [115] by developing a model of tumour growth integrating two compartments within the cell cycle: one for cells in phases G_1 and S and another for cells in G_2 and M. They based their model on experimental data to represent the effect on tumour growth of an antimitotic agent (curacin A), that prevents cells from dividing. They modeled the blockade of cells in the G_2/M phase of the cell cycle according to the treatment dose, and distinguished resistant cells from sensitive ones. This model was able to predict a minimum dose of treatment able to stop growth of both kinds of cells. To analyse the effect on tumour growth of another anticancer treatment (mercaptopurine) according to varying degrees of cell resistance, Panetta et al. [117] modified the model introduced by Kozusko et al. [80] by distinguishing phases G_0/G_1, S and G_2/M.

3.2 PDE Models with Spatial Dynamics for Tumour Growth and Drug Effects

ODE models presented above do not integrate any spatial dimension. They were historically developed to explain in vitro tumour growth curves. Obviously in vivo tumour growth depends on its environment. For instance, it depends on the mechanical properties of the supporting tissue, on the local quantity of nutrients and oxygen, on the local concentrations of pro and anti-growth chemical factors, etc. PDE models integrating spatial dynamics are thus better suited than ODE models for the design of realistic models of tumour growth. Greenspan [63] was the first author to take into account the spatial dynamics of tumour cells and oxygen through the simplifying hypothesis of a spherical symmetry of the diseased tissue (tumour spheroid). Several models are based on his [32, 33, 43, 45, 101, 127].

Reaction-diffusion PDEs are best suited to describe the space and time evolution of the concentrations of chemical substances and of cell densities. Such equations allow to take into account the interactions of a diffusing chemical molecule or of a population of cells with its environment.

Thus Swanson et al. [131–133] used the classical KPP-Fisher model that is frequently used to represent the spatial progression of so-called "travelling waves" (see [103] for details), to develop a model of brain tumour growth that takes into account tumour cell proliferation and diffusion

$$\frac{\partial p}{\partial t} = \nabla.(D(x)\nabla p) + \rho p(1-p), \tag{6}$$

where p is the tumour cell density (that depends on space and time), $D(x)$ the diffusion rate (that depends on space), ρ the net proliferation rate. In fact, its linearised form around the origin

$$\frac{\partial p}{\partial t} = \nabla.(D(x)\nabla p) + \rho p \tag{7}$$

is sufficient to describe tumour progression (and it has an analytic solution if $D(x)$ is constant). The difficulty here resides in the identification of the diffusion coefficient $D(x)$, which is in fact far from constant, since it depends on the nature of cerebral matter, grey or white, and the brain is not known to possess a simple spatial structure.

In [132], Swanson et al. modelled the action of a chemotherapy by introducing in Eq. (6) a linear death term that also depends on space and time. They investigated drug delivery according to the tissue heterogeneities of the brain (white or grey matter).

Competition between cells for the gain of space and nutrients influences tumour growth. Thus, on the basis of a Lotka-Volterra type model [109, 110], Gatenby et al. [59, 60] modelled competition between healthy and tumour cells to phenomenologically represent the mutual negative influences of the populations on each other. They highlighted the limitations of the clinical cytotoxic strategies that solely focus on killing tumour cells and not on preserving healthy cell populations from toxic side effects of the anticancer drugs.

Cell proliferation and cell death induce changes in the tumour volume. This phenomenon has to be taken into account to model tumour cell growth in a more mechanistic way. It is usually done by adding a transport term in the left-hand side of the reaction-diffusion equation

$$\frac{\partial p}{\partial t} + \nabla.(\mathbf{v}p) = \nabla.(D\nabla p) + f(p), \tag{8}$$

where f represents the reaction term and $\mathbf{v}$ the velocity of the transport movement. The velocity $\mathbf{v}$ can be determined by using Darcy's law and, for instance, assumptions on the total amount of cells.

Authors generally consider that the population of tumour cells is submitted to a growth signal representing all growth signals (inhibitors or promoters). The equations governing the evolution of the density (or mass) of tumour cells and of the concentration of chemicals are derived by applying the principle of mass conservation to each species. The common form of such equations is

$$\frac{\partial p}{\partial t} + \nabla.(\mathbf{v}p) = \nabla.(D_p\nabla p) + \alpha_p(c,p) - \delta_p(c,p), \tag{9}$$

$$\frac{\partial c}{\partial t} = \nabla.(D_c\nabla c) + \alpha_c(c) - \delta_c(c,p), \tag{10}$$

where p is the density of tumour cells, c the concentration of the chemical (nutrients, oxygen, ...), D_p the diffusion rate of tumour cells, α_p represents their proliferation rate, δ_p their spontaneous death rate, D_c is the diffusion rate of the chemical, α_c represents its production rate, δ_c its degradation rate. Because tumour cell proliferation and death depend on the concentration of the chemical, functions α_p and δ_p depend, for instance linearly, on the concentration of the chemical and on the tumour cell density. The function α_c depends, for instance linearly, on the concentration of the chemical and the function δ_c depends both on the chemical concentration and on the density of tumour cells to model, for instance, the consumption of the chemical substance by tumour cells. The same kind of equation as Eq. (10) can be used to describe the evolution of the concentration of drug u.

As already mentioned in Sect. 2, cancer therapies can have different effects on tumour growth. To model the effect of a drug-inducing tumour cell death (cytotoxic), one can add in the right-hand term of Eq. (9) a death term; thus the equation governing the density of tumour cells submitted to the effect of a cytotoxic drug may be given by

$$\frac{\partial p}{\partial t} + \nabla.(\mathbf{v}p) = \nabla.(D_p\nabla p) + \alpha_p(c,p) - \delta_p(c,p) - K_{\text{cyto}}(u,p), \tag{11}$$

where K_{cyto} is a positive function that depends on the drug concentration and on the tumour cell density, Eq. (10) remaining unchanged.

Instead, to model the effect on tumour growth of an anti-angiogenic therapy that will reduce oxygen supply, one can add such a decay term in the right-hand term of Eq. (10), which then becomes

$$\frac{\partial c}{\partial t} = \nabla.(D_c\nabla c) + \alpha_c(c) - \delta_c(c,p) - K_{\text{angio}}(u,c), \tag{12}$$

where K_{angio} is a positive function that depends on the drug concentration and on the chemical concentration; in this case, Eq. (9) remains unchanged.

Thus, in [73], Jackson et al. considered a model with two kinds of cells differing by their sensitivity to a cytotoxic treatment: one cell type was less sensitive than the other one. They assumed that the tumour was a spheroid, thus reducing the dimension from three to one, using radial symmetry. The drug fate was modelled through the variations of its tissue concentration, via a term of blood-to-tissue transfer (the drug concentration in blood being prescribed by the therapy scheduling). The authors compared the tumour response to an equal amount of drug administered either by bolus injection or by continuous infusion. Jackson based herself on this work to develop a model of the action of an anti-cancer agent (doxorubicin) on tumour growth [72]. This model is composed of a submodel of tumour growth coupled to a three-compartment submodel of intratumour drug concentration (extracellular space, intracellular fluid space, nucleus space) and to a submodel of the plasma concentration of the drug. The intracellular action of the drug on tumour cells is modelled through a Hill-type function. This model allows to study the tumour response to repeated rounds of chemotherapy.

In [112], Norris et al. investigated the effects of different drug kinetics (linear vs. Michaelis-Menten kinetics) and different drug schedules (single infusion vs. repeated infusions) on tumour growth.

Frieboes et al. [57] developed a mathematical model of tumour drug response that takes into account the local concentration of drug and nutrients. The authors considered two cell phenotypes, viable and dead tumour cells, and supposed that their mitosis and apoptosis rates depended on the nutrients and drug concentration. This model was calibrated on in vitro cultures of breast cancer cells.

More mechanistic (i.e., more molecular than purely phenomenological) models have been used to take into account details of the angiogenic process. Endothelial cells that constitute the blood vessel wall migrate towards a gradient of a chemoattractant substance secreted by quiescent (or hypoxic) tumour cells (this movement is termed chemotaxis). Continuous models of angiogenesis usually take into account the density of endothelial cells and the tissue concentration of the chemoattractant substance. They are based on the following equations

$$\frac{\partial m}{\partial t} = \nabla.(D_m \nabla m) + \alpha_m m - \chi_m \nabla.(m \nabla w) - \delta_m m, \tag{13}$$

$$\frac{\partial w}{\partial t} = \nabla.(D_w \nabla w) + \alpha_w(q) w - \delta_w w, \tag{14}$$

where m denotes the density of endothelial cells, D_m their diffusion rate, α_m their proliferation rate, χ_m their chemotaxis rate, δ_m their death rate, w the concentration of the chemoattractant substance, D_w its diffusion rate, δ_w its production rate that depends on the density of quiescent tumour cells q, δ_w its degradation rate. Such models can also include other kinds of cells or chemical substances, see for instance [34] for a review of angiogenesis models.

The coupling of angiogenesis models to tumour growth models is usually done via the concentration in oxygen (or nutrients), the time and space evolution of which is given by a reaction-diffusion PDE based on the fact that oxygen is delivered by the vasculature and mostly consumed by tumour cells. This coupling enables to describe in a more realistic way the effects of anti-angiogenic therapies on tumour vasculature and thus on tumour growth.

Thus, Sinek et al. [128] based on the model of vascular tumour growth developed by Zheng et al. [141] and on experimental data to develop a model of tumour growth and vascular network coupled to a multi-compartment pharmacokinetic-pharmacodynamic (PK-PD) model. Their purpose was to analyse the effect on tumour growth of two anti-cancer drugs, doxorubicin and cisplatin (compartments of the PK-PD model were drug-specific). They concluded that drug and oxygen heterogeneities, possibly due to irregularities of the vasculature, can impact drug efficacy on tumour cells. Kohandel et al. [79] proposed a model that also takes into account tumour cells, the tumour vascular network and oxygen to investigate the effect on tumour growth of different schedules of single and combined radiotherapy and anti-angiogenic therapy.

3.3 Phase-Structured Cellular Automata for the Cell Division Cycle and Drug Effects

Drug effects on tumour growth can also be investigated by means of phase-structured cellular automata to represent the cell division cycle. Cellular automata enable to describe individual cancer cell evolution within a population of cells. Thus Altinok et al. developed a cellular automaton for the cell cycle [3–6]. This automaton does not take into account molecular events but phenomenologically describes cell cycle progression. The states of this automaton correspond to the phases of the cell cycle. Transition between two states of the automaton correspond to cell progression through the cell cycle, or exit from the cell cycle, and are supposed to respect some prescribed rules. For instance each phase of the cell cycle is supposed to be characterised by a mean duration and a variability in order to take into account inter-cell variability that can appear within a population. This model enables to study, on a whole population of cells, the impact of the variability in the duration of the cell cycle phases on cell desynchronisation through the cell cycle.

Such modelling is motivated by the fact that one way to optimise pharmacological treatments in cancer, taking into account of the cell division cycle on which tissue proliferation relies, is to take advantage of the control that circadian clocks are known to exert on it. Such treatments are termed chronotherapies of cancer [87–92]. In order to investigate the effects of chronotherapy on the growth of a tumour cell population, Altinok et al. coupled this cellular automaton with a model of the circadian clock through kinases known to induce or inhibit the transition from G_2 to M. For instance, in [3, 4], the authors were interested in the action of

5-fluorouracil (5-FU), an anticancer drug known to block cells in the S phase. They modelled the effects of this drug by increasing the probability that cells submitted to 5-FU while in S phase exit from the cell cycle at the next G_2/M transition. They compared the cytotoxic efficacy of continuous administration of 5-FU and of several chronomodulated therapies that differed from their administration peak time. Later, in [5], Altinok et al. analysed the cytotoxic effects of 5-FU chronotherapies according to their administration peak time and to the cell cycle mean duration. As they did for 5-FU, Altinok et al. also investigated the effects of oxaliplatin chronomodulated therapies on tumour cells. Contrary to 5-FU, oxaliplatin is an anticancer agent that is not phase-specific. Therefore the authors modelled the effects of oxaliplatin in a non phase-specific way, by increasing the probability for exposed cells of exiting the cell cycle at the next checkpoint (G_1/S or G_2/M transitions).

3.4 *Physiologically Structured PDE Models for the Cell Cycle and Drug Effects*

Time and space are not the only two variables on which tumour growth depends. In fact, tumour growth also depends on the physiological properties of cancer cells, that can be for instance age of the cells (i.e., time since the last cell division), mass or volume of the cells, or their DNA content. To take this phenomenon into account, the McKendrick PDE framework is the best suited

$$\begin{cases} \dfrac{\partial n}{\partial t}(a,t) + \dfrac{\partial}{\partial a}[g(a)n(a,t)] + d(a)n(a,t) = 0 & (t > 0, a > a_{\min}), \\ n(a_{\min},t) = \displaystyle\int_{a_{\min}}^{\infty} \beta(s)n(s,t)\mathrm{d}s & (t > 0), \\ n(a,0) = n_0(a) & (a > a_{\min}), \end{cases} \tag{15}$$

where $n(a,t)$ is the density of tumour cells with the characteristic a (age, mass, volume, DNA content, etc.) at time t, g is the tumour growth rate, d is the death rate, β is the tumour cell birth rate, $a_{\min} \geq 0$ is the minimum value of a. Note that g, d, β depend on a.

Physiologically structured cell population dynamics models have been extensively studied in the last 25 years, see e.g. [7–9, 14, 17, 20, 27, 40, 41, 65, 69, 71, 74, 103, 123, 138]. For instance, Iwata et al. [71] developed a model of the dynamics of the colony size distribution of metastatic tumours, assuming that both primary and metastatic tumour growth depended on the size of the tumour. The authors proposed a Gompertz equation to model the primary tumour growth and a McKendrick type equation to model the evolution of the colony size distribution of metastases. Kheifetz et al. [74] proposed a model for tumour cell age distribution to investigate tumour cell dynamics under periodic age-specific chemotherapy. Hinow et al. [69] developed an age-structured PDE model to investigate the cytotoxic and cytostatic effects on tumour growth of a cancer drug, lapatinib, on the basis of biological experiments. The authors distinguished between proliferative and non-proliferative cells and assumed that only proliferative cells were ageing.

One of us with his co-workers [27,40,41] considered a multiphase age-structured PDE model in which they introduced time dependency of the parameters (death rate, transition rate from one phase of the cell cycle to the next one). They investigated the effects of a circadian control on the tumour growth rate with and without a periodic cell cycle phase specific chronotherapy. More details about [27] can be found in Sect. 7.

Basse et al. [16, 18, 19] developed a phase- and size-structured model of a cell population submitted to paclitaxel, a cancer agent that induces mitotic arrest of the cell cycle and cell death. The size of the cell was considered as determined by its DNA content and some of the model parameters were determined by fitting experimental flow cytometry data. On the basis of the work of Spinelli et al. [130], Basse et al. [20] developed a phase- and age-structured model of a cell population submitted to a chemotherapy. They considered several cancer agents and assumed that these chemotherapies affected tumour cell population dynamics by modifying cell cycle phase transition functions or by killing cells in the mitotic phase.

Webb [138] proposed a both age- and size-structured model for normal and tumour cell dynamics under chemotherapy, on the basis of the McKendrick model with an additional transport term. He supposed that the two population differed by their mean cell cycle duration, which was longer for tumour cells, and modelled the effects of the chemotherapy by a time periodic death term. His aim was to take the resonance phenomenon into account to determine optimal period of the cancer treatment in order to induce the lowest tumour growth rate and the highest normal cell population growth rate.

Finally, we present some examples of models that are not actually physiologically structured PDE models but that derive from them.

Ubezio and co-workers [94,95,104,135] based themselves on an age- and phase-structured PDE model to develop a discrete age-structured model of cell cycle describing the time evolution of the number of cells of age a at time t in the phases G_1, S and G_2/M of the cell cycle. This model also takes into account the inter-cell variability in phase duration. The potential effects of drug (blocking cells in G_1 or in G_2, etc) were modelled through separate parameters. Thanks to this model and experimental data, Montalenti et al. [104] investigated the effects of several doses of cisplatin on ovarian carcinoma cells. Although cisplatin is known to block cells mostly (but not only, since as an alkylating agent inducing double strand breaks throughout the cell cycle, it is not phase-specific) in G_2, they also analysed the effects of cisplatin on cells in G_1. In [94,95], based on experimental data, Lupi et al. investigated the effects of topotecan and melphalan, respectively, on ovarian cells. Later, Ubezio et al. [135] deepened the previous works of their team by examining the effects of five drugs (doxorubicin, cisplatin, topotecan, paclitaxel and melphalan) on ovarian cancer cells using several doses.

Delay differential models can also be viewed as deriving from age- structured PDE models since they can be obtained by integrating PDEs along characteristics. Thus Bernard et al. [23] proposed a model composed of delayed differential equations to model tumour and normal cell population dynamics in the phases of the cell cycle under circadian control and chemotherapy. They compared the efficacy and toxicity of constant and chronomodulated schedules of 5-FU, a phase-specific drug used in the treatment of colorectal cancer.

3.5 Mixed Models, Both Spatially and Physiologically Structured

We call "mixed models", models that include both spatial and physiological dynamics. Such models are useful to investigate spatial changes induced by a phase-specific chemotherapy combined or not with an antiangiogenic agent. This kind of models has not been highly developed. Bresch and co-workers [25,30,31] developed a multiscale model of tumour growth that includes cell age in the proliferative phases of the cell cycle and tissue motion of tumour cells. On the basis of the model developed by Bresch et al., coupled with an angiogenesis model, one of us and her co-workers [25] investigated the effects of an innovative antiangiogenic drug on tumour vasculature and hence on tumour growth. This multiscale model takes into account some molecular events such as cell cycle dynamics and cell receptor binding. This model could be coupled to a model of phase-specific drug, such as 5-FU, to analyse tumour and endothelial cell dynamics under drug infusion. It could also be interesting to determine optimal drug schedules that would maximise tumour cell death under constraints of minimising endothelial cell death to ensure drug delivery to tumour cells (remember that endothelial cells are cells that constitute the vessel wall, see Sect. 3.4 for details).

Alarcón et al. [2] proposed a more complex multiscale model of vascular tumour growth that integrates tissue, cell and intracellular scales. For instance, this model accounts for vascular network, blood flow, cell–cell interaction, cell-cycle, VEGF production and integrates several kinds of models (ODE models, cellular automata, etc). The authors investigated the effects of low and high concentrations of protein p27 on the dynamics of tumour and normal cell populations.

4 Control and Its Missions: Representing the Action of Drugs

In the previous section, we have described some dynamic models for cell populations frequently used in the study of cancer growth and treatments. These models can thus be seen as controlled dynamic systems with drug effects as their control functions. Various examples of such drug effects have been given in Sect. 2. Introducing pharmacokinetics (i.e., evolution of concentrations) for the drugs chosen produces additional equations to the cell population dynamic model, and their pharmacodynamics (i.e., actual drug actions) modify this cell dynamics according to the target and to the effect of the drugs. Then, optimisation of cancer treatments can be represented as an optimal control problem on this controlled dynamic system. In this section, we first discuss how the drug infusions are taken into account in the model, then we give examples of objective functions and constraints considered in the literature on the treatment of cancers.

4.1 Classes of Control Functions: What Is Fixed and What May Be Optimised

We introduce a vector space X, called the state space. At each time $t \in \mathbb{R}_+$, the state of the system is $x(t) \in X$. This variable lists all the data necessary to represent the system. It should at least contain the number (or density) of cells for each type considered. The state may have coordinates for healthy and cancer cells and for each phase considered. In a PDE model, the state may also distinguish between ages or between locations of cells. In the PDE case, X has infinite dimension. The state should also contain concentrations of drugs in each compartment of the pharmacokinetic model.

We denote by u the control function, $u : \mathbb{R}_+ \to U$. It represents the (multi)drug infusion schedule time by time, one coordinate per drug. The dynamics of the biological system can thus be written as

$$\dot{x}(t) = f(t, x(t), u(t)),$$

where $f : \mathbb{R}_+ \times X \times U \to X$. Given a control $u(\cdot)$, under standard hypothesis on the dynamics f, the state is uniquely defined and we will denote the state variable associated with the control $u(\cdot)$ by $x_u(\cdot)$. Examples of such functions are given in Sect. 3, for instance, in Eqs. (1)–(4), (11)–(12) and (15).

Alternatively, instead of a control function, one may consider simpler predefined infusion schemes with only a small number of control parameters. Such infusion schemes may represent either a simple model for an early study or a consequence of technical constraints such as the fact that oral drugs can only be administered at fixed hours (at meal time for instance). Then $u \in \mathbb{R}^m$ is a set of parameters and the dynamics is $\dot{x}(t) = f_u(t, x(t))$. Examples of such parameters are the period of a periodic scheme [114, 138] or the phase difference between a circadian clock and the time of drug infusion initiation [5] (see also Sect. 3).

4.2 Objective Functions: Measuring the Output

An optimisation problem consists in maximising or minimising a given real-valued objective function that models the objective we want to reach.

The main purpose of a cancer treatment is to minimise the number of cancer cells. When the model takes into account the number of cancer cells directly [5, 15, 49, 56, 75, 85, 118, 136], the objective function is simply the value of the coordinate of the state variable corresponding to the number of cancer cells at a time T, T being either fixed or controlled. To take into account the drug effects, Swierniak et al. [75, 134] defined a performance index to minimise the number of tumour cells at the end of the treatment while minimising the cumulated drug dose (viewed as a measure of the cumulated drug effects on healthy cells).

The optimisation problem can also be formalised as the minimisation of the asymptotic growth rate of the cancer cell population [27, 115, 138]. Hence, the number of cancer cells will increase more slowly, or even eventually decrease. We will present this approach in a linear frame (hence controlling eigenvalues) in Sect. 7.

Alternatively, in [134], Swierniak et al. discussed the problem of maximising both the final number of normal cells and the cumulated drug effects on tumour cells. They concluded that this approach led to optimisation principles similar to those developed to solve the problem of minimising both the final number of tumour cells and the cumulated drug effects on healthy cells.

4.3 Constraints, Technological and Biological, Static or Dynamic

Toxicity Constraints

A critical issue in cancer treatment is due to the fact that drugs usually exert their effects not only on cancer cells but also on healthy cells. A simple way to minimise the number of cancer cells is to deliver a huge quantity of drug to the patient, who is however then certainly exposed at high lethal risk. In order to avoid such "toxic solutions", one may set constraints in the optimisation problem, which thus becomes an optimisation problem under constraints.

Putting an upper bound on the drug instantaneous flow [56] and/or on the total drug dose is a simple way to prevent too high a toxicity for a given treatment. A bound on total dose may also represent a budget limit for expensive drugs [85].

However, fixed bounds on drug doses are not dynamic, i.e., they do not take into account specificities of the patient's metabolism and response to the treatment, other than by adapting daily doses to fixed coarse parameters such as body surface or weight (as is most often the case in the clinic so far). In order to get closer to actual toxicity limits, and hoping for a better result, it is possible to consider instead a lower bound on the number of healthy cells, as in [15]. In the same way, using a

Malthusian growth model, where growth exponents are the targets of control, such a constraint becomes a lower bound on the asymptotic growth rate of the healthy cell population [138].

In the same way, a drug used in a treatment must reach a minimal concentration at the level of its target (which blood levels reflect only very indirectly) to produce therapeutic effects. Classically, clinical pharmacologists are accustomed to appreciating such efficacy levels by lower threshold blood levels, that are themselves estimated as functions of pharmacokinetic parameters such as first and second half-life times and distribution volume of the drug, with confidence interval estimates for a general population of patients. As in the case of toxicity, a more dynamic view is possible, by considering drug levels that decrease the number of cancer cells, that is, which yield a negative growth rate in the cancer cell population.

This leads to the definition of admissible sets for drug infusion flows, the union of $\{0\}$ and of a therapeutic range containing the infusion levels that are at the same time efficient and not too toxic (such a constraint is considered in [136]). Those admissible sets are rather difficult to take into account, however, as they lead to complex combinatorial problems.

An approach that is consequently often chosen (see [85] for instance) is to forget this constraint in the model and to a posteriori check that the optimal drug infusion schedules found are high enough to be efficient when they are nonzero.

That may be an elementary reason why so-called bang-bang controls (i.e., all-or-none) are of major interest in chemotherapy optimisation: they are defined as controls such that at each time, either the drug infusion flow is the smallest possible (i.e., 0), or it is the highest possible. Even though it is now easy to use in the clinic (and also in ambulatory conditions) programmable pumps that may deliver drug flows according to any predefined schedule with long-lasting autonomy, solutions to optimisation problems often turn out to be bang-bang (tap open-tap closed).

But solutions to optimisation problems in cancer chemotherapy are not always bang-bang, when considerations other than on simple parallel growth of the two populations are taken into account, and this includes competition, when the two populations are in contact, e.g. in the bone marrow normal haematopoietic and leukaemic cells, or when both populations are submitted to a common—but differently exerted—physiological control, such as by circadian clocks [15].

Another interesting approach, relying on two models, one of them including the cell division cycle [115], and putting the optimal control problem with toxicity constraints, is developed in [54]. The optimal control problem is solved by using the industrial software gPROMS®.

Drug Resistance

Whereas therapeutic efficacy and limitation of toxic adverse effects are the first concern when dealing with chemotherapy, the frequent development of drug resistances in the target cancer cell populations is certainly the second bigger issue in the clinic. The development of such resistances may come from overexpression in individual cells of defence mechanisms as an exaggeration of physiological phenomena, such as are ABC transporters (the P-gp, or P-glycoprotein, being its most known representant), but they may also result, at least as likely, in proliferative populations encompassing mitoses, from mutations yielding more fit, i.e., resistant in the presence of drug, subpopulations.

A classical solution to this problem is to forbid too low drug concentrations that are supposed to create environmental conditions favourable to the development of more fit drug resistant cell populations without killing them, as is also the case, for instance, in antibiotherapy with bacteria. Nevertheless, other, more recent, arguments to support an opposite view, have been put forth: assuming that there exists a resistant cell population at the beginning of the treatment, or that it may emerge during the treatment, then delivering high drug doses often produces the effect to kill all sensitive cells, giving a comparative fitness advantage to resistant cells, that subsequently become very hard to eradicate. Thus a paradoxical solution has been proposed, at least in slowly developing cancers: killing just enough cancer cells to limit tumour growth, but letting enough of these drug sensitive cancer cells to oppose by competition for space the thriving of resistant cells, that are supposed to be less fit, but just the same, usually slowly, will invade all the tumour territory if no opponents are present [58, 61]. Indeed, such free space left for resistant tumour cells to thrive, when high drug doses have been administered with the naive hope to eradicate all cancer cells, may result in the rise of tumours that escape all known therapeutics, a nightmare for physicians which is unfortunately too often a clinical reality. Hence the proposed strategy to avoid high doses, that are able to kill all sensitive cells, and to only *contain* tumour growth by keeping alive a minimal population of drug-sensitive tumour cells.

Both those constraints, toxicity and resistance, can be considered as part of the objective function by setting the objective to be a balance between two objectives. For instance, Kimmel and Swierniak in [75] proposed to minimise a linear combination of the number of cancer cells and of the total drug dose. This yields an unconstrained optimisation problem, that has a simpler resolution, while still taking into account the diverging goals of minimising the number of cancer cells and keeping the number of healthy cells high enough.

But whereas cancer and healthy cells are two quite distinct populations, with growth models that may easily be distinguished and experimentally identified by their parameters, it is more difficult to take into account the evolutionary lability (i.e., the genomic instability) and heterogeneity of cancer cell populations with respect to mutation-selection towards drug resistance, according to evolution mechanisms that are not completely elicited. Note that *acquired* (as opposed to *intrinsic*, i.e., genetically constitutive) drug resistance may result as well from individual cell adaptation (enhancement of physiological mechanisms) as from genetic mutations, both under the pressure of a drug-enriched environment, as discussed in [42]. In this respect, acquired resistance may be reversible, if no mutation has initiated the mechanism, or irreversible, and it is likely irreversible in the case of intrinsic resistance.

Ideally, the optimal solution of a therapeutic control problem should take into account both the drug resistance (using evolutionary cell population dynamics) and the toxicity constraints, but these constraints have usually been treated separately so far. Whereas the difficult problem of drug resistance control is certainly one of our concerns in a cell Darwinian perspective, in the sequel we shall present only results for the (easier) toxicity control problem.

5 Identification of Parameters: The Target Model and Drug Effects

5.1 Methods of Parameter Identification

In the many works dedicated to modelling pharmacological control of tumour growth and its optimisation that have been published in the last 40 years, when the issue of confronting a theoretical optimisation method with actual data has been tackled, quite different attitudes have been displayed. When identifying parameters of a biological model, one may use different methods, according to the nature of considered experimental data, their precision and reliability, and, also of course according to the scientific background of people in charge of identification. One may distinguish between at least three types of methods, all of which, to yield the best estimation of the parameters at stake usually rely on least squares minimisation, otherwise said minimisation of a L^2 distance between experimental quantitative observations and numerical features of the model, either direct outputs, such as cell numbers, or computed statistical parameters, such as mean cell cycle times.

Probabilistic Methods

The first method is based on the theory of parameter estimation in statistical models, and supposes that a probability measure, depending on a set of statistical parameters, e.g., mean and variance of a probability density function (p.d.f.), is a priori given in a space of constitutive parameters of the model, e.g., coefficients in a set of differential equations. In its simplest form, estimation will result from the minimisation of the L^2 distance between a model p.d.f. and a corresponding observed histogram, yielding with precision a best set of parameters for the p.d.f. It may also result from more elaborated principles, such as maximum likelihood estimation (including the use of computational algorithms of the expectation-maximisation (EM) type, with or without the assumption of an underlying Markov chain), see the statistical literature on the subject, e.g., [86, 137] for a general presentation (To this class of methods may also be related attempts to characterise by its statistical properties a chaotic deterministic system, as studied for instance in [82], when no actual model is given of the system, which is only supposed to have trajectories converging towards a chaotic attractor—on which they are dense—an attractor which by definition is endowed with an invariant ergodic measure.).

Control Science and Dynamical Systems

The second method comes from signal processing and control science. It is applied to the representation of dynamic systems by state-space models and it relies on the fact that the system is given by a set of ODEs, which may be converted by Laplace transform to the study of transfer functions, i.e., the system in the frequency domain. Such systems may be studied by their responses to input excitations to better characterise them. Presentations of such identification methods may be found in [93, 137].

Inverse Problems

The third method, inverse problem solving, belongs to the domain of PDEs. The models under consideration are close to the physical world and the method can comprise almost all situations, but requires specific studies for each case and nontrivial mathematics. The general principle is that observations of the real system represented by a PDE model correspond to an ill-posed problem, i.e., that the system of PDEs as it is given cannot be identified in a unique manner from the observations. Nevertheless, small regularisations (such as Tikhonov's), i.e., small modifications of the underlying differential operator, make the problem well-posed, i.e., amenable to the identification of its parameters in a unique manner. For a general presentation, see [76]. Recent developments on physiologically structured models may also be found in [50–52, 64, 120].

5.2 Parameters in Macroscopic Models of Tumour Growth

In macroscopic models of tumour growth, parameter identification most often relies on imagery techniques, mainly radiological or MRI, as in [131] for brain tumour growth. But it is also possible to obtain tumour growth curves representing three-dimensional growth by using a method which may seem very coarse, but which has not found any really better competitor so far. It consists in growing a tumour (homograft or xenograft, i.e., of the same animal species, or of another) under the skin or on the skin of a laboratory rodent and measuring everyday by using a caliper diameters in three dimensions (one longitudinal and two orthogonal transverse) of the tumour, which is protected by the skin coating when the tumour is subcutaneous. It is possible only when the tumour is already palpable under the skin (or visible when it is on the skin), which excludes avascular tumours and generally involves histologically heterogeneous, but physically (i.e., in density with respect to water) homogeneous tumours. This allows an approximate estimation of the tumour mass, assumed to be proportional to the number of tumour cells, from its approximate volume and keeps the animal alive (until tumour weight reaches 10 % of the animal body weight, at which point the animal is sacrificed for obvious ethical reasons). Coarse as it may seem, this method is still widely used because of its simplicity, see e.g., [35, 61, 124]. In a macroscopic (whole body) perspective, it should in principle also be possible to relate tumour growth with blood concentrations of biomarkers such as ACE, CA19.9 or PSA, known to be elevated in cancer or even only in tissue hyperplasia but they are not very specific; even for more specific antigenic biomarkers, models relating these quantities to actual cell population increase, to our knowledge, are still wanted. A recent review on this topic may be found in [113].

5.3 Parameters in Cell and Molecular-Based Models of Tissue Growth

At the cell population level, more easily quantifiable data have been recorded to identify model parameters in structured PDE models. They are linked to cell population samples, either with observation on a global population, such as given by flow cytometry (or FACS, for fluorescence activated cell sorting) [16–20,94,135], or by observations on individual cells and statistics performed on the sampled population of individual cells marked with fluorescent proteins, such as FUCCI [125] in [26, 27]. These methods require previous cell staining, e.g. with propidium iodide (for flow cytometry), or hybridisation of intracellular proteins with external fluorescent proteins (for FUCCI). In all cases, in the presence of a stationary (i.e., asymptotic) distribution of cells may constitute the basis for applying an inverse problem method, as shown in [12, 51]. A difference between these two sorts of experimental data (direct cell population or reconstruction of population with previous individual cell recordings) is that flow cytometry is a snapshot on a population of cells that are destroyed by the sorting process, whereas FUCCI investigates individual living cells without destroying them, a richer experimental situation.

5.4 Measurement of Pharmacodynamic Effects

Evaluation of the effects of a treatment using an anticancer drug involves measurements with and without treatment, which obviously is not ethically possible in most clinical situations. Nevertheless, in the case of low-grade gliomas [132, 133], where tumours can evolve very slowly during many years, and for which it is known that no actually efficient therapy exists (in particular neurosurgery may be more detrimental than beneficial to the patient) therapeutists may unfortunately, but non unethically, find themselves in a situation of mere observers. In this case, macroscopic images of tumours may be used, providing parameter estimation without treatment, which may then serve as a basis for comparisons with treated gliomas. The observations are always radiological or MRI. But in cell cultures and in animals, such experimental observations are of course much easier and may allow comparing parameters of interest evaluated by using any of the methods mentioned above in different situations: cell cultures, fresh blood samples from patients, tumour growth curves [11, 18–20, 35, 116] and it is then possible to propose optimised treatments based on these estimations [10, 15, 35].

6 Therapeutic Optimisation Procedures

In the previous sections we have described the drugs used in chemotherapy, various cell population dynamic models and the objectives and constraints considered in chemotherapy optimisation problems that have been published. Those three topics can be seen as components of an optimisation model. In order to get quantitative results, the parameters of this model should be estimated by using the techniques presented in Sect. 5. Then, one has to choose an optimisation procedure to solve the optimisation problem considered.

When choosing an optimisation procedure, one first needs to identify what are the optimisation variables. For chemotherapy optimisation, there are two main situations: either the optimisation variables are some parameters of a predefined infusion scheme or they are the infusion scheme itself, represented by a time-dependent control function $u(t)$ (cf. Sect. 4.1).

6.1 Graphic Optimisation

"Graphic optimisation" simply consists in plotting the value of the objective for all admissible points. It is a very simple scheme and the only requirement for its success is that the admissible set must have a nonempty interior. It also provides graphics to present the result.

Graphic optimisation suits particularly the case of a predefined infusion scheme with only a few parameters. For instance, this technique was used by Webb [138] and Panetta and Adam [115]. These authors considered models of the McKendrick type (see Sect. 3) and they searched for the best period of periodic drug infusions, i.e., the period of predefined drug infusion schemes that minimises the growth rate of cancer cells. Altinok et al. [5] proposed a cellular automaton model controlled by two predefined infusion schedules of drugs where the parameter is the phase difference between a circadian clock and the drug infusion.

The drawback of this method is that when the number of parameters grows, the time necessary for the resolution of the problem grows exponentially. Moreover, graphics are less practical when the dimension exceeds 3. The classical solution is, rather, to make use of a more evolved optimisation algorithm. However, if one needs an evolved optimisation algorithm anyway, one might as well consider an optimal control problem, in which the whole infusion schedule is the optimisation variable.

6.2 Pontryagin's Maximum Principle

An optimal control problem is an optimisation problem where the objective is a function of the state variables $x(\cdot) : \mathbb{R}_+ \to \mathbb{R}^n$ of a dynamical system and of the controlled variables $u(\cdot) : \mathbb{R}_+ \rightarrow \mathbb{R}^m$ that control the dynamical system. It can be written formally as

$$\begin{aligned} \min_{u(\cdot),T} \int_0^T f^0(t,x(t),u(t))\mathrm{d}t + g^0(T,x(T)) \\ \dot{x}(t) = f(t,x(t),u(t)), \quad \forall t \in [0,T] \\ u(t) \in U_t, \quad \forall t \in [0,T] \\ x(0) = x_0 \,, \quad x(T) \in M_1 \end{aligned} \tag{16}$$

Here, f^0 is the cost function and g^0 is the final cost. The final time T can be either fixed or be part of the control. The dynamical system is represented by function f, which gives the evolution of x and is controlled by u. At each time, the control may be subject to constraints represented by the set U_t and the set M_1 is a subset of $\mathbb{R}^n$ representing conditions on the final state. If we replace the constraints $x(0) = x_0$ and $x(T) \in M_1$ by $x(T) = x(0)$, we have a T-periodic optimal control problem (see [28] for more precision on the consequences of this model).

A major tool of optimal control is Pontryagin's maximum principle [121]. It gives necessary optimality conditions for the optimal trajectories. We denote the Hamiltonian of the system by

$$H(t,x,p,p^0,u) = \sum_{i=1}^{n} p^i f_i(t,x,u)\rangle + p^0 f^0(t,x,u),$$

where $p = (p^1,\ldots,p^n) \in \mathbb{R}^n$ and $p^0 \in \mathbb{R}$. If $u(\cdot)$ associated with the trajectory $x(\cdot)$ is an optimal control on $[0,T]$, then there exists a continuous application $p(\cdot)$ called the adjoint vector and a nonpositive number p^0 such that for almost all $t \in [0,T]$,

$$\dot{x}(t) = \frac{\partial H}{\partial p}(t,x(t),p(t),p^0,u(t)) \,, \quad \dot{p}(t) = -\frac{\partial H}{\partial x}(t,x(t),p(t),p^0,u(t)) \tag{17}$$

and we have the maximisation condition for almost all $t \in [0,T]$,

$$H(t,x(t),p(t),p^0,u(t)) = \max_{v \in U_t} H(t,x(t),p(t),p^0,v). \tag{18}$$

If in addition, the final time to reach the target M_1 is not fixed, we have the condition

$$\max_{v \in U_T} H(T,x(T),p(T),p^0,v) = -p^0 \frac{\partial g^0}{\partial t}(T,x(T))$$

and if M_1 is manifold of $\mathbb{R}^n$ with a tangent space $\mathrm{T}_x M_1$ at x, we have

$$p(T) - p^0 \frac{\partial g}{\partial t}(T,x(T)) \perp \mathrm{T}_{x(T)} M_1.$$

Under the conditions of the Pontryagin's maximum principle given in Eqs. (17) and (18), we have

$$\frac{\mathrm{d}}{\mathrm{d}t}H(t,x(t),p(t),p^0,u(t)) = \frac{\partial H}{\partial t}(t,x(t),p(t),p^0,u(t))$$

and thus if f, f^0 and U_t do not depend on t, then H does not depend on t and $\max_{v \in U_t} H(t,x(t),p(t),p^0,v)$ is constant.

Thanks to Pontryagin's maximum principle one is often able to determine the optimal control as a function of the adjoint vector. Nevertheless, the adjoint vector is not easy to compute. It is defined through its value at the terminal point, $p(T)$, and solutions to the associated boundary value problem are difficult to compute, nor need they be unique.

For Pontryagin's maximum principle to be applicable, the cell population model must be a set of ODEs controlled by drug infusions, as presented in Sect. 3.1. The authors generally minimise the number of cancer cells at final time with a bound on the instantaneous drug flow. The total dose is either constrained to be bounded or is part of the objective, a smaller dose improving the objective. When the information provided by Pontryagin's maximum principle is enough to know the optimal control, as in [49, 56, 75, 85], it gives the control, i.e., the solution to the problem, in an explicit formula, without any discretisation. This is then a very valuable information. Unfortunately, for most optimal control problems, and optimal control arising from chemotherapy problems are not an exception, we do not have enough information and we have to use a numerical algorithm to solve the problem.

6.3 *Numerical Methods for Optimal Control Problems*

Two classes of numerical methods exist for optimal control problems, namely indirect methods, also called shooting methods, and direct methods.

Shooting Method

The shooting method is based on the observation that, if ever we knew the value $p_0 = (p_0^0, p_0^1, \dots, p_0^n)$ of the adjoint state at the initial point, we could get the optimal controls time by time. Thus we define the shooting function $G(p_0)$ such that $G(p_0) = 0$ if and only if $p(T)$ satisfies the final conditions (recall that T is the final time). The shooting method simply consists in solving the equation $G(p_0) = 0$, with variable p_0, for instance by a Newton method.

A variant of the shooting method was used in [85] for chemotherapy optimisation. Ledzewicz et al. considered two drugs that act on a Gompertzian model: one is an anti-angiogenic, which controls the carrying capacity of the tumour and the other is a cytotoxic drug, which controls a death term. The pharmacodynamics of the drugs was modelled by linear differential equations. The authors apply Pontryagin's maximum principle on this model with the objective of minimising the quantity of tumour cells and as constraints an upper bound on the drug instantaneous flow and an upper bound on drug total dose. The authors obtained the optimal control as a function of the adjoint vector. The optimal control reaches the dose bound when a function related to the adjoint is nonzero and it follows a singular curve when this function vanishes along an interval. They then used a shooting method to construct the optimal control as a feedback function from these adjoint-vector dependent singular curves. Algorithmic details are given in [99].

In general, shooting methods give very precise results but the structure of commutations, given by studying Pontryagin's maximum principle, must be known in advance for them to be efficient. When this structure is unknown, one can still perform direct methods, which we describe next.

Direct Methods

Direct methods consist of a total discretisation of the control problem and then of solving the finite dimensional optimisation problem obtained. The discretisation of an optimal control problem results in an optimisation problem with a large number of variables. The theory of differentiable optimisation is the classical tool for such problems [24, 29, 111]. However, in order to overcome the limits of differentiable optimisation, some authors use stochastic algorithms to solve the discretised problem. We next give some examples of these techniques in the context of chemotherapy.

Gradient algorithm

When the problem is formulated without any state constraint, one can use the gradient algorithm, as in [118]. The authors proposed a cell-cycle-dependent model written with one ODE by cell-cycle phase. They controlled the transition and death rates and optimised a linear combination of the number of cancer cells and of the total dose of drugs. The gradient algorithm starts here with an initial control strategy u_0 and the associated trajectory x_{u_0}. It consists in successive improvements of the discretised objective $F^0(u) = \sum_{l=0}^{N} f^0(t_l, x_u(t_l), u(t_l)) + g^0(T, x_u(T))$ by

$$u_{k+1} = P_U(u_k - \alpha \nabla F^0(u_k)),$$

where U is the set of admissible controls and α is a length step chosen in order to guarantee a sufficient decrease of the objective, for instance with an Armijo or Wolfe line search rule. When computing the gradient of the objective with respect to the control, there appears an adjoint vector which is a discrete version of the adjoint vector in Pontryagin's maximum principle.

Uzawa algorithm

An optimal control problem with K constraints is a problem of the form of problem (16) where we add constraints $\int_0^T f^i(t,x(t),u(t))\mathrm{d}t + g^i(T,x(T)) \leq 0$ for $i = 1,\ldots,K$. For such problems, direct methods are particularly suited and the discretised optimal control problem can be solved by the Uzawa algorithm.

We denote $F^i(u) = \sum_{l=0}^{N} f^i(t_l, x_u(t_l), u(t_l)) + g^i(T, x_u(T))$ and we introduce the Lagrangian

$$L(u,\lambda) = F^0(u) + \sum_{i=1}^{K} \lambda^i F^i(u),$$

where λ is a vector with one coordinate by state constraint called a Lagrange multiplier. At a given iterate (u_k, λ_k), we solve

$$u_{k+1} = \arg\min_u L(u, \lambda_k)$$

by a nonconstrained optimisation algorithm, as is the gradient algorithm, and then we compute

$$\lambda_{k+1}^i = \max(0, \lambda_k^i + \alpha F^i(u_{k+1})), \quad \forall i \in \{1,\ldots,K\}$$

where α is an appropriate step size. If the constraint is an equality constraint instead of an inequality constraint, we accept nonpositive values for λ and we do not perform the maximum against 0.

Basdevant et al. used the Uzawa algorithm in [15] to solve the problem of minimising the number of cancer cells while maintaining the number of healthy cells over a tolerability threshold. They modelled the cell population dynamics and the action of the drug by a set of coupled differential equations.

In [27], we solved the problem of minimising the asymptotic growth rate of the cancer cell population while keeping the asymptotic growth rate of the healthy cell population over a prescribed threshold; see a sketch of the method and of its results below in Sect. 7. We modelled the cell population dynamics by a McKendrick model physiologically controlled by a circadian clock, considering a phase-dependent drug acting on transitions. We firstly discretised the problem and then solved it by using a Uzawa algorithm with augmented Lagrangian. That is to say, we replaced the Lagrangian by

$$L_c(u,\lambda) = F^0(u) + \frac{1}{2c} \sum_{i=1}^{K} (\max(0, \lambda^i + cF^i(u))^2 - (\lambda^i)^2)$$

Compared to the classical Lagrangian, the augmented Lagrangian has better convergence and stability properties for a small computational cost.

Other differentiable optimisation algorithms may also be used, depending on the properties of the problem at stake. In general, all these algorithms give a local optimal solution quickly but they do not give any guarantee that the control solution produced is a global optimum. In order to overcome this drawback, some authors chose to use stochastic algorithms instead.

Stochastic algorithms

Stochastic algorithms are algorithms that use a random number generator to find the optimal solutions of a given problem. These random numbers are used to explore the admissible control set with the hope that the optimal controls will eventually be hit. Each stochastic optimisation algorithm is a compromise between focusing on good solutions and letting enough freedom to exploration in order not to miss the global optimum. See [68, 129] for more details on this subject.

In [1], Agur et al. considered an age-structured cell cycle model with deterministic cycle phase lengths. The drug under consideration is toxic for cells in one of the phases only. They considered a composite objective function that takes into account the number of cancer and healthy cells in the end and a measure of the survival of the patient. They assumed that a patient survives if at no time the number of healthy cells falls below a threshold. The authors compared three versions of simulated annealing. They first defined the neighbourhood of every point of the admissible set, that is, at every point, they defined the possible ways to go to another point. This neighbourhood should be large enough to give freedom to the algorithm but not too large because otherwise the computational cost of searching the neighbourhood would be dissuasive. Then simulated annealing gives the rule for the acceptance or rejection of a neighbour, which gets stricter when a parameter, call the temperature, decreases. In theory, if the temperature is decreased properly, the iterates converge to an optimum of the problem. In practice, convergence may be desperately slow. The other two heuristics presented in the paper do the same work but with simplified rules, that do not guarantee convergence to an optimum but have smaller computational costs.

Villasana et al. proposed in [136] an ODE model with three types of cells: cancer cells in interphase (i.e., G_1, S and G_2), cancer cells in mitosis phase (M) and healthy cells. Each type of cells has a particular dynamic and there are interactions between them. They considered a combination of a cytotoxic and of a cytostatic drug and they wanted to minimise the number of cancer cells while keeping the number of healthy cells above a threshold. They used the covariance matrix adaptation evolutionary strategy (CMA-ES) to solve this problem. This is an algorithm based on probabilistic mutations of the current iterates and on a selection of the best ones [68]. The covariance matrix adaptation is a way to give the mutations directions for them to be more effective.

7 Focus: Cancer Therapeutics to Control Long-Term Cell Population Behaviour in Structured Cell Population Models

7.1 Linear and Nonlinear Models

We have presented in [26, 27] a method based on the control of eigenvalues in an age-structured model, yielding the numerical solution of an optimal control problem in the context of cancer chronotherapeutics, and we sum up below some of its results as regards system modelling, identification, and theoretical therapeutic optimisation. To this goal, we used an age-structured cell population model, since our aim was to represent the action of cytotoxic anticancer drugs, which always act onto the cell division cycle in a proliferating cell population. The model chosen, of the McKendrick type [102], is linear. This may be considered as a harsh simplification to describe biological reality, which involves nonlinear feedbacks to represent actual growth conditions such as population size limitation due to space scarcity. Nonetheless, having in mind that linear models in biology are just linearisations of more complex models (for instance considering the fact a first course of chemotherapy will most often kill enough cells to make room for a non space-limited cell population to thrive in the beginning) we think that it is worth studying population growth and its asymptotic behaviour in linear conditions and thus analyse it using its growth (or Malthus) exponent. This first eigenvalue of the linear system may be considered as governing the asymptotic behaviour, at each point where it has been linearised, of a more complex nonlinear system, as described in [21, 22].

7.2 Age-Structured Models for Tissue Proliferation and Its Control

We know that circadian clocks [87–92] normally control cell proliferation, by gating at checkpoints between cell cycle phases (i.e., by letting cells pass to the next phase only conditionally). We also know that circadian clock disruption has been reported to be a possible cause of lack of physiologically control on tissue proliferation in cancer [91], a fact that we will represent in our model to distinguish between cancer and healthy cell populations.

The representation of the dynamics of the division cycle in proliferating cell proliferations by physiologically structured PDEs is thus a natural frame to model proliferation in cell populations, healthy or tumour. The inclusion in such proliferation models of targets for its control, physiological (circadian) and pharmacological (by drugs supposed to act directly on checkpoints), allows to develop mathematical methods of their analysis and therapeutic control [26, 27, 36], in particular for cancer chronotherapeutics, i.e., when the drug control is made 24 h-periodic to take advantage of favourable circadian times.

Physiologically structured cell population dynamics models have been extensively studied in the last 20 years, see Sect. 3.4 for some examples. We consider here typically age-structured cell cycle models, in which the cell division cycle is divided into I phases (classically 4: G_1, S, G_2 and M), and the variables are the densities $n_i(t,x)$ of cells having age x at time t in phase i. Equations read

$$\begin{cases} \dfrac{\partial n_i(t,x)}{\partial t} + \dfrac{\partial n_i(t,x)}{\partial x} + d_i(t,x)n_i(t,x) + K_{i\to i+1}(t,x)n_i(t,x) = 0, \\ n_{i+1}(t,0) = \displaystyle\int_0^\infty K_{i\to i+1}(t,x)n_i(t,x)\mathrm{d}x, \\ n_1(t,0) = 2\displaystyle\int_0^\infty K_{I\to 1}(t,x)n_I(t,x)\mathrm{d}x \end{cases} \tag{19}$$

together with an initial condition $(n_i(t=0,.))_{1\le i\le I}$. This model was first introduced in [39] and further studied in other publications, among which [40, 41]. In this model, in each phase i, cells are ageing with constant speed 1 (transport term), they die with rate d_i or with rate $K_{i\to i+1}$ go to next phase, in which they start with age 0. To represent the effect of circadian clocks on phase transitions [91], one may consider time-periodic coefficients d_i and $K_{i\to i+1}$, the period being in this case 24 h.

7.3 Basic Facts About Age-Structured Linear Models

One of the most important facts about linear models is the trend of their solutions to exponential growth. The study of the growth exponent, first eigenvalue of the system, is therefore crucial. Solutions to system (19) satisfy (if the coefficients are time-periodic, or stationary) $n_i(t,x) \sim C^0 N_i(t,x)\mathrm{e}^{\lambda t}$ [119], where N_i are defined by

$$\begin{cases} \dfrac{\partial N_i(t,x)}{\partial t} + \dfrac{\partial N_i(t,x)}{\partial x} + \Big(\lambda + d_i(t,x) + K_{i\to i+1}(t,x)\Big)N_i(t,x) = 0, \\ N_{i+1}(t,0) = \displaystyle\int_0^\infty K_{i\to i+1}(t,x)N_i(t,x)\mathrm{d}x, \\ N_1(t,0) = 2\displaystyle\int_0^\infty K_{I\to 1}(t,x)N_I(t,x)\mathrm{d}x, \\ N_i > 0, \quad N_i(t+T,.) = N_i(t,.), \quad \displaystyle\sum_i \int_0^T \int_0^\infty N_i(t,x)\mathrm{d}x\mathrm{d}t = 1 \end{cases} \tag{20}$$

with $T-$periodic coefficients.

We focus now on the case of stationary phase transition coefficients $(K_{i\to i+1}(t,x) = K_{i\to i+1}(x))$ and we do not consider death rates ($d_i = 0$). Note that if one considers constant nonzero death rates, the problem does not change, only the eigenvalue λ is then in fact $\lambda + d$, as one can see in the equations of system (20). As shown in [39], the first eigenvalue λ is then solution of the following equation, which in population dynamics is referred to, in the 1-phase case ($I = 1$) with no death term, as Euler-Lotka's equation

$$\frac{1}{2} = \prod_{i=1}^{I} \int_0^{+\infty} K_{i\to i+1}(x) \mathrm{e}^{-\int_0^x K_{i\to i+1}(\xi)\mathrm{d}\xi} \mathrm{e}^{-\lambda x}\, \mathrm{d}x \tag{21}$$

Integrating the first equation of System (19) along its characteristics [119], we can in the stationary case with no death rate derive the formula

$$n_i(t+x,x) = n_i(t,0)\mathrm{e}^{-\int_0^x K_{i\to i+1}(\xi)\mathrm{d}\xi}.$$

This can be interpreted in the following way: the probability for a cell which entered phase i at time t to stay for at least an age duration x in phase i is given by

$$P(\tau_i \geq x) = \mathrm{e}^{-\int_0^x K_{i\to i+1}(\xi)\mathrm{d}\xi}$$

The time τ_i spent in phase i is thus a random variable on $\mathbb{R}_+$, with probability density function f_i given by

$$\mathrm{d}P_{\tau_i}(x) = f_i(x)\mathrm{d}x = K_{i\to i+1}(x).\mathrm{e}^{-\int_0^x K_{i\to i+1}(\xi)\mathrm{d}\xi}\mathrm{d}x$$

or equivalently:

$$K_{i\to i+1}(x) = \frac{f_i(x)}{1 - \int_0^x f_i(\xi)\mathrm{d}\xi}. \tag{22}$$

7.4 FUCCI (Cell Cycle) Reporters to Identify Model Parameters

FUCCI (for fluorescence ubiquitination-based cell cycle indicator) is a recent cell imaging technique that allows tracking progression within the cell cycle of an individual cell [125] after hybridisation with fluorescent proteins of cell cycle phase characteristic indicators (geminin and Cdt1) involved in the ubiquitination, i.e., natural degradation, of the actual proteic determinants of evolution in the cell cycle. We used FUCCI data that consisted in time series of the intensity of red and green fluorescence emitted by individual NIH 3T3 cells (mouse embryonic fibroblasts) within an in vitro homogeneous population proliferating without any control in a liquid medium. This allowed us to measure the time an individual cell spent in the G_1 phase and in the phases $S/G_2/M$ of the cell cycle. Our data consisted of cell cycle phase durations from 55 proliferating cells. Note that is fully justified in our case to assume $K_{i\to i+1}(t,x) = K_{i\to i+1}(x)$ since these experimental conditions correspond to cells proliferating in a completely independent manner, without any communication

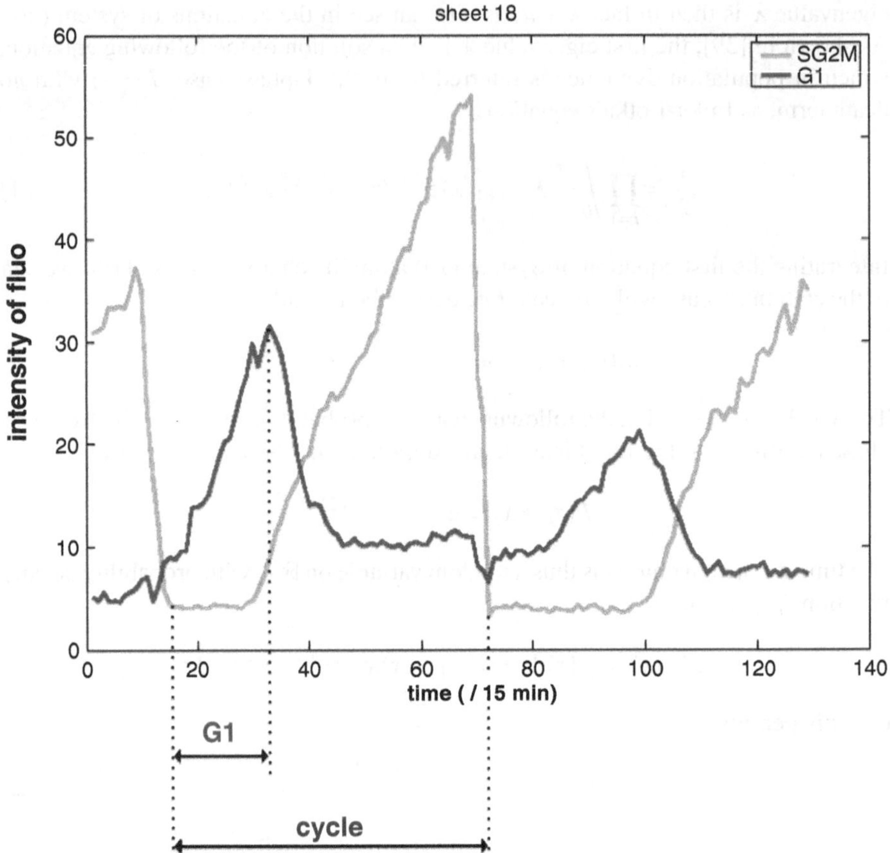

Fig. 1 Graphic method used to determine the durations of the whole cell cycle and of G_1 phase. The duration of phase $S/G_2/M$ was deduced by subtracting the duration of phase G_1 from the duration of the cell cycle

nor external control on their proliferation. It is noteworthy that, though we deal with PDEs, our method is simpler than methods involving inverse problem solving (which nevertheless have been used on comparable situations, for instance in [51]), and this is due to the fact that using the FUCCI reporter technique we have access to precisely defined data in individual cells, with the counterpart that quite few individual proliferating cells have been recorded. A graph representing a time series from an individual cell and the method used to record phase durations is presented on Fig. 1.

We used these experimental data to identify the parameters of our model by fitting shifted Gamma distributions $f(x) = \rho^{-k}(\Gamma(k))^{-1}(x-a)^{k-1}e^{-(x-a)/\rho}$ on $[a,+\infty[$ to frequencies of appearance of G_1 and $S/G_2/M$ durations within the population (recall that variable x stands for age in each one of the two phases). These Gamma distributions were approximations of the probability density functions of

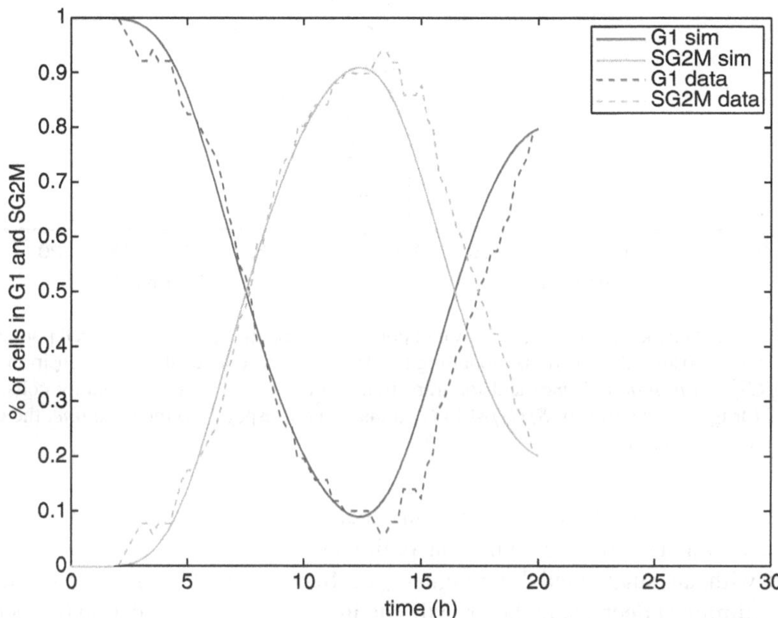

Fig. 2 Time evolution of the percentages of cells in the phases G_1 (*red* or *deep grey*) and $S/G_2/M$ (*green* or *light grey*) from biological data (*dashed line*) and from numerical simulations (*solid line*). Our model results in a reasonably good fit to biological data

the random variable corresponding to the time spent in G_1 and $S/G_2/M$. This enabled us to determine the expression of the transitions rate according to the formula (22). We then compared the solutions of the system with cell recordings that had previously been synchronised "by hand", i.e., all recordings were artificially made to start simultaneously at the beginning of G_1 phase. The result is shown on Fig. 2. Note that using an inverse problem method—see Sect. 5—instead of ours could have consisted here in determining the parameters of the model, i.e., $K_{i\to i+1}$ transition functions, by minimising an L^2 distance between this experimental data curve and a theoretical, parameter-dependent curve representing these data.

7.5 *Optimising Eigenvalues as Objective and Constraint Functions*

We then used combined time-independent data on phase transition functions, obtained from experimental identification of the parameter functions $K_{i\to i+1}(t,x) = \kappa_i(x)$ in the uncontrolled model, with cosine-like functions representing the periodic

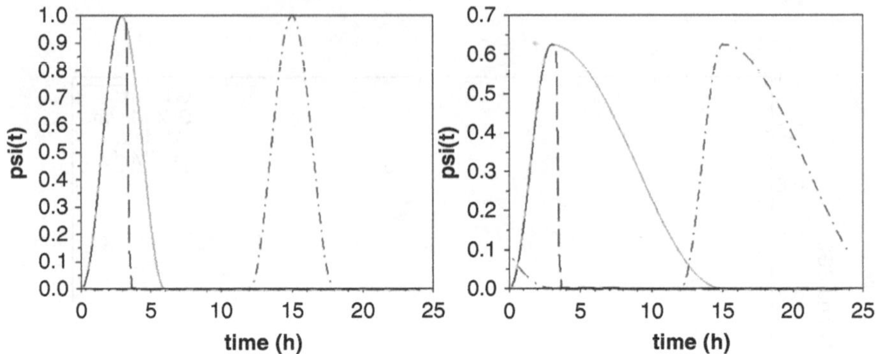

Fig. 3 Drug and circadian controls, healthy cell population case (**a**, *left*) and cancer cell population case (**b**, *right*). Cosine-like functions modelling the drug and circadian controls for transition from G_1 to $S/G_2/M$ (*dash-dotted line*) and for transition from $S/G_2/M$ to G_1 in healthy cells. The "natural" (drug-free) control for $S/G_2/M$ to G_1 transition corresponds to the *solid line*, the drug-induced one to the *dashed line*

control on these transitions by circadian clocks, together with free-running drug infusion regimens. The drug infusion regimens were optimised using a Uzawa method with an augmented Lagrangian (see Sect. 6.3 or [29] for algorithmical details), aiming at decreasing the growth rate in a cancer cell population (objective) while preserving the same in a healthy cell population (constraint) by maintaining it over a prescribed threshold. The idea is the same as in [15], except that we deal here with cell population growth exponents instead of cell numbers.

We considered two cell populations, that we called cancer cells and healthy cells. In these simulations, we took into account cell death via a constant death rate, the same for both populations. We made the two cell populations only differ by their circadian control function ψ and we assumed that there was no interaction between the two populations, healthy and cancer. We took for this circadian control a continuous piecewise cosine-like function for each phase (Fig. 3a). We assumed that cancer cell populations still obey circadian control at these main checkpoints but less faithfully, and we modelled their behaviour by a looser answer to the circadian control signal (Fig. 3b).

Transitions from one phase to the other are described by the transition rates $K_{i \to i+1}(t,x)$. We took them with the form

$$K_{i \to i+1}(t,x) = \kappa_i(x)\psi_i(t)(1 - g_i(t)),$$

where $\kappa(x)$ is the transition rate of the cell without circadian control identified from FUCCI data, $\psi_i(t)$ is the natural circadian control modelled by a cosine-like function and $g_i(t)$ is the effect at the cell level of the drug infusion at time t on the transition rate from phase i to phase $i+1$. No drug corresponds to $g_i(t) = 0$, a transition-blocking infusion corresponds to $g_i(t) = 1$. We assumed that the drug has the same effect on both populations, which couples their behaviours through the drug infusions.

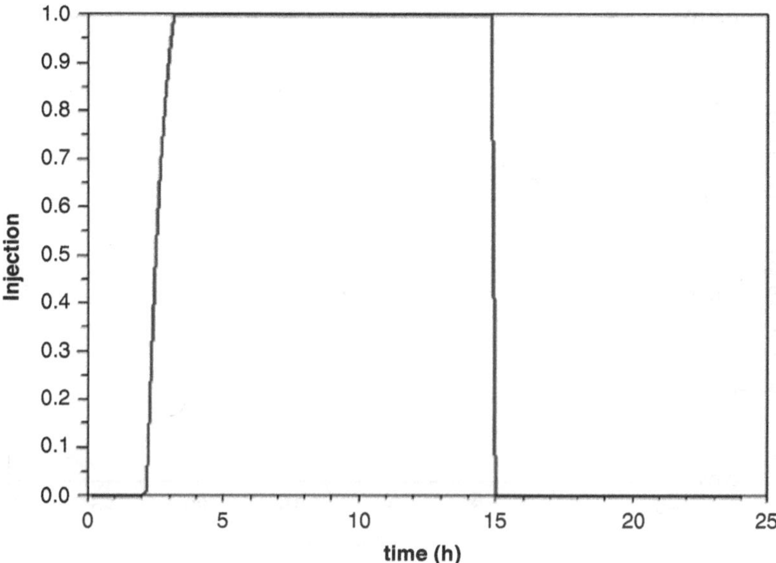

Fig. 4 Locally optimal periodic drug infusion strategy (function g_2, see text for details) found by the optimisation algorithm

We solved the constrained optimisation problem [26,27] with Uzawa's algorithm. After convergence of the algorithm, we get the locally optimal strategy, shown in our case on Fig. 4, defining on $[0;24]$ the 24 h-periodic function g_2 (assuming that $g_1 = 0$, that is here mimicking the action of an anticancer drug—as 5-fluorouracil—active in S phase only). It can be seen that it mainly consists in forbidding transitions when healthy cells do no change phase, thus harming cancer cells only.

The result of the optimised drug infusion regimen rates is shown in Fig. 5, where it can be seen that the asymptotic growth rate of cancer cells, initially positive and higher than the one of healthy cells, has been rendered negative by the periodic treatment exerted on transition rates while the new growth rate of healthy cells, though moderately affected by the treatment, remains positive.

Note that the FUCCI technology only enables us to distinguish between cells in G_1 and $S/G_2/M$, without distinction between S, G_2 and M. However, we may note that the method used in [27] to identify phase transitions relies in fact on the probability distribution of durations of phases. Since the duration of the phase M is known to be most of the time very short, with almost zero variability within cell populations, it would be legitimate to consider it as fixed, as 1 h, say, and that the recorded variability of $S/G_2/M$ is in fact the variability of S/G_2. Thus, we could have considered that we were dealing in this identification process with a transition function from S/G_2 to M instead of the one from $S/G_2/M$ to G_1. In this case, the transition function from M to G_1 could have been modelled by an indicator function, representing the fact that mitotic cells divide and enter the G_1 phase once they are old enough. Under these assumptions, we could have applied our optimisation

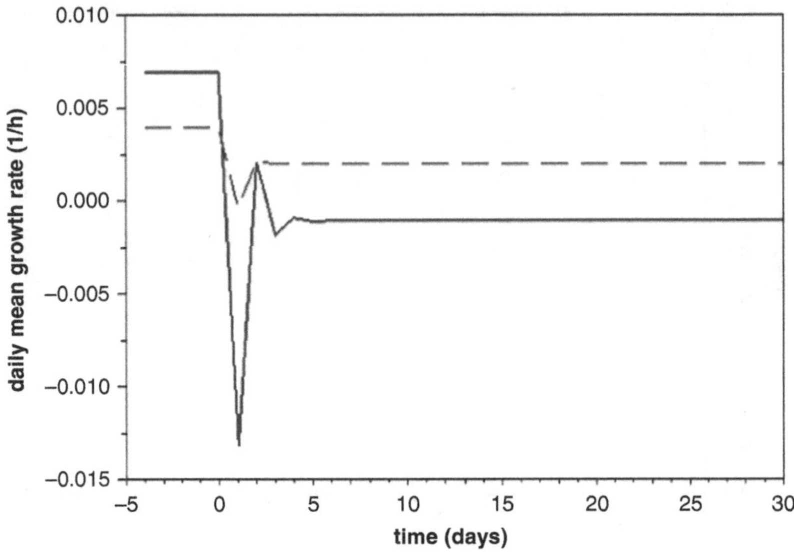

Fig. 5 Daily mean growth rates for cancer (*solid line*) and healthy cells (*dashed line*) when starting drug infusion at time 0. After a 10-day transient phase, the drug-forced biological system stabilizes towards the expected asymptotic growth rate

problem to an age-structured model accounting for three phases of the cell cycle, G_1, S/G_2 and M. As a first step in our analysis, we preferred keeping the model as simple as possible, in the absence of actual knowledge of the duration of M phase, but taking it as fixed, for instance to 1 h, such considerations might be developed in future works to assume that we have, thanks to FUCCI reporters, accessed the main two checkpoints, G_1/S and G_2/M.

8 Future Prospects: Multi-target Multi-drug Delivery Optimisation in Physiologically Structured Cell Population Models

From a therapeutic point of view, we have only represented the action of one drug acting on cell cycle phase transitions (on only one transition, G_2/M, if it acts on S phase). We know indeed that, in particular via p53, drugs that create DNA damage ultimately act by blocking the cell cycle at checkpoints, mainly G_1/S and G_2/M, and this only subsequently sends the cell into apoptosis, therefore cell cycle transition checkpoints are the most accurate targets to represent the effect of most cytotoxic drugs. The action of spindle poisons, such as taxans and vinca alcaloids, that block the cell cycle in M phase (but not at G_1/S or G_2/M), might also be represented by using the same formalism.

Now, in everyday oncology, most treatments use combinations of drugs that exert their action in synergy on different targets. These drugs may act on phase transitions only, but they may also act by inhibition of growth factor receptors (such as cetuximab, or tyrosine kinase inhibitors), impinging on the speed v_1, depending on age x at which G_1 phase is run through. In this case, one may use, as in [69], an extended version of model (19), where

$$\frac{\partial}{\partial t}n_1(t,x)+\frac{\partial}{\partial x}(v_1(x)n_1(t,x))+d_1(t,x)n_1(t,x)+K_{1\to 2}(t,x)n_1(t,x)=0 \quad (23)$$

Another possibility would be to introduce a non-proliferating, or quiescent, phase G_0 exchanging cells with G_1 and to represent the action of growth inhibitors by a control of these exchanges.

In Sect. 7, summing up [27], and following [15], we have focused on mathematical models of tissue growth having in mind only the problem of limiting drug toxicity to healthy cell populations to optimise cancer treatments. In the future, making available models of the emergence of drug-resistant cell subpopulations under drug pressure in a cell Darwinian perspective, we will simultaneously tackle at the cell population level the constraints of drug resistance in tumour cells and of toxicity to healthy tissues, to propose globally efficient combined therapies using at least two complementary drugs.

In a multiscale perspective, integrating a representation of the vasculature around a cancer cell population will also allow us to represent and optimise the action of combined therapies associating cytotoxic and antiangiogenic drugs, as in [49, 84].

To be relevant for actual clinical applications, models based on the representation of evolving structured populations will also need to be integrated in a whole-body level, from the infusion of drugs into the central compartment of general blood circulation until the actions they exert at the peripheral sites on proliferating cell populations. This has partly been done, but still without control, and in the case of an avascular tumour, in [31]. Moreover, to take into account in a dynamic way the constraint of limiting unwanted toxicity to healthy tissues, cell population growth models will have to separately represent both tumour and healthy cell populations, with therapeutic control, wanted or not, exerted on both, and possibly with competition between the two populations, as is the case for space in the bone marrow between leukaemic and normal haematopoietic cells at different stages of their maturation.

Whole-body integration of different spatial scales of description of pharmacological control on tumour and on healthy cell populations should certainly go as down as possible at the single cell level, including for instance nucleocytoplasmic transport to model the control by p53 of the cell cycle in case of damage to the DNA, as produced by cytotoxic drugs, and it must necessarily contain an intermediate tissue level as the main level of description, for cancer is certainly a tissue disease, which may be controlled only at the tissue, i.e., at the cell population level. Then higher levels of integration: whole-body, and for the individualisation of treatments (in particular adaptation of whole-body model parameters to clusters of patients), population of individuals, must be considered, as sketched in [38].

Apart from toxicity issues, at this intermediate level (the tissue), will also be considered in models presently under construction [42] the possible evolution of cells towards drug resistance, which is the other main problem encountered in cancer therapeutics. To prevent or overcome the emergence of resistant cell subpopulations, it is usually better to make use of several anticancer drugs acting on different targets, to avoid as much as possible strategies used by cancer cells, which, due to their genomic instability, easily adapt by mutations to single-drug therapies. This has been the case for instance with imatinib, a drug that has completely changed the prognosis of CML, but has nevertheless, after being considered as a miracle drug, also to be confronted to the issue of resistance in CML cells [126]. Future optimisation problems in cancer therapeutics will have to take into account as a constraint, given the possibility to induce drug resistance, to limit it, as much as possible, not necessarily by a complete eradication of tumour cells, but, as sketched in [58, 61], more realistically by its containment, and this will likely more easily done by using combinations of therapies than by using monotherapies.

9 Discussion and Conclusion

We have presented in this chapter firstly a brief review of models of cancer that have been used or may be used to tackle the general problem of therapeutic optimisation in oncology. As sketched elsewhere [37, 38], theoretic drug delivery optimisation is the last step of therapeutic optimisation, which must rely firstly on an accurate representation of the behaviour of targets (wanted and unwanted) without treatment and on the changes the means of action of the physician—drugs—exert on them.

The point of view we have adopted here may indeed be considered as complementary to the one of molecular biologists, pharmacologists and systems theoreticians [77, 78] who seek to control the cell division cycle at the single cell level by targeted drugs that are hoped-for blockers of intracellular pathways essential in cancer proliferation. Either by targeting a "hub" in the network by a single drug (e.g., imatinib in CML targeting BCR-Abl chimeric tyrosine kinase [126]) or by combining drugs that can hit complementary pathways, they search for "druggable" proteins that can be hit to arrest the cell cycle. A typical and recent example may be found in [122]. However, this approach, obviously valuable to provide new weapons in the war against cancer, by its nature cannot take into account the constraints linked to toxicity or drug resistance issues, which must be considered at the cell population level in a whole-body drug delivery optimisation perspective. This molecular biology approach should also be completed by whole-body pharmacokinetic-pharmacodynamic molecular modelling to represent the fate of drugs in the organism, as sketched above and in [37, 38]. In other words, these approaches may be thought of as understanding the target and the weapon, whereas, to stay in this metaphor, optimisation of drug control is training to shoot in all conditions to safely reach the target.

Focusing on cell population models, for cancer is never the problem of a single cell, we thus advocate in this chapter the interest of using structured cell population dynamics, to be further integrated in a multiscale setting, in the optimisation of drug delivery in oncology. From intracellular molecular dynamics to human populations, aiming at getting closer to actual clinical applications, we clearly have still hard work ahead, both in modelling and model analysis and in experimental identification and validation. Various therapeutic optimisation methods have been reviewed in their principles, and we have shown, focusing even more on linear population growth for cancer and for healthy cells, how it is possible to choose one, adapted to the model under consideration. The question of therapeutic optimisation in cancer is vast, and it may be treated in quite different manners, which have to be adapted to the particular clinical problem at stake. Nevertheless, modelling the target, the means of control, and taking account of the known clinical issues, there is still room for mathematical developments to pave the way for optimisation methods that will be able to face always more clinical challenges, all the more so as more links will be developed between mathematicians and clinicians.

Acknowledgments Access to data mentioned in Subsection 7.4 has been provided to us by G. van der Horst's lab in Erasmus Medical Centre (Rotterdam, The Netherlands); it was supported by a grant from the European Research Area in Systems Biology (ERASysBio+) and FP7 to the French National Research Agency (ANR) #ANR-09-SYSB-002-004 for the research network Circadian and Cell Cycle Clock Systems in Cancer (C5Sys) coordinated by Francis Lévi (Villejuif, France).

References

1. Agur, Z., Hassin, R., Levy, S.: Optimizing chemotherapy scheduling using local search heuristics. Oper. Res. **54**, 829–846 (2006)
2. Alarcón, T., Byrne, H., Maini, P.: A multiple scale model for tumor growth. Multiscale Model. Simul. **3**, 440–475 (2005)
3. Altinok, A., Lévi, F., Goldbeter, A.: A cell cycle automaton model for probing circadian patterns of anticancer drug delivery. Adv. Drug Deliv. Rev. **59**, 1036–1010 (2007)
4. Altinok, A., Lévi, F., Goldbeter, A.: Optimizing temporal patterns of anticancer drug delivery by simulations of a cell cycle automaton. In Bertau, M., Mosekilde, E., Westerhoff, H. (eds.) Biosimulation in Drug Development, pp. 275–297. Wiley-VCH Verlag GmbH & Co. KGaA, Weinheim, Germany (2008)
5. Altinok, A., Lévi, F., Goldbeter, A.: Identifying mechanisms of chronotolerance and chrono-efficacy for the anticancer drugs 5-fluorouracil and oxaliplatin by computational modeling. Eur. J. Pharm. Sci. **36**, 20–38 (2009)
6. Altinok, A., Gonze, D., Lévi, F., Goldbeter, A.: An automaton model for the cell cycle. Interface Focus **1**, 36–47 (2011)
7. Arino, O.: A survey of structured cell population dynamics. Acta. Biotheor. **43**, 3–25 (1995)
8. Arino, O., Kimmel, M.: Comparison of approaches to modeling of cell population dynamics. SIAM J. Appl. Math. **53**, 1480–1504 (1993)
9. Arino, O., Sanchez, E.: A survey of cell population dynamics. J. Theor. Med. **1**, 35–51 (1997)
10. Ballesta, A., Clairambault, J., Dulong, S., Lévi, F.: Theoretical optimization of irinotecan-based anticancer strategies in case of drug-induced efflux. Appl. Math. Lett. **24**, 1251–1256 (2011)

11. Ballesta, A., Dulong, S., Abbara, C., Cohen, B., Okyar, A., Clairambault, J., Levi, F.: A combined experimental and mathematical approach for molecular-based optimization of irinotecan circadian delivery. PLoS Comp. Biol. **7**, e1002143 (2011)
12. Banks, H., Sutton, K.L., Thompson, W.C., Bocharov, G., Doumic, M., Schenkel, T., Argilaguet, J., Giest, S., Peligero, C., Meyerhans, A.: A new model for the estimation of cell proliferation dynamics using cfse data. J. Imunol. Meth. **373**, 143–160 (2011)
13. Barbolosi, D., Iliadis, A.: Optimizing drug regimens in cancer chemotherapy: a simulation study using a pk-pd model. Comput. Biol. Med. **31**, 157–172 (2001)
14. Barbolosi, D., Benabdallah, A., Hubert, F., Verga, F.: Mathematical and numerical analysis for a model of growing metastatic tumors. Math. Biosci. **218**, 1–14 (2009)
15. Basdevant, C., Clairambault, J., Lévi, F.: Optimisation of time-scheduled regimen for anti-cancer drug infusion. Math. Model. Numer. Anal. **39**, 1069–1086 (2006)
16. Basse, B., Baguley, B.C., Marshall, E.S., Joseph, W.R., van Brunt, B., Wake, G., Wall, D.J.N.: A mathematical model for analysis of the cell cycle in cell lines derived from human tumors. J. Math. Biol. **47**, 295–312 (2003)
17. Basse, B., Baguley, B., Marshall, E., Wake, G., Wall, D.: Modelling the flow cytometric data obtained from unperturbed human tumour cell lines: Parameter fitting and comparison. Bull. Math. Biol. **67**, 815–830 (2005)
18. Basse, B., Baguley, B.C., Marshall, E.S., Joseph, W.R., van Brunt, B., Wake, G., Wall, D.J.N.: Modelling cell death in human tumour cell lines exposed to the anticancer drug paclitaxel. J. Math. Biol. **49**, 329–357 (2004)
19. Basse, B., Baguley, B.C., Marshall, E.S., Wake, G.C., Wall, D.J.N.: Modelling cell population growth with applications to cancer therapy in human tumour cell lines. Prog. Biophys. Mol. Biol. **85**, 353–368 (2004)
20. Basse, B., Ubezio, P.: A generalised age- and phase-structured model of human tumour cell populations both unperturbed and exposed to a range of cancer therapies. Bull. Math. Biol. **69**, 1673–1690 (2007)
21. Bekkal Brikci, F., Clairambault, J., Perthame, B.: Analysis of a molecular structured population model with polynomial growth for the cell cycle. Math. Comput. Model. **47**, 699–713 (2008)
22. Bekkal Brikci, F., Clairambault, J., Ribba, B., Perthame, B.: An age-and-cyclin-structured cell population model for healthy and tumoral tissues. J. Math. Biol. **57**, 91–110 (2008)
23. Bernard, S., Čajavec Bernard, B., Lévi, F., Herzel, H.: Tumor growth rate determines the timing of optimal chronomodulated treatment schedules. PLoS Comput. Biol. **6**(3), e1000712 (2010)
24. Bertsekas, D.: Nonlinear Programming. Athena Scientific, Nashua, NH (1995)
25. Billy, F., Ribba, B., Saut, O., Morre-Trouilhet, H., Colin, T., Bresch, D., Boissel, J.-P., Grenier, E., Flandrois, J.-P.: A pharmacologically based multiscale mathematical model of angiogenesis and its use in investigating the efficacy of a new cancer treatment strategy. J. Theor. Biol. **260**(4), 545–562 (2009)
26. Billy, F., Clairambault, J., Fercoq, O., Gaubert, S., Lepoutre, T., Ouillon, T.: Proliferation in cell population models with age structure. In: Proceedings of ICNAAM 2011, Kallithea Chalkidis (Greece), pp. 1212–1215. American Institute of Physics (2011)
27. Billy, F., Clairambault, J., Fercoq, O., Gaubert, S., Lepoutre, T., Ouillon, T., Saito, S.: Synchronisation and control of proliferation in cycling cell population models with age structure. Math. Comp. Simul. in press, (2012)
28. Bittanti, S., Guardabassi, G.: Optimal periodic control and periodic systems analysis - an overview. In: 25th IEEE Conference on Decision and Control, pp. 1417–1423 (1986)
29. Bonnans, J., Gilbert, J., Lemaréchal, C., Sagastizábal, C.: Numerical Optimization – Theoretical and Practical Aspects. Universitext. Springer, Berlin (2006)
30. Bresch, D., Colin, T., Grenier, E., Ribba, B., Saut, O.: A viscoelastic model for avascular tumor growth. Special issue Discrete Continuous Dyn. Syst. 101–108 (2009)
31. Bresch, D., Colin, T., Grenier, E., Ribba, B., Saut, O.: Computational modeling of solid tumor growth: the avascular stage. SIAM J. Sci. Comput. **32**, 2321–2344 (2010)

32. Byrne, H.M., Chaplain, M.A.: Growth of nonnecrotic tumors in the presence and absence of inhibitors. Math. Biosci. **130**(2), 151–181 (1995)
33. Byrne, H.M., Chaplain, M.A.: Growth of necrotic tumors in the presence and absence of inhibitors. Math. Biosci. **135**(15), 187–216 (1996)
34. Byrne, H., Alarcón, T., Owen, M., Webb, S., Maini, P.: Modelling aspects of cancer dynamics: A review. Phil. Trans. Roy. Soc. A **364**, 1563–1578 (2006)
35. Clairambault, J.: Modelling oxaliplatin drug delivery to circadian rhythm in drug metabolism and host tolerance. Adv. Drug Deliv. Rev. **59**, 1054–1068 (2007)
36. Clairambault, J.: A step toward optimization of cancer therapeutics. physiologically based modelling of circadian control on cell proliferation. IEEE-EMB Mag. **27**, 20–24 (2008)
37. Clairambault, J.: Modelling physiological and pharmacological control on cell proliferation to optimise cancer treatments. Math. Model. Nat. Phenom. **4**, 12–67 (2009)
38. Clairambault, J.: Optimising cancer pharmacotherapeutics using mathematical modelling and a systems biology approach. Personalized Medicine **8**, 271–286 (2011)
39. Clairambault, J., Laroche, B., Mischler, S., Perthame, B.: A mathematical model of the cell cycle and its control. Technical report, Number 4892, INRIA, Domaine de Voluceau, BP 105, 78153 Rocquencourt, France (2003)
40. Clairambault, J., Gaubert, S., Lepoutre, T.: Comparison of Perron and Floquet eigenvalues in age structured cell division models. Math. Model. Nat. Phenom. **4**, 183–209 (2009)
41. Clairambault, J., Gaubert, S., Lepoutre, T.: Circadian rhythm and cell population growth. Math. Comput. Model. **53**, 1558–1567 (2011)
42. Clairambault, J., Hochberg, M., Lorenzi, T., Lorz, A, Perthame, B.: Populational adaptive evolution, chemotherapeutic resistance and multiple anti-cancer therapies. Math. Model. Numer. Anal. accepted, 2012.
43. Cui, S., Friedman, A.: Analysis of a mathematical model of the effect of inhibitors on the growth of tumor. Math. Biosci. **164**, 103–137 (2000)
44. Cunningham, D., Humblet, Y., Siena, S., Khayat, D., Bleiberg, H., Santoro, A., Bets, D., Mueser, M., Harstrick, A., Verslype, C., Chau, I., Van Cutsem, E.: Cetuximab monotherapy and cetuximab plus irinotecan in irinotecan-refractory metastatic colorectal cancer. N. Engl. J. Med. **351**, 337–345 (2004)
45. Deakin, A.S.: Model for the growth of a solid in vitro tumor. Growth **39**(1), 159–165 (1975)
46. d'Onofrio, A.: Rapidly acting antitumoral antiangiogenic therapies. Phys. Rev. E Stat. Nonlin. Soft Matter Phys. **76**(3 Pt 1), 031920 (2007)
47. d'Onofrio, A., Gandolfi, A.: Tumour eradication by antiangiogenic therapy: analysis and extensions of the model by Hahnfeldt et al. (1999). Math. Biosci. **191**, 159–184 (2004)
48. d'Onofrio, A., Gandolfi, A.: A family of models of angiogenesis and anti-angiogenesis anti-cancer therapy. Math. Med. Biol. **26**(1), 63–95 (2009)
49. d'Onofrio, A., Ledzewicz, U., Maurer, H., Schättler, H.: On optimal delivery of combination therapy for tumors. Math. Biosci. **222**, 13–26 (2009)
50. Doumic, M., Perthame, B., Zubelli, J.: Numerical solution of an inverse problem in size-structured population dynamics. Inverse Problems **25**, 045008 (25pp) (2009)
51. Doumic, M., Maia, P., Zubelli, J.: On the calibration of a size-structured population model from experimental data. Acta Biotheor. **58**, 405–413 (2010)
52. Doumic, M., Hoffmann, M., Reynaud, P., Rivoirard, V.: Nonparametric estimation of the division rate of a size-structured population. SIAM J. Num. Anal. **50**(2), 925–950 (2012)
53. Druker, B., Talpaz, M., Resta, D., Peng, B., Buchdunger, E., Ford, J., Lydon, N., Kantarjian, H., Capdeville, R., Ohno-Jones, S., Sawyers, C.: Efficacy and safety of a specific inhibitor of the bcr-abl tyrosine kinase in chronic myeloid leukemia. N. Engl. J. Med. **344**, 1031–1037 (2001)
54. Dua, P., Dua, V., Pistikopoulos, E.: Optimal delivery of chemotherapeutic agents in cancer. Comp. Chem. Eng. **32**, 99–107 (2008)
55. Ergun, A., Camphausen, K., Wein, L.M.: Optimal scheduling of radiotherapy and angiogenic inhibitors. Bull. Math. Biol. **65**(3), 407–424 (2003)

56. Fister, K.R., Panetta, J.C.: Optimal control applied to cell-cycle-specific cancer chemotherapy. SIAM J. Appl. Math. **60**(3), 1059–1072 (2000)
57. Frieboes, H.B., Edgerton, M.E., Fruehauf, J.P., F.Rose, R.A.J., Worrall, L.K., Gatenby, R.A., Ferrari, M., Cristini, V.: Prediction of drug response in breast cancer using integrative experimental/computational modeling. Canc. Res. **69**, 4484–4492 (2009)
58. Gatenby, R.: A change of strategy in the war on cancer. Nature **459**, 508–509 (2009)
59. Gatenby, R., Gawlinski, E.: A reaction-diffusion model of cancer invasion. Canc. Res. **56**, 5745–5753 (1996)
60. Gatenby, R., Maini, P.K., Gawlinski, E.: Analysis of tumor as an inverse problem provides a novel theoretical framework for understanding tumor biology and therapy. Appl. Math. Lett. **15**, 339–345 (2002)
61. Gatenby, R., Silva, A., Gillies, R., Friden, B.: Adaptive therapy. Canc. Res. **69**, 4894–4903 (2009)
62. Gompertz, B.: On the nature of the function expressive of the law of human mortality, and on a new mode of determining the value of life contingencies. Philos. Transact. Roy. Soc. Lond. **115**, 513–585 (1825)
63. Greenspan, H.: Models for the growth of a solid tumor by diffusion. Stud. Appl. Math. **52**, 317–340 (1972)
64. Gyllenberg, M., Osipov, A., Päivärinta, L.: The inverse problem of linear age-structured population dynamics. J. Evol. Equat. **2**, 223–239 (2002)
65. Gyllenberg, M., Webb, G.F.: A nonlinear structured population model of tumor growth witsh quiescence. J. Math. Biol. **28**, 671–694 (1990)
66. Haferlach, T.: Molecular genetic pathways as therapeutic targets in acute myeloid leukemia. Hematology **2008**, 400–411 (2008), Am. Soc. Hematol. Educ. Program.
67. Hahnfeldt, P., Panigrahy, D., Folkman, J., Hlatky, L.: Tumor development under angiogenic signaling: a dynamical theory of tumor growth, treatment response, and postvascular dormancy. Canc. Res. **59**, 4770–4775 (1999)
68. Hansen, N.: The CMA evolution strategy: A comparing review. In Lozano, J., Larraaga, P., Inza, I., Bengoetxea, E. (eds.) Towards a New Evolutionary Computation. Advances in Estimation of Distribution Algorithms, pp. 75–102. Springer, Berlin (2006)
69. Hinow, P., Wang, S., Arteaga, C., Webb, G.: A mathematical model separates quantitatively the cytostatic and cytotoxic effects of a her2 tyrosine kinase inhibitor. Theor. Biol. Med. Model. **4**, 14 (2007). doi:10.1186/1742-4682-4-14
70. Iliadis, A., Barbolosi, D.: Optimizing drug regimens in cancer chemotherapy by an efficacy-toxicity mathematical model. Comput. Biomed. Res. **33**, 211–226 (2000)
71. Iwata, K., Kawasaki, K., Shigesada, N.: A dynamical model for the growth and size distribution of multiple metastatic tumors. J. Theor. Biol. **203**, 177–186 (2000)
72. Jackson, T.: Intracellular accumulation and mechanism of action of doxorubicin in a spatio-temporal tumor model. J. Theor. Biol. **220**, 201–213 (2003)
73. Jackson, T., Byrne, H.: A mathematical model to study the effects of drug resistance and vasculature on the response of solid tumors to chemotherapy. Math. Biosci. **164**, 17–38 (2000)
74. Kheifetz, Y., Kogan, Y., Agur, Z.: Long-range predictability in models of cell populations subjected to phase-specific drugs: growth-rate approximation using properties of positive compact operators. Math. Model. Meth. Appl. Sci. **16**, 1155–1172 (2006)
75. Kimmel, M., Świerniak, A.: Control theory approach to cancer chemotherapy: Benefiting from phase dependence and overcoming drug resistance. In: Friedman, A. (ed.) Tutorials in Mathematical Biosciences III. Lecture Notes in Mathematics, vol. 1872, pp. 185–221. Springer, Berlin (2006)
76. Kirsch, A.: An Introduction to the Mathematical Theory of Inverse Problems, 2nd edn. Springer, New York (2011)
77. Kitano, H.: Cancer as a robust system: Implications for anticancer therapy. Nat. Rev. Canc. **3**, 227–235 (2004)
78. Kitano, H.: A robustness-based approach to systems-oriented drug design. Nat. Rev. Drug Discov. **6**, 202–210 (2007)

79. Kohandel, M., Kardar, M., Milosevic, M., Sivaloganathan, S.: Dynamics of tumor growth and combination of anti-angiogenic and cytotoxic therapies. Phys. Med. Biol. **52**, 3665–3677 (2007)
80. Kozusko, F., Chen, P., Grant, S.G., Day, B.W., Panetta, J.C.: A mathematical model of in vitro cancer cell growth and treatment with the antimitotic agent curacin a. Math. Biosci. **170**(1), 1–16 (2001)
81. Laird, A.: Dynamics of tumour growth. Br. J. Canc. **13**, 490–502 (1964)
82. Lasota, A., Mackey, M.: Chaos, Fractals, and Noise: Stochastic Aspects of Dynamics, 2nd edn. Springer, New York (1994)
83. Ledzewicz, U., Schttler, H.: Optimal controls for a model with pharmacokinetics maximizing bone marrow in cancer chemotherapy. Math. Biosci. **206**, 320–342 (2007)
84. Ledzewicz, U., Maurer, H., Schaettler, H.: Optimal and suboptimal protocols for a mathematical model for tumor anti-angiogenesis in combination with chemotherapy. Math. Biosci. Eng. **8**, 307–323 (2011)
85. Ledzewicz, U., Maurer, H., Schaettler, H.: Optimal and suboptimal protocols for a mathematical model for tumor anti-angiogenesis in combination with chemotherapy. Math. Biosci. Eng. **8**(2), 307–323 (2011)
86. Lehmann, E., Casella, G.: Theory of point estimation. Springer Texts in Statistics, 2nd edition, Springer, New York (1998)
87. Lévi, F.: Cancer chronotherapeutics. Special issue Chronobiol. Int. **19**, 1–19 (2002)
88. Lévi, F.: Chronotherapeutics: the relevance of timing in cancer therapy. Canc. Causes Contr. **17**, 611–621 (2006)
89. Lévi, F.: The circadian timing system: A coordinator of life processes. implications for the rhythmic delivery of cancer therapeutics. IEEE-EMB Magazine **27**, 17–20 (2008)
90. Lévi, F., Altinok, A., Clairambault, J., Goldbeter, A.: Implications of circadian clocks for the rhythmic delivery of cancer therapeutics. Phil. Trans. Roy. Soc. A **366**, 3575–3598 (2008)
91. Lévi, F., Okyar, A., Dulong, S., Innominato, P., Clairambault, J.: Circadian timing in cancer treatments. Ann. Rev. Pharmacol. Toxicol. **50**, 377–421 (2010)
92. Lévi, F., Schibler, U.: Circadian rhythms: Mechanisms and therapeutic implications. Ann. Rev. Pharmacol. Toxicol. **47**, 493–528 (2007)
93. Ljung, L.: System Identification - Theory for the User, 2nd edn. PTR Prentice Hall, Upper Saddle River, N.J. (1999)
94. Lupi, M., Matera, G., Branduardi, D., D'Incalci, M., Ubezio, P.: Cytostatic and cytotoxic effects of topotecan decoded by a novel mathematical simulation approach. Canc. Res. **64**, 2825–2832 (2004)
95. Lupi, M., Cappella, P., Matera, G., Natoli, C., Ubezio, P.: Interpreting cell cycle effects of drugs: the case of melphalan. Canc. Chemother. Pharmacol. **57**, 443–457 (2006)
96. Martin, R.: Optimal control drug scheduling of cancer chemotherapy. Automatica **28**, 1113–1123 (1992)
97. Martin, R.B., Fisher, M.E., Minchin, R.F., Teo, K.L.: Low-intensity combination chemotherapy maximizes host survival time for tumors containing drug-resistant cells. Math. Biosci. **110**, 221–252 (1992)
98. Martin, R.B., Fisher, M.E., Minchin, R.F., Teo, K.L.: Optimal control of tumor size used to maximize survival time when cells are resistant to chemotherapy. Math. Biosci. **110**, 201–219 (1992)
99. Maurer, H., Büskens, C., Kim, J.-H.R., Kaya, C.Y.: Optimization methods for the verification of second order sufficient conditions for bang-bang controls. Optim. Contr. Appl. Meth. **26**(3), 129–156 (2005)
100. Mauro, M.J., O'Dwyer, M., Heinrich, M.C., Druker, B.J.: STI571: A paradigm of new agents for cancer therapeutics. J. Clin. Oncol. **20**, 325–334 (2002)
101. McElwain, D., Ponzo, P.: A model for the growth of a solid tumor with non-uniform oxygen consumption. Math. Biosci. **35**, 267–279 (1977)
102. McKendrick, A.: Applications of mathematics to medical problems. Proc. Edinburgh Math. Soc. **54**, 98–130 (1926)

103. Metz, J., Diekmann, O.: The dynamics of physiologically structured populations. In: Lecture Notes in Biomathematics, vol. 68. Springer, New York (1986)
104. Montalenti, F., Sena, G., Cappella, P., Ubezio, P.: Simulating cancer-cell kinetics after drug treatment: application to cisplatin on ovarian carcinoma. Phys. Rev. E **57**, 5877–5887 (1998)
105. Morgan, D.: The Cell Cycle: Principles of Control. Primers in Biology series. Oxford University Press, Oxford (2006)
106. Murray, J.: Optimal control for a cancer chemotherapy problem with general growth and loss functions. Math. Biosci. **98**, 273–287 (1990)
107. Murray, J.: Some optimal control problems in cancer chemotherapy with a toxicity limit. Math. Biosci. **100**, 49–67 (1990)
108. Murray, J.: The optimal scheduling of two drugs with simple resistance for a problem in cancer chemotherapy. IMA J. Math. Appl. Med. Biol. **14**, 283–303 (1997)
109. Murray, J.: Mathematical Biology. I: An Introduction, 3rd edn. Springer, New York (2002)
110. Murray, J.: Mathematical Biology. II: Spatial Models and Biomedical Applications, 3rd edn. Springer, New York (2003)
111. Nocedal, J., Wright, S.: Numerical Optimization. Springer, New York (1999)
112. Norris, E., King, J., Byrne, H.: Modelling the response of spatially structured tumours to chemotherapy: Drug kinetics. Math. Comput. Model. **43**, 820–837 (2006)
113. Nowsheen, S., Aziz, K., Panayiotidis, M.I., Georgakilas, A.G.: Molecular markers for cancer prognosis and treatment: have we struck gold? Canc. Lett. (2011) Published on line, November 2011, DOI:10.1016/j.canlet.2011.11.022.
114. Panetta, J.C.: A mathematical model of breast and ovarian cancer treated with paclitaxel. Math. Biosci. **146**(2), 89–113 (1997)
115. Panetta, J., Adam, J.: A mathematical model of cell-specific chemotherapy. Math. Comput. Model. **22**, 67 (1995)
116. Panetta, J.C., Kirstein, M.N., Gajjar, A.J., Nair, G., Fouladi, M., Stewart, C.F.: A mechanistic mathematical model of temozolomide myelosuppression in children with high-grade gliomas. Math. Biosci. **186**, 29–41 (2003)
117. Panetta, J.C., Evans, W.E., Cheok, M.H.: Mechanistic mathematical modelling of mercaptopurine effects on cell cycle of human acute lymphoblastic leukaemia cells. Br. J. Canc. **94**(1), 93–100 (2006)
118. Pereira, F., Pedreira, C., de Sousa, J.: A new optimization based approach to experimental combination chemotherapy. Frontiers Med. Biol. Engng. **6**(4), 257–268 (1995)
119. Perthame, B.: Transport equations in biology. Frontiers in Mathematics Series. Birkhäuser, Boston (2007)
120. Perthame, B., Zubelli, J.: On the inverse problem for a size-structured population model. Inverse Problems **23**, 1037–1052 (2007)
121. Pontryagin, L.S., Boltyanski, V.G., Gamkrelidze, R.V., Mishchenko, E.F.: The mathematical theory of optimal processes. Interscience Publishers, New York (1962) Translated from the Russian by K.N. Trirogoff.
122. Prenen, H., Tejpar, S., Cutsem, E.V.: New strategies for treatment of kras mutant metastatic colorectal cancer. Clin. Canc. Res. **16**, 2921–2926 (2010)
123. Ribba, B., Saut, O., Colin, T., Bresch, D., Grenier, E., Boissel, J.P.: A multiscale mathematical model of avascular tumor growth to investigate the therapeutic benefit of anti-invasive agents. J. Theor. Biol. **243**, 532–541 (2006)
124. Ribba, B., Watkin, E., Tod, M., Girard, P., Grenier, E., You, B., Giraudo, E., Freyer, G.: A model of vascular tumour growth in mice combining longitudinal tumour size data with histological biomarkers. Eur. J. Canc. **47**, 479–490 (2011)
125. Sakaue-Sawano, A., Ohtawa, K., Hama, H., Kawano, M., Ogawa, M., Miyawaki, A.: Tracing the silhouette of individual cells in S/G2/M phases with fluorescence. Chem. Biol. **15**, 1243–48 (2008)
126. Shah, N., Tran, C., Lee, F., Chen, P., Norris, D., Sawyers, C.: Overriding imatinib resistance with a novel ABL kinase inhibitor. Science **305**, 399–401 (2004)

127. Shymko, R.M.: Cellular and geometric control of tissue growth and mitotic instability. J. Theor. Biol. **63**(2), 355–374 (1976)
128. Sinek, J.P., Sanga, S., Zheng, X., Frieboes, H.B., Ferrari, M., Cristini, V.: Predicting drug pharmacokinetics and effect in vascularized tumors using computer simulation. J. Math. Biol. **58**, 485–510 (2009)
129. Spall, J.C.: Introduction to stochastic search and optimization: estimation, simulation, and control. Wiley, Hoboken (2003)
130. Spinelli, L., Torricelli, A., Ubezio, P., Basse, B.: Modelling the balance between quiescence and cell death in normal and tumour cell populations. Math. Biosci. **202**, 349–370 (2006)
131. Swanson, K., Alvord, E., Murray, J.: A quantitative model for differential motility of gliomas in grey and white matter. Cell Prolif. **33**, 317–329 (2000)
132. Swanson, K., Alvord, E., Murray, J.: Quantifying efficacy of chemotherapy of brain tumors with homogeneous and heterogeneous drug delivery. J. Neurosurg. **50**, 223–237 (2002)
133. Swanson, K.R., Alvord, E.C., Murray, J.D.: Virtual brain tumours (gliomas) enhance the reality of medical imaging and highlight inadequacies of current therapy. Br. J. Canc. **86**, 14–18 (2002)
134. Swierniak, A., Polanski, A., Kimmel, M.: Optimal control problems arising in cell-cycle-specific cancer chemotherapy. Cell Prolif. **29**, 117–139 (1996)
135. Ubezio, P., Lupi, M., Branduardi, D., Cappella, P., Cavallini, E., Colombo, V., Matera, G., Natoli, C., Tomasoni, D., D'Incalci, M.: Quantitative assessment of the complex dynamics of G1, S, and G2-M checkpoint activities. Canc. Res. **69**, 5234–5240 (2009)
136. Villasana, M., Ochoa, G., Aguilar, S.: Modeling and optimization of combined cytostatic and cytotoxic cancer chemotherapy. Artif. Intell. Med. **50**(3), 163–173 (2010)
137. Walter, E., Pronzato, L.: Identification of parametric models from experimental data. In: Communications and Control Engineering Series, 2nd edn. Springer, London (1997)
138. Webb, G.: Resonance phenomena in cell population chemotherapy models. Rocky Mountain J. Math. **20**(4), 1195–1216 (1990)
139. Webb, G.: A cell population model of periodic chemotherapy treatment. Biomedical Modeling and Simulation, pp. 83–92. Elsevier, Netherlands (1992)
140. Webb, G.: A non linear cell population model of periodic chemotherapy treatment. Recent Trends Ordinary Differential Equations, Series in Applicable Analysis 1, pp. 569–583. World Scientific, Singapore (1992)
141. Zheng, X., Wise, S.M., Cristini, V.: Nonlinear simulation of tumor necrosis, neo-vascularization and tissue invasion via an adaptive finite-element/level-set method. Bull. Math. Biol. **67**, 211–259 (2005)

Tumor Development Under Combination Treatments with Anti-angiogenic Therapies

Urszula Ledzewicz, Alberto d'Onofrio, and Heinz Schättler

1 Introduction

Tumors are a family of high-mortality diseases, each differing from the other, but all exhibiting a derangement of cellular proliferation and characterized by a remarkable lack of symptoms [52] and by time courses that, in a broad sense, may be classified as nonlinear. As a consequence, despite the enormous strides in prevention and, to a certain extent, cure, cancer is one of the leading causes of death worldwide, and, unfortunately, is likely to remain so for many years to come [4, 53].

Phenomenal progress in the field of molecular biology has qualitatively suggested how the macroscopic complexity of tumor behavior reflects the intricacy of its underlying deregulating microscopic biochemical mechanisms. At an inter-cellular level, considering tumor cell populations as ecosystems [57], further sources of complexity arise from its internal cell-to-cell cooperative and competitive interactions [41]. Additional interactions, which are critically relevant for the survival of a cancer, are its relationships with external populations, such as blood vessels, lymphatic vessels and with the cells of the immune system. Moreover, the responses of tumor cells (TCs) to these interactions are characterized by a considerable evolutionary ability via changes by means of mutations to enhance

U. Ledzewicz (✉)
Department of Mathematics and Statistics, Southern Illinois University Edwardsville, Edwardsville, IL, 62026, USA
e-mail: uledzew@siue.edu

A. d'Onofrio
Department of Experimental Oncology, European Institute of Oncology, 20139 Milan, Italy
e-mail: alberto.donofrio@ifom-ieo-campus.it

H. Schättler
Department of Electrical and Systems Engineering, Washington University, St. Louis, MO, 63130, USA
e-mail: hms@wustl.edu

U. Ledzewicz et al., *Mathematical Methods and Models in Biomedicine*, Lecture Notes on Mathematical Modelling in the Life Sciences, DOI 10.1007/978-1-4614-4178-6_11,

their survival in a hostile environment. Summarizing, cancer is a disease with at times anti-intuitive behavior whose macroscopic time course reflects intra-cellular and inter-cellular phenomena that are strongly nonlinear and time varying.

In this framework, methods of modern mathematics, such as the theory of finite and infinite dimensional dynamical systems, can play an important role in better understanding, preventing, and treating the family of bio-physical phenomena collectively called cancer. As Bellomo and Maini stressed in [3]: “the heuristic experimental approach, which is the traditional investigative method in the biological sciences,” and in medicine, “should be complemented by the mathematical modeling approach. The latter can be used as a hypothesis-testing and indeed, hypothesis-generating tool which can help to direct experimental research. In turn, the results of experiments help to refine the modeling. The goal of this research is that, by iterating back and forth between experiment and theory, we eventually arrive at a deeper conceptual understanding of how the highly nonlinear processes in biology interact. *The ultimate goal in the clinical setting is to use mathematical models to help design therapeutic strategies* [our emphasis]”.

In this chapter, we describe some of the fundamental principles that underlie the mathematical modeling of the evolution of a tumor that need to be taken into account in any treatment approach. Clearly, no attempt can be made to be comprehensive in a short chapter and for this reason we focus on one particular topic, combination therapies that involve anti-angiogenic treatments. We start with a brief discussion of various mathematical models for tumor growth which form the basis on which any kind of treatment needs to be imposed. Of these, by far the most important ones are chemo- and radiotherapies and we discuss their benefits and shortcomings. Anti-angiogenic therapies target the vasculature of a developing tumor and in combination with these traditional treatment approaches provide a two-pronged attack on both the cancer cells and the vasculature that supports them. Starting with general models that capture the characteristic features of these treatment approaches, we lead over to more detailed models as they are needed to optimize treatment protocols and discuss the implications of our mathematical analysis.

2 Phenomenological Models of Tumor Growth

A phenomenological model that describes the growth of a population of cells may be written in the general form

$$\dot{p} = pR(p), \tag{1}$$

where p is the size (measured as volume, number of cells, density of cells, etc.) of the population and $R(\cdot)$ models the net proliferation rate, i.e., the difference between the proliferation rate of the tumor cells, $\Pi(p)$, and their death and apoptosis rate $M(p)$. Ideally, these rates are constant and then the growth law is a pure exponential [68]. In the biological reality, however, the rate $R(p)$ normally is a decreasing function of p. Indeed, due to limited amounts of nutrients, the proliferation rate is a decreasing function of the population size whereas the rate $M(p)$ is an increasing function and, typically, the net effect will be positive. We shall use mainly the net proliferation rate since generally it is difficult to infer Π and M separately from external growth measurements. If $\Pi(0) > M(0)$, then it is easy to show that for all $p(0) > 0$ it follows that $\lim_{t\to+\infty} p(t) = K$, where K is the unique solution of the equation $\Pi(p) = M(p)$. This value K, called the *carrying capacity*, then represents the maximum sustainable size of the population. Unfortunately, in the vast majority of cases, the value K well exceeds values compatible with the life of the host.

Initially, for a small tumor size $p(0)$, an *exponential* growth law is appropriate for tumor growth. For $p \ll K$ higher-order terms can be neglected and approximatively it holds that

$$\dot{p} \simeq R(p(0))p. \tag{2}$$

However, as the tumor grows, these neglected terms matter. One of the most commonly used laws to describe tumor growth is the *Gompertz law* [68],

$$R(p) = a - b\ln(p), \qquad a > b > 0. \tag{3}$$

The parameter a represents a baseline proliferation rate, while b summarizes the effects of mutual inhibitions between cells and competition for nutrients; it sometimes is called the growth retardation factor. Normalizing $p(0) = 1$, the tumor size then becomes

$$p(t) = \exp\left(\frac{a}{b}(1 - \mathrm{e}^{-bt})\right), \tag{4}$$

which has a typical double exponential structure. The normalized carrying capacity is $K = \exp\left(\frac{a}{b}\right)$ and thus it is convenient to rewrite $R(p)$ in the form $R(p) = \xi \ln(K/p)$, with the coefficient ξ a growth parameter that determines the rate of convergence of p to K.

The Gompertz law belongs to the class of phenomenological growth models that are based on competition between processes associated with proliferation and death. The number of such models is amazingly large, and, as another prolific example, we only mention the ubiquitous generalized logistic law,

$$R(p) = a\left(1 - \left(\frac{p}{K}\right)^{\nu}\right), \qquad a > 0, \quad \nu > 0, \tag{5}$$

notwithstanding the existence of various other models (e.g., see, [15, 17, 41]).

Indeed, both the Gompertz and logistic models were generalized in many ways, and the recurrent question which model is more realistic [33] has no correct answer. Since populations of cancer cells of different types and/or in different conditions may behave very differently, it should not be surprising that models of cancer growth can be so diversified. Indeed, any macroscopic growth law has to mirror a set of phenomena that occur at the cellular scale including metabolic processes and intercellular interactions that vary considerably from case to case [35, 41]. Concerning specifically Gompertz-like models, as pointed out in [39, 68], all growth laws that produce a relative growth rate $\frac{p'}{p}$ that tends to ∞ as the size p tends to zero are clearly not adequate to describe the growth of small aggregate tumors whose doubling time, a quantity related to a complex set of biological processes such as cell division cycle and apoptosis, cannot be arbitrarily small. The Gomp-ex law by Wheldon [68]

$$\dot{p} = \begin{cases} p(a - b\ln(C)) & \text{if } 0 < p < C, \\ p(a - b\ln(p)) & \text{if } p \geq C \end{cases} \tag{6}$$

provides a modification that uses the Gompertz law above a certain threshold C, but uses a simple exponential growth for smaller tumor sizes. Even if the tumor size is measured in terms of cells, this stabilizes the ratio $\frac{p'}{p}$ for small populations. For another generalization of the Gompertz law, see [49].

All of these models were obtained by qualitative reasoning and then, for specific cases, validated by means of data fitting, e.g., [36, 37]. Some of them are remarkably successful in the process of data-based validation. It is thus natural to ask to what extent these models reproduce at a large degree of approximation finer microscopical details, for example, of intercellular inhibitions. A second natural question then arises as to whether these models can be unified in some general framework. Can each of them be considered as a particular instance of some metamodel? Among the few works aimed at introducing a mechanistic theory that links macroscopic phenomenological models to microscopic interactions and parameters, we cite the simple, yet plausible model in [35] which is based on the realistic hypothesis of long range interactions between cells in a population whose "structure is fractal."This approach, significantly extended by one of us in [41], allows to show that *apparently contradictory growth models* (logistic, generalized logistic, Gompertzian, exponential, von Bertanlaffy, power law, del Santo-Guiot) *are simply macroscopic different manifestations of a common physical microscopic framework*. In other words, different values of the parameters of the microscopic law result in different analytical laws for $R(p)$. Thus, while one of these models may be more appropriate depending on a specific medical situation, in principle they all become viable alternatives in the investigation of the development of a tumor under treatment.

3 Cancer Chemotherapy

Cancer chemotherapy, and notwithstanding significant and very interesting recent developments towards novel cancer therapies such as immunotherapy, to this day remains the elective non-surgical choice for treatment of tumors. Strictly speaking, chemotherapy merely indicates the use of a chemical to cure a disease, especially due to proliferating pathogens such as bacteria, tumor cells, etc. However, chemotherapy has a so important clinical role in oncology that in the common usage of language the word chemotherapy nowadays uniquely denotes *anti-tumor chemotherapy*. The use of chemotherapies in oncology, indeed, has been one of the major steps forward in the so-called "war against cancer" [4]. There not only exists a huge body of experimental and clinical literature that has been produced in the last 60 years, but chemotherapy also triggered a large amount of theoretical research in connection with its apparently simple translation into mathematical models (e.g., [1, 23–25, 51, 54, 58–61, 64, 65, 68]). Also, and unlike in other fields of biomedicine, on this topic there has been a small, but important interplay between theoretical and experimental-clinical scientists [14, 38, 67].

Broadly speaking, we may consider two major classes of actions of chemo-therapeutic drugs: *cytostatic*, where the chemical decelerates or blocks the tumor cells' proliferation, and *cytotoxic*, where the agent kills the neoplastic cell. This, of course, is a highly idealized description and often it is not clear how to classify the actions of a specific drug. For example, paclitaxel, still one of the more commonly used drugs in chemotherapy, binds to tubulin which locks microtubules in place and thus prevents cell-duplication. In principle, this is a blocking action. However, generally, a drug that prevents the further duplication of cancer cells indefinitely is considered cytotoxic even when it does not induce apoptosis.

The main adverse effects of chemotherapy are due to the fact that drugs are rarely selective to identify tumor cells, but, especially in the first stages of modern chemotherapy, target all or at least large classes of proliferating cells. The mechanism of action for these drugs is to interfere with one or more biochemical pathways and thus the more the targeted pathway is specific to cancer cells, the less severe side effects are. Since its first use, it has been plain that for this reason—the scarce selectivity of chemotherapeutic agents—a number of serious side effects are related to the use of cytotoxic chemicals to cure tumors. They simply also kill a more or less wide range of physiologically proliferating cells important for life.

Even when side effects are limited, a high number of failures of chemotherapy due to both *intrinsic* and *acquired drug resistance* plagues this treatment approach. Cancer cells often are genetically unstable and, coupled with high proliferation rates, this leads to significantly higher mutation rates than in healthy cells [14]. If a mutated cell has a biochemical structure that invalidates the mechanism of attack of the chemotherapeutic agent, these cells have acquired drug resistance. Indeed, the response of tumor cells to chemotherapy is characterized by a considerable evolutionary ability to enhance the cell survival in an environment that is becoming hostile. Moreover, and more importantly, because of the tremendous heterogeneity of cancer cells, often small sub-populations of cells are intrinsically not sensitive to the treatment ab initio (intrinsic resistance). In this case, as the sensitive cells are killed by the treatment, a tiny fraction of remaining, intrinsically resistant tumor cells can grow to become the dominant remaining population leading to the failure of therapy, possibly only after many years of seeming remission of the cancer. Considerable research efforts thus have been, and still are being devoted to finding means to overcome drug resistance [13]. Tumor anti-angiogenic therapy falls into the realm of these procedures [20, 21] and, for this reason, combinations of chemotherapy with anti-angiogenic treatments offer synergistic advantages.

We briefly extend the growth model considered in the previous section to include chemotherapy. In this model we assumed that $M(0) < \Pi(0)$ since a negative net proliferation rate $R(p) < 0$ implies the self-extinction of the neoplasm, a case only of interest for immunogenic tumors. But a negative net proliferation rate is exactly what chemotherapeutic agents aim at and thus this has relevance in the theoretical analysis of chemotherapy where the agents either reduce the proliferation rate or increase the death rate of the neoplastic cells. When a drug is delivered to a human or an animal host, two different types of processes take place called *pharmacokinetics (PK)* and *pharmacodynamics (PD)*: pharmacokinetics determines the density of the drug in the blood, i.e., *what the body does to the drug*, and pharmacodynamics models the effects the drugs have, *what the drug does to the body*. If we administer a drug whose density profile in the blood is $c(t)$, in many cases it is considered realistic [68] to assume that the number of cells killed per time unit is proportional to $c(t)p(t)$, i.e., the pharmacodynamic model is linear in both the concentration c and p. This hypothesis is called the *linear log-kill hypothesis*, and it modifies the basic growth model to become

$$\dot{p} = pR(p) - \varphi cp \tag{7}$$

with φ a positive parameter. For example, a simple model of chemotherapy assuming a Gompertz law for unperturbed growth can be written as

$$\dot{p}(t) = -\xi p(t) \ln\left(\frac{p(t)}{p_\infty}\right) - \varphi c(t)p(t), \tag{8}$$

where p_∞ denotes the (constant) carrying capacity. In a more general setting, one can assume that the pharmacodynamics still is linear in p, but nonlinear in c, say

$$\dot{p} = pR(p) - H(c)p. \tag{9}$$

If the therapy is delivered by constant continuous infusion therapy, after some initial transient, it is reasonable to assume that $c(t) \approx C$, and in this case the tumor can be eradicated if $R(0) < H(C)$. In the important case of periodic therapy, $c(t+T) = c(t)$, this eradication condition becomes $R(0) < \langle H(c) \rangle$, where $\langle f \rangle$ denotes the average of a periodic function f over one period.

4 Modeling Radiotherapeutic Treatments

The log-kill model used in Eq. (8) can also be considered a rudimentary approximation for including effects of radiotherapy when only first-order killing effects are considered. While such a linear model is reasonable for a cytotoxic agent, it is, however, only a crude approximation for the effects of radiotherapy. It is more realistic to assume that the damage to DNA made by the effects of ionization radiation consists of a linear component that corresponds to a simultaneous break in both DNA strands caused by a single particle and a quadratic term that accounts for two adjacent breaks on different chromosomes caused by two different particles. This is the so-called linear-quadratic (LQ) model [12, 66, 68] that has become the accepted model in radiotherapy. The damage of radiation to the tumor can then be modelled in the form

$$-p(t)\left(\varphi+\beta\int_0^t w(s)\exp\left(-\rho(t-s)\mathrm{d}s\right)\right)w(t), \tag{10}$$

where w represents the radiation dose rate and φ, β and ρ are positive constants with φ and β related to the tumor LQ parameters and ρ the tumor repair rate. A small tumor repair rate implies a larger influence of the integral term that describes the secondary, i.e., quadratic effects, and thus a greater effectiveness of the therapy, while large repair rates imply that the integral can be neglected. Note that the integral term in parenthesis in Eq. (10) is simply the solution to the first-order linear equation

$$\dot{r}=-\rho r+w, \qquad r(0)=0. \tag{11}$$

Mathematically, the structure of the overall model becomes more transparent if we replace the integral by this differential equation. Briefly, in case of a constant dose rate $w(t)=W$, the eradication condition becomes

$$R(0)<\varphi W+(\beta/\rho)W^2, \tag{12}$$

whereas for periodically delivered therapies it is

$$R(0)<\varphi\left\langle w(t)\right\rangle+\beta\left\langle w(t)r(t)\right\rangle. \tag{13}$$

5 Tumor Angiogenesis and Solid Vascularized Tumors

Chemotherapy targets the main characteristic of tumor cells, their proliferative derangement. However, as already mentioned, tumor cells show a vast array of microscopic and macroscopic interactions with other cellular populations. As a consequence, the study of these phenomena may open the way for the creation of new therapies. Folkman [9, 10] already stressed in the early seventies that the development of a vascular network inside the tumor mass becomes necessary to support tumor growth. Indeed, primary solid tumors and metastases require the formation of new blood vessels in order to grow beyond 1–2 mm^3. Folkman named this process neo-angiogenesis. It is sustained by various mechanisms—tumors may coopt existing vessels, may induce the formation of new vessels from preexisting ones or may exploit endothelial precursors originating from the bone marrow [11]. Tumor angiogenesis is a complex process driven by *pro-angiogenic* factors that are being released as the tumor cells lack a full level of nutrients [70]. Interestingly enough, tumor cells also release *anti-angiogenic* chemicals that modulate the growth of the vessel network. In this way, a solid tumor deploys a sophisticated strategy to control its own growth. Folkman suggested that inhibiting the development of the tumoral vessel network could be a powerful way to control, in turn, the neoplastic growth via the reduction of nutrients supply. He termed this new kind of therapy *anti-angiogenic therapy*.

Angiogenic inhibitors are commonly classified [22] as *direct inhibitors* which act on the endothelial cells and inhibit their proliferation and migration or induce their apoptosis, or as *indirect inhibitors* that block the production of angiogenic factors by malignant cells, or as mixed agents that target both endothelial and malignant cells. Most angiogenic inhibitors are cytostatic inhibiting the formation of new blood vessels. Some of the direct inhibitors have a cytotoxic action that induce a rapid destruction of existing blood vessels. Various anti-angiogenic drugs have undergone clinical development in recent years, and some of them have led to improvement in overall survival or disease-free survival in various clinical scenarios. Since the therapy targets healthy cells, namely the endothelial cells forming blood vessels, that are far more genetically stable than tumor cells, anti-angiogenic agents are far less subject to drug resistance [20]. *Per se*, this way of controlling the tumor burden appears intriguing and there is evidence from experimental work that inhibiting angiogenesis may induce tumor regression and sometimes cure [50].

Modeling the interplay between tumor growth and the development of its vascular network, as well as the action of angiogenic inhibitors, is an important step that could help to plan effective anti-angiogenic therapies and a large number of mathematical models have been proposed, e.g., [1, 2, 16, 34, 42, 43, 47]. Quite interestingly, Folkman himself and his coworkers formulated a simple, but largely influential mathematical model in [16] that describes the vascular phase of tumor growth assuming that this growth is strictly controlled by the dynamics of the vascular network and that the vascular dynamics is the result of the opposite influence of pro-angiogenic and anti-angiogenic factors produced by the tumor itself. This model provides a framework to portray the effects of anti-angiogenic therapies, and it was successful in fitting experimental data on the growth and response to different anti-angiogenic drugs for Lewis lung carcinomas implanted in mice.

The appreciation of the role of angiogenesis in tumor development has led Folkman and his coworkers to introduce the concept of a varying carrying capacity, $q(t)$, defined as the tumor size potentially sustainable by the existing vascular network at a given time [16]. This carrying capacity may be assumed proportional to the extent of the actual tumor vasculature. Making the carrying capacity in Eq. (8) variable, and following [16] in modeling, the sophisticated, tightly controlled strategy for the production of vessels reduces to the following dynamical system for tumor size and carrying capacity under chemotherapy:

$$\dot{p} = -\xi p \ln\left(\frac{p}{q}\right) - \varphi c p, \tag{14}$$

$$\dot{q} = bp - dp^{\frac{2}{3}} q - \mu q. \tag{15}$$

Here bp models the growth stimulated by the pro-angiogenic factors, $dq^{2/3}$ is a variable loss rate "constant" due to the endogenous anti-angiogenic factors produced autonomously by tumor cells, μ is the natural loss rate constant of the vasculature, and c denotes the concentration of the chemotherapeutic agent. This model was obtained under a series of simplifying assumptions that include spherical symmetry of the tumor, a fast degradation of pro-angiogenic factors and a slow degradation of inhibitory factors. The dynamics of anti-angiogenic factors reflects their more systemic effects and leads to an interaction term between the surface area of the spheroid and the vasculature of the form $dp^{\frac{2}{3}}q$, whereas the dynamics of the pro-angiogenic factors suggests the term bp. A mathematical analysis of this model was presented in [42], focusing on the tumor eradication under regimens of continuous or periodic anti-angiogenic therapy; the problem of determining optimal treatment schedules for a given amount of inhibitors has been solved in [26].

By relaxing the assumptions made in [16], and also by considering more general laws of tumor growth, the above model was generalized in [44] assuming that the specific growth rate of the tumor, $\frac{\dot{p}}{p}$, and the specific birth rate of vessels depend on the ratio between the carrying capacity and the tumor size. Since the ratio $\frac{q}{p}$ may be interpreted as proportional to the tumor vessel density, the second assumption agrees with the model proposed by Agur et al. [1]. In the absence of therapy, the model proposed in [44] takes the form

$$\dot{p} = pF\left(\frac{q}{p}\right), \tag{16}$$

$$\dot{q} = q\left(\beta\left(\frac{q}{p}\right) - I(p) - \mu\right), \tag{17}$$

where the growth function $F : (0,\infty) \to \mathbb{R}$ is strictly increasing and satisfies

$$-\infty \leq \lim_{\rho\to 0^+} F(\rho) < 0, \quad F(1) = 0, \quad \text{and} \quad 0 < \lim_{\rho\to+\infty} F(\rho) \leq +\infty.$$

The stimulation term $\beta : (0,\infty) \to \mathbb{R}$ is strictly decreasing and satisfies $\beta(+\infty) = 0$ and $\beta(1) > \mu$. It may be unbounded, like in the biologically important case of power laws, $\beta(\rho) = b\rho^{-\delta}$, $\delta > 0$, or bounded such as

$$\beta(\rho) = \frac{\beta_M}{1 + k\rho^n}, \qquad n \geq 1.$$

In this case, $\lim_{\rho\to 0^+} \beta(\rho)$ is finite and β is a decreasing Hill function. Also, combinations of the above two expressions are allowed. The inhibition term is a strictly increasing function $I : [0,\infty) \to [0,\infty)$ that satisfies $I(0) = 0$ and $\lim_{p\to+\infty} I(p) = +\infty$. Equations (16) and (17) together provide a general mathematical framework within which the time evolution of solid vascularized tumors can be analyzed.

6 Combinations of Chemo- and Anti-angiogenic Therapies of Vascularized Tumors

Anti-angiogenic therapy is an indirect approach that only limits the tumor's support mechanism without actually killing the cancer cells. Therefore it is only natural, and this has been observed consistently, that therapeutic effects are only temporary and, in the absence of further treatment, the tumor will grow back once treatment is halted. Thus tumor anti-angiogenesis is not efficient as a stand-alone or monotherapy treatment, but it needs to be combined with other mechanisms like traditional chemotherapy or radiotherapy treatments that kill cancer cells. In this context, it is worth noting that tumors differ from normal tissues also in density, topology and functionality of their vessel network. Tumor vasculature is characterized by a remarkable degree of intricacy as well as by a variety of disfunctionalities. Since the neovessel network that brings nutrients to the tumor is also the route to deliver chemotherapeutic drugs, R.K. Jain hypothesized that the preliminary delivery of a vessel disruptive anti-angiogenic agent, by "pruning" the vessel network, may regularize it with beneficial consequences for the successive delivery of cytotoxic chemotherapeutic agents [18, 19]. If treatment schedules are optimized to minimize the tumor volume, such a structure of protocols is confirmed as optimal [48] and our results support this hypothesis.

Experimental studies on mouse models and clinical trials [6,7] showed that some cell cytotoxic agents (e.g., cyclophosphamide) also have significant anti-angiogenic effects. In [40], this effect was modelled for a chemotherapeutic monotherapy that also has a vessel disruptive action. On the other hand, here we are interested to fully assess the effects of a combined therapy when three classes of drugs are co-present: (a) an anti-angiogenic agent u having effects of vessel disruption, (b) a chemotherapeutic agent v which may or may not have effects of vessel disruption or inhibition, and (c) an anti-angiogenic agent w inhibiting the proliferation of the tumor vessels. Generalizing the model in [48], these effects are included in the following equations:

$$\dot{p} = pF\left(\frac{q}{p}\right) - \varphi v p, \tag{18}$$

$$\dot{q} = q\left(\theta(w,v)\cdot\beta\left(\frac{q}{p}\right) - I(p) - \mu - \gamma u - \eta v\right). \tag{19}$$

Here $\theta = \theta(w,v)$ is a function that takes values in the interval $[0,1]$ and is decreasing in both variables. We also assume that $0 \leq \eta \leq \varphi$ since, for biological reasons, the log-kill effect on the carrying capacity, if it exists, is not the prevalent one. However, this assumption has no consequences on the asymptotic behavior of the solutions of the proposed system. The model considered in [48] was the special case $\beta(\rho) = \frac{b}{\rho}$ so that $q\beta(q/p) = bp$.

We first analyze the behavior of the tumor and its vessels in the absence of therapies or under continuous infusion therapies (CITs) of infinite temporal length, i.e., for $v(t) \equiv V \geq 0$, $u(t) \equiv U \geq 0$ and $w(t) = W \geq 0$. If $F(\infty)$ is small enough, then the tumor can in fact be eradicated by anti-angiogenic action alone.

Lemma 6.1. *If $\varphi V > F(\infty)$, then, in the limit $t \to \infty$, the tumor is eradicated,* $\lim_{t\to+\infty} p(t) = 0$.

The p-nullcline, $\dot{p} = 0$, is given by $q = A(V)p$ where $A(V) = F^{-1}(\varphi V)$ and, setting $\dot{q} = 0$, we obtain that

$$q = Q(p) = p\beta^{-1}\left(\frac{I(p)+\mu+\gamma U+\eta V}{\theta(W,V)}\right). \tag{20}$$

It is then straightforward to prove the following proposition:

Proposition 6.1. *Under continuous infusion therapies, $U \geq 0$ and $V \geq 0$, if*

$$\theta(W,V)\beta\left(A(V)\right) > (\mu+\gamma U+\eta V), \tag{21}$$

then there exists a unique, non-null, globally asymptotically stable equilibrium point $EQ = (p_e(U,V,W), q_e(U,V,W))$ that satisfies

$$q_e(U,V,W) = p_e(U,V,W)A(V) \tag{22}$$

and

$$p_e(U,V,W) = I^{-1}\left[\theta(W,V)\beta\left(A(V)\right) - (\mu+\gamma+\eta V)\right]. \tag{23}$$

Moreover, the orbits of the system are bounded and the set

$$\Omega_{(U,V,W)} = \left\{(p,q) \in \mathbb{R}^2_+ : q \leq M = \max_{p\in[0,p_e]} Q(p) \text{ and } 0 \leq p \leq \frac{M}{A(V)}\right\}$$

is positively invariant and attractive.

Thus, in case of infinitely long therapies, in principle it is possible to eradicate the tumor under suitable constraints on the drug density in the blood. A first condition to reach this target has been illustrated in Lemma 6.1, but it is simply the translation to the angiogenic setting of the eradication constraint $R(0) < H(C)$ from the chemotherapy setting. Here we are interested in results that genuinely relate to the tumor–vessel interaction, and we also would like to show possible synergies between chemotherapy and the anti-angiogenic therapies. This leads to the following proposition:

Proposition 6.2. *Under continuous infusion therapy, if*

$$\theta(W,V)\beta\left(A(V)\right) \leq \mu+\gamma U+\eta V \tag{24}$$

then, independently from the initial burden of the tumor and of the vessels (i.e., for all initial conditions $(p(0),q(0)) \in \mathbb{R}_+^2$), the tumor is eradicated, $\lim_{t\to+\infty}(p(t),q(t)) = (0^+,0^+)$. *Moreover, under generic time varying therapies, a sufficient condition for eradication is that*

$$\theta(w_{\min},v_{\min})\beta\left(A(v_{\min})\right) - (\mu + \gamma u_{\min} + \eta v_{\min}) \leq 0, \tag{25}$$

where $u_{\min}$, $v_{\min}$ *and* $w_{\min}$ *denote the minimal values during therapy.*

It is interesting to notice the various implications of condition Eq. (24) in case of combinations of a non-null chemotherapy ($V > 0$) with anti-angiogenic therapies. We start with the biologically interesting case when no anti-angiogenic agents are present [69], i.e., $U = W = 0$. In this case, the model studies the effects of the inclusion of a varying carrying capacity to a continuous infusion chemotherapy. In the classical setting, with constant carrying capacity $K > 0$,

$$\dot{p} = pF\left(\frac{K}{p}\right) - \varphi V p, \tag{26}$$

it follows that chemotherapy can never eradicate (at least, in theory) the tumor in case of an unbounded value $F(\infty)$. However, with the inclusion of the dynamics of the vessels, there exists a threshold V^* such that the tumor can be eradicated if $V \geq V^*$. It is easily determined from Eq. (24) with $W = U = 0$:

$$\theta(W,V^*)\beta\left(A(V^*)\right) - \mu - \eta V^* = 0. \tag{27}$$

Note that, provided that $\mu > 0$, the threshold also exists in case $\eta = 0$. Of course, since μ usually is small and satisfies $\mu \ll b$, such a threshold is very large. On the other hand, if $\mu = \eta = 0$, then there cannot be eradication, even if we add a proliferation inhibiting effects, since $\theta(W,V)\beta\left(A(V)\right) > 0$.

7 Beyond Linear Models of Chemotherapy in Vascularized Tumors

For many solid tumors, a log-kill law to model the effects of cytotoxic drugs might be oversimplified. Indeed, the efficacy of a blood-borne agent on the tumor cells will depend on its actual concentration at the cell site, and thus it will be influenced by the geometry of the vascular network and by the extent of blood flow. The efficacy of a drug will be higher if vessels are close to each other and sufficiently regular to permit a fast blood flow; it will be lower if vessels are distanced, or irregular and tortuous so to hamper the flow. To represent these phenomena in a simple form, in [45, 46] it has been assumed that the drug action to be included in the equation for $\dot{p}$ is dependent on the vessel density, i.e., in our model on the ratio $\rho = q/p$.

If $v(t)$ is the concentration of the agent in blood, we assume that its effectiveness is modulated by an unimodal non-negative function $\gamma(\rho)$ with $\gamma(0) = 0$. This leads to the new equation

$$\dot{p} = pF\left(\frac{q}{p}\right) - \gamma\left(\frac{q}{p}\right)v(t)p, \tag{28}$$

while the equation for q is unchanged. The above change deeply impacts the behavior of the dynamical system. It is not difficult to show that, under constant continuous infusion therapy, the model allows that the tumor-vessels system is multi-stable. This multi-stability may have both beneficial and detrimental "side effects" [45, 46]. On one hand, multi-stability allowed [45] to explain the pruning effect as a change of attractor of a tumor under chemotherapy. For example, the temporary delivery of anti-angiogenic therapy before an uninterrupted chemotherapy may move an orbit in the (p,q) space from the basin of attraction of a locally stable equilibrium with a large tumor size to the basin of attraction of another equilibrium with a small tumor size. On the other hand, the actual drug concentration profiles are affected by large bounded stochastic fluctuations, so that—as shown in [46]—the tumor volume may undergo detrimental noise-induced transitions from small to large equilibrium sizes. These stochastic transition phenomena might thus explain some cases of resistance to chemotherapy as due to non-genetic mechanisms.

8 Optimal Protocols for Combined Anti-angiogenic and Chemotherapies

In view of the high cost and limited availability of anti-angiogenic agents and because of harmful side effects of cytotoxic drugs, it is not feasible to give indefinite administrations of agents. The practically relevant question is how an a priori given, limited amount of anti-angiogenic and chemotherapeutic agents is best administered. Clearly, while anti-angiogenic agents are mostly limited because of their cost, chemotherapeutic agents must be limited because of their toxic side effects. For optimization problems, it is no longer possible to keep the model general, but a specific choice needs to be made for the growth function F and all the other functional terms that define the model. Here, for sake of definiteness, we consider the following model for tumor anti-angiogenesis that is based on [16]:

[AC] for a free terminal time T, minimize the objective $J(u) = p(T)$ subject to the dynamics

$$\dot{p} = \xi p \ln\left(\frac{q}{p}\right) - \varphi p v, \qquad p(0) = p_0, \tag{29}$$

$$\dot{q} = bp - dp^{\frac{2}{3}}q - \mu q - \gamma q u - \eta q v, \qquad q(0) = q_0, \tag{30}$$

$$\dot{y} = u, \qquad y(0) = 0, \tag{31}$$

$$\dot{z} = v, \qquad z(0) = 0, \tag{32}$$

over all piecewise continuous (respectively, Lebesgue measurable) functions $u : [0,T] \to [0,u_{\max}]$ and $v : [0,T] \to [0,v_{\max}]$ for which the corresponding trajectory satisfies the endpoint constraints $y(T) \leq y_{\max}$ and $z(T) \leq z_{\max}$.

The constants $u_{\max}$ and $v_{\max}$ denote the maximum dose rates of the anti-angiogenic and cytotoxic agents, respectively, and Eqs. (31) and (32) limit the overall amounts of each drug given to $y_{\max}$ and $z_{\max}$. In this formulation, for the moment, the dosages u and v are identified with their concentrations. We comment on the changes that occur to the optimal solutions if a standard linear model for the pharmacokinetics of the drugs is included in Sect. 10. Variations of this model are considered, for example, in [30, 56, 62, 63].

It follows from the dynamics (in fact, in greater generality) that for any admissible control pair (u,v) defined on $[0,\infty)$, solutions to this dynamical system exist for all times and remain positive. Some properties of optimal controls have been established for a more general formulation in [48], but complete solutions require a specification of the dynamics. We also denote by **[A]** the special case of problem **[AC]** that corresponds to an anti-angiogenic monotherapy. This problem is obtained from the formulation above by simply setting $z_{\max} = 0$ which eliminates the chemotherapeutic agent. For the monotherapy problem **[A]**, a complete solution in form of a regular synthesis of optimal controlled trajectories has been given in [26] and the significance of this solution lies in the fact that the optimal solution for the combination therapy problem indeed does build upon this synthesis. This is a nontrivial feature which does not hold for solutions to optimal control problems for nonlinear systems in general, but seems to be prevalent for the models combining anti-angiogenic treatments with chemotherapy and also radiotherapy. We therefore start with a brief review of the optimal synthesis for the monotherapy problem.

An optimal *synthesis* of controlled trajectories acts like a GPS system. It provides *a full "road map" of how optimal protocols look like depending on the current conditions of the state variables in the problem, both qualitatively and quantitatively*. Given any particular point (p,q) that represents the tumor volume and the current value of the carrying capacity, and any value y that represents the amount of inhibitors that have already been used up, equivalently, the amount remaining to be used, it tells how to choose the control u. Figure 1, for specific parameter values that have been taken from [16], gives a two-dimensional rendering of such a synthesis for the monotherapy problem when the variable y has been omitted. The actual numerical values of the parameters are not important for our presentation here since, indeed, the optimal synthesis looks qualitatively identical regardless of these numerical values [26]. In this chapter, we do not pursue quantitative results, but our aim is to describe *robust qualitative results* about optimal controls that are transferable to other models and give insights about the structure of optimal solutions that can be useful in more general situations as well. In fact, given our theoretically optimal analytical solution, for a typical initial condition, a straightforward Matlab code just takes seconds to compute the optimal control and corresponding trajectory. Since we have a full understanding of the theoretically

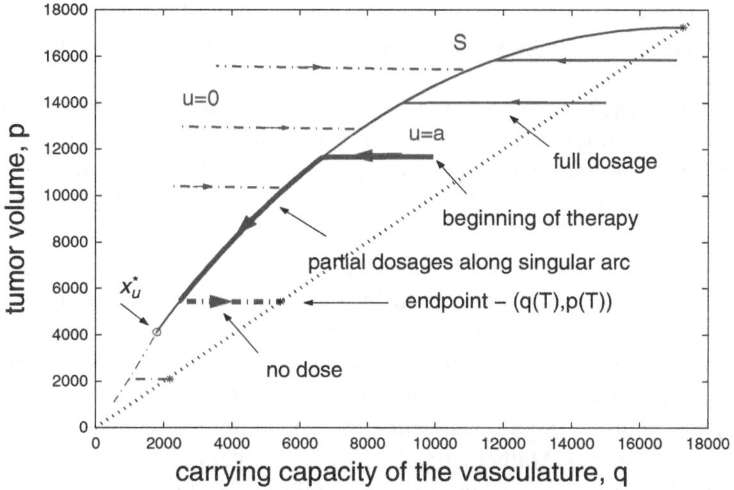

Fig. 1 A 2-dimensional projection of the optimal synthesis for the monotherapy model **[A]** into (p,q)-space

optimal solution, no minimization algorithm needs to be invoked, but the procedure simply consists in evaluating a piecewise defined, albeit somewhat complicated function.

We briefly describe the structure of optimal controlled trajectories. First of all, note that we plot the tumor-volume p vertically and the carrying capacity q horizontally in Fig. 1. This simply better visualizes a decrease or increase in the primary cancer volume. The anchor piece of the synthesis is an optimal singular arc $\mathscr{S}$ shown in blue. This is a unique curve defined in (p,q)-space along which the *best tumor reductions* occur and optimal controls follow this curve whenever the data allow. That is, if (p,q) happens to lie on $\mathscr{S}$, and if angiogenic inhibitors y are still available, then the optimal control consists in giving a specific time-varying dosage that makes the system stay on this curve, i.e., makes the singular arc invariant. There is a unique control that has this property, the corresponding singular control, and as long as its values lie in the admissible range $[0,u_{\max}]$, this is the optimal control. For the model by Hahnfeldt et al., there is a unique point on the singular arc when the control reaches its saturation limit $u_{\max}$ which is denoted by x_u^* in Fig. 1. Mathematically, the structure of the optimal synthesis near a saturating singular arc is well understood [26, 55], but for simplicity of presentation we limit our discussions here to cases when saturation does not occur. In this case, once (p,q) lies on $\mathscr{S}$, the optimal protocol then simply consists in giving these singular dosages until all inhibitors are exhausted. At that time, the state of the system lies in the region $\mathscr{D}_+ = \{(p,q) : p > q\}$ and, because of aftereffects, the tumor volume will still be decreasing until it reaches the diagonal $\mathscr{D}_0 = \{(p,q) : p = q\}$ of the system. This behavior is preprogrammed in the properties of the Gompertz growth law which makes the tumor volume decrease in $\mathscr{D}_+$ regardless of the actual control being used and is true for all growth models F that are strictly increasing and satisfy $F(1) = 0$.

If the state (p,q) does not lie on the singular arc and inhibitors are still available, then the optimal policy simply consists in getting to this singular arc in the "best" possible way as measured by the objective. If $p < q$, and initially this clearly is the medically relevant case, then this is done by simply giving a full dose treatment until the singular arc is reached and then the control switches to the singular regime. In such a case, since the carrying capacity is high, immediate action needs to be taken and it would not be optimal to let the tumor grow further. These are the trajectories in Fig. 1 that are shown by solid green curves. Note that these curves are almost horizontal and thus show little tumor reduction. Of course, the beneficial effect of this full dose segment is that it prevents the tumor increase that otherwise would have occurred. Significant reductions in tumor volume only arise as the singular arc is reached. Mathematically, it is no problem to include in the solution initial conditions when $p > q$, but, from the medical side, this case is less interesting. Here the optimal solution consists of giving no anti-angiogenic agents, but to wait until the system reaches the singular arc as the carrying capacity increases and then to start treatment when the singular arc is reached by once more following the singular regime. Intuitively, inhibitors are put to better use along this curve than if they would have been applied directly. For small tumor volumes (e.g., those that lie below the saturation level x_u^* for the singular control,) anti-angiogenic agents are always given at maximum dose. The minimum tumor volumes are realized when, after all inhibitors have been exhausted, the trajectory corresponding to $u = 0$ crosses the diagonal after termination of treatment.

The interesting feature of this synthesis is the relative simplicity and full robustness of the resulting optimal controls once the singular curve $\mathscr{S}$ is known. Furthermore, not only for the model considered here, but also for various of its modifications [8, 56], it is possible to determine this singular arc and its corresponding control analytically by means of well-known procedures in geometric optimal control theory which makes the construction of this synthesis and the resulting computations of optimal protocols a worthwhile endeavor [26, 30].

Below, we give the explicit formulas for the singular arc and control. They equally apply to the monotherapy problem **[A]** and to the combination therapy problem **[AC]** when no chemotherapeutic drugs are administered.

Proposition 8.1 ([26]). *If an optimal control* u_* *is singular on an interval* (α,β) *and* $v \equiv 0$ *on* (α,β), *then the corresponding trajectory* (p_*,q_*) *follows a uniquely defined singular curve* $\mathscr{S}$ *which, defining* $x = \frac{p}{q}$, *can be parameterized in the variables* (p,x) *in the form*

$$dp^{\frac{2}{3}} + bx(\ln x - 1) + \mu = 0 \tag{33}$$

with x *in some interval* $[x_1,x_2] \subset (0,\infty)$. *The singular control* $u_{\sin}$ *that keeps this curve* $\mathscr{S}$ *invariant is given as a feedback function of* p *and* q *in the form*

$$\gamma u_{\sin}(p,q) = \Psi(p,q) = \left(\xi \ln\left(\frac{p}{q}\right) + b\frac{p}{q} + \frac{2}{3}\xi \frac{d}{b}\frac{q}{p^{1/3}} - (\mu + p^{2/3})\right) \tag{34}$$

or, equivalently, using Eq. (33), in terms of x by

$$\gamma u_{\sin}(x) = \left(\frac{1}{3}\xi + bx\right)\ln x + \frac{2}{3}\xi\left(1 - \frac{\mu}{bx}\right). \tag{35}$$

There exists exactly one connected arc on the singular curve $\mathcal{S}$ along which the singular control is admissible, i.e., satisfies the bounds $0 \leq u_{\sin}(x) \leq u_{\max}$. This arc is defined over an interval $[x_\ell^*, x_u^*]$ where x_ℓ^* and x_u^* are the unique solutions to the equations $u_{\sin}(x_\ell^*) = 0$ and $u_{\sin}(x_u^*) = u_{\max}$. At these points the singular control saturates at the control limits $u = 0$ and $u = u_{\max}$.

An important feature of this solution is that it becomes the basis for the optimal solution of the combination therapy problem **[AC]**. Indeed, for a typical initial condition with $p < q$, optimal controls for the combination therapy problem have the following structure: *optimal controls for the anti-angiogenic agent follow the optimal angio-monotherapy and then, at a specific time, chemotherapy becomes active and is given in one full dose session.* The formulas for the singular control and singular arc need to be adjusted to the presence of chemotherapy, but in this case it is not possible that both controls are singular simultaneously. More specifically, we have the following result:

Proposition 8.2 ([48]). *If the optimal anti-angiogenic dose rate u is singular on an open interval I, then the chemotherapeutic agent v is bang-bang on I with at most one switching on I from $v = 0$ to $v = v_{\max}$ and we have the following relation between the controls u and v:*

$$\gamma u_{\sin}(t) + (\eta - \varphi)v(t) = \Psi(p(t), q(t)) \tag{36}$$

with Ψ defined by Eq. (34). Given v, this determines the anti-angiogenic dose rate with a jump-discontinuity where chemotherapy becomes active.

This structure allows to set up a minimization problem over a 1-dimensional parameter τ that denotes the time when chemotherapy becomes active. We illustrate this for an initial condition (p_0, q_0) with $p_0 < q_0$ where the anti-angiogenic inhibitor will immediately be applied at full dose. In principle, the time τ when chemotherapy is activated can lie anywhere in the interval of definition. For example, if the amount $z_{\max}$ of chemotherapeutic agents is high, then it is possible that chemotherapy already becomes active along the interval when the anti-angiogenic dose is at maximum. Analogously, if this amount is very low, it is possible that this activation will only occur after all anti-angiogenic inhibitors have been used up. Typically, however, this time τ will lie somewhere in the interval where the anti-angiogenic dosage follows the singular monotherapy structure and this is illustrated in Fig. 2a.

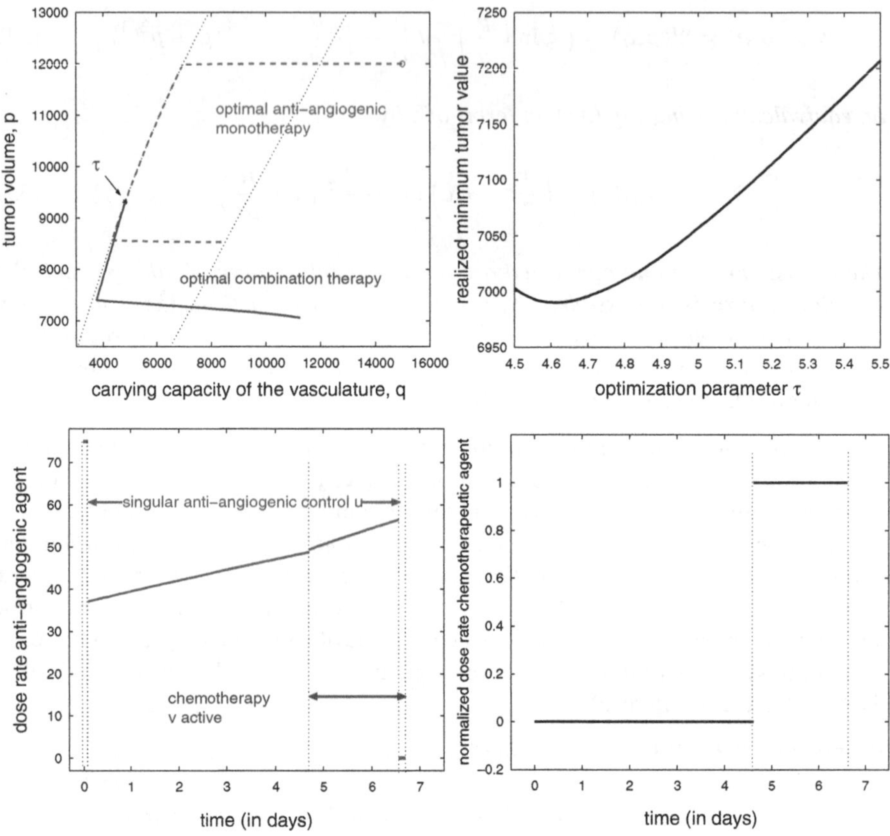

Fig. 2 Optimal combination therapy: (**a**, *top left*) structure of controlled trajectory depending on τ, (**b**, *top right*) value of the objective as a function of τ, (**c**, *bottom left*) optimal anti-angiogenic agent u and (**d**, *bottom right*) optimal chemotherapeutic agent v

It is not difficult to compute the value of the objective as the parameter τ in this family of trajectories is varied and Fig. 2b gives a representative graph of this value as a function of τ. Figures 2c, d show the corresponding optimal controls u and v.

It is interesting to note that this optimization leads to the conclusion that the anti-angiogenic agent is applied first with the chemotherapeutic agent to follow in one full dose session later on. In a clear sense, the optimal control solution points out a specific "path" that should be followed in order to obtain the best possible tumor reductions. Even in the combination therapy model, this path is closely linked with the optimal singular arc from the monotherapy problem as shown by Eq. (34). It lies in the region where the tumor volume p is higher than its carrying capacity q, but there exists a specific relation between these variables and clearly q is not pushed to zero too fast, but a definite balance between these two variables is maintained along the optimal solution. Since the vascular network of the tumor is needed to deliver the chemotherapeutic agents, this perfectly makes sense. Although no "pruning" aspects have been taken into account in the model, it appears that the optimal solutions precisely suggest such a behavior: give anti-angiogenic agents at full dose rate until an "optimal" relation between the tumor volume and its carrying capacity has been established, then maintain this relation through specially designed partial dose rates and, at the right time, apply full dose chemotherapy while still maintaining the optimal relation between p and q through the administration of anti-angiogenic agents. This is in agreement with R. K. Jain's already mentioned hypothesis that *the preliminary delivery of anti-angiogenic agents may regularize a tumor's vascular network with beneficial consequences for the successive delivery of cytotoxic chemotherapeutic agents* [18, 19]. Even for combinations of anti-angiogenic therapy with radiotherapy, a similar feature seems to exist.

9 Combination of Anti-angiogenic Treatment with Radiotherapy

Model **[AC]** has to be modified in order to include effects of radiotherapy on a vascularized tumor. Indeed, as seen in Sect. 4, one has to deal with nonlinear delayed cytotoxic effects on tumor cells, to which one has also to add the radiation damage to the carrying capacity q. Of course, the effects of radiation on tumor cells, endothelial cells and on healthy cells are not equal and thus need to be modelled by separate linear-quadratic equations with different sets of parameters. This leads to similar formulations, but in spaces of varying dimension. Mathematically, this generates different behaviors since the structure of singular controls very much depends on the existing degrees of freedom [29]. In the literature, often the effects of radiation therapy on the tumor cells and its vasculature are modelled by one equation (for example, see [8], which also gives numerical values that are based on [5]) and here, for sake of definiteness and simplicity we take this approach as well. As before, we then include separate states y and z that keep track of the total amounts of anti-angiogenic agents given, respectively, the total damage done by radiotherapy on healthy tissue. This damage is measured in terms of its biologically equivalent dose (BED). We then arrive at the following 6-dimensional optimal control formulation:

[AR] for a free terminal time T, minimize the objective $J(u,v) = p(T)$ subject to the dynamics

$$\dot{p} = \xi p \ln\left(\frac{q}{p}\right) - (\varphi + \beta r) pw, \qquad p(0) = p_0, \tag{37}$$

$$\dot{q} = bp - dp^{\frac{2}{3}} q - \gamma q u - (\eta + \delta r) qw, \qquad q(0) = q_0, \tag{38}$$

$$\dot{r} = -\rho r + w, \qquad r(0) = 0, \tag{39}$$

$$\dot{y} = u, \qquad y(0) = 0, \tag{40}$$

$$\dot{z} = (1 + \alpha s)w, \qquad z(0) = 0, \tag{41}$$

$$\dot{s} = -\sigma s + w, \qquad s(0) = 0, \tag{42}$$

over all Lebesgue measurable functions $u : [0,T] \to [0,u_{\max}]$ and $w : [0,T] \to [0,w_{\max}]$ for which the corresponding trajectory satisfies the end-point constraints $y(T) \le y_{\max}$ and $z(T) \le z_{\max}$.

The coefficients are assumed constant and generally are positive; $u_{\max}$, $w_{\max}$ represent maximum dose rates at which the agents can be administered and $y_{\max}$ and $z_{\max}$ limit the total amounts of the respective agents to be given. A medically reasonable selection for all parameter values is given in Table 1 in [8]. The main difference between model **[AR]** and the model in [8] is that we dropped the so-called early-tissue constraint which prevents an overestimation of the damage done to the early tissue. Also, rather than distinguishing between separate variables r_p and r_q that model the damage done to p and q, here, for simplicity, we only use one variable r as the numerical values given in [8] for the coefficients in these equations agree. Furthermore, we consider a continuous time version of radiotherapy that is not necessarily given in fractionated doses. We shall comment below on how such a treatment protocol can be derived from our version.

The addition of the radio-therapy terms has no structural effects on the anti-angiogenic treatment and if the control u is singular, then regardless of the form of the radio-therapy schedule, we have the following direct extension of Proposition 8.2 for model **[AR]**:

Proposition 9.1 ([29]). *If the optimal anti-angiogenic dose rate u follows a singular control on an open interval I and if the radiotherapy dose rate is given by w, then we have the following relation between the controls u and w:*

$$\gamma u_{\sin}(t) + [(\eta + \delta r) - (\varphi + \beta r)]\, w(t) = \Psi(p(t), q(t)) \tag{43}$$

with Ψ the function defined in Eq. (34) for the optimal anti-angiogenic monotherapy. Given w, this determines the anti-angiogenic dose rate.

Like for combinations with chemotherapy, also in this case there is an immediate and mathematically simple extension of the formula that determines the optimal singular anti-angiogenic dose rate to the more structured and more complicated mathematical model that describes the combination treatment with radiotherapy. However, now the structure of the second control is very different. In fact, it typically (if the bounds on the dose rates permit) is singular as well. The controls u and w are said to be *totally singular* on an open interval I if they are singular simultaneously. This is not optimal for the combination therapy model **[AC]** with chemotherapy, but it is the defining structure for the combination of anti-angiogenic therapy with radiotherapy. For this we need a second equation that links $u_{\sin}$ with $w_{\sin}$.

Proposition 9.2 ([29]). *If the optimal anti-angiogenic dose rate u and the radiotherapy dose rate w both follow singular regimens u_{sin} and w_{sin} on an open interval I, then, in addition to Eq. (43), a second relation of the form*

$$B(x(t))u_{\text{sin}}(t) + w_{\text{sin}}(t) = A(x(t)) \tag{44}$$

holds on I where A and B are smooth functions that only depend on the dynamics of the system and can be determined analytically.

Overall, $(u_{\text{sin}}, w_{\text{sin}})$ thus are the solutions of a 2×2 system of linear equations whose coefficients are determined solely by the equations defining the dynamics of the system. It is possible to give explicit expressions for the functions A and B and thus also for the controls. However, these formulas depend on the second derivatives of the terms in the dynamics and they are long and unwieldy. On the other hand, given any particular value (p,q) of the state and specified values of the parameters, it is not difficult to compute these coefficients A and B numerically and solve for the controls.

Figure 3 gives an example of a totally singular anti-angiogenic dose rate u and a radiotherapy dose rate w that have been computed in this way for parameter values taken from [8]. Part (a) shows the graph of the radiation schedule if no upper limit on the dose rate is imposed. If we set the radiation limit to $w_{\max} = 5$, then this upper bound is initially exceeded and part (b) shows the control that has been computed by saturating this schedule at $w_{\max}$. Since Eq. (43) is valid regardless of the structure of w, the calculations easily adjust. The corresponding graph of the singular control u is given in part (c). Part (d) shows the corresponding trajectory. Note that this trajectory is almost linear. In fact, whenever the anti-angiogenic control u follows a singular regimen, then the quotient $\frac{p}{q}$ follows the simple dynamics

$$\frac{\mathrm{d}}{\mathrm{d}t}\left(\frac{p}{q}\right) = \frac{2}{3}\xi\frac{d}{b}p^{\frac{2}{3}} \tag{45}$$

and, along this simulation, the right-hand side only varies between 0.03 and 0.09, i.e., is almost constant.

The controls given in this figure were not computed to be optimal, but they only illustrate totally singular controls for a combination of anti-angiogenic and radiotherapy. Based on our theoretical analysis, it is clear that these controls will play an essential part in the structure of optimal protocols. This is seconded by the structure of optimal protocols computed in [8] where all the solutions given have exactly this structure, but no hard limits on the dosage rates were imposed. In order to solve the overall optimal control problem, however, it is necessary to take these constraints into account and then to establish the structure of optimal controls before and after the singular segments. Different from the monotherapy problem described earlier, in this case there exists a vector field whose integral curves are the trajectories for totally singular controls everywhere, not just on some lower-dimensional surface. However, it matters which of these trajectories is taken. Research on determining an optimal synthesis is ongoing.

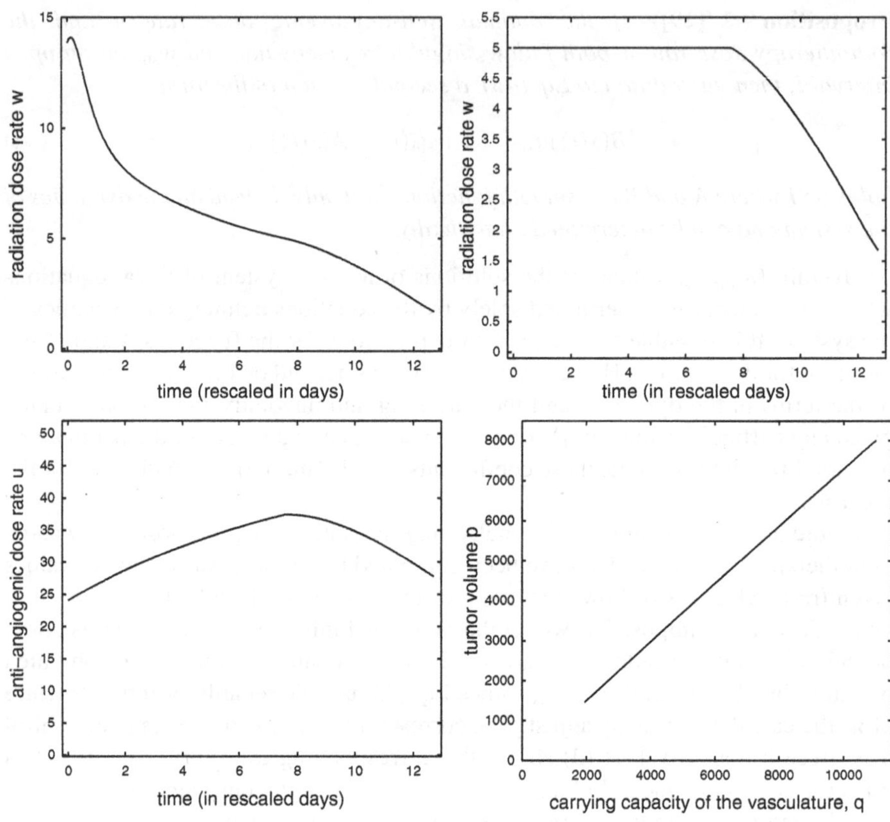

Fig. 3 Combination with radiotherapy: (**a**, *top left*) the unsaturated singular radiation dose rate w, (**b**, *top right*) the radiation dose rate w with upper limit $w_{\max}$ enforced, (**c**, *bottom left*) corresponding singular anti-angiogenic agent u and (**d**, *bottom right*) corresponding trajectory (p,q)

10 Suboptimal Protocols and Pharmacokinetic Equations

In all these optimal control problems, singular controls and their corresponding trajectories are the most important feature that determine the structure of optimal protocols. This is true both for the monotherapy case where a complete theoretical solution for the optimal control problem has been given [26] and for combination therapies where similar complete solutions are currently still elusive. But in these cases, explicit formulas for the singular controls allow to utilize numerical procedures to determine optimal solutions for given initial conditions and parameters. However, singular controls are defined in terms of the variables p and q (and also r and s in the case of the radiation dose) and thus are not practically realizable.

Yet, these optimal solutions have bi-fold relevance for designing practical protocols for these therapies: (1) Clearly, the optimal solutions define benchmark values to which other protocols—simple, heuristically chosen and implementable—can be compared and thus they determine a measure for how close to optimal a given general protocol is. (2) Equally important, the structure of optimal controls directly indicates simple, piecewise constant, and thus easily realizable protocols that approximate the optimal solutions well, so-called *suboptimal* protocols. In the papers [27,31], an extensive analysis of suboptimal protocols for the anti-angiogenic monotherapy problem was undertaken, and it was shown that piecewise constant suboptimal protocols with a very small number of switchings exist that are able to replicate the optimal values within 1 %. Similar results are valid for the modification of the underlying model by Ergun et al. given in [8]. Since optimal controls for the chemotherapeutic agent in combination with anti-angiogenic treatments are bang-bang, there is no need to approximate these and thus these results directly carry over to these problems.

These excellent approximation properties remain valid if the model is made more realistic by including pharmacokinetic equations for the anti-angiogenic and chemotherapeutic agents [32]. The models considered so far identify their dosages with their concentrations in the plasma. The controls u and v, as they were used, actually represent the concentrations of these agents and linear terms of the type $-\gamma q u$ model the pharmacodynamics of the drugs. If a standard linear exponential growth/decay model is added for the concentrations of the agents, e.g., $\dot{c} = -\alpha c + u$, then indeed there exist qualitative changes in the optimal solutions that are due to the fact that the so-called intrinsic order of the singular control changes from 1 to 2 [28]. These make the solutions even more complicated from a mathematical point of view. However, the added complexity disappears if only suboptimal protocols are considered. Then, as in the simplified models, the structure of the theoretically optimal controls immediately suggests how to choose excellent simple approximating protocols [32]. On the level of suboptimal realizable protocols, the simplified modeling that ignores a linear pharmacokinetic model for the therapeutic agents is fully justified.

Similar investigations are ongoing for the model including radiotherapy. Radiation doses are commonly administered in daily fractionated doses (of short durations) and this does not agree with the model considered above. Mathematically, this leads to a more complex, hybrid optimization problem that generally is solved with numerical methods. These methods and solutions, however, do not provide much insight into the underlying principles. A continuous-time formulation, as it was presented here, gives this information about the structure of optimal controls (and why they look the way they do) and this is what makes it rather straightforward (e.g., by averaging) to compute approximating fractionated doses. But our investigations on this topic are still in their preliminary stages.

11 Conclusion and Discussion

For various treatment strategies, we have considered optimal control problems to minimize the tumor volume when the overall amounts of agents are limited a priori. This may be simply because only a limited amount of the agents is available, like in the case of anti-angiogenic inhibitors which still are very expensive and thus are only used in limited quantities, or it may be because these treatments have severe side effects that need to be limited and thus a priori decisions are made to limit the total amount of drugs or radiation to be given, a standard medical approach. Then the question how to schedule this agents in time arises naturally. Here we have considered treatment strategies that combine anti-angiogenic therapies with the classical approaches of chemo- and radiotherapy. Our main conclusions are that singular controls (which can be computed analytically using Lie-derivative based calculations) are at the center of optimal solutions for both the mono- and combination therapy treatments. Although these controls are feedback functions, and thus cannot be directly applied, they point the way to simple realizable approximations that are excellent.

From a general biomedical point of view, our results suggest that in order to optimally treat a highly dynamical disease such as a cancer, a highly dynamical schedule of drug delivery may be needed. This remark, which could seem trivial from a mathematical point of view, has some deep implications in medical oncology. Indeed, the paradigm of dynamical drug scheduling requires a substantial rethinking of the concept of clinical trial, which in its current form is largely linked to a static vision of tumors.

Acknowledgements We would like to thank an anonymous referee for his careful reading of our chapter and several suggestions that we incorporated into the final version. The research of A. d'Onofrio has been done in the framework of the Integrated Project "p-medicine—from data sharing and integration via VPH models to personalized medicine," project identifier: 270089, which is partially funded by the European Commission under the 7th framework program. The research of U. Ledzewicz and H. Schättler has been partially supported by the National Science Foundation under collaborative research grant DMS 1008209/1008221.

References

1. Agur, Z., Arakelyan, L., Daugulis, P., Ginosar, Y.: Hopf point analysis for angiogenesis models. Discrete Continuous Dyn. Syst. B **4**, 29–38 (2004)
2. Anderson, A.R.A., Chaplain, M.A.: Continuous and discrete mathematical models of tumor-induced angiogenesis. Bull. Math. Biol. **60**, 857–899 (1998)
3. Bellomo, N., Maini, P.K.: Introduction to "special issue on cancer modelling'. Math. Models Meth. Appl. Sci. **15**, iii (2005)
4. Boyle, P., d'Onofrio, A., Maisonneuve, P., Severi, G., Robertson, C., Tubiana, M., Veronesi, U.: Measuring progress against cancer in Europe: has the 15% decline targeted for 2000 come about? Ann. Oncol. **14**, 1312–1325 (2003)

5. Brenner, D.J., Hall, E.J., Huang, Y., Sachs, R.K.: Optimizing the time course of brachytherapy and other accelerated radiotherapeutic protocols. Int. J. Radiat. Oncol. Biol. Phys. **29**, 893–901 (1994)
6. Browder, T., Butterfield, C.E., Kraling, B.M., Shi, B., Marshall, B., O'Reilly, M.S., Folkman, J.: Antiangiogenic scheduling of chemotherapy improves efficacy against experimental drug-resistant cancer. Canc. Res. **60**, 1878–86 (2000)
7. Colleoni, M., Rocca, A., Sandri, M.T., Zorzino, L., Masci, G., Nole, F., Peruzzotti, G., Robertson, C., Orlando, L., Cinieri, S., de Braud, F., Viale, G., Goldhirsch, A.: Low-dose oral methotrexate and cyclophosphamide in metastatic breast cancer: antitumour activity and correlation with vascular endothelial growth factor levels. Ann. Oncol. **13**, 73–80 (2002)
8. Ergun, A., Camphausen, K., Wein, L.M., Optimal scheduling of radiotherapy and angiogenic inhibitors, Bull. of Math. Biology, **65**, 407–424 (2003)
9. Folkman, J.: Tumor angiogenesis: Therapeutic implications. New Engl. J. Med. **295**, 1182–1196 (1971)
10. Folkman, J.: Antiangiogenesis: New concept for therapy of solid tumors. Ann. Surg. **175**, 409–416 (1972)
11. Folkman, J.: Opinion - Angiogenesis: an organizing principle for drug discovery? Nature Rev. Drug. Disc. **6**, 273–286 (2007)
12. Fowler, J.F.: The linear-quadratic formula and progress in fractionated radiotherapy, Br. J. Radiol. **62**, 679–694 (1989)
13. Frame, D.: New strategies in controlling drug resistance. J. Manag. Care Pharm. **13**, 13–17 (2007)
14. Goldie, J.H., Coldman, A.J.: A mathematic model for relating the drug sensitivity of tumors to their spontaneous mutation rate. Canc. Treat. Rep. **63**, 1727–1733 (1979)
15. Guiot, C., Degiorgis, P.G., Delsanto, P.P., Gabriele, P., Deisboecke, T.S.: Does tumor growth follow a 'universal law'? J. Theor. Biol. **225**, 147–151 (2003)
16. Hahnfeldt, P., Panigrahy, D., Folkman, J., Hlatky, L.: Tumor development under angiogenic signaling: A dynamical theory of tumor growth, treatment response, and postvascular dormancy. Canc. Res. **59**, 4770–4775 (1999)
17. Hart, D., Shochat, E., Agur, Z.: The growth law of primary breast cancer as inferred from mammography screening trials data. Br. J. Canc. **78**, 382–387 (1999)
18. Jain, R.K., Normalizing tumor vasculature with anti-angiogenic therapy: a new paradigm for combination therapy. Nat. Med. **7**, 987–989 (2001)
19. Jain, R.K., Munn, L.L.: Vascular normalization as a rationale for combining chemotherapy with antiangiogenic agents. Princ. Practical Oncol. **21**, 1–7 (2007)
20. Kerbel, R.S.: A cancer therapy resistant to resistance. Nature **390**, 335–336 (1997)
21. Kerbel, R.S.: Tumor angiogenesis: Past, present and near future. Carcinogensis **21**, 505–515 (2000)
22. Kerbel, R., Folkman, J.: Clinical translation of angiogenesis inhibitors. Nat. Rev. Canc. **2**, 727–739 (2002)
23. Kimmel, M., Swierniak, A.: Control theory approach to cancer chemotherapy: benefiting from phase dependence and overcoming drug resistance. In: Tutorials in Mathematical Bioscences III: Cell Cycle, Proliferation, and Cancer, Lecture Notes in Mathematics, vol. 1872, pp. 185–221. Springer, Berlin (2006)
24. Ledzewicz, U., Schättler, H.: Optimal bang-bang controls for a 2-compartment model in cancer chemotherapy. J. Optim. Theor. Appl. - JOTA **114**, 609–637 (2002)
25. Ledzewicz, U., Schättler, H.: Analysis of a cell-cycle specific model for cancer chemotherapy, J. Biol. Syst. **10**, 183–206 (2002)
26. Ledzewicz, U., Schättler, H.: Anti-angiogenic therapy in cancer treatment as an optimal control problem, SIAM J. Contr. Optim. **46**(3), 1052–1079 (2007)
27. Ledzewicz, U., Schättler, H.: Optimal and suboptimal protocols for a class of mathematical models of tumor anti-angiogenesis. J. Theor. Biol. **252**, 295–312 (2008)
28. Ledzewicz, U., Schättler, H.: Singular controls and chattering arcs in optimal control problems arising in biomedicine. Contr. Cybern. **38**, 1501–1523 (2009)

29. Ledzewicz, U., Schättler, H.: Multi-input optimal control problems for combined tumor anti-angiogenic and radiotherapy treatments. J. Optim. Theor. Appl. - JOTA **153**, 195–224 (2012), doi:10.1007/s10957-011-9954-8, published online: 16 November 2011
30. Ledzewicz, U., Munden, J., Schättler, H.: Scheduling of anti-angiogenic inhibitors for Gompertzian and logistic tumor growth models. Discrete Continuous Dyn. Syst. Ser. B **12**, 415–439 (2009)
31. Ledzewicz, U., Marriott, J., Maurer, H., Schättler, H.: Realizable protocols for optimal administration of drugs in mathematical models for novel cancer treatments. Math. Med. Biol. **27**, 157–179 (2010), doi:10.1093/imammb/dqp012
32. Ledzewicz, U., Maurer, H., Schättler, H.: Minimizing tumor volume for a mathematical model of anti-angiogenesis with linear pharmacokinetics. In: Diehl, M., Glineur, F., Jarlebring, E., Michiels, W. (eds.) Recent Advances in Optimization and its Applications in Engineering, pp. 267–276. Springer, Berlin (2010)
33. Marusic, M., Bajzer, A., Freyer, J.P., Vuk-Povlovic, S.: Analysis of growth of multicellular tumor spheroids by mathematical models. Cell Prolif. **27**, 73ff (1994)
34. McDougall, S.R., Anderson, A.R.A., Chaplain, M.A.: Mathematical modelling of dynamic adaptive tumour-induced angiogenesis: Clinical implications and therapeutic targeting strategies. J. Theor. Biol. **241**, 564–589 (2006)
35. Mombach, J.C.M., Lemke, N., Bodmann, B.E.J., Idiart, M.A.P.: A mean-field theory of cellular growth. Europhys. Lett. **59**, 923–928 (2002)
36. Norton, L.: A Gompertzian model of human breast cancer growth. Canc. Res. **48**, 7067–7071 (1988)
37. Norton, L., Simon, R.: Growth curve of an experimental solid tumor following radiotherapy. J. Nat. Canc. Inst. **58**, 1735–1741 (1977)
38. Norton, L., Simon R.: The Norton-Simon hypothesis revisited. Canc. Treat. Rep. **70**, 163–169 (1986)
39. d'Onofrio, A.: A general framework for modelling tumor-immune system competition and immunotherapy: Mathematical analysis and biomedial inferences. Physica D **208**, 202–235 (2005)
40. d'Onofrio, A.: Rapidly acting antitumoral antiangiogenic therapies. Phys. Rev. E Stat. Nonlin. Soft Matter Phys. **76**, 031920 (2007)
41. d'Onofrio, A.: Fractal growth of tumors and other cellular populations: Linking the mechanistic to the phenomenological modeling and vice versa. Chaos, Solitons and Fractals **41**, 875–880 (2009)
42. d'Onofrio, A., Gandolfi, A.: Tumour eradication by antiangiogenic therapy: Analysis and extensions of the model by Hahnfeldt et al. Math. Biosci. **191**, 159–184 (2004)
43. d'Onofrio, A., Gandolfi, A.: The response to antiangiogenic anticancer drugs that inhibit endothelial cell proliferation, Appl. Math. and Comp. **181**, 1155–1162 (2006)
44. d'Onofrio, A., Gandolfi, A.: A family of models of angiogenesis and anti-angiogenesis anti-cancer therapy. Math. Med. Biol. **26**, 63–95 (2009)
45. d'Onofrio, A., Gandolfi, A.: Chemotherapy of vascularised tumours: role of vessel density and the effect of vascular "pruning". J. Theor. Biol. **264**, 253–265 (2010)
46. d'Onofrio, A., Gandolfi, A.: Resistance to anti-tumor chemotherapy due to bounded-noise transitions. Phys. Rev. E. **82**, (2010) Art.n. 061901, doi:10.1103/PhysRevE.82.061901
47. d'Onofrio, A., Gandolfi, A., Rocca, A.: The dynamics of tumour-vasculature interaction suggests low-dose, time-dense antiangiogenic schedulings. Cell Prolif. **42**, 317–329 (2009)
48. d'Onofrio, A., Ledzewicz, U., Maurer, H., Schättler, H.: On optimal delivery of combination therapy for tumors. Math. Biosci. **222**, 13–26 (2009), doi:10.1016/j.mbs.2009.08.004
49. d'Onofrio, A., Fasano, A., Monechi, B.: A generalization of Gompertz law compatible with the Gyllenberg-Webb theory for tumour growth Math. Biosci. **230**(1), 45–54 (2011)
50. O'Reilly, M.S., Boehm, T., Shing, Y., Fukai, N., Vasios, G., Lane, W.S., Flynn, E., Birkhead, J.R., Olsen, B.R., Folkman, J.: Endostatin: An endogenous inhibitor of angiogenesis and tumour growth. Cell **88**, 277–285 (1997)

51. Panetta, J.C.: A mathematical model of breast and ovarian cancer treated with paclitaxel. Math. Biosci. **146**, 89–113 (1997)
52. Peckham, M., Pinedo, H.M., Veronesi, U.: Oxford Textbook of Oncology. Oxford Medical Publications, New York (1995)
53. Quinn, M.J., d'Onofrio, A., Moeller, B., Black, R., Martinez-Garcia, C., Moeller, H., Rahu, M., Robertson, C., Schouten, L.J., La Vecchia, C., Boyle, P.: Cancer mortality trends in the EU and acceding countries up to 2001. Ann. Oncol. **14**, 1148–1152 (2003)
54. Ribba, B., Marron, K., Agur, Z., Alarcïn, T., Maini, P.K.: A mathematical model of Doxorubicin treatment efficacy for non-Hodgkin's lymphoma: Investigation of the current protocol through theoretical modelling results. Bull. Math. Biol. **67**, 79–99 (2005)
55. Schättler, H., Jankovic, M.: A synthesis of time-optimal controls in the presence of saturated singular arcs. Forum Math. **5**, 203–241 (1993)
56. Schättler, H., Ledzewicz, U., Cardwell, B.: Robustness of optimal controls for a class of mathematical models for tumor anti-angiogenesis. Math. Biosci. Eng. **8**, 355–369 (2011), doi:10.3934/mbe.2011.8.355
57. Sole, R.V.: Phase transitions in unstable cancer cell populations. Eur. J. Phys. B **35**, 117–124 (2003)
58. Skipper, H.E.: On mathematical modeling of critical variables in cancer treatment (goals: Better understanding of the past and better planning in the future). Bull. Math. Biol. **48**, 253–278 (1986)
59. Swan, G.W.: Role of optimal control in cancer chemotherapy. Math. Biosci. **101**, 237–284 (1990)
60. Swierniak, A.: Optimal treatment protocols in leukemia - modelling the proliferation cycle. In: Proceedings of 12th IMACS World Congress, vol. 4, pp. 170–172, Paris (1988)
61. Swierniak, A.: Cell cycle as an object of control. J. Biol. Syst. **3**, 41–54 (1995)
62. Swierniak, A.: Direct and indirect control of cancer populations. Bull. Pol. Acad. Sci. **56**, 367–378 (2008)
63. Swierniak, A.: Comparison of six models of antiangiogenic therapy. Appl. Math. **36**, 333–348 (2009)
64. Swierniak, A., Kimmel, M., Smieja, J.: Mathematical modeling as a tool for planning anticancer therapy, Eur. J. Pharmacol. **625**, 108–121 (2009)
65. Swierniak, A., Ledzewicz, U., Schättler, H.: Optimal control for a class of compartmental models in cancer chemotherapy. Int. J. Appl. Math. Comput. Sci. **13**, 357–368 (2003)
66. Thames, H.D., Hendry, J.H.: Fractionation in Radiotherapy. Taylor and Francis, London (1987)
67. Ubezio, P., Cameron, D.: Cell killing and resistance in pre-operative breast cancer chemotherapy. BMC Canc. **8**, 201ff (2008)
68. Wheldon, T.E.: Mathematical Models in Cancer Research. Hilger Publishing, Boston-Philadelphia (1988)
69. Wijeratne, N.S., Hoo, K.A.: Understanding the role of the tumour vasculature in the transport of drugs to solid cancer tumors. Cell Prolif. **40**, 283–301 (2007)
70. Yu, J.L., Rak, J.W.: Host microenvironment in breast cancer development: Inflammatory and immune cells in tumour angiogenesis and arteriogenesis. Breast Canc. Res. **5**, 83–88 (2003)

Saturable Fractal Pharmacokinetics and Its Applications

Rebeccah E. Marsh and Jack A. Tuszyński

1 Introduction

In this chapter we discuss an application of fractal kinetics under steady state conditions to model the enzymatic elimination of a drug from the body. A one-compartment model following fractal Michaelis–Menten kinetics under a steady state is developed and applied to concentration-time data for the cardiac drug mibefradil in dogs. The model predicts a fractal reaction order and a power law asymptotic time-dependence of the drug concentration. A mathematical relationship between the fractal reaction order and the power law exponent is derived. The goodness-of-fit of the model is assessed and compared to that of four other models suggested in the literature. The proposed model provides the best fit to the data. In addition, it correctly predicts the power law shape of the tail of the concentration-time curve. The new fractal reaction order can be explained in terms of the complex geometry of the liver, the organ responsible for eliminating the drug. Furthermore, we investigate the potential for identifying global characteristics in the pharmacokinetics of the anticancer drug paclitaxel. An analysis of data in the literature yields both clearance curves and dose-dependency curves that are accurately described by power laws with similar exponents.

Pharmacokinetics is the study of the absorption, distribution, metabolism, and eventual elimination (ADME) of a drug from the body [19]. It is fundamental in developing dosing regimes, predicting the behavior of new drugs, and estimating their therapeutic effects. For numerous compounds, such as anesthetics, cardiac drugs, and chemotherapeutic drugs, quantitative knowledge about the interaction of the drug with the body is of vital importance to its successful therapeutic applications. Pharmacological data usually consist of discrete values of the concentration of a drug in the plasma or blood as a function of time. A plot of these values

R.E. Marsh • J.A. Tuszyński (✉)
Department of Physics, University of Alberta, Edmonton, AB, Canada, T6G 2J1
e-mail: rmarsh@ualberta.ca; jackt@ualberta.ca

U. Ledzewicz et al., *Mathematical Methods and Models in Biomedicine*, Lecture Notes on Mathematical Modelling in the Life Sciences, DOI 10.1007/978-1-4614-4178-6_12,

generates a concentration-time curve, $C(t)$, that first rises as absorption of the drug dominates (invasion) and then decreases after a maximum concentration value, $C_{\max}$, is reached (elimination). This decline may be relatively short, taking place over minutes or hours, or may last for several days, and it is mainly governed by the rate of elimination of the drug from the body and drug distribution including such processes as tissue binding. The goal of pharmacokinetic modeling is to use these curves to describe, compare, and predict a drug's course in the body, as well as to determine optimum dosing regimes, potential toxicity, and possible drug–drug interactions. One of the main difficulties is to combine organ-level modeling with molecular-level modeling that provides insights into drug distribution and binding with proteins and lipids. This is still a key challenge in pharmacokinetic modeling and misconceptions have occurred in the past due to the difficulties in combining these two vastly different temporal and spatial scales [37]. Drug distribution processes cover complicated molecular and cellular details of the hepatic elimination processes of drugs. Sensitivity analysis has shown that complex models of hepatic elimination cannot be identified on the basis of whole body drug disposition data (plasma concentration-time curves) making this aspect in need of appropriate model development.

While classical compartmental models are the most common type of pharmacokinetic mathematical models and they can provide adequate agreement with clinical pharmacokinetic data sets, they often fail to provide a good fit to the tail regions, where non-exponential time-dependence can occur that is better fit by power laws or gamma functions [52, 64]. Since all data sets are finite in size, they can always be fit with a sufficiently large number of compartments and an associated large number of adjustable parameters associated with a chosen basis set of functions (e.g. exponentials). However, this does not address the fundamental origin of the frequently encountered non-exponential behavior in pharmacokinetics. A link has been made previously [9] between concentration-time curves with power-law tails and fractal kinetics. We believe that nonlinearities in pharmacokinetic models are of crucial importance. Therefore, to provide a clear contrast, we first introduce standard ideas in pharmacokinetics which are based on linearity.

In general, a system is considered to be linear when its output is directly proportional to its input. The concept of linearity in the body's handling of a drug is very important; it implies that the concentration of the drug as well as any derived parameters scale easily with both dose and time. The associated kinetic processes can be described by a set of linear differential equations that follow the superposition principle (which states that the whole is equal to the sum of its individual components).

According to traditional pharmacokinetic models, the body can be divided into compartments, and a drug's journey between two connected compartments is described by a rate coefficient [39]. In a linear model, these rate coefficients k are assumed to be constant. The concentration in each compartment can be described by the following differential equation:

$$\frac{\mathrm{d}C(t)}{\mathrm{d}t} = kC(t). \tag{1}$$

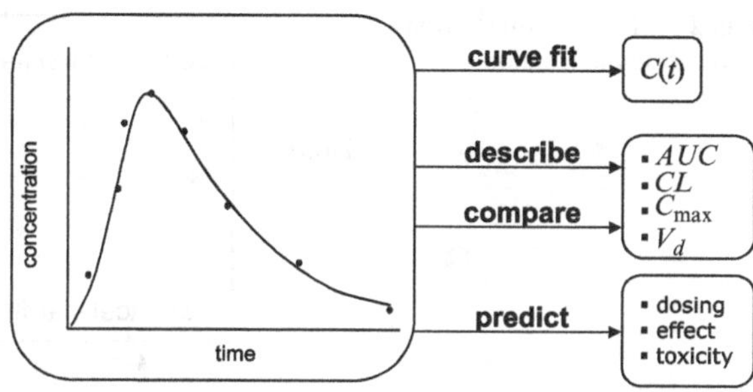

Fig. 1 A typical pharmacokinetic curve for $C(t)$

Integrating it produces:

$$C(t) = C_0 e^{-kt}. \tag{2}$$

Thus for multiple simultaneous processes, a series of coupled differential equations is obtained as stated above.

Nonlinear drug–organ interactions, however, cannot be described adequately by classical compartment models. As observed in numerous sets of clinical data, the concentration-time curve shows a non-exponential time-dependence, at least asymptotically. Furthermore, observation of anomalous kinetics in the experimental data also suggests that the kinetic reactions should occur on or within fractal media with time-dependent kinetic rate coefficients [1, 33].

Nonlinearity usually suggests that the underlying physical and chemical relationships are complicated. Among the various types of nonlinearities identified in pharmacokinetic systems, a couple have been shown to follow a power law relationship, signifying that scaling laws and universality may be present across a wide range of patient characteristics as well as types and stage of disease. Clinically, nonlinearity in the behavior of a drug can complicate the design of dosage regimes as well as the prediction of the toxicity and effectiveness of the drug. The types of nonlinearity that are discussed in the literature fall into two categories: dose-dependence and time-dependence.

Dose-proportionality is a concept common in toxicity experiments or dose escalation studies, which are primarily conducted early in the research of a new drug. The response of patients to different doses of the drug is measured. If the pharmacokinetic parameters are unchanged with changes in the dose, the pharmacokinetics of the system are said to be dose-independent over the therapeutic range studied. If a doubling of the dose of a drug produces a doubling in one or more pharmacokinetic parameters, the system is considered to be dose-proportional and linear. If the parameter values increase or decrease by a factor other than two, the system is considered to be dose-dependent and nonlinear. Typically, the parameters that are tested versus the dose are the area under the plasma-concentration curve (*AUC*), which is often referred to as the systemic exposure, the maximum plasma concentration ($C_{\max}$), and the clearance (*CL*), which is a measure of the volume of the blood that is cleared of drug per unit time (see Fig. 1).

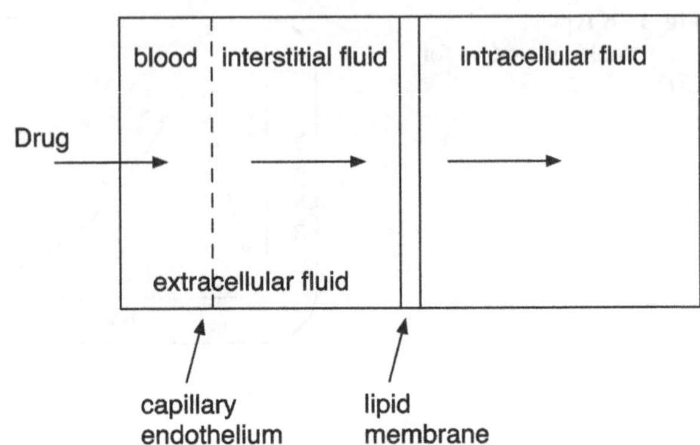

Fig. 2 Schematic illustration of the drug distribution processes

Time-dependent pharmacokinetics is present when the pharmacokinetic parameters vary with time due to actual physical alterations in the body, whether in the form of physiological changes related to the body's circadian rhythms (called chronopharmacokinetics) or chemically induced changes brought about by the introduction of the drug into the body [38]. The potential causes of time-dependence fall into the categories of absorption and elimination parameters, metabolism, plasma binding, renal or hepatic clearance, overall systemic clearance, and enzyme activity [38].

Both dose-dependence and time-dependence can be present in the same system (i.e. the pharmacokinetic parameters describing a system can vary both in time and with dose), and in fact, the sources proposed for both are similar. The causes of dose-dependent pharmacokinetics may include oral absorption, tissue distribution, protein binding in both the plasma and tissues, and elimination [40].

Of particular significance is the role of drug distribution processes throughout the body which we briefly discuss below. Distribution processes are illustrated in Fig. 2 and they represent movement of drug molecules from the blood to and from the tissues. Distribution is determined by: (a) partitioning across various membranes, (b) binding to tissue components, (c) binding to blood components (RBC, plasma protein), and (d) physiological volumes.

All of the fluid in the body, in which a drug can be dissolved, can be roughly divided into three compartments:

1. Intravascular (blood plasma found within blood vessels): $V = 3\,\mathrm{L}$ (4 % of body weight)
2. Interstitial (fluid surrounding cells): $V = 9\,\mathrm{L}$ (13 % body weight)
3. Intracellular (fluid within cells, i.e. cytosol): $V = 28\,\mathrm{L}$ (41 % of body weight)

The distribution of a drug into these compartments is dictated by its physical and chemical properties. Many drugs bind to plasma proteins in the blood stream and plasma protein binding limits distribution. Consequently, a drug that binds plasma protein diffuses less efficiently than a drug that does not.

To address the shortcomings of homogeneously stirred compartment models, several mathematical formulations have been proposed. Liver clearance plays a particularly important role in drug elimination kinetics, hence it garnered special attention in liver clearance models that can be grouped into three main classes: (a) well-stirred models, (b) parallel tube models, and (c) distributed tube models. We briefly compare and contrast these three models and their outcomes based on an earlier meta-analysis [56].

The Well-Stirred Model is the simplest model mathematically but it does not correspond to liver anatomy. Here the kinetic equation for the drug concentration in the liver, $C(t)$, is given by

$$V\frac{\mathrm{d}C(t)}{\mathrm{d}t} = Q(C_{\mathrm{in}} - C) - fCL_{\mathrm{int}}C, \tag{3}$$

where Q is the rate of blood flow, CL_{int} is the internal clearance rate and C_{in} is the drug concentration at the entrance to the liver. At steady state we find that

$$C = \frac{QC_{\mathrm{in}}}{Q + fCL_{\mathrm{int}}} = C_{\mathrm{in}}F, \tag{4}$$

where the coefficient F is termed bioavailability which is given by

$$F = \frac{1}{1 + \dfrac{fCL_{\mathrm{int}}}{Q}} = \frac{1}{1+X}, \tag{5}$$

where $X = fCL_{\mathrm{int}}/Q$. Total hepatic clearance is then defined as

$$CL_H = Q(1-F) = \frac{QX}{1+X}. \tag{6}$$

In the Parallel Tube Model each tube (blood vessel) satisfies a plug flow equation given by

$$S\frac{\mathrm{d}C}{\mathrm{d}t} + Q\frac{\mathrm{d}C}{\mathrm{d}x} = -\rho(x)g(C), \tag{7}$$

where $\rho(x)$ describes enzyme distribution and g is an enzyme kinetics Michaelis–Menten term. The linear kinetics in this model simplifies to

$$Q\frac{\mathrm{d}C}{\mathrm{d}z} = \frac{-fCL_{\mathrm{int}}}{L}C, \tag{8}$$

with the introduction of the moving variable $z = x - (Q/S)t$, such that its solution is governed by exponential decay

$$C(z) = C_{\mathrm{in}}\exp\left(\frac{-zx}{L}\right). \tag{9}$$

Hence, we find for bioavailability that

$$F = \exp\left(\frac{-fCL_{\mathrm{int}}}{Q}\right). \tag{10}$$

This is the simplest mode which captures sinusoid structure of transport and metabolism in the liver but has been met with only mixed success in predicting physiological data.

The Distributed Tube Model assumes that the distribution of tubes satisfies a statistical profile regarding their geometries and enzyme densities. The variance in this distribution, ε^2, depends on both path length and enzyme statistics and hence is substrate dependent. It can be measured from a quadratic fit of $\log(F)$ to $1/Q$. By a series expansion one finds

$$F = < F_i > = < \mathrm{e}^{-xi} > = \mathrm{e}^{-<xi>}\left(1 + \frac{1}{2} < x_i{}^2 > + \dots\right), \tag{11}$$

which results in the following relation

$$F = \exp\left(\frac{-fCL_{\mathrm{int}}}{Q}\right)\left(1 + \frac{1}{2}\varepsilon^2\left(\frac{fCL_{\mathrm{int}}}{Q}\right)^2\right), \tag{12}$$

where $\varepsilon^2 = \Sigma_i[(CL_{\mathrm{int},i}/CL_{\mathrm{int}})^2/(Q_i/Q)]$ and the index i runs over all individual tubes.

Comparative studies of these models [56] revealed that the choice of a liver model has most important effects for drugs with high clearance. All models have reasonable correlation for most drugs at low clearance, but exhibit a systematic discrepancy at high clearance, especially for the well-stirred model. The distributed model corrects this discrepancy without altering low clearance behavior.

At a more fundamental level of model development, agent-based and fractal-geometry-based models have been proposed and investigated. For example, a stochastic random walk model for the drug molecules was studied [67], a model of convective-diffusive transit behavior in the liver [52], a gamma-distributed drug residence time model [61], transient fractal kinetics studies [1], and fractal Michaelis–Menten kinetics [45] were all undertaken to come up with a more realistic description of physiological processes involving drug molecules.

To take into account the organ heterogeneity and simulate enzyme kinetics in disordered media, lattice models have been introduced by investigators. Berry [7] performed Monte Carlo simulations of a Michaelis–Menten reaction on a two-dimensional lattice with a varying density of obstacles to simulate the barriers to diffusion caused by biological membranes. He found that fractal kinetics resulted at high obstacle concentrations. Kosmidis et al. [35] performed Monte Carlo simulations of a Michaelis–Menten enzymatic reaction on a two-dimensional percolation lattice at criticality. They found that fractal kinetics emerged at long times.

A network model of the liver was developed [10] consisting of a square lattice of vascular bonds connecting two types of sites that represent either sinusoids or hepatocytes. Random walkers explored the lattice at a constant velocity and were removed with a set probability from hepatocyte sites. To simulate different pathological states of the liver, random sinusoid or hepatocyte sites were removed. For a lattice with regular geometry, it was found that the number of walkers decayed according to an exponential relationship. For a percolation lattice with a fraction p of the bonds removed, the decay was found to be exponential for high trap concentrations but transitioned to a stretched exponential at low trap concentrations. The models described above are all basic random walk models, and the lattices are abstract representations of the geometry of the exploration space. Ideally, one would strive for the incorporation of agent-based lattice models with anatomically correct fractal-like models of specific organs such as the liver, kidneys, lungs, etc.

2 Nonlinearity in Pharmacokinetics

As stated above, nonlinear pharmacokinetics exists when the parameters are dose- or time- dependent [52]. With dose-dependence, an increase in the administered dose results in a disproportionate increase in the absorbed dose. The most common type of dose-dependence discussed in the literature follows Michaelis–Menten kinetics, where the clearance of a drug changes with concentration due to saturation of the drug action sites. References to time-dependent nonlinearity are much less frequent, although Levy [38] lists the following possible sources: absorption and elimination parameters, systemic clearance, enzymatic metabolic activity, plasma binding, renal clearance, and cerebrospinal fluid drug concentration. It is also worth mentioning that both dose and time dependences can be present simultaneously. Because the body is a complex system, the observed concentration values are the end product of many intricate interactions.

2.1 *Power Laws and Fractals*

Complex systems often pose major challenges to applied mathematicians who model them using various simplifications. Fortunately, sometimes, as is the case with pharmacokinetic processes taking place in the human body, they can be described by reasonably simple mathematics due to the presence of scaling rules which result in power laws. Power laws have been found across scientific disciplines including ecology, biology, economics, chemistry, physiology, and physics [20]. They are given by the simple formula:

$$g(x) = Ax^{\alpha}. \tag{13}$$

By taking the log of both sides, we obtain:

$$\log g(x) = \log A + \alpha \log x. \tag{14}$$

Thus when plotted on a log–log scale, this type of function produces a straight line with slope α. Power laws are attractive because they are scale-invariant. Multiplying the x-variable by a factor of a merely changes the constant of proportionality:

$$g(x) = A(ax^{\alpha}) = (aA)x^{\alpha}. \tag{15}$$

The shape of such a system is the same irrespective of the scale. Because the characteristics of the system remain the same as we zoom in or out of the physical space occupied by the system, this allows us to extrapolate from shorter time or space intervals to longer ones and vice versa.

Fractals are objects that have non-integer power exponents. They can be described by the following equation:

$$L(d) \propto d^{1-D}, \tag{16}$$

where D is the fractal dimension of the system and can be thought of as a measure of the plane-filling properties of a structure [43, 44].

Fractals describe systems behaving under constraints. For example, the circulatory system consists of a series of bifurcating vessels, and the time it takes for a circulating drug to reach its target will depend on the confines of its path. Similarly, most organs in the body are complex structures and the diffusion of a drug across them will be limited by the available surfaces. Because their structure persists down to smaller and smaller scales, fractals have the unique ability to fill the available space as efficiently as possible.

Power laws and fractals can be generated in many different ways and can exist in both space and time. That is, a system characteristic can scale either with length or with time. In pharmacokinetics, for example, the concentration of a drug can depend on the path it must follow through the body or on characteristic times such as diffusion or residence rates. It should be noted that both types of fractals can exist simultaneously in a system. In the sections that follow, we will consider the calcium antagonist drug mibefradil, whose elimination is governed by the fractal geometry of its site of metabolism, as well as the chemotherapeutic drug paclitaxel, whose elimination is proposed to be governed by its residence time within the cells.

2.2 Power Laws in Pharmacokinetics

Power laws have been applied in the analysis of both dose- and time-dependences in pharmacokinetic systems. When a drug is injected into the body via a single bolus intravenous dose, the resulting plasma concentration time curve is called a clearance curve. While traditionally clearance curves are fit by a sum of exponential terms, some groups have shown that they can also be fit by a single negative power according to $y(t) = at^{-\gamma}$, by a gamma function according to $y = at^{-\alpha}e^{-\beta t}$, or by two sequential negative powers [52, 66]. Within a given range, while the same curves can be fit within experimental error by a sum of negative exponentials, the converse is not true [66]. Different explanations for these power-law and gamma-function fits have been proposed. Because the drugs examined by Norwich and Siu [52] were predominately eliminated through the liver, they developed a model based on the anatomy of the liver and the flow of blood through it. Their convection–diffusion equations for the functional unit of the liver, the acinus, generated approximate solutions with gamma and power functions. Wise [66] explained the power law fits in terms of a heterogeneous distribution of drug particles, where each particle cycles through the system a number of times by series of random walks with drift.

Dose-proportionality can be identified using a simple plot of the pharmacokinetic parameter as a function of the dose. The graph will be a straight line with a zero intercept if the parameter is linearly proportional to dose. As a better diagnostic tool, the "power model" was proposed [24]

$$y = \alpha D^{\beta}, \tag{17}$$

where y is a pharmacokinetic parameter and D is the dose. When $\log y$ is plotted as a function of $\log D$, the slope of the line will be equal to the parameter β. Two scenarios are discussed in this chapter: $\beta = 0$ (dose-independence) and $\beta = 1$ (dose-dependence). The study [24] found that the slopes analyzed were internally consistent, and they were compared to an expected value of one. We propose to build on this model by expanding the definition of the exponent to include fractional values.

2.3 *Compartmental Models*

To understand and predict drug behavior through the body, compartmental models are commonly used in pharmacokinetics [28]. In principle, each organ should be represented by a separate compartment (see Fig. 3 for a schematic illustration) that is assumed to contain a *homogenously distributed* concentration of drug molecules undergoing a set of chemical kinetic reactions. In practice, models with a few hypothetical compartments have been used with variable success rates. The compartment's input/output is typically assumed to be governed by *linear* kinetic processes with *constant* rate coefficients.

A compartment is defined by the number of drug molecules having the same probability of undergoing a set of chemical kinetic processes. The exchange of drug molecules between compartments is described by kinetic rate coefficients. The classical compartmental model is based on two main assumptions: (a) each compartment is homogeneous (i.e. there is instantaneous mixing) and (b) the kinetic

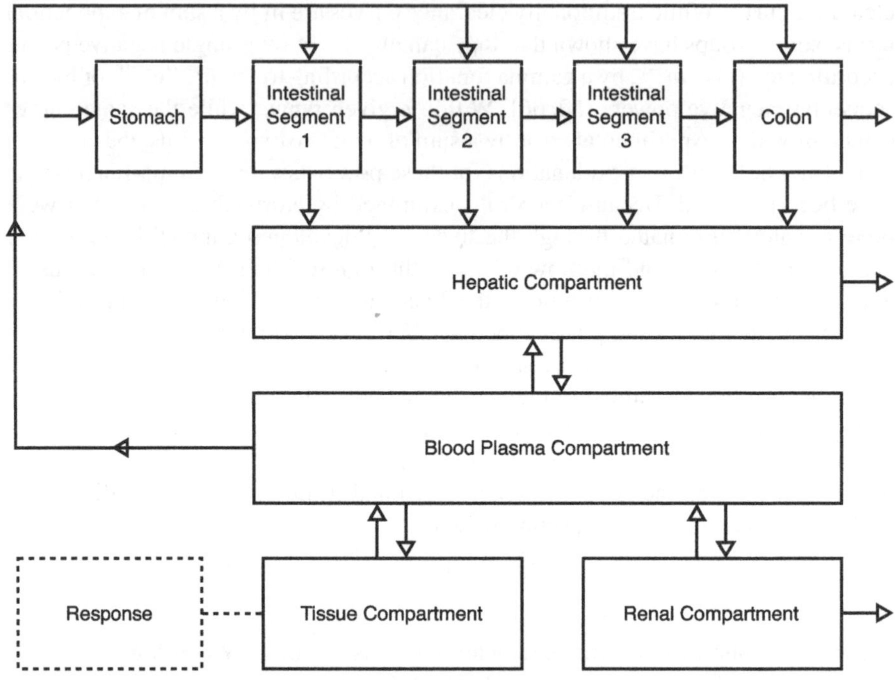

Fig. 3 Schematic illustration of a multi-compartment physiologically-based model

rate coefficients are all constant, such that the fraction of drug transferred between any two compartments is constant in time. The system is described by coupled first-order differential equations whose solutions take the form of a sum of terms that are exponential in time.

Within the pharmacokinetic model, the body is represented with a number of compartments depending on the availability of parameter values and a specific application of the model. Each compartment has a given volume. Of particular interest is the "Tissue Compartment" which represents all body tissues not otherwise accounted for in the model. The compound concentrations in this compartment determine the intensity of physiologic effects. This is shown in Fig. 3 by the dashed connection to the adjacent "Response" box. The movement of blood that carries with it drug molecules between compartments is shown with arrows. Initial doses enter the stomach if the drug is orally available, while the colon, liver, and kidneys are sites of potential elimination of the drug from the system.

Classical kinetics is based on the law of mass action, which states that the rate v of a chemical reaction is directly proportional to the product of the concentrations of the N reactants each raised to the power n_i:

$$v = k \prod_{i=1}^{N} C_i^{n_i}, \tag{18}$$

where C_i is the concentration of reactant i and k is the kinetic rate coefficient in units of inverse time. The law of mass action "mass-balance" assures the equality between the amounts of material that leave one compartment and the amounts that enter into another. The reaction order n_i is the number of concentration terms that must be multiplied together to get the rate of the reaction [57]. For a single step, n_i is typically equal to the molecularity, which is the number of molecules that are altered during the reaction. When only one molecule is modified, the reaction is given by

$$v = kC. \tag{19}$$

The movement of the drug molecules between compartments is proportional to the volume rate of fluid flow between the connected compartments and the concentration of the compound in the originating compartment, and inversely proportional to the volume of the original compartment. Given an initial distribution of the compounds in the various compartments, the model is integrated forwards in time, using a 4th order Runge–Kutta algorithm, with variable time-steps.

2.4 *Enzyme Kinetics and the Michaelis–Menten Equation*

The rate of enzyme-catalyzed reactions can deviate from those predicted by classical kinetics. At high concentrations, saturation of the enzymes limits the maximum reaction rate that can be achieved, while at low concentrations, the rate of formation of the enzyme-substrate complex becomes significant and the reaction becomes dependent on the substrate concentration [13].

Consider the reaction:

$$E + S \underset{k_{-1}}{\overset{k_1}{\rightleftharpoons}} ES \xrightarrow{k_2} E + P, \tag{20}$$

where E, S, ES, and P represent the enzyme, substrate, enzyme-substrate complex, and product, respectively. If we denote the concentration of the substrate as C, the concentration of the enzyme-substrate as x, and the total concentration of enzymes as e_0, the system is described by the following ordinary differential equations:

$$\frac{\mathrm{d}x}{\mathrm{d}t} = k_1(e_0 - x)C - (k_{-1} + k_2)x, \tag{21}$$

$$\frac{\mathrm{d}p}{\mathrm{d}t} = k_2 x. \tag{22}$$

Using the Briggs–Haldane treatment [57] to simplify the problem, a quasi-steady-state assumption is made where the concentration of the substrate-enzyme complex is taken to be constant, i.e. $\mathrm{d}x/\mathrm{d}t = 0$. Therefore,

$$k_1(e_0 - x)C - (k_{-1} + k_2)x = 0. \tag{23}$$

Collecting the terms in x and rearranging gives:

$$x = \frac{k_1 e_0 C}{k_{-1} + k_2 + k_1 C}. \tag{24}$$

Using the fact that the rate of the reaction is $v = k_2 x$ gives

$$v = \frac{k_2 e_0 C}{\frac{k_{-1} + k_2}{k_1} + C}. \tag{25}$$

Finally, denoting $v_{\max} = k_2 e_0$ and $K_{\mathrm{M}} = (k_{-1} + k_2)/k_1$ gives the Michaelis–Menten equation:

$$v = \frac{v_{\max} C}{k_M + C}. \tag{26}$$

The parameter $v_{\max}$ is the maximum velocity of the reaction, and the Michaelis–Menten constant K_{M} is the substrate concentration at half the maximum velocity.

In the low-concentration case, where $C \ll K_{\mathrm{M}}$, Eq. (26) reduces to

$$v = \frac{v_{\max}}{K_{\mathrm{M}}} C, \tag{27}$$

which describes first-order kinetics with $k = v_{\max}/K_{\mathrm{M}}$. In the high-concentration case, where $C \gg K_{\mathrm{M}}$, Eq. (26) becomes

$$v = v_{\max}, \tag{28}$$

which is steady-state kinetics with a constant reaction rate.

2.5 *Asymptotics of the Concentration-Time Curve*

The solution to a compartmental model with constant coefficients takes the form of a linear superposition of exponential terms, and the resulting concentration-time curve exhibits an exponentially decaying tail. However, there is evidence that the concentration-time curves of many drugs exhibit long-time power law tails of the form

$$C(t) \sim t^{-\gamma} \qquad \text{for } t > T, \tag{29}$$

where T marks the time of the onset of the tail. Negative power laws were first applied, empirically, to describe the washout of bone-seeking radioisotopes [3, 6, 51]. Subsequently, other types of clearance curves have been fit by a single power law, two sequential power laws, or the gamma function $y(t) = at^{-\alpha}e^{-\beta t}$ [61]. Different explanations for these fits have been proposed, including a stochastic random walk model based on the cycling of molecules in and out of the plasma [48], a set of convection–diffusion equations for transit in the liver, and gamma-distributed drug residence times [65]. The authors explain the power law behavior by hepatic processes although tissue distribution may be playing an equally important role as shown in [62].

In this chapter, we introduce a model based on fractal kinetics with an anomalous reaction order as a physiologically based mechanism that generates power law tails using only one compartment. This model, which may be a gross oversimplification of the actual situation in the human body, is naturally interpreted in terms of the anatomy and physiology of the liver, the organ of drug elimination.

2.6 Transient Fractal Kinetics

Anacker and Kopelman [2] found that reactions that occur on or within fractal media exhibit anomalous kinetics that do not follow the classical mass-action form. Specifically, the kinetic rate coefficient is time-dependent [32]:

$$k = k_0 t^{-h}, \tag{30}$$

where

$$h = 1 - \frac{d_s}{2}. \tag{31}$$

The quantity d_s is the spectral dimension that describes the path of a random walker within the medium [50]. The classical case corresponds to $d_s = 2$. Equations (30) and (31) have been supported by experiments of trapping and binary reactions on the Sierpinski gasket, percolation clusters, and lattices with disordered transition rates [4, 41, 44] to name but a few examples. While Eq. (30) applies to diffusion-limited reactions on fractals, it also applies to any situation for which $h > 0$.

Equation (30) has been incorporated into pharmacokinetics through both non-compartmental and compartmental models. The former includes the homogeneous–heterogeneous distribution model introduced by Macheras [42] to quantify the global and regional characteristics of blood flow to organs. The latter includes the fractal compartmental model [16] in which a classical compartment was used to represent the plasma while a fractal compartment was used to represent the liver. In this formalism, the rate of elimination from the liver is given by

$$v = k_0 t^{-h} C. \tag{32}$$

Simulations of the model showed that h plays a significant role in determining the shape of the concentration-time curve [9].

Several attempts have been made to incorporate Eq. (30) into the Michaelis–Menten equation to describe concentration-dependent reactions that occur in spatially constrained conditions. Berry [7] made a substitution producing the formula

$$v = \frac{v_{\max} C}{K_{M0} t^h + C}. \tag{33}$$

Using Monte Carlo simulations on a 2D lattice to model enzyme reactions in low-dimensional media, it was found [7] that h increases independently with increasing obstacle density on the lattice and increasing initial substrate concentration. Kosmidis et al. [35] also performed Monte Carlo simulations and found that Eq. (33) holds mainly when the initial substrate concentration is high, either through an intravenous bolus administration or a high rate of absorption. In addition, Eq. (33) was incorporated into a one-compartment model [35]. Simulations performed by Aranda et al. [4] also confirm these results but suggest that K_{M0} is characterized by multifractality and hence a set of fractal exponents.

2.7 Steady State Fractal Kinetics

As seen above, the effect of complex geometry on the rate of transient reactions produces an anomalous kinetic rate coefficient. Anacker and Kopelman [2] demonstrated that under steady state conditions, however, the effect of the geometry is manifested as an anomalous reaction order. They showed that Eq. (19) should be replaced by the effective rate equation

$$v = kC^X, \tag{34}$$

where X is a fractal reaction order related to the spectral dimension of the random walk. For example,

$$X = \begin{cases} 1 + \dfrac{2}{d_s} & \text{for } A + A \text{ reactions,} \\[2ex] 1 + \dfrac{4}{d_s} & \text{for } A + B \text{ reactions.} \end{cases} \tag{35}$$

These equations have been confirmed using Monte Carlo simulations. Anacker et al. [2] found that $X = 2.44$ for the 2D Sierpinski gasket and $X = 2.01$ (as expected) for the homogeneous cubic lattice. Klymko and Kopleman [32] found that for bimolecular reactions in solids, X ranged from the homogeneous value of 2 up to a value of 30. Newhouse and Kopelman [50] found values of $X \approx 5$ for ensembles of 10×10 islands and $X \approx 15$ for ensembles of 5×5 islands. Therefore, as available space becomes more finely divided, as in the example a fractal dust like the Cantor set [44], $d_s \to 0$ and therefore $X \to \infty$.

A form of concentration-dependent fractal kinetics was developed by López-Quintela and Casado [41], who proposed the following scaling relationship:

$$k_i^{\mathrm{eff}} = A_i C^{1-d_{\mathrm{f}}} \qquad 0 \leq d_{\mathrm{f}} \leq 1, \tag{36}$$

where d_{f} is the fractal dimension of the space. By applying this equation to $v_{\max}$, the formula was obtained:

$$v = \frac{v_{\max}^{\mathrm{eff}} C^{2-d_{\mathrm{f}}}}{K_{\mathrm{M}}^{\mathrm{eff}} + C}, \tag{37}$$

where $v_{\max}^{\mathrm{eff}}$ and $K_{\mathrm{M}}^{\mathrm{eff}}$ are new constants. For $d_{\mathrm{f}} = 1$, the classical Michaelis–Menten equation is recovered, and as $d_{\mathrm{f}} \to 0$, the complexity of the reaction becomes more and more important. Heidel and Maloney [26] performed an analytical exploration of this equation, and Ogihara [53] applied it to model carrier-mediated transport under heterogeneous conditions.

A seemingly different approach to concentration-dependent fractal kinetics is the "power-law formalism" developed by Savageau [57], expressed through the generalized mass-action representation:

$$\frac{\mathrm{d}C_i}{\mathrm{d}t} = \sum_{k=1}^{r} \alpha_{ik} \prod_{j=1}^{n} C_j^{g_{ijk}} - \sum_{k=1}^{r} \beta_{ik} \prod_{j=1}^{n} C_j^{h_{ijk}}, \tag{38}$$

where α and β are the kinetic rate coefficients and g and h are the kinetic rate orders associated with each reactant. The equations for the power-law formalism are complicated and Savageau admits that this model works best for large series of reactions rather than one or more reactions catalyzed by only one enzyme [57]. Savageau justifies this formalism by showing that for homodimeric reactions, its equations are equivalent to the fractal kinetics equations. However, this equivalence has yet to be proven for any other reactions due to the complexity of the equations. In principle, it is possible that Eq. (38) can be obtained by summing over several Michaelis–Menten reactions.

To summarize, any reaction for which $h > 0$ or $X > n$ is referred to as following fractal-like kinetics. In this chapter, an alternative formulation of dose-dependent fractal kinetics is proposed based on fractal reaction orders under steady state conditions.

2.8 Model Solutions

In a strict sense, a steady state regime means that the concentration of the reactant is constant in time. One way in which this can be achieved is if the concentration of drug molecules is much greater than the concentration of enzymes, even if the local concentration values vary considerably. Even in the presence of drug elimination, a steady state can be maintained due to the recycling of drug molecules by the circulatory system. If the environment is heterogeneous, the system is described by the equations:

$$\frac{\mathrm{d}x}{\mathrm{d}t} = k_1(e_0 - x)C^X - (k_{-1} + k_2)x, \tag{39}$$

$$\frac{\mathrm{d}p}{\mathrm{d}t} = k_2 x. \tag{40}$$

Therefore, the Michaelis–Menten equation becomes

$$v = \frac{v_{\max} C^X}{K_{\mathrm{M}} + C^X}. \tag{41}$$

It can be noted that Eq. (41) has the same form as the Hill equation that describes the response of the system as a function of the drug concentration. Incorporating this formula into a one-compartment model with an intravenous infusion yields

$$\frac{\mathrm{d}C}{\mathrm{d}t} = -\frac{v_{\max} C^X}{k_M + C^X} + \frac{i(t)}{V_{\mathrm{d}}}, \tag{42}$$

where $i(t)$ is the infusion rate in units of mass/time and V_{d} is the volume of distribution in units of volume. To investigate the asymptotics of Eq. (42), we consider the model post-infusion. For high concentrations (those occurring at or above K_{M}):

$$\frac{\mathrm{d}C}{\mathrm{d}t} = -v_{\max}. \tag{43}$$

For low concentrations (those occurring far below K_{M}):

$$\frac{\mathrm{d}C}{\mathrm{d}t} = -\frac{v_{\max}}{K_{\mathrm{M}}} C^X. \tag{44}$$

Integrating Eq. (44) leads to the asymptotic power law behavior

$$C(t) \sim t^{1/(1-x)}. \tag{45}$$

Comparing to Eq. (29) yields the relationship

$$\gamma = \frac{1}{1 - X}, \tag{46}$$

or

$$X = 1 - \frac{1}{\gamma}. \tag{47}$$

Note that Eqs. (45)–(47) are undefined for $X = 1$, since this value corresponds to the classical model with an exponential tail, which is inconsistent with a power law.

The fact that the proposed steady state model predicts long-time power law behavior provides a point of comparison with other models. For example, the Michaelis–Menten model predicts an exponential tail, and the transient fractal and fractal Michaelis–Menten equations predict stretched exponential tails of the form $C(t) \sim \exp(at^{1-h})$.

3 Applications

3.1 Mibefradil: Spatially Induced Nonlinearity

Mibefradil is a calcium antagonist that was developed to reduce ventricular fibrillation [58]. It is orally administered, and its major site of elimination is the liver. Studies done on chronically instrumented dogs [58] concluded that observed nonlinear pharmacokinetics are due to dose- and time-dependent reduction of hepatic clearance of the drug. Fuite et al. [16] proposed that the source of this reduction is the fractal geometry of the liver. The circulation in the liver can be divided into the macro-circulation (including the hepatic artery and the hepatic and portal veins) and the micro-circulation (consisting of the portal vein, hepatic arterioles, and the sinusoids) [8]. The microvasculature of the liver consists of vessels that bifurcate towards smaller and smaller daughter vessels. In fact, the vessels supplying the liver, lungs, kidney, and heart have been found to exhibit scaling relationships for branch diameter, branch length, pressure, and radius-to-length ratios [5, 21, 22]. Javanaud estimated the fractal dimension of the liver to be $d_f \approx 2$ [29].

Due to differences in the global and regional nature of blood flow to organs, Macheras developed a homogeneous–heterogeneous distribution model [42]. The homogeneous conditions are considered "well-stirred," and the heterogeneous conditions near tissues are considered "under-stirred." Figure 4 illustrates this composite system.

While the homogeneous portions of the circulatory system can be described using conventional kinetics, regional areas such as those feeding the liver are fractal and thus should display fractal kinetics. Essentially, the geometry of a surface where a chemical process is taking place affects the rate at which the process can occur. This reaction rate can be expressed as follows [25]:

$$k(t) \propto t^{-(1-d_S/2)}. \tag{48}$$

The theoretical model [16] combining well-stirred Euclidean and fractal compartments led to analytical solutions for the time evolution of the drug concentration which were obtained using perturbation analysis. The equations were then tested

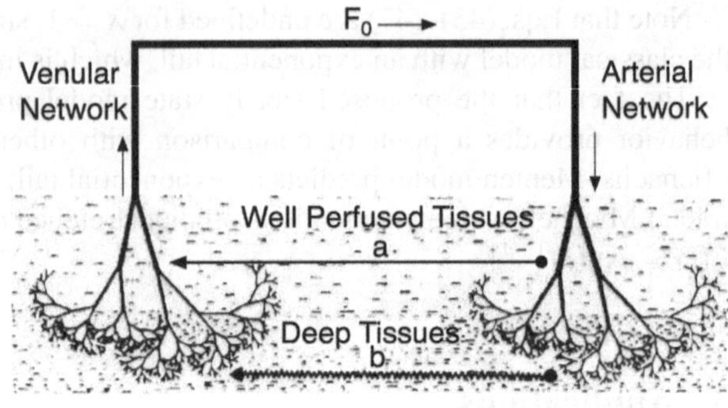

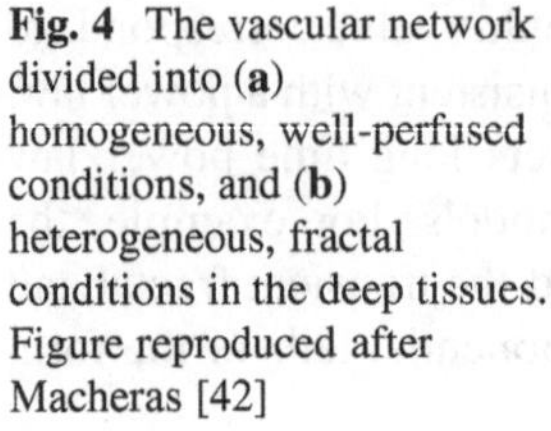
Fig. 4 The vascular network divided into (**a**) homogeneous, well-perfused conditions, and (**b**) heterogeneous, fractal conditions in the deep tissues. Figure reproduced after Macheras [42]

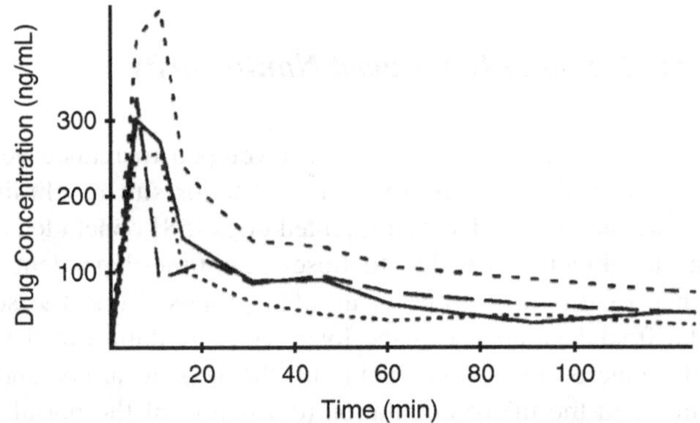

Fig. 5 The early time behavior of the concentration-time data for four different dogs [58]. Concentration-time data were obtained for the cardiac drug mibefradil in four dogs. The dogs received a dose of 1 mg/kg of mibefradil infused over 10 min

using mibefradil data [58]. The results for four dogs are shown in Fig. 5. The spectral dimension for the dog liver estimated using the model was between 1.778 and 1.914, which compares favorably with the measured value of 2.0 by Javanaud [29].

The value and standard deviation of the power law tail exponent γ were calculated from the concentration-time curves using linear regression analysis of the log-transformed data.

The various models were fit to the data using a parameter optimization method based on a simulated annealing (SA) algorithm implemented in C++. The SA algorithm [31] minimizes an objective function through an efficient exploration of the parameter space. All downhill moves are accepted and selective uphill moves are allowed according to the Metropolis algorithm and an effective temperature. At the start of the annealing process, the temperature is relatively high compared to the standard deviation of the objective function, and the probability of accepting an uphill move is high. Hence, the random walk is able to explore a wide area of parameter space without getting trapped in local minima. As the temperature is decreased, the algorithm is able to focus on the most promising areas and locate the global minimum.

Table 1 Summary of models for the enzyme-mediated kinetics of drug elimination

Model	Abbreviation	Reaction rate
Fractal	F	$k_0 t^{-h} C$
Michaelis–Menten	MM	$v_{\max} C/(K_M + C)$
Transient fractal Michaelis–Menten	FMM	$v_{\max} C/(K_{M0} t^h + C)$
López-Quintela fractal Michaelis–Menten	LQC	$v^{\mathrm{eff}}_{\max} C^{2-d_f}/(K^{\mathrm{eff}}_M C^{2-d_f})$
Steady state fractal Michaelis–Menten	SSFMM	$v_{\max} C^X/(K_m + C^X)$

Table 2 Slope γ of the log(concentration) versus log(time) curve between $t = 30\,\mathrm{min}$ and $t = 1{,}440\,\mathrm{min}$

Dog	γ	R^2
1	−0.702 (0.028)	0.991
2	−0.464 (0.049)	0.927
3	−0.597 (0.024)	0.989
4	−0.705 (0.066)	0.943

Values are given as mean (standard deviation)

The SA algorithm has many advantages over other optimization methods. It is largely independent of the starting values, it can escape local minima through selective uphill moves, and the underlying function need not be continuous. The SA method has been found to be superior to the simplex method, the Adaptive Random Search, the quasi-Newton algorithm, and the Levenberg–Marquardt algorithm in finding the optimum of continuous functions [11, 15, 23]. The objective function was chosen to be the weighted residual sum of squares (WRSS) [17]:

$$\mathrm{WRSS} = \sum_{i=1}^{n} \frac{(C_i - \hat{C}_i)^2}{\hat{C}_i^2}, \tag{49}$$

where $\hat{C}_i$ denotes the predicted value of C_i based on the given model. The goodness-of-fit of each model was assessed using the Akaike information criterion (AIC), which takes into account WRSS as well as the number of model parameters, N_{par}, and the number of data points, N_{obs} [17]:

$$\mathrm{AIC} = N_{\mathrm{obs}} \ln(\mathrm{WRSS}) + 2N_{\mathrm{par}}. \tag{50}$$

A lower value indicates a better fit. Five one-compartment models were fit to the data sets, and they are summarized in Table 1.

The shape of the tail was determined for the four data sets and was found to be a straight line on a log–log plot, indicating a power law relationship. The values for the power law exponent γ are listed in Table 2. The power law tail extends over three

Table 3 One-compartment parameters for the drug concentration in the jugular vein of dogs following a 10-min infusion of 1 mg/kg of mibefradil

		Value				
Model	Parameter	Dog 1	Dog 2	Dog 3	Dog 4	Mean
MM	v_{max} (ng/mL/min)	327	4,699	4,737	4,375	
	K_M (ng/mL)	96,593	101,139	100,131	101,046	
	V_d (L)	10.5	0.00369	0.00361	0.00405	
	WRSS	7.27	11.0	11.0	11.0	
	AIC	31.8	37.2	37.2	37.2	35.9
LQC	v_{max}^{eff} (ng/mL/min)	587	847	363	623	
	K_M^{eff} (ng/mL)	5,323	6,961	4,702	5,693	
	V_d (L)	7.21	8.45	4.64	5.67	
	$2-D$	2.00	2.00	2.00	2.00	
	WRSS	2.68	4.06	4.15	3.05	
	AIC	20.8	26.2	26.5	22.5	24.0
F	k_0 (/min)	1.01	1.16	1.03	1.21	
	V_d (L)	4.80	4.31	4.95	3.55	
	h	0.999	0.999	0.998	0.998	
	WRSS	2.56	5.33	3.74	4.14	
	AIC	18.2	27.8	23.2	24.5	23.4
FMM	v_{max} (ng/mL/min)	4,358	3,306	2,486	4,170	
	K_{M0} (ng/mL)	4,623	3,638	2,709	4,401	
	V_d (L)	6.04	11.0	10.1	12.0	
	h	1.00	1.00	1.00	1.00	
	WRSS	2.01	1.82	2.01	2.01	
	AIC	17.1	15.8	17.1	17.1	16.8
SSFMM	v_{max} (ng/mL/min)	3,575	8,201	3,548	3,817	
	K_M (ng/mL)	5,217	799	6,778	7,098	
	V_d (L)	1.30	2.39	16.5	9.54	
	X	2.56	3.35	2.74	2.61	
	WRSS	0.845	0.263	0.219	0.544	
	AIC	5.99	−9.38	−11.7	0.0797	−3.75

orders of magnitude in time, and the goodness-of-fit represented by the R^2 value is greater than 0.9 for every dog. This result indicates that the SSFMM model is an appropriate model for the data.

The results from the model fits are listed in Table 3. The SSFMM model provides the best fit to all data sets, while the MM and LQC models perform the worst. Furthermore, the reaction orders of $2-d_f$ in the LQC model yield values of zero for the fractal dimension d_f, essentially eliminating the fractal nature of the model. In the case of the F and FMM models, the exponent h takes the maximum value of 1; however, the fit is not substantially improved.

The values for X determined from the model fit were compared to those calculated from the power law tail exponent γ using Eq. (46), and the results are

Table 4 Comparison between the values for the fractal reaction order X predicted from the slope γ and obtained from the model fit

	X	
Dog	Predicted from γ	Model value
1	2.42 (0.10)	2.56
2	3.16 (0.33)	3.35
3	2.68 (0.11)	2.74
4	2.41 (0.22)	2.61

listed in Table 4. The values agree within error for all but Dog 1. Figure 6 panel (a) shows the power law tail for Dog 3, and Fig. 6 panel (b) shows the same data fit by the SSFMM model.

According to Eq. (44), the existence and onset of the power law tail correlate with the value of K_M, the power law behavior only holds for $C \ll K_M$. The values estimated for K_M range from 800 ng/mL to 7,000 ng/mL and are between 30 % and 90 % higher than the maximum plasma concentrations (556.1–1,400 ng/mL). Therefore, the power law tails are observable because the dose of mibefradil given to the dogs in this study leads to plasma concentrations well below saturation levels. Furthermore, Eq. (44) can be interpreted alternatively in terms of a concentration-dependent v_{max} of the form $v_{max}^{eff} = v_{max}C^{X-1}$. When the approximate Eq. (44) was used instead of Eq. (42), it resulted in similar parameter values as the SSFMM model but with a poorer fit to the rise of the curve.

3.2 *Paclitaxel: Temporally Induced Nonlinearity*

One of the biggest challenges in clinical oncology is to estimate the optimum dose of an anticancer drug for a given patient. Such a calculation normally combines the physiological characteristics of the patient with the results of preclinical investigations and clinical trials. Chemotherapeutic drugs offer a unique opportunity for pharmacokinetic studies. In the early stages of drug development, a drug is typically administered at high doses and through a wide range of doses in order to establish toxicity and tolerance. For such experiments to be clinically useful, we need to be able to take a limited set of measurements and extrapolate from them an optimum dose for a specific patient. In the simplest case, a doubling of the dose would result in a doubling of the systemic exposure to the drug. However, a dose of drugs in a patient is a confined many-body system that can exhibit multiple degrees of complexity. In many cases, the doubling of a dose produces either less than or more than double the effect. While we already have the tools to extrapolate in the linear case, here we strive to develop adequate tools for situations that involve nonlinearities.

Paclitaxel is a chemotherapeutic drug derived from the European Yew tree bark. It consists of a taxane nucleus with three rings, and it is poorly water soluble [27]. The current formulation is a 6 mg/ml solution in a solvent consisting

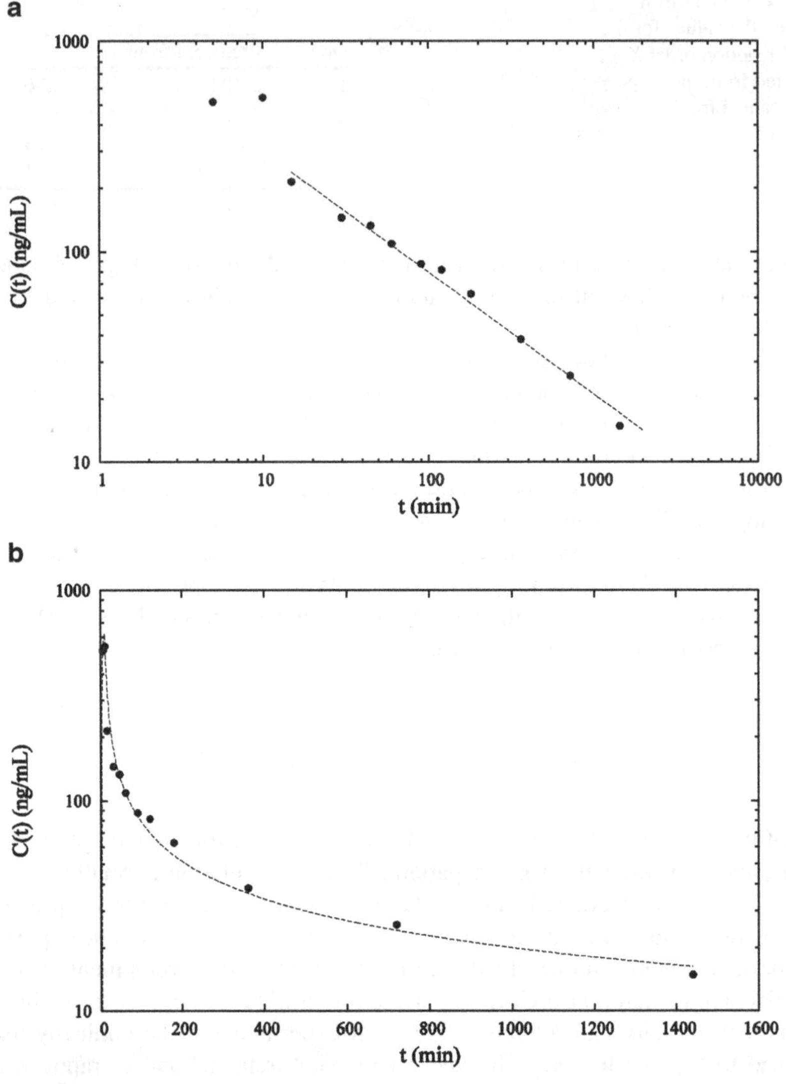

Fig. 6 Concentration-time curves for mibefradil data for Dog 3. (**a**) Log–log plot showing the long-time power law tail from 30 min to 1,440 min. The *dashed line* is the regression line with $\gamma = -0.597 \pm 0.024$. (**b**) The same data but the *dashed line* now represents the best-fit curve found using the steady state fractal model with $X = 2.74$

of 50 % polyoxyethylated castor oil (Cremophor EL) and 50 % dehydrated alcohol (USP) [60]. It is usually administered by intravenous infusion. Paclitaxel binds preferentially to microtubules, which are cellular components necessary for mitosis, intracellular transport, maintenance of cell shape, and cellular motility. It has been found to strongly inhibit cell replication by blocking cells in the G2 mitotic phase by

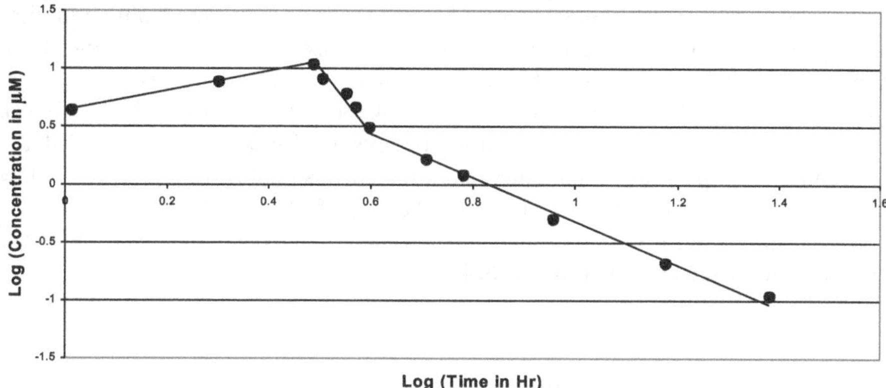

Fig. 7 A typical log–log plot of concentration versus time for a 3-h infusion of 225 mg/m^2 of paclitaxel, measured over a 24-h period. The biphasic elimination can be clearly seen. Data were taken from [30]

stabilizing microtubules [54]. Paclitaxel is well suited for pharmacokinetic studies because it is administered at high doses, it has a long residence time in the body, and it has been the subject of many dose escalation studies. It has shown a consistent disproportional increase in the maximum plasma concentration (C_{max}) and the area under the curve (AUC) [30]. Following the end of the infusion, a biphasic elimination has been noted [27].

In a meta-study of data from the literature, not only biphasic elimination, but also a distinct power law relationship for each phase has been found. The first phase begins immediately after the end of the infusion and consists of a rapid decline lasting approximately 1 h. The subsequent prolonged terminal phase can last up to 72 h. Significantly, although the data span a wide range of age, sex, type of cancer, total dose, and infusion time, the power exponents for the two elimination phases seem to be approximately equal to -3.3 and -1.6, respectively. In addition, the time period corresponding to uptake during the infusion also seems to obey a power law, but with a much larger range of exponents. Figure 7 shows a representative log–log plot for concentration-time data of paclitaxel.

Various reasons have been proposed in the literature for the source of the nonlinearity of paclitaxel pharmacokinetics. The predominant explanation being offered is the high binding of paclitaxel molecules to the Cremophor EL vehicle and plasma proteins [30, 59]. However, a related drug docetaxel, which does not have Cremophor EL, has also shown nonlinear behavior [68]. In the study, 100 mg/m^2 of docetaxel was administered during a 6-h infusion. We found an exponent of -1.34 ($R^2 = 0.9830$) for the period of 8–30 h. In addition, pharmacokinetic studies of ABI-007, a Cremophor-free formulation of paclitaxel, demonstrated similar nonlinear behavior. The data from one experiment [49] with ABI-007 produced an exponent of -1.60 ($R^2 = 0.9469$), while in another study [14] the area under the curve scaled with dose with an exponent of -1.61 ($R^2 = 0.9477$).

Several experiments provide support for this theory. Using data from an in vitro study looking at the cytotoxicity of paclitaxel and docetaxel in three cell lines [55], a power law relationship was found between the IC_{50} and the exposure time. Interestingly, for one cell line, the exponent was found to be -1.62 ($R^2 = 0.9957$). In addition, in another in vitro study of tumor cells in a supportive medium [36], the concentration of paclitaxel in the medium after 4 h was related through a power law to the initial concentration with an exponent of -1.15 ($R^2 = 0.998$).

The results from our study and a more in-depth analysis have been published earlier where details can be found [46, 47].

4 Conclusion and Discussion

Biological systems are complex, but the advantage of such systems is the relatively simple mathematics that seems to consistently emerge from their behavior. It is well known that the liver has a complex geometry. The blood vessels supplying it are arranged as a fractal tree [29], its cellular network has fractal properties [18], and the perfusion of blood at the terminal branches is heterogeneous [63]. Both transient and steady state reactions occurring within such spaces can exhibit anomalous behavior. For transient reactions, it is assumed that there is a random distribution of reactants [34]. Therefore, anomalous kinetics in the transient case strictly results from the decreased efficiency of random walkers in exploring their irregular space (quantified by d_s). In the steady state regime, however, there is an influx of molecules. In regular geometries, this influx can cause a net stirring effect [34]; however, in fractal and confined geometries, self-stirring is inefficient. The spaces are instead characterized by large fluctuations in the local concentration and an increasing segregation of molecules. This effect has been reported for reaction-diffusion phenomena in physical systems [12]. As a result, under steady state conditions, the distribution of molecules is partially ordered due to the influx of molecules, and the non-randomness reduces the reaction probabilities and consequently the reaction rate. To summarize, transient fractal kinetics occur in well-stirred heterogeneous media while steady state kinetics occur in poorly stirred heterogeneous media. Here, the term heterogeneity refers to the geometry of the environment.

In the case of steady state fractal kinetics, Klymko and Kopelman [32] interpreted non-integer values of X as characteristic of a microscopically heterogeneous medium that is best described as a collection of kinetically independent clusters. The kinetic rate coefficients are then kinetic averages taken over domains of different sizes and local concentration. This interpretation is consistent with the studies that reported high X values for reactions occurring on clusters and islands [50]. A similar model can be developed for the liver, the organ predominantly responsible for the elimination of drugs from the body. The metabolic enzymes are located in the liver cells, called hepatocytes, which are organized around the terminal supply vessels. Each set of vessels and their surrounding cells are called a sinusoid. Not only does each sinusoid have a different number and distribution of hepatocytes, it receives a different portion of the blood flow. Therefore, the liver can be considered as a network of clusters of sinusoids. Because X increases as the size of the clusters decreases [50], $X = 1$ means that the liver acts as a homogeneous, well-mixed compartment and $X > 1$ indicates segmentation and a lack of mixing.

This interpretation is consistent with a model proposed by Weiss [63], who described the transit times in the liver as being determined by both the micro-mixing and macro-mixing processes. He suggested two models at different ends of the spectrum: (a) a distributed model in which the sinusoids are parallel and there is complete segregation of the pathways, and (b) a dispersion model in which the sinusoids are interconnected and there is perfect micro-mixing. Considering our results in this framework, X provides a quantitative measure of the degree of micro-mixing between sections of the liver and locates the model somewhere between Weiss's two extreme models.

This chapter describes applications of fractal kinetics under steady state conditions to pharmacokinetics. We have demonstrated that a steady state fractal Michaelis–Menten equation best describes the elimination of the drug mibefradil from dogs. Furthermore, it accounts for the long-time power law behavior of the concentration through the inclusion of a fractal reaction order, X. This anomalous reaction order suggests that the liver, the organ of elimination for mibefradil, is best treated as a collection of clusters of sinusoids. The higher the value of X, the less mixing that occurs between adjacent sinusoid clusters. Mibefradil is an orally administered calcium antagonist with a relatively short elimination time, while paclitaxel is an intravenously administered anticancer drug that can remain in the body for over three days. Nevertheless, the elimination curves of both drugs can be accurately described using a similar power law relationship.

We conclude that transient fractal kinetics is appropriate for describing reactions that occur within well-mixed heterogeneous environments, while steady state fractal kinetics better describes reactions that occur in under-stirred heterogeneous spaces. The latter can occur due to the continuous influx of drug molecules through recycling in the circulatory system.

References

1. Anacker, L.W., Kopelman, R.: Fractal chemical kinetics: Simulations and experiments. J. Chem. Phys. **81**, 6402–6403 (1984)
2. Anacker, L.W., Kopelman, R.: Steady-state chemical kinetics on fractals: Segregation of reactants. Phys. Rev. Lett. **58**, 289–291 (1987)
3. Anderson, J., Osborn, S.B., Tomlinson, R.W., Weinbren, I.: Some applications of power law analysis to radioisotope studies in man. Phys. Med. Biol. **18**, 287–295 (1963)
4. Aranda, J.S., Salgado, E., Muñoz-Diosdado A.: Multifractality in intracellular enzymatic reactions. J. Theor. Biol. **240**, 209–217 (2006)
5. Bassingthwaighte, J., Liebovitch, L.S., West, B.J.: Fractal Physiology. Oxford University Press, New York (1994)

6. Bassingthwaighte, J.B., Beard, D.A.: Fractal 15O-labeled water washout from the heart. Circ. Res. **77**, 1212–1221 (1995)
7. Berry, H.: Monte carlo simulations of enzyme reactions in two dimensions: Fractal kinetics and spatial segregation. Biophys. J. **83**, 1891–1901 (2002)
8. Campra, J.L., Reynolds, T.B.: The hepatic circulation. In: Arias, I.M., Popper, H., Schachter, D., Shafritz, D.A. (eds) The Liver: Biology and Pathobiology. Raven Press, New York (1982)
9. Chelminiak, P., Marsh, R.E., Tuszyński J.A., Dixon, J.M., Vos, K.J.E.: Asymptotic time dependence in the fractal pharmacokinetics of a two-compartment model. Phys. Rev. E Stat. Nonlin. Soft. Matter Phys. **72**, 031903 (2005)
10. Chelminiak, P., Dixon, J.M., Tuszyński J.A., Marsh, R.E.: Application of a random network with a variable geometry of links to the kinetics of drug elimination in healthy and diseased livers. Phys. Rev. E Stat. Nonlin. Soft. Matter Phys. **73**, 051912 (2006)
11. Corana, A., Marchesi, M., Martini, C., Ridella, S.: Minimizing multimodal functions of continuous variables with the "simulated annealing" algorithm. ACM Trans. Math. Softw. **13**, 262–280 (1987)
12. Cornell, S., Droz, M., Chopard, B.: Role of fluctuations for inhomogeneous reaction-diffusion phenomena. Phys. Rev. A **44**, 4826–4832 (1991)
13. Cornish-Bowden, A.: Fundamentals of Enzyme Kinetics, Rev edn. Portland Press, London (1995)
14. Damascelli, B., Cantù, G., Mattavelli, F., Tamplenizza, P., Bidoli, P., Leo, E., Dosio, F., Cerrotta, A.M., Di Tolla, G., Frigerio, L.F., Garbagnati, F., Lanocita, R., Marchianò, A., Patelli, G., Spreafico, C., Tichà, V., Vespro, V., Zunino, F.: Intraarterial chemotherapy with polyoxyethylated castor oil free paclitaxel, incorporated in albumin nanoparticles (ABI-007): Phase II study of patients with squamous cell carcinoma of the head and neck and anal canal: Preliminary evidence of clinical activity. Cancer **92**, 2592–2602 (2001)
15. Eftaxias, A., Font, J., Fortuny, A., Fabregat, A., Stüber, F.: Nonlinear kinetic parameter estimation using simulated annealing. Comput. Chem. Eng. **26**, 1725–1733 (2002)
16. Fuite, J., Marsh, R., Tuszyński, J.A.: Fractal pharmacokinetics of the drug mibefradil in the liver. Phys. Rev. E Stat. Nonlin. Soft. Matter Phys. **66**, 021904 (2002)
17. Gabrielsson, J., Weiner, D.: Pharmacokinetic and Pharmacodynamic Data Analysis: Concepts and Applications, 2nd edn. Swedish Pharmaceutical Press, Stockholm (1997)
18. Gaudio, E., Chaberek, S., Montella, A., Pannarale, L., Morini, S., Novelli, G., Borghese, F., Conte, D., Ostrowski, K.: Fractal and Fourier analysis of the hepatic sinusoidal network in normal and cirrhotic rat liver. J. Anat. **207**, 107–115 (2005)
19. Gibaldi, M., Perrier, D.: Pharmacokinetics, 2nd edn. Marcel Dekker, New York (1982)
20. Gisiger, T.: Scale invariance in biology: Coincidence or footprint of a universal mechanism? Biol. Rev. Camb. Philos. Soc. **76**, 161–209 (2001)
21. Glenny, R.W., Robertson, H.T.: Fractal properties of pulmonary blood flow: Characterization of spatial heterogeneity. J. Appl. Physiol. **69**, 532–545 (1990)
22. Glenny, R.W., Robertson, H.T.: Fractal modeling of pulmonary blood flow heterogeneity. J. Appl. Physiol. **70**, 1024–1030 (1991)
23. Goffe, W.L., Ferrier, G.D., Rogers, J.: Global optimization of statistical functions with simulated annealing. J. Econometrics **60**, 65–99 (1994)
24. Gough, K., Hutchinson, M., Keene, O., Byrom, B., Ellis, S., Lacey, L., McKellar, J.: Assessment of dose proportionality: Report from the statisticians in the pharmaceutical industry/pharmacokinetics UK joint working party. Drug Inform. J. **29**, 1039–1048 (1995)
25. Havlin, S., Ben-Avraham, D.: Diffusion in disordered media. Adv. Phy. **36**, 695–798 (1987)
26. Heidel, J., Maloney, J.: An analysis of a fractal Michaelis–Menten curve. J. Aust. Math. Soc. Ser. B **41**, 410–422 (2000)
27. Huizing, M.T., Misser, V.H., Pieters, R.C., ten Bokkel Huinink, W.W., Veenhof, C.H., Vermorken, J.B., Pinedo, H.M., Beijnen, J.H.: Taxanes: A new class of antitumor agents. Canc. Invest **13**, 381–404 (1995)
28. Jacquez, J.: Compartmental Analysis in Biology and Medicine, 3rd edn. BioMedware, Ann Arbor MI (1996)

29. Javanaud, C.: The application of a fractal model to the scattering of ultrasound in biological media. J. Acoust. Soc. Am. **86**, 493–496 (1989)
30. Kearns, C.M., Gianni, L., Egorin, M.J.: Paclitaxel pharmacokinetics and pharmacodynamics. Semin. Oncol. **22**, 16–23 (1995)
31. Kirkpatrick, S., Gelatt, C.D. Jr, Vecchi, M.P.: Optimization by simulated annealing. Science **220**, 671–680 (1983)
32. Klymko, P.W., Kopelman, R.: Heterogeneous exciton kinetics: Triplet naphthalene homofusion in an isotopic mixed crystal. J. Phys. Chem. **86**, 3686–3688 (1982)
33. Kopelman, R.: Rate processes on fractals: Theory, simulations, and experiments. J. Stat. Phys. **42**, 185–200 (1986)
34. Kopelman, R.: Fractal reaction kinetics. Science **241**, 1620–1626 (1988)
35. Kosmidis, K., Karalis, V., Argyrakis, P., Macheras, P.: Michaelis–Menten kinetics under spatially constrained conditions: Application to mibefradil pharmacokinetics. Biophys. J. **87**, 1498–1506 (2004)
36. Kuh, H.J., Jang, S.H., Wientjes, M.G., Au, J.L.: Computational model of intracellular pharmacokinetics of paclitaxel. J. Pharmacol. Exp. Ther. **293**, 761–770 (2000)
37. Landaw, E.M., Katz, D.: Comments on mean residence time determination. J. Pharmacokinet. Biopharm. **13**, 543–547 (1985)
38. Levy, R.H.: Time-dependent pharmacokinetics. Pharmacol. Ther. **17**, 383–397 (1982)
39. Levy, R.H., Bauer, L.A.: Basic pharmacokinetics. Ther. Drug Monit. **8**, 47–58 (1986)
40. Lin, J.H.: Dose-dependent pharmacokinetics: Experimental observations and theoretical considerations. Biopharm. Drug Dispos. **15**, 1–31 (1994)
41. López-Quintela, M.A., Casado, J.: Revision of the methodology in enzyme kinetics: A fractal approach. J. Theor. Biol. **139**, 129–139 (1989)
42. Macheras, P.: A fractal approach to heterogeneous drug distribution: Calcium pharmacokinetics. Pharm. Res. **13**, 663–670 (1996)
43. Mandelbrot, B.B.: Fractals: Form, Chance, and Dimension. W.H. Freeman, San Francisco (1977)
44. Mandelbrot, B.B.: The Fractal Geometry of Nature. W.H. Freeman, San Francisco (1982)
45. Marsh, R.E., Tuszyński, J.A.: Fractal Michaelis–Menten kinetics under steady state conditions, application to mibefradil. Pharmaceut. Res. **23**, 2760–2767 (2006)
46. Marsh, R.E., Tuszyński, J.A., Sawyer, M.B., Vos, K.J.E.: Emergence of power laws in the pharmacokinetics of paclitaxel due to competing saturable processes. J. Pharm. Pharm. Sci. **11**, 77–96 (2008)
47. Marsh, R.E., Tuszyński, J.A., Sawyer, M., Vos, K.J.E.: A model of competing saturable kinetic processes with application to the pharmacokinetics of the anticancer drug paclitaxel. Math. Biosci. Eng. **8**, 325–354 (2011)
48. Marshall, J.H.: Calcium pools and the power function. In: Bergner, P.E., Lushbaugh, C.C. (eds.) Compartments, Pools, and Spaces in Medical Physiology. USAEC Division of Technical Information, Oak Ridge TN (1967)
49. McLeod, H.L., Kearns, C.M., Kuhn, J.G., Bruno, R.: Evaluation of the linearity of docetaxel pharmacokinetics. Canc. Chemother. Pharmacol. **42**, 155–159 (1998)
50. Newhouse, J.S., Kopelman, R.: Reaction kinetics on clusters and islands. J. Chem. Phys. **85**, 6804–6806 (1986)
51. Norris, W.P., Tyler, S.A., Brues, A.M.: Retention of radioactive bone-seekers. Science **128**, 456–462 (1958)
52. Norwich, K.H., Siu, S.: Power functions in physiology and pharmacology. J. Theor. Biol. **95**, 387–398 (1982)
53. Ogihara, T., Tamai, I., Tsuji, A.: Application of fractal kinetics for carrier-mediated transport of drugs across intestinal epithelial membrane. Pharm. Res. **15**, 620–625 (1998)
54. Pazdur, R., Kudelka, A.P., Kavanagh, J.J., Cohen, P.R., Raber, M.N.: The taxoids: Paclitaxel (Taxol) and docetaxel (Taxotere). Canc. Treat. Rev. **19**, 351–386 (1993)

55. Riccardi, A., Servidei, T., Tornesello, A., Puggioni, P., Mastrangelo, S., Rumi, C., Riccardi, R.: Cytotoxicity of paclitaxel and docetaxel in human neuroblastoma cell lines. Eur. J. Canc. **31A**, 494–499 (1995)
56. Ridgway, D., Tuszyński, J.A., Tam, Y.K.: Reassessing Models of Hepatic Extraction. J. Biol. Phys. **29**, 1–21 (2003)
57. Savageau, M.A.: Development of fractal kinetic theory for enzyme-catalysed reactions and implications for the design of biochemical pathways. BioSystems **47**, 9–36 (1998)
58. Skerjanec, A., Tawfik, S., Tam, Y.K.: Mechanisms of nonlinear pharmacokinetics of mibefradil in chronically instrumented dogs. J. Pharmacol. Exp. Ther. **278**, 817–825 (1996)
59. Sparreboom, A., van Tellingen, O., Nooijen, W.J., Beijnen, J.H.: Nonlinear pharmacokinetics of paclitaxel in mice results from the pharmaceutical vehicle Cremophor EL. Canc. Res. **56**, 2112–2115 (1996)
60. Vaishampayan, U., Parchment, R.E., Jasti, B.R., Hussain, M.: Taxanes: An overview of the pharmacokinetics and pharmacodynamics. Urology **54**, 22–29 (1999)
61. Weiss, M.: Use of gamma distributed residence times in pharmacokinetics. Eur. J. Clin. Pharmacol. **25**, 695–702 (1983)
62. Weiss, M.: Importance of tissue distribution in determining drug disposition curves. J. Theor. Biol. **103**, 649–52 (1983)
63. Weiss, M.: A note on the interpretation of tracer dispersion in the liver. J. Theor. Biol. **184**, 1–6 (1997)
64. Wise, M.E.: The evidence against compartments. Biometrics **27**, 262 (1971)
65. Wise, M.E.: Interpreting both short- and long-term power laws in physiological clearance curves. Math. Biosci. **20**, 327–337 (1974)
66. Wise, M.E.: Negative power functions of time in pharmacokinetics and their implications. J. Pharmacokinet. Biopharm. **13**, 309–346 (1985)
67. Wise, M.E., Osborn, S.B., Anderson, J., Tomlinson, R.W.S.: A stochastic model for turnover of radiocalcium based on the observed power laws. Math. Biosci. **2**, 199–224 (1968)
68. van Zuylen, L., Gianni, L., Verweij, J., Mross, K., Brouwer, E., Loos, W.J., Sparreboom, A.: Inter-relationships of paclitaxel disposition, infusion duration and cremophor EL kinetics in cancer patients. Anticancer Drugs **11**, 331–337 (2000)

A Mathematical Model of Gene Therapy for the Treatment of Cancer

Alexei Tsygvintsev, Simeone Marino, and Denise E. Kirschner

1 Introduction

Cancer is a major cause of death worldwide, resulting from the uncontrolled growth of abnormal cells in the body. Cells are the body's building blocks, and cancer starts from normal cells. Normal cells divide to grow in order to maintain cell population equilibrium, balancing cell death. Cancer occurs when unbounded growth of cells in the body happens fast. It can also occur when cells lose their ability to die. There are many different kinds of cancers, which can develop in almost any organ or tissue, such as lung, colon, breast, skin, bones, or nerve tissue. There are many known causes of cancers that have been documented to date including exposure to chemicals, drinking excess alcohol, excessive sunlight exposure, and genetic differences, to name a few [38]. However, the cause of many cancers still remains unknown. The most common cause of cancer-related death is lung cancer. Some cancers are more common in certain parts of the world. For example, in Japan, there are many cases of stomach cancer, but in the USA, this type of cancer is pretty rare [49]. Differences in diet may play a role. Another hypothesis is that these different populations could have different genetic backgrounds predisposing them to cancer. Some cancers also prey on individuals who are either missing or have altered genes as compared to the mainstream population. Unfortunately, treatment

A. Tsygvintsev (✉)
UMPA, ENS de Lyon, 46, allée d'Italie, 69364 Lyon Cedex 07, France
e-mail: alexei.tsygvintsev@ens-lyon.fr

S. Marino • D.E. Kirschner
Department of Microbiology and Immunology, University of Michigan Medical School, 6730 Med. Sci. Bldg. II, Ann Arbor, MI 48109-0620, USA
e-mail: simeonem@umich.edu; kirschne@umich.edu

U. Ledzewicz et al., *Mathematical Methods and Models in Biomedicine*, Lecture Notes on Mathematical Modelling in the Life Sciences, DOI 10.1007/978-1-4614-4178-6_13,

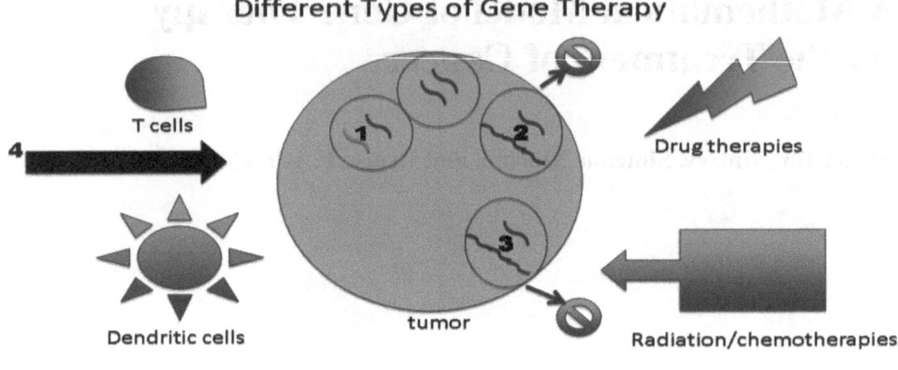

Fig. 1 Gene therapy and immunotherapy treatments. As denoted by the numbers in the figure: (1) Replace missing or mutated genes with healthy genes. (2) Insert genes into tumors that act like suicide bombs once they are turned on by drugs (3) Insert genes that make tumors more susceptible to treatments such as chemotherapy and radiotherapy. (4) Augment the immune response to cancers by enhancing the ability of immune cells, such as T cells and dendritic cells, to fight cancer cells

of cancer is still in its infancy, although there are some successes when the cancer is detected early enough. To begin to address these important issues, in this work we will focus solely on genetic issues related to cancer so that we can explore a new treatment area known as gene therapy as a viable approach to treatment of cancer.

Genes are located on chromosomes inside all of our cells and are made of DNA. Humans have approximately 35,000 genes. Gene therapy is an experimental treatment currently being tested in clinical trials that involves introducing additional genetic material (either DNA or RNA) into cells to fight cancer in a few different ways. There are several gene therapy approaches that are being explored. First, scientists are attempting to use gene therapy to replace missing or mutated genes with healthy genes (for example, p. 53, [14]). Second, scientists are attempting to put genes into tumors that act like suicide bombs once they are turned on by drugs that are administered to the patient [53]. Similar to the suicide genes, a third approach is to insert genes that make tumors more susceptible to treatments such as chemotherapy and radiotherapy. And finally, gene therapy is being used to improve the immune response to cancers by enhancing the ability of immune cells, such as T cells, to fight cancer cells [42]. Figure 1 summarizes these different gene therapies. Not surprising, gene therapy research has continued to include other genetics manipulations of gene expression through delivery of modified genes, short pieces of RNA known as siRNA [5], naked plasmid DNA, and even viruses as vectors for delivery of genetic material into cells. To reduce the risks of side effects, specific tissues and cell types must be targeted. Not only is the type of gene therapy best suited for each case not known a priori, the choice of which gene to target, the dose and timing of treatments all must be determined. A recent review summarizes advances in gene therapy and also highlights opportunities for systems biology and mathematical modeling to synergize efforts with experimentalists and clinicians to push cancer research forward [23].

Mathematical modeling has been instrumental in the past 50 years in helping decipher different aspects of complex systems in biology. In particular, mathematical modeling has had an impact on our understanding of cancer biology and treatment (cf. [4, 24, 25] for excellent reviews). We begin by briefly reviewing existing models designed specifically for capturing tumor-immune dynamics, one of which forms the basis for our current work. As a first step to exploring the use of gene therapy on the tumor–immune interaction during cancer, we will apply a simple mathematical model to explore the dynamics of these different types of gene therapies, with the goal of predicting optimal combinations of approaches leading to clearance of a tumor. We present the model and its analysis (both analytical and numerical) and offer some conclusions.

2 Brief Review of Mathematical Models Describing Tumor-Immune Dynamics

For the past 40 years, mathematical models have been developed describing many aspects of cancer from tumor growth dynamics (cf. [11,22,32]), angiogenesis, and vascularization (cf. [36,41,52]), to the topic of immune response to tumors. Since the work herein will be focused solely on tumor-immune dynamics, we briefly review work in this area.

Tumor-immune models have been around since the early 1990s and have evolved to capture very complex aspects of the immune response as knowledge of the molecular dynamics of immunity has increased. For example, an important aspect of immunity is the recognition of non-self, or foreign antigens. Specialized antigen presenting cells (known as macrophages and dendritic cells) present foreign antigen to immune cells, such as T cells, to train them to respond and clear the foreign material (like bacteria and viruses). Of course, since tumor cells began as self, or non-foreign host cells, the level of antigenicity of the tumor may be weak as the human immune system is trained to not kill self. Mathematical models of tumor-immune interactions that have explored dynamics at this scale are [27,34,40,47,51]. Recently, Joshi et al. [26] develop a new mathematical model to capture immunotherapy that involves the antigen presentation pathway and its role in tumor–immune dynamics. Other models focus only on therapy as well as on boosting immunity [3,6,8–10,16–18,21,30,48]. Immune competition models have been studied [15, 34, 35], which focus on the dynamics between host immune cells and tumors. These types of studies have their origin in the Lotka–Volterra models established almost 100 years ago.

2.1 Lotka–Volterra Models for Tumor-Immune Interactions

The idea to use the qualitative theory of ordinary differential equations in mathematical biology reaches back to 1920s when Lotka and Volterra formulated a simple mathematical model in population dynamics theory. A good summary published in 1997 by Adam and Bellomo [1] presents a summary of early work regarding this approach to tumor-immune dynamics, and much of the original work on this was done by Kutznetsov [31] and colleagues. We review it briefly.

Let $y(t)$ be the population of predator and $x(t)$ is population of its prey (for example, one can imagine populations of wolves and rabbits in a forest). Assuming that numbers $x(t), y(t)$ are big enough and that the predator and prey populations are homogeneous, one can view them as continuous functions of time. Let $\Delta x(t) = x(t + \Delta t) - x(t)$ and $\Delta y(t) = y(t + \Delta t) - y(t)$ be small variations of populations during a certain period of time Δt. Taking $\Delta t = 1$ (for example 1 day) one can replace $\Delta x(t)$, $\Delta y(t)$ by their derivatives, i.e. write $\dot{x}(t), \dot{y}(t)$ instead. The Lotka–Volterra equations are given by

$$\begin{cases} \dot{x} = ax - bxy, \\ \dot{y} = -cy + dyx, \end{cases} \tag{1}$$

where a, b, c, d are some positive numbers.

The linear positive term ax in the first equation (prey) corresponds to exponential growth; the negative predation term, $-bxy$, describes the rate prey are lost and is proportional to number of prey and predators in mass action form. In the second equation (predator), the negative linear term $-cy$ corresponds to natural death, as prey will not survive without prey, and $+dxy$ describes the growth of the predator population proportional to prey and number of predators. The simple form of Lotka–Volterra (LV) system is remarkable. It allows for investigation of the quantitative and qualitative behavior for all of its solutions both analytically and numerically. First, no chaotic behavior is possible according to Poincaré-Bendixon theorem, and, asymptotically, every non-periodic solution either goes to a fixed point or approaches a limit cycle. Simple analysis shows that most solutions of LV system are periodic, i.e. the population numbers $x(t), y(t)$ are oscillating around a certain equilibrium state $x(t) = x^*, y(t) = y^*$. The stable, stationary solution is (0,0).

In 1994 Kuznetsov et al. [31] applied Lotka–Volterra ideas to cancer modeling, where $E(t)$ represents the effector immune cells (predators) and $T(t)$ the tumor cells (prey). The equations, which are similar to the LV system, are written as follows:

$$\begin{cases} \dot{E} = s + p\dfrac{ET}{g+T} - mET - dE, \\ \\ \dot{T} = aT(1-bT) - nET, \end{cases} \tag{2}$$

where s,p,g,m,d,a,n,b are positive parameters.

Here the exponential growth of T in the second equation was replaced by a more realistic one in logistic form: $aT(1-bT)$ (originally due to Verhulst, 1838), where b^{-1} is the maximal carrying capacity for tumor cells and a is the maximal growth rate. The term $-nET$ describes the loss of tumor cells due to the presence of immune cells. In the first equation s is normal immune cell growth, which is n cell death with d the loss rate; $-mET$ describes the decay of E cells due to interacting with tumor cells in a mass action way. The term $p\frac{ET}{g+T}$ represents Mchaelis–Meten growth of the immune response in response to tumors.

The Kuznetsov equations describe several important features and allow us to make predictions that are relevant for understanding cancer immunotherapy. The paper by Kuznetsov et al. [31] establishes existence of long period oscillations of tumor that agrees with recurrent clinical manifestations of certain human leukemias. In addition, the model predicts the existence of a critical level of E-cells in the body below which the tumor growth cannot be controlled by the immune response. It describes qualitatively the "escape" phenomena in which low doses of tumor cells can escape immune defenses and grow into a large tumor, whereas larger doses of tumor cells are eliminated.

The Kuznetsov model was generalized by Kirschner and Panetta in 1998 [28]. The idea was to introduce a third population (concentration) of effector molecules known as *cytokines*, which are information signaling molecules used extensively in intercellular communication by the immune system. Below we describe briefly the Kirschner–Panetta equations. Tumor cells are tracked as a continuous variable as they are large in number and are generally homogeneous; their concentration is denoted by $T(t)$. Immune cells (called effector cells) are also large in number and represent those cells that have been stimulated and are ready to respond to the foreign matter (known as antigen); their concentration is denoted by $E(t)$. Finally, effector molecules are represented as a concentration $C(t)$. These are self-stimulating, positive feedback proteins for effector cells that produce them. The equations that describe the interactions of these three state variables are referred herein as the Kirschner–Panetta (KP) system:

$$\begin{cases} \dfrac{dE}{dt} = cT - \mu_2 E + \dfrac{p_1 EC}{g_1 + C} + s_1 & \text{(3a)} \\ \dfrac{dT}{dt} = r_2 T(1-bT) - \dfrac{aET}{g_2 + T} & \text{(3b)} \\ \dfrac{dC}{dt} = \dfrac{p_2 ET}{g_3 + T} + s_2 - \mu_3 C & \text{(3c)} \end{cases}$$

In Eq. (3a), the first term represents stimulation by the tumor to generate effector immune cells. The parameter c is known as the antigenicity of the tumor. Since tumor cells begin as self, c represents how different the tumor cells are from self cells (i.e., how foreign). The second term in Eq. (3a) represents natural death and the third is the proliferative enhancement effect of the cytokine IL-2. In Eq. (3b), the first term is a logistic growth term for tumor growth and the second is a clearance term by the immune effector cells. In the final Eq. (3c), IL-2 is produced by effector cells (in a Michaelis–Menten fashion) and decays via a known half-life (third term).

To capture a novel treatment approach (still in use in some clinical settings), KP introduced three terms into their models. The first one is *Adoptive cellular immunotherapy* (ACI), representing the introduction of immune cells into cancer patients that have been stimulated to have specific anti-tumor activity [42, 44–46]. T cells, also known as lymphocytes, produce cytokines that are either self-stimulating or can stimulate (or shut down) other cells. ACI is usually performed in conjunction with large amounts of IL-2. There are two types of immune cells that are cultured for this purpose: (1) LAK-(lymphokine-activated killer cells): cells taken from host and then stimulated with activating factors. These cells are then injected back to patient. (2) TIL-(tumor infiltrating lymphocytes): Immune cells are taken from patient, and grown with high concentrations of IL-2 before injected back to the patient.

In the KP model, s_1 represents the treatment terms of introducing LAK and TIL cells to the tumor site of a patient. The second term, s_2, is a treatment term that represents administration of the cytokine IL-2 by a physician to a patient, to again stimulate effector cell growth and proliferation.

The KP system can exhibit chaotic behavior. The typical example of chaotic behavior is the system of Lorenz (1963) representing a so called strange attractor. The complete qualitative analysis of KP equations is much harder than the conventional Kuznetsov model. Nevertheless, Kirschner and Panetta, using stability analysis and modern bifurcation theory, classified representative behaviors of solutions and stability of cancer-free equilibrium states. Description of oscillations with long time dormant periods of illness was described in order to complete the studies by Kuznetsov.

Arciero and colleagues [5] extended the KP equation by including a suppressive cytokines known as TGF-β and also a simple type of gene therapy known as siRNA [39]. The use of siRNA is an early type of gene therapy where short-interfering RNA fragments interfere with the expression of a specific gene and modify behaviors in a cell. In addition, they added TGF-β to the model, which is a cytokine that acts to suppress immunity by inhibiting activation of effector cells and reducing antigenicity of tumors. It also stimulates tumor growth by promoting tumor vascularization. Their model predicts that increasing the rate of TGF-β production for reasonable values of tumor antigenicity enhanced tumor growth and its ability to escape host detection. siRNA treatment focused on the gene expression for TGF-β: it acts to suppress TGF-β production by targeting the messenger RNA that codes for TGF-β. Reducing TGF-β helped to rescue these negative effects to the host. Another group also recently explored the development of a microRNA-mRNA for the purpose of gene network regulation in tumors [2]. In an additional paper, Burden and colleagues [12] explored optimal control methods for determining the best treatment strategy based on the KP equations. They designed a control functional to maximize numbers of effector cells and interleukin-2 concentration while minimizing numbers of tumor cells. In 2008, Banerjee [7] proposed a delay version of KP equations where the Eq. (3a) was replaced by a new one:

$$\frac{\mathrm{d}E(t)}{\mathrm{d}t} = cT(t) - \mu_2 E(t) + \frac{p_1 E(t-\tau)C(t-\tau)}{g_1 + C(t)} + s_1 \tag{4}$$

with the rest of the KP system [Eq. (3)] remaining the same. The introduction of a time delay, $\tau > 0$, corresponds to the delay that occurs between the production of a cytokine production, and its downstream binding and activation action on host effector cells. In that work, Banerjee analyzed the local stability of the cancer free equilibrium in the presence of the delay using semi-numerical bifurcation methods.

All of the above applications of dynamical system theory were studied using a similar approach: investigation of local stability of solutions by linear approximation (i.e., nonlinear equations are replaced by linear ones in a suitable regions of phase space). However, non-linear phenomena have a much greater complexity and require analysis on a global level. Very few generalized methods have been developed; as such, usually each nonlinear system must be studied individually. The main difficulty is the presence of free, non-numerical parameters in the system as clearly exemplified in the KP equations. Parameters (13 of them for KP model) such as antigenicity, c, or maximal growth ratio, a, are not known experimentally. These parameters are mostly composite parameters that phenomenologically represent a set of biological mechanisms in a simple way. To describe different qualitative scenarios of the model when performing a stability analysis without using numerical values for the model parameters is a critical task. Thus, we consider the pairing of numerical and analytical methods as the best approach to gain as much information as possible. Finally, In 2009, Kirschner and Tsygvintsev [29] performed a global analysis of the KP system using the generalized Lyapunov method. They derived sufficient conditions that guarantee asymptotic convergence as $t \to +\infty$ of $T(t)$ to 0. For a "virtual" patient, that would imply complete clearance of cancer once a corresponding therapy is adopted. Another result of [29] was to analytically prove the existence of host self-regulation of cytokine levels that never exceed certain critical values. See also [13, 20] for further discussion.

3 A Gene Therapy Model

The problem with the use of LAK and TIL cells as described above is that only about half of the TILs that are typically generated are reactive to tumors [50]. Thus, the ability to genetically engineer TIL cells that are directed against tumor-specific antigens is a key objective. Recently, this was attempted in a small clinical trial [43]

Table 1 Parameter values for the model (5)

Name	Definition	Baseline (units)	Range
μ_2	Half-life of effector cells E	0.03 (1/time)	0.03
p_3	Proliferation rate of E	0.1245 (1/time)	0.1245
f	Half-sat for E proliferation term	10^{-3} (cells)	$[10^{-5}, 1]$
$s_1(t)$	Immunotherapy term	1 (cells/time)	$[10^{-2}, 10^2]$
$c(t)$	Cancer antigenicity	0.05 (1/time)	$[10^{-3}, 0.5]$
$r_2(t)$	Cancer growth rate	0.18 (1/time)	$[10^{-1}, 2]$
b	Cancer cell capacity (logistic growth)	10^{-9} (1/cells)	10^{-9}
$a(t)$	Cancer clearance term	1 (1/time)	$[10^{-2}, 10^2]$
g_2	Half-saturation, for cancer clearance	10^5 (cells)	10^5

and a small percentage of patients had complete tumor regression. In this study, Rosenberg and colleagues took a blood sample from each patient and transferred genes into T cells inducing each cell to produce specialized T-cell receptors (TCR). These cells are then transferred back into the patient. In the body, T cells produce TCRs on their outer membrane and the TCRs recognize and attach to certain molecules found on the surface of the tumor cells. Finally, signaling through the TCRs activates T cells to attack and kill the tumor cells. To explore these studies further we will build on the KP model.

First, to simplify the model we can remove the IL-2 Eq. (3c). We replace the IL-2 saturation term in Eq. (3a) with a self-proliferation term, i.e. $p_1E/(E+f)$. The idea that the proliferation rate of effectors may be a decreasing function of effectors has been explored by d'Onofrio et al. [19]. To capture the effects of gene therapy (see Fig. 1) we must allow for the immune parameters of the model, i.e. a and c, to be step functions. Antigenicity, c, will signal stronger to the immune system during gene therapy and the clearance of tumor cells, a, will be strongly enhanced after gene therapy. Finally the source term representing TIL cells, $s_1(t)$ should be time dependent. We can also combine this with a self-limiting gene therapy treatment for tumors, which affects the growth rate of the tumor, r_2, by allowing it to be a step function that decreases its growth rate. The new equations are:

$$\begin{cases} \dot{E} = c(t)T - \mu_2 E + p_3 \dfrac{E}{E+f} + s_1(t), \\ \\ \dot{T} = r_2(t)T(1-bT) - a(t)\dfrac{ET}{T+g_2}. \end{cases} \tag{5}$$

It is this model that we analyze both analytically and numerically in the next sections. We define the parameters of system (5), their values as well as their ranges of variation in Table 1. They are mostly based on previously published data (cf.,[5, 7]).

4 Stability Conditions for Nonautonomous Gene Therapy Model

In this section we derive conditions for global stability of the cancer free state ($T = 0$) for the Gene Therapy model (5). First, we investigate the second equation of system (5) independently from the first equation. Thus, we consider $r_2(t)$, $a(t)$ and $e(t) = E(t)$ as *arbitrary* positive data functions:

$$\dot{T} = r_2(t)T(1-bT) - a(t)\frac{e(t)T}{T+g_2}. \tag{6}$$

The only biological plausible solutions should satisfy the condition $T(t) \in [0, b^{-1}]$. Moreover, as easily seen from the second equation of system (5), the interval $[0, b^{-1}]$ is dynamically invariant under the flow. Our first aim is to derive conditions on the functions $r_2(t), a(t)$ and $e(t)$, which would imply asymptotical global stability of the cancer free equilibrium state $T = 0$.

Theorem 4.1. *Let one of the following two conditions holds*

Condition 1: *There exist $t_0 > 0$ and $\varepsilon > 0$ such that*

$$\frac{a(t)e(t)}{r_2(t)} > g_2 + \frac{(1-bg_2)^2}{4b} + \frac{\varepsilon}{r_2(t)}, \quad \forall t \geq t_0 \tag{7}$$

or

Condition 2: $g_2 > b^{-1}$ *and there exist $t_0 > 0$ and $\varepsilon > 0$ such that*

$$\frac{a(t)e(t)}{r_2(t)} > g_2 + \frac{\varepsilon}{r_2(t)}, \quad \forall t \geq t_0. \tag{8}$$

Then every solution of Eq. (6) satisfies $\lim_{t\to+\infty} T(t) = 0$ with exponential convergence.

Proof. We write Eq. (6) in the form:

$$\dot{T} = \frac{r_2(t)T}{T+g_2}V(T), \tag{9}$$

where V is a quadratic polynomial with respect to T given by:

$$V(T) = -bT^2 + (1-bg_2)T + g_2 - \frac{a(t)e(t)}{r_2(t)}, \tag{10}$$

with discriminant D as follows:

$$D = (1-bg_2)^2 + 4b\left(g_2 - \frac{a(t)e(t)}{r_2(t)}\right). \tag{11}$$

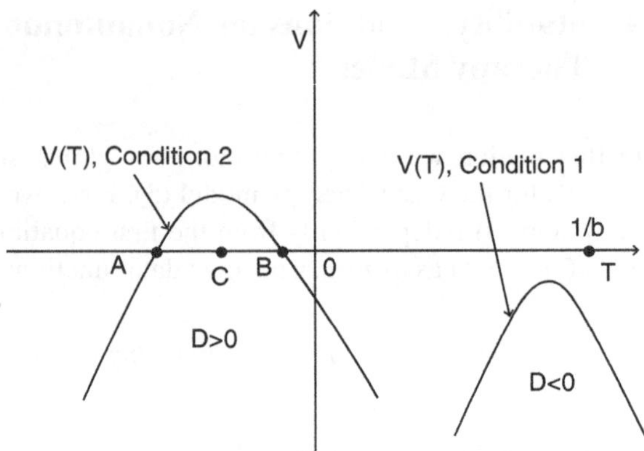

Fig. 2 The quadratic function $V(T)$

Condition 1 is equivalent to $D < -4b\varepsilon/r_2(t)$. Thus,

$$r_2(t)V(T) < -\varepsilon, \tag{12}$$

for all T, since the leading coefficient of $V(T)$ is negative (i.e., $-b < 0$, see Fig. 2). The Eq. (9) can be written, for every fixed solution $T(t)$, in the form:

$$\dot{T} = -\delta_T(t)T, \tag{13}$$

where $\delta_T(t) = -r_2(t)V(T)/(T+g_2)$. Since $T(t)$ is bounded above by b^{-1} and because of Eq. (12), the inequality $\delta_T(t) > \delta_0 > 0$ will hold starting from a certain moment. That completes the proof of the exponential convergence of $T(t)$ to 0 as $t \to +\infty$.

Now, let us assume that Eq. (10) have two real roots A and B, $A < B$. The quadratic polynomial function $V(T)$ has unique extremum given by $C = \frac{(1-bg_2)}{2b} < b^{-1}$, corresponding to maximal value of $V(T)$. Also, $V(b^{-1}) = -\frac{a(t)e(t)}{r_2(t)} < 0$. Both roots are negative, i.e. $A, B < 0$ if and only if $C < 0$ and $V(0) < 0$. We assume that $r_2(t)V(0) < -\varepsilon$ for certain positive ε. The same arguments of Condition 1 can be applied. □

We note that Conditions 1 and 2 are also necessary for *global* stability of the state $T = 0$: if both are not satisfied, it is always possible to choose parameter values that return a solution for which $T(t)$ is not converging to 0. As it follows from the analysis above, key parameters governing the stability of cancer free equilibrium state $T = 0$ are included in the function:

$$S(t) = \frac{a(t)E(t)}{r_2(t)}. \tag{14}$$

In order to stabilize or completely eliminate the cancer, we suggest the choice of functions in system (5) that force $S(t)$ to be uniformly bounded from zero for all values of time. Here, different paths can be proposed. We can adjust the external source of effector cells $s_1(t)$ every time $S(t)$ starts to decrease or, alternatively, the functions $r_2(t), a(t)$ can be made E-dependent in a way that the stability condition $S(t) > S_0$ holds until complete eradication of the tumor is achieved. Below we propose conditions which do not involve effector cells, $E(t)$ explicitly. In the next theorem, T_0 plays a key role: if T falls below T_0, the cancer is assumed cleared.

Theorem 4.2 (Main Stability Theorem). *Let the following condition be satisfied for all $t \geq t_0$ with some constants $t_0 \geq 0$, $\varepsilon > 0$, $\sigma > 0$, $\beta > 0$ and $T_0 \in (0, b^{-1})$*

$$\begin{cases} \mu_2(\varepsilon + \theta r_2(t))^2 + (\mu_2 f - p_3 - s_1(t) - c(t)T_0)(\varepsilon + \theta r_2(t))a(t) \\ \qquad\qquad -(s_1(t) + fc(t)T_0 - \beta)a^2(t) < 0 \\ \dfrac{\varepsilon + \theta r_2(t)}{a(t)} \quad \text{is a nonincreasing function of time} \\ fc(t)T_0 + s_1(t) > \sigma > 0 \end{cases} \tag{15}$$

where

Case a: $\quad \theta = g_2 + \dfrac{(1 - bg_2)^2}{4b}$

or

Case b: $\quad \theta = g_2$ *and* $g_2 > b^{-1}$

The following statements hold:

Case I (partial clearance). For every solution $(T(t), E(t))$ of Eq. (5) given by initial condition $(T(t_1), E(t_1))$, $t_1 \geq t_0$ with $T(t_1) > T_0$ the function $T(t)$ will reach in finite time the value T_0.

Case II (complete clearance). If condition (15) is satisfied with $T_0 = 0$, then for all solutions of the system (5)

$$\lim_{t \to +\infty} T(t) = 0. \tag{16}$$

Proof. We write the first equation of the system (5) as follows

$$\dot{E} = \frac{K(E)}{E + f}, \tag{17}$$

where K is quadratic polynomial with respect to E given by

$$K = \tilde{\alpha}E^2 + \tilde{\beta}E + \tilde{\gamma} \tag{18}$$

and $\tilde{\alpha} = -\mu_2$, $\tilde{\beta} = c(t)T(t) + p_3 + s_1(t) - \mu_2 f$, $\tilde{\gamma} = fc(t)T(t) + fs_1(t)$.

The conditions (7) and (8) can be all expressed, for suitable real positive number θ in the following form

$$E(t) - h(t) > 0 \tag{19}$$

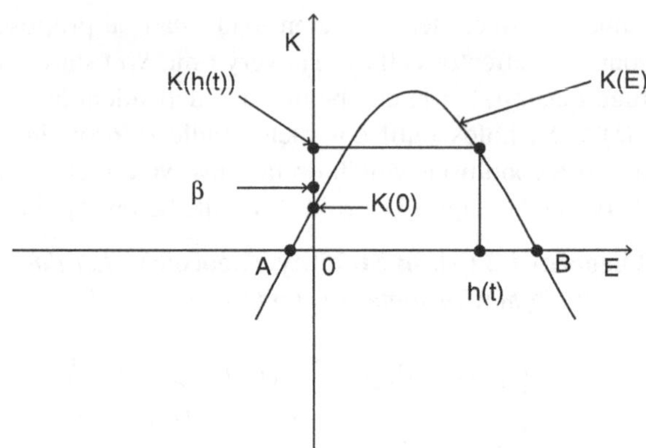

Fig. 3 The quadratic function $K(E)$

with

$$h(t) = \frac{\varepsilon + \theta r_2(t)}{a(t)}. \tag{20}$$

Checking the discriminant of quadratic polynomial $K(E)$ one proves that $K(E) = 0$ has always two real roots $A < 0$, $B > 0$ of opposite signs such that $\tilde{\gamma} > 0$. Since the leading coefficient $\tilde{\alpha} < 0$ the quadratic polynomial $K(E)$ is positive in the interval (A,B) and negative outside. The first inequality of Eq. (15), in the case $T > T_0$, is equivalent to condition $K(h(t)) > \beta \Rightarrow h(t) \in [0,B)$ (see Fig. 3).

At the same time $K(0) = \tilde{\gamma} = fc(t)T(t) + s_1(t)f > \sigma$. This implies that one will have $\dot{E}(t) > \beta_0/(E+f)$ with

$$\beta_0 = \min(\sigma, \beta) \tag{21}$$

once the inequality (19) is violated, forcing $E(t)$ to increase. Since $h(t)$ is non-increasing function, the inequality (19) will be satisfied for certain $t = t^*$ and will hold then for all $t \geq t^*$. As follows from the proof of Theorem 4.1 and Eq. (13), for $t \geq t^*$ the function $T(t)$ will decrease till it takes the value $T = T_0$. If $T_0 = 0$ (Case II) then one uses Theorem 4.1 again to derive Eq. (16). □

One can interpret the significance of the inequality $fc(t)T_0 + s_1(t) > \sigma > 0$ in Eq. (15) as follows. Once antigenicity is switched off, i.e. $c(t) = 0$, treatment, $s_1(t)$, should be nonzero and vice versa. In the case of complete tumor clearance ($T_0 = 0$), the treatment term should be always positive above a certain level. Indeed, partial clearance does not exclude future regressing via the "escape" effect.

5 Numerical Simulations

Because the analytical results hold only for very small regions of parameter space,we would like to explore the gene therapy model more fully. To this end, we will apply statistical sampling techniques and numerical analysis to the system.

Sufficient conditions (15) of the Main Stability Theorem 4.2 imply large ranges for the four *treatment parameters* $c(t)$, $s_1(t)$, $a(t)$, $r_2(t)$, which are directly related to the four treatment strategies in Fig. 1. The values of initial conditions have been varied between 1 and 10^5 for the populations of effector and tumor cells, $E(0)$ and $T(0)$. Global stability conditions (15) hold for any initial conditions (we use $E(0) = C(0) = 10^3$ for our baseline run). Sufficient conditions (15) are tested numerically by solving system (5) in Matlab using *ode15s* (a solver for stiff systems). Since conditions (15) are sufficient, we combine techniques from uncertainty and sensitivity analysis (see [33] for a review) to efficiently and comprehensively investigate treatment combinations and how they might affect cancer progression.

Regions of the parameter space where cancer is cleared are searched by sampling the parameter space in the ranges defined in Table 1. We only vary the four *treatment parameters*, while all others are kept constant at their baseline values (see *Baseline* column in Table 1). Samples are generated from uniform distributions and the sampling scheme used is known as Latin hypercube sampling (LHS) [37]. LHS scheme comprises three main steps: (1) definition of probability density functions to use as a priori distributions for the parameters under analysis, (2) number N of samples to perform and (3) independent sampling of each parameter. The last step assumes that each parameter distribution is divided into N subintervals of equal probability and that the sampling is preformed without replacement. The accuracy of LHS is comparable to simple random sampling schemes but more efficient (i.e., with a significant reduction in the number of samples needed). In our study we use a sample size of 10,000 and tested numerically the impact of combining only constant treatment strategies, although conditions (15) are also valid for time-varying inputs.

5.1 Sensitivity Analysis as a Way to Determine Optimal Parameters for Treatment

In conjunction with uncertainty analysis, we use a generalized correlation coefficient (partial rank correlation coefficient, PRCC) to guide our understanding of which treatment parameter(s) contribute most to drive cancer proliferation or clearance (our model outcomes). PRCC is one of the most popular sensitivity indexes used for the analysis of deterministic models [33]. PRCCs results can be interpreted as a degree of correlation between input and output variability: PRCCs vary between -1 and 1 and can be applied to any nonlinear monotonic relationship. A test of significance is also available: only PRCCs that are significantly different from zero are shown in this study. In order to select an optimal combination of treatments, a pairwise comparison between PRCCs has been performed by a generalized z-test (see p. 183 in [33]) and a ranking of the treatments is generated. Uncertainty and sensitivity analysis results are shown in Table 2. We review these techniques and

Table 2 Uncertainty and sensitivity analysis results

Name	Definition	PRCC	Ranking
$a(t)$	Cancer clearance term	−0.1993	1st
$s_1(t)$	Immunotherapy term	−0.1061	2nd
$c(t)$	Cancer antigenicity	−0.0814	3rd
$r_2(t)$	Cancer growth rate	−0.0791	3rd

PRCC values significantly different from 0 (with $p < 0.01$) PRCCs ranking based on generalized z-test ($p < 0.05$)

others in [33]; our Matlab scripts to perform LHS, PRCC as well as other uncertainty and sensitivity analysis techniques are available online at http://malthus.micro.med.umich.edu/lab/usanalysis.html.

Table 2 shows how all four parameters have PRCCs that are statistically different from zero (with $p < 0.01$): not surprisingly they are all negatively correlated with cancer cell count (i.e., increasing their values from the baseline, decreases cancer cell count). Two treatment parameters, $a(t)$ and $s_1(t)$ (cancer clearance and immunotherapy terms, respectively), have the highest impact on reducing cancer cells. The other two parameters (i.e., $c(t)$ and $r_2(t)$) have similar PRCCs (they share the same ranking since they are not statistically different from each other), so they are equally effective in reducing cancer cell count. Figure 4 shows scatter plots of parameters versus cancer-cell counts, resulting from our extensive uncertainty analysis with an LHS scheme of 10,000 samples. We classify the outputs in four groups: complete clearance (green dots, no cancer cells), partial clearance (blue dots, cancer cell count below the initial condition $T(0) = 10^3$), small growth (red dots, cancer cell count between 10^3 and 10^6), and large growth (black dots, cancer cell count above 10^6). Cases where conditions (15) are satisfied are included in the "green" region of Fig. 4. There is clearly a synergy between immunotherapy and cancer clearance terms ($s_1(t)$ and $a(t)$): both must be large to achieve complete clearance (green). High values for $s_1(t)$ or $a(t)$ are always associated with lower cancer cell counts, but no correlation can be inferred between either of these two parameters and antigenicity ($c(t)$) or cancer growth rate ($r_2(t)$). Below we show two examples of numerical simulations leading to complete clearance, when conditions (15) are either satisfied or not (Fig. 5). Clearance is usually achieved fast when conditions (15) are satisfied.

6 Conclusion and Discussion

Using mathematical models to explore important problems in biology is an ever-increasing tool towards shedding light on these complex systems. Cancer modeling has had a recent and successful history of making predictions that can assist in hypothesis generation leading to experimental and perhaps clinical verification. For example, gene therapy is a relatively young idea in treatment of diseases,

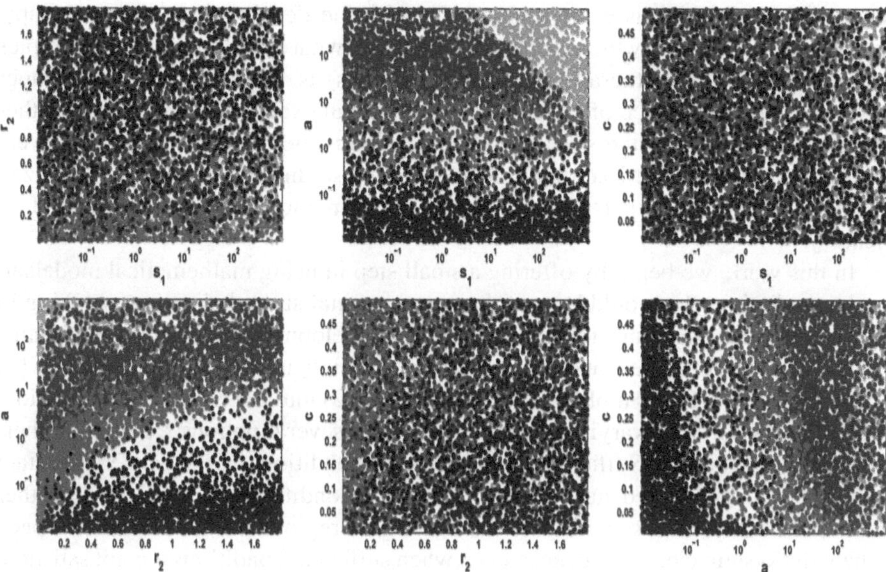

Fig. 4 Treatment combinations. Scatter plot of the numerical solution of system (5). The values plotted correspond to the cancer cell counts at day 2000 and they are classified by color: *green*-clearance (no cancer), *blue*-partial clearance (cancer below the initial condition of 10^3), *red*-small growth (from 10^3 to 10^6), *black*-large growth (above 10^6). The x and y axes represent the six combinations of the four *treatment parameters* $c(t)$, $s_1(t)$, $a(t)$, $r_2(t)$ as sampled in the LHS described in the *Numerical simulations* section. We vary all four inputs simultaneously (sampling from uniform distributions within their respective ranges) and keep the rest of the parameters constant to the baseline values shown in Table 1 (*Baseline* column)

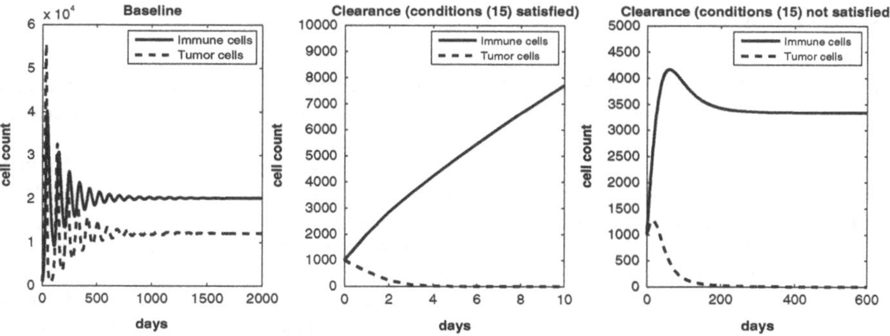

Fig. 5 Numerical solutions of system (5). Shown are plots for the baseline simulation (Parameters are chosen from the *Baseline* column in Table 1), for clearance when conditions (15) are satisfied (values of the treatment parameters are: $s_1(t) = 764.5072$, $r_2(t) = 0.0023$, $a(t) = 38.0040$, $c(t) = 0.3710$), and clearance where conditions (15) are not satisfied (values of the treatment parameters are: $s_1(t) = 10^2$,$r_2(t) = 0.0523$, $a(t) = 2$, $c(t) = 0.05$). The rest of the parameters are set to baseline values shown in Table 1

the practice of which is even younger. As with the development of any therapy, questions relating to which gene to target, or what combination of therapies can be used (immunotherapy plus gene therapies) is important. A recent paper reviewed the importance of pairing high-throughput experimental studies together with computational systems biology studies to help determine the optimal answers to these questions [23]. Excitingly, these types of studies can lead to personalized medical treatment, which one would expect from medicine in the twenty first century.

In this work, we begin by offering a small step in using mathematical models to make predictions that could be useful to experimentalists and clinicians working in the area of tumor–immune interactions and the development of treatment protocols. To this end, we simplified an existing model describing tumor-immune dynamics [7] by merging the effector molecule equation (for IL-2) into the effector cell equation, and allowing for time-varying inputs representing several options for immunotherapy and gene therapy. Sufficient global stability conditions of the cancer-free state were derived and tested numerically. Since the conditions are sufficient, further numerical analysis was performed to investigate regions of the parameter space where the system clears the cancer, even when sufficient conditions are not satisfied. Our results suggest that the source term of TIL cells, s_1, in combination with the cancer clearance term, a, provide the optimal treatment combination: high levels of both will clear the tumor. Further investigation is necessary to establish whether this is a viable immunotherapy/gene therapy option in the clinical setting. We are working now on deriving necessary conditions for the stability of the cancer-free state for the model system (5) in the general time-dependent therapy case.

References

1. Adam, J.A., Bellomo, N.: A Survey of Models for tumor-immune system dynamics, Birkhauser Series on Modeling and Simulation in Science, Engineering and Technology. Birkhauser, Boston (1997)
2. Aguda, B.D., Kim, Y., Piper-Hunter, M.G., Friedman, A., Marsh, C.: MicroRNA regulation of a cancer network: Consequences of the feedback loops involving miR-17-92, E2F and Myc. Proc. Natl. Acad. Sci. USA **105**, 19678–19683 (2008)
3. Ambrosi, D., Bellomo, N., Preziosi, L.: Modelling tumor progression, heterogeneity, and immune competition. J. Theor. Med. **4**, 51–65 (2002)
4. Andasari, V., Gerisch, A., Lolas, G., South, A.P., Chaplain, M.A.: Mathematical modeling of cancer cell invasion of tissue: Biological insight from mathematical analysis and computational simulation. J. Math. Biol. **63**(1), 141–71 (2011)
5. Arciero, J., Jackson, T., Kirschner, D.: A mathematical model of tumor- immune evasion and siRNA treatment. Discrete Continuous Dyn. Syst. Ser. B **4**, 39–58 (2004)
6. Arlotti, L., Gamba, A., Lachowicz, M.: A kinetic model of tumor/immune system cellular interactions. J. Theor. Med. **4**, 39–50 (2002)
7. Banerjee, S., Immunotherapy with Interleukin–2: A Study Based on Mathematical Modeling, Int. J. Appl. Math. Comput. Sci. **18**(3), 389–398 (2008)
8. Bellomo, N., Delitala, M.: From the mathematical kinetic, and stochastic game theory to modelling mutations, onset, progression and immune competition of cancer cells. Phys. Life Rev. **5**, 183–206 (2008)

9. Bellomo, N., Preziosi, L.: Modelling and mathematical problems related to tumor evolution and its interaction with the immune system. Math. Comput. Model. **32**, 413–452 (2000)
10. Bellomo, N., Bellouquid, A., Delitala, M.: Mathematical topics on the modelling complex multicellular systems and tumor immune cells competition. Math. Models Methods Appl. Sci. **14**, 1683–1733 (2004)
11. Bodnar, M., Forys, U.: Three types of simple DDEs describing tumor growth. J. Biol. Syst. **15**, 453–471 (2007)
12. Burden, T., Ernstberger, J., Fister, K.R.: Optimal control applied to immunotherapy. Discrete Continuous Dyn. syst. 4, 135–146 (2004)
13. Caravagna, G., D'Onofrio, A., Milazzo, P., Barbuti, R.: Tumour suppression by immune system through stochastic oscillations. J. theor. biol. **265**(3), 336–345 (2010)
14. Chang, Z., Song, J., Gao, G., Shen, Z.: Adenovirus-mediated p53 gene therapy reverses resistance of breast cancer cells to adriamycin. Anticancer Drugs **22**(6), 556–62 (2011)
15. Chaplain, M., Matzavinos, A.: Mathematical modelling of spatio-temporal phenomena in tumour immunology. In: Friedman, A. (ed.) Tutorials in Mathematical Biosciences III: Cell Cycle, Proliferation, and Cancer. Lecture Notes in Mathematics, vol. 1872, pp. 131–183. Springer, Berlin (2006)
16. De Angelis, E., Delitala, M., Marasco, A., Romano, A.: Bifurcation analysis for a mean field modelling of tumor and immune system competition. Math. Comput. Model. **37**, 1131–1142 (2003)
17. Delitala, M.: Critical analysis and perspectives on kinetic (cellular) theory of immune competition. Math. Comput. Model. **35**, 63–75 (2002)
18. de Pillis, L., Gu, W., Radunskaya, A.: Mixed immunotherapy and chemotherapy of tumors: Modeling, applications and biological interpretations. J. Theor. Biol. **238**, 841–862 (2006)
19. D'Onofrio, A.: The role of the proliferation rate of effectors in the tumor-immune system competition. Math. Model. Meth. Appl. Sci. **16**(8), 1375–1401 (2006)
20. d'Onofrio, A., Gatti, F., Cerrai, P., Freschi, L.: Delay-induced oscillatory dynamics of tumour–immune system interaction. Math. Comput. Model. **51**, 572–591 (2010)
21. Forys, U.: Marchuks model of immune system dynamics with application to tumour growth. J. Theor. Med. **4**, 85–93 (2002)
22. Friedman, A., Kim, Y.: Tumor cells proliferation and migration under the influence of their microenvironment. Math. Biosci. Eng. **8**(2), 371–83 (2011)
23. Gabhann, F.M., Annex, B.H., Popel, A.S.: Gene therapy from the perspective of systems biology. Curr. Opin. Mol. Ther. **12**(5), 570–577 (2010)
24. Gatenby, R.A., Maini, P.: Modelling a new angle on understanding cancer. Nature **420**(6915), 462 (2002)
25. Gatenby, R.A., Maini, P.K.: Mathematical oncology: Cancer summed up. Nature **421**(6921), 321 (2003)
26. Joshi, B., Wang, X., Banerjee, S., Tian, H., Matzavinos, A., Chaplain, M.A.: On immunotherapies and cancer vaccination protocols: A mathemati cal modelling approach. J. Theor. Biol. **259**(4), 820–827 (2009)
27. Kelly, C., Leek, R., Byrne, H., Cox, S., Harris, A., Lewis, C.: Modelling macrophage infiltration into avascular tumours. J. Theor. Med. **4**, 21–38 (2002)
28. Kirschner, D., Panetta, J.: Modeling immunotherapy of the tumor-immune interaction, J. Math. Biol. **37**, 235–252 (1998)
29. Kirschner, D., Tsygvintsev, A.: On the global dynamics of a model for tumor immunotherapy. J. Math. Biosci. Eng. Vol. **6**(3), 573–583 (2009)
30. Kronik, N., Kogan, Y., Vainstein, V., Agur, Z.: Improving alloreactive CTL immunotherapy for malignant gliomas using a simulation model of their interactive dynamics. Cancer Immunol. Immunother. **57**, 425–439 (2008)
31. Kuznetsov, V., Makalkyn, I., Taylor, M., Perelson, A.: Nonlinear dynamics of immunogenic tumors: Parameter estimation and global bifurcation analysis, Bull. Math.Biol. **56**(2), 295–321 (1994)

32. Macklin, P., McDougall, S., Anderson, A.R., Chaplain, M.A., Cristini, V., Lowengrub, J.: Multiscale modelling and nonlinear simulation of vascular tumour growth. J. Math. Biol. **58**(4–5), 765–98 (2009)
33. Marino, S., Hogue, I.B., Ray, C.J., Kirschner, D.E.: A methodology for performing global uncertainty and sensitivity analysis in systems biology. J. Theor. Biol. **254**, 178–196 (2008) PMID:18572196
34. Matzavinos, A., Chaplain, M.: Travelling-wave analysis of a model of the immune response to cancer. C.R. Biol. **327**, 995–1008 (2004)
35. Matzavinos, A., Chaplain, M., Kuznetsov, V.: Mathematical modelling of the spatio-temporal response of cytotoxic T-lymphocytes to a solid tumour. IMA J. Math. Med. Biol. **21**, 1–34 (2004)
36. McDougall, S.R., Anderson, A.R., Chaplain, M.A.: Mathematical modelling of dynamic adaptive tumour-induced angiogenesis: Clinical implications and therapeutic targeting strategies. J. Theor. Biol. **241**(3), 564–589 (2006)
37. McKay, M.D., Beckman, R.J., Conover, W.J.: A Comparison of three methods for selecting values of input variables in the analysis of output from a computer code. Technometrics **21**(2), 239–245 (1979)
38. Moscow, J.A., Cowan, K.H.: Biology of cancer. In: Goldman, L., Ausiello, D. (eds. Cecil. Med. 23rd edn. Saunders Elsevier, Philadelphia (2007), chap 187
39. Nana-Sinkam, S.P., Croce, C.M.: MicroRNAs as therapeutic targets in cancer. Transl. Res. **157**(4), 216–225 (2011)
40. Owen, M., Sherratt, J.: Mathematical modelling of macrophage dynamics in tumours. Math. Model. Meth. Appl. Sci. **9**, 513–539 (1999)
41. Owen, M.R., Alarcon, T., Maini, P.K., Byrne, H.M.: Angiogenesis and vascular remodelling in normal and cancerous tissues. J. Math. Biol. **58**(4–5), 689–721 (2009)
42. Rabinowich, H., Banks, M., Reichert, T.E., Logan, T.F., Kirkwood, J.M., Whiteside, T.L.: Expression and activity of signaling molecules n T lymphocytes obtained from patients with metastatic melanoma before and after interleukin 2 therapy. Clin. Canc. Res. **2**, 1263–1274 (1996)
43. Robbins, P.F., Morgan, R.A., Feldman, S.A., Yang, J.C., Sherry, R.M., Dudley, M.E., Wunderlich, J.R., Nahvi, A.V., Helman, L.J., Mackall, C.L., Kammula, U.S., Hughes, M.S., Restifo, N.P., Raffeld, M., Lee, C.-C.R., Levy, C.L., Li, Y.F., El-Gamil, M., Schwarz, S.L., Laurencot, C., Rosenberg, S.A.: Tumor regression in patients with metastatic synovial cell sarcoma and melanoma using genetically engineered lymphocytes reactive with NY-ESO-1. J. Clin. Oncol. **29**, 917–924 (2011)
44. Rosenberg, S.A., Lotze, M.T.: Cancer immunotherapy using interleukin-2 and interleukin- 2-activated lymphocytes. Ann. Rev. Immunol. **4**, 681–709 (1986)
45. Rosenberg, S.A., Yang, J.C., Topalian, S.L., Schwartzentruber, D.J., Weber, J.S., Parkinson, D.R., Seipp, C.A., Einhorn, J.H., White, D.E.: Treatment of 283 consecutive patients with metastatic melanoma or renal cell cancer using high-dose bolus interleukin 2. JAMA **271**, 907–913 (1994)
46. Rosenstein, M., Ettinghousen, S.E., Rosenberg, S.A.: Extravasion of intravascular uid mediated by the systemic administration of recombinant interleukin 2. J. Immunol. **137**, 1735–1742 (1986)
47. Sherratt, J., Perumpanani, A., Owen, M.: Pattern formation in cancer. In: Chaplain, M., Singh, G., McLachlan, J. (eds.) On Growth and Form: Spatio- temporal Pattern Formation in Biology. Wiley, New York (1999)
48. Szymanska, Z. (2003). Analysis of immunotherapy models in the context of cancer dynamics. Appl. Math. Comput. Sci. 13, 407–418.
49. Thun, M.J.: Biology of cancer. In: Goldman L, Ausiello D (eds.) Cecil. Med. 23rd edn. Saunders Elsevier, Philadelphia, Pa (2007), chap 185
50. Tran, K.Q., Zhou, J., Durflinger, K.H., et al.: Mini- mally cultured tumor-infiltrating lymphocytes display optimal characteristics for adoptive cell therapy. J. Immunother. **31**, 742–751 (2008)

51. Webb, S., Sherratt, J., Fish, R.: Cells behaving badly: A theoretical model for the Fas/FasL system in tumour immunology. Math. Biosci. **179**, 113–129 (2002)
52. Wu, F.T., Stefanini, M.O., Mac Gabhann, F., Kontos, C.D., Annex, B.H., Popel, A.S.: A systems biology perspective on sVEGFR1: Its biological function, pathogenic role and therapeutic use. J. Cell Mol. Med. **14**(3), 528–52 (2010)
53. Zhao, Y., Lam, D.H., Yang, J., Lin, J., Tham, C.K., Ng, W.H., Wang, S.: Targeted suicide gene therapy for glioma using human embryonic stem cell-derived neural stem cells genetically modified by baculoviral vectors. Gene Therapy **19**, 189–200 (2012)

Part V
Epidemiological Models

Epidemiological Models with Seasonality

Avner Friedman

1 Introduction

Epidemiology is the branch of medicine that deals with incidence, distribution, and control of diseases in a population. At the basic level the population is divided into susceptible, exposed, infected, and recovered compartments. However, often infection is caused not only by exposed or infected individuals but also by other species, such as mosquitos in the case of malaria, or waste water in the case of cholera. In attempting to model the transmission of the disease one has to take into account the facts that infection rates may vary among different populations (due, for instance, to those who underwent vaccination and those who did not), as well as from one season to another. In this chapter we focus on seasonality-dependent diseases and ask the question whether initial infection of one or a small number of individuals will cause the disease to spread or whether the disease will die out. To answer this question we invoke the concept of the basic reproduction number, a number which is easy to compute in the case of seasonality-independent diseases, but difficult to compute in the case of diseases with seasonality.

The basic reproduction number R_0 is an important concept in epidemiology. In a healthy susceptible population, any small infection will die out if $R_0 < 1$, but may persist and become endemic if $R_0 > 1$. If we denote by J the Jacobian matrix about the disease free equilibrium (DFE) and by λ the eigenvalue of J with largest real part, then $R_0 < 1$ if $\mathrm{Re}\{\lambda\} < 0$ and $R_0 > 1$ if $\mathrm{Re}\{\lambda\} > 0$; R_0 is the norm, or the spectral radius, of the matrix operator J. For epidemiological models with ω-periodic coefficients, R_0 is defined as the spectral radius of a certain linear operator L in a Banach space of ω-periodic functions. Here again the DFE is asymptotically stable if $R_0 < 1$ and unstable if $R_0 > 1$. However, it is generally

A. Friedman (✉)
Department of Mathematics, The Ohio State University, Columbus, OH 43210, USA
e-mail: afriedman@math.osu.edu

U. Ledzewicz et al., *Mathematical Methods and Models in Biomedicine*, Lecture Notes on Mathematical Modelling in the Life Sciences, DOI 10.1007/978-1-4614-4178-6_14,

difficult to compute R_0 in this case, and, in particular, to determine when R_0 is less than 1 or larger than 1. In the present chapter we develop general methods to determine when $R_0 < 1$ and when $R_0 > 1$. However, for the sake of clarity we shall first apply the method to a special case of waterborne diseases and then, in the final section of this chapter, we extract from this special case the general features of our methods and give some other examples.

Consider a dynamical system in $\mathbb{R}^n$

$$\frac{dx}{dt} = f(x, \gamma), \tag{1}$$

where $\gamma = (\gamma_1, \ldots, \gamma_k)$ varies in a k-dimensional parameter space Ω, and suppose that x^0 is a stationary point independent of γ, i.e.,

$$f(x^0, \gamma) = 0 \text{ for all } \gamma \in \Omega\,. \tag{2}$$

We denote the eigenvalues of the Jacobian matrix $(\partial f/\partial x)(x^0, \gamma)$ by $\lambda_i(\gamma)$ and arrange them so that

$$\mathrm{Re}\{\lambda_n(\gamma)\} \leq \mathrm{Re}\{\lambda_{n-1}(\gamma)\} \leq \cdots \leq \mathrm{Re}\{\lambda_2(\gamma)\} \leq \mathrm{Re}\{\lambda_1(\gamma)\}\,.$$

If $\mathrm{Re}\{\lambda_1(\gamma)\} < 0$, then x^0 is asymptotically stable, and if $\mathrm{Re}\{\lambda_1(\gamma)\} > 0$, then x^0 is unstable. In epidemiological models there is a special interest in the steady state x^0 which represents DFE. Associated with x^0 is the concept of the basic reproduction number R_0, and it is shown that if $R_0 < 1$ then the DFE x^0 is asymptotically stable (so that $\mathrm{Re}\{\lambda_1\} < 0$) and if $R_0 > 1$ then the DFE x^0 is not stable (so that $\mathrm{Re}\{\lambda_1\} > 0$).

Consider next a nonautonomous system

$$\frac{dx}{dt} = f(x^0, \gamma(t)), \tag{3}$$

where $\gamma(t)$ is ω-periodic, and assume, as before, that Eq. (2) holds for all $\gamma{=}\gamma(t) \in \Omega$. For an epidemiological model of the form Eq. (3) one can still define the concept of the basic reproduction number R_0 [12] and again show that if $R_0 < 1$ then the DFE is asymptotically stable, whereas if $R_0 > 1$ then the DFE is not stable. In the autonomous case Eq. (1) R_0 is the spectral radius of a matrix defined in terms of some submatrices of $(\partial f/\partial x)(x^0, \gamma)$. In the nonautonomous case Eq. (3), R_0 is the spectral radius of a linear integral operator on ω-periodic functions with a kernel which is defined in terms of some submatrices of $(\partial f/\partial x)(x^0, \gamma(t))$.

Suppose we express the basic reproduction number for the DFE x^0 of Eq. (1) as a function

$$R_0 = R_0(\gamma_1, \ldots, \gamma_k)\,,$$

and then define $R_0(t) = R_0(\gamma_1(t), \ldots, \gamma_k(t)$, $[R_0] = R_0(\langle\gamma_1\rangle, \ldots, \langle\gamma_k\rangle)$ where $\langle\gamma_i\rangle = \frac{1}{\omega}\int_0^{\omega} \gamma_i(t)\mathrm{d}t$.

Then the question arises whether $[R_0]$ can be used to estimate the basic reproduction number R_0 of Eq. (10) for the same DFE x^0.

Examples of epidemiological models for periodically occurring diseases were given in [1,2,4,8–10,12], and the references therein. In [1,12] there are examples where $R_0 > [R_0]$, and in [8] there is an example where $R_0 < [R_0]$. For a malaria model [3] it was shown that

$$\text{if } [R_0] < 1 \text{ then } R_0 < 1, \tag{4a}$$

$$\text{if } [R_0] > 1 \text{ then } R_0 > 1\,. \tag{4b}$$

This means that $[R_0]$ can be used to determine the stability or instability of the DFE for Eq. (3). Such a result is very useful since it is much more difficult to compute R_0 than to compute $[R_0]$.

In this chapter we prove the statements in Eqs. (4a) and (4b) in the context of a waterborne disease model. Some of the arguments overlap with those in [3]. In the concluding section we discuss the general features of our proofs and illustrate how the same methods can be applied to other disease models.

2 The Basic Reproduction Number for Autonomous Systems

A fundamental concept in epidemiological studies of infectious diseases is the concept of the basic reproduction number, R_0. It is defined as the expected number of secondary infections produced by an infective individual in a completely healthy but susceptible population. If $R_0 < 1$ then the disease will die out, whereas if $R_0 > 1$, then the disease will not die out and will persist. For autonomous epidemic models, $R_0 < 1$ implies that all the eigenvalues of the Jacobian matrix about the DFE have negative real parts; if $R_0 > 1$, then at least one eigenvalue has a positive real part.

A general autonomous compartmental model was developed in [11]. The model includes m disease compartments with population densities $x_1,\ldots,x_m$ and $n-m$ non-disease compartments with population densities $x_{m+1},\ldots,x_n$. The dynamical system has the form

$$\frac{\mathrm{d}x_i}{\mathrm{d}t} = \mathscr{F}_i(x) - \mathscr{V}_i(x) \qquad (1 \le i \le m), \tag{5a}$$

$$\frac{\mathrm{d}x_j}{\mathrm{d}t} = g_j(x) \qquad (m+1 \le j \le n), \tag{5b}$$

where x varies in the space

$$\mathbb{R}^n_+ = \{x = (x_1,\ldots,x_n),\ x_i \ge 0 \text{ for all } i\}.$$

We define

$$\mathbb{R}^{n,0}_{+} = \{x \in \mathbb{R}^n_+,\ x_i = 0 \text{ for } i = 1,\dots,m\}$$

and make the following assumptions:

(A1) $\mathscr{F}_i, \mathscr{V}_i$ and g_j and their first x-derivatives are continuous functions in R^n_+, for all i, j.
(A2) $\mathscr{F}_i = \mathscr{V}_i = 0$ on $\mathbb{R}^{n,0}_+$, for all i.
(A3) $\mathscr{F}_i \geq 0$, for all i.
(A4) $\mathscr{V}_i \leq 0$ whenever $x_i = 0$; i is any number $1,2,\dots,m$.
(A5) $\sum\limits_{i=1}^{m} \mathscr{V}_i \geq 0$.
(A6) The disease free system $\frac{dx_j}{dt} = g_j(0,\dots,0,x_{m+1},\dots,x_n)$ has a unique equilibrium point $x^0 = (0,\dots,0,x^0_{m+1},\dots,x^0_n)$ which is globally asymptotically stable.

We refer to x^0 as the DFE. We introduce the Jacobian matrices

$$F = (F_{ij}) = \left(\frac{\partial \mathscr{F}_i(x^0)}{\partial x_j}\right)_{1\leq i,j\leq m}, \quad V = (V_{ij}) = \left(\frac{\partial \mathscr{V}_i(x^0)}{\partial x_j}\right)_{1\leq i,j\leq m}$$

and assume that
(A7) V is nonsingular.

Then V^{-1} and FV^{-1} are nonnegative matrices [11].

We denote by $\rho(A)$ the spectral radius of a bounded linear operator A; if A is a matrix, viewed as a linear operator in $\mathbb{R}^n$, then $\rho(A)$ is the maximum absolute value of the eigenvalues of A.

Theorem 2.1 ([11]). *Under the assumptions (A1)–(A7) there holds:*

$$R_0 = \rho(FV^{-1}).$$

Since R_0 is the principal eigenvalue of the nonnegative matrix FV^{-1} and thus also of its transpose, there exists an eigenvector $\lambda = (\lambda_1,\dots,\lambda_m)$ in $\mathbb{R}^m_+$ of $(FV^{-1})^{\mathrm{T}}\lambda = R_0\lambda$, so that

$$\sum_{i=1}^{m} F_{ij}\lambda_i = R_0 \sum_{i=1}^{m} V_{ij}\lambda_i\,. \tag{6}$$

Using Eq. (6) one can show, under some additional conditions, that if $R_0 < 1$ then x^0 is *globally* asymptotically stable for the system Eqs. (5a) and (5b). The proof uses a Liapounov function of the form

$$h(t) = \sum_{i=1}^{m} (\lambda_i + \varepsilon\mu_i)x_i(t) \tag{7}$$

for appropriate parameters μ_i and ε positive and sufficiently small.

Since $h'(t) < 0$ for all $t > 0$, by the LaSalle invariance principle [7] it follows that $h(t) \to 0$ as $t \to \infty$. This easily leads to the conclusion that $x(t) \to x^0$ as $t \to \infty$, so that x^0 is a globally asymptotically stable equilibrium.

A function of the form Eq. (7) is a Liapounov function if

$$\mathscr{V}_i = \mathscr{V}_i(x^*), \; \mathscr{F}_i(x) \leq \mathscr{F}_i(x_0^*)$$

for all

$$x^* = (x_1, \ldots, x_m) \in \mathbb{R}_+^m, \; x_0^* = (x^*, \, x_{m+1}^0, \ldots, x_n^0) \,,$$

and

$$\mathscr{F}_i(x_0^*) - \mathscr{V}_i(x^*) \leq \sum_{j=1}^{m} x_j \frac{\partial}{\partial x_j}(\mathscr{F}_i(x_0^*) - \mathscr{V}_j(x^*)), \; \sum_{i=1}^{m} V_{ij}\lambda_i > 0.$$

Similarly one can show, under somewhat different conditions, that if $R_0 > 1$ then there is a function $h(t)$ of the form Eq. (7) which satisfies:

$$h'(t) \geq \delta h(t) \; (\delta > 0)$$

provided $\sum_{i=j} x_i(t) < \eta$ where η is sufficiently small and t is sufficiently large depending on $h(0)$. This instability result of the DFE can be used, with the aid of Horn's lemma [5] (see also [4]) to deduce the existence of an (endemic) equilibrium point $\bar{x}$, $\bar{x} \neq 0$, provided the system (5a) and (5b) has a compact convex invariant set with x^0 in its interior.

3 The Basic Reproduction Number for Nonautonomous Systems

Consider now a nonautonomous system of the same structure as in Eqs. (5a) and (5b),

$$\frac{\mathrm{d}x_i}{\mathrm{d}t} = \mathscr{F}_i(t,x) - \mathscr{V}_i(t,x) \qquad (1 \leq i \leq m), \tag{8a}$$

$$\frac{\mathrm{d}x_j}{\mathrm{d}t} = g_j(t,x) \qquad (m+1 \leq j \leq n)\,. \tag{8b}$$

Following [12] (see also [2, 13]) we assume:

(B1) $\mathscr{F}_i$, $\mathscr{V}_i$ and g_j and their first x-derivatives are continuous functions in $\mathbb{R} \times \mathbb{R}_+^n$, for all i, j.

(B2) $\mathscr{F}_i = \mathscr{V}_i = 0$ on $\mathbb{R} \times \mathbb{R}_+^{n,0}$, for all i.

(B3) $\mathscr{F}_i \geq 0$, for all i.

(B4) $\mathscr{V}_i \leq 0$ on $\mathbb{R} \times \mathbb{R}_+^n$ if $x_i = 0$; i is any number $1, 2, \ldots, m$.
(B5) $\mathscr{F}_i$, $\mathscr{V}_i$, g_j are ω-periodic in t, for all i, j.
(B6) There exists a disease-free periodic solution $x^0(t) = (0, 0, \ldots, 0,\ x^0_{m+1}(t), \ldots,$ $x^0_n(t)$ of Eqs. (8a) and (8b).
We introduce the Jacobian matrices

$$G(t) = \left(\frac{\partial g_i(t, x^0(t))}{\partial x_j} \right)_{m+1 \leq i, j \leq n},$$

$$F(t) = \left(\frac{\partial \mathscr{F}_i(t, x^0(t))}{\partial x_j} \right)_{1 \leq i, j \leq m}, \qquad V(t) = \left(\frac{\partial \mathscr{V}_i(t, x^0(t))}{\partial x_j} \right)_{1 \leq i, j \leq m}.$$

Let $Y(t, s)$ denote the solution of

$$\frac{\mathrm{d}Y}{\mathrm{d}t} = -V(t)Y, \quad Y(s, s) = I,$$

where I is the unit matrix, and set $Y(\omega) = Y(\omega, 0)$.

We assume that $x^0(t)$ is linearly asymptotically stable by imposing the condition

(B7) $\rho(G(\omega)) < 1$.

We also assume that the internal evolution in the infectious disease compartments due to death is dissipative, that is,

(B8) In $\rho(Y(\omega)) < 1$.

Denote by C_ω the Banach space of ω-periodic functions from $\mathbb{R}$ into $\mathbb{R}^m$ equipped with the maximum norm, and denote by C_ω^+ the subspace of functions with values in $\mathbb{R}_+^m$. We introduce the bounded linear operator

$$(L\phi)(t) = \int_0^\infty Y(t, t-a) F(t-a) \phi(t-a) \mathrm{d}a, \ t \in \mathbb{R}$$

for $\phi \in C_\omega^+$. The basic reproduction number R_0 is defined as the spectral radius of the operator L,

$$R_0 = \rho(L) . \tag{9}$$

This definition coincides with the definition of R_0 given in Sect. 2 when the system (8a) and (8b) is autonomous [12].

Theorem 3.1 ([12]). *Under the assumptions (B1)–(B8), $x^0(t)$ is asymptotically stable if $R_0 < 1$, and unstable if $R_0 > 1$.*

In the sequel we consider nonautonomous systems (8a) and (8b) where the dependence on t is in terms of parameters $\gamma_1(t), \ldots, \gamma_k(t)$. We associate with Eqs. (8a) and (8b) the autonomous system (5a) and (5b) with parameters $\gamma_1, \ldots, \gamma_k$ and DFE x^0 independent of t.

We denote by ω the common period of the functions $\gamma_j(t)$, and set

$$\langle z \rangle = \frac{1}{\omega} \int_0^{\omega} z(t) \mathrm{d}t$$

for any ω-periodic function $z(t)$. Suppose R_0 for Eqs. (5a) and (5b) is given as a function

$$R_0 = R_0(\gamma_1, \ldots, \gamma_k) \tag{10}$$

and set

$$\begin{aligned} R_0(t) &= R_0(\gamma_1(t), \ldots, \gamma_k(t)) , \\ [R_0] &= R_0(\langle \gamma_1 \rangle, \ldots, \langle \gamma_k \rangle) . \end{aligned} \tag{11}$$

If

$$F = \mathrm{diag}(F_1, \ldots, F_m),\ V = \mathrm{diag}(V_1, \ldots, V_m) ,$$

then

$$R(t) = \max_{1 \le i \le m} \frac{F_i(t)}{V_i(t)} .$$

As shown in [12], in this case

$$R_0 = \max_{1 \le i \le m} \frac{\langle F_i \rangle}{\langle V_i \rangle} ,$$

so that

$$R_0 = [R_0] ; \tag{12}$$

but, in general, $R_0 \neq [R_0]$. For example, in a model of tuberculosis it was shown, in [8], that $R_0 < [R_0]$ for R_0 and $[R_0]$ when both vary in an interval containing 1, as shown in Fig. 1a, and in a model of Dengue fever it was shown, in [12], that $R_0 > [R_0]$ for R_0 and $[R_0]$ in an interval containing 1; see Fig. 1b.

Since it is much harder to compute R_0 than to compute $[R_0]$, the question arises whether one can use $[R_0]$ to estimate R_0. We would especially want to derive such estimates when $[R_0]$ or R_0 are equal to (or near to) the value 1, a value associated with stability/instability of the DFE. If we can show, for example, that

$$[R_0] < 1 \text{ implies that } R_0 < 1 \tag{13}$$

and

$$[R_0] > 1 \text{ implies that } R_0 > 1 , \tag{14}$$

then it would follow that Eq. (12) holds when either side of the equation is equal to 1. This situation is illustrated in Fig. 2.

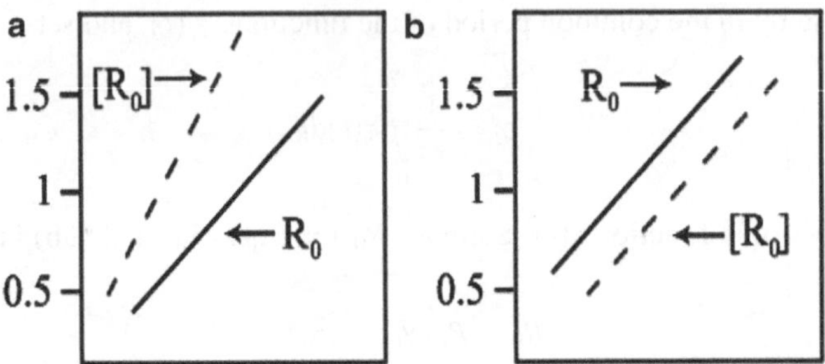

Fig. 1 Comparison between R and $[R_0]$ for some range of a parameter in the model

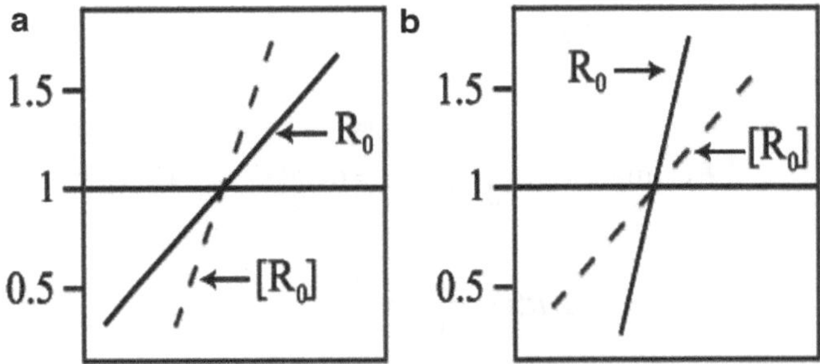

Fig. 2 Example showing that R_0 coincides with $[R_0]$ when $[R_0] = 1$

For a malaria model with periodic coefficients it was proved in [3] that if $[R_0] < 1$ ($[R_0] > 1$) then the DFE is globally asymptotically stable (unstable); this is equivalent to the assertions (13) and (14). In the present chapter we establish Eqs. (13) and (14) for a model of waterborne diseases with periodic coefficients. Our approach is similar to [3], but with some technical differences. In the concluding section of the chapter we summarize the general features of our methods, and briefly give additional examples.

4 Statement of Results for a Waterborne Disease Model

Waterborne diseases are diseases which are transmitted through water. They include, for instance, diarrhea and cholera, typhoid fever, hepatitis A, enteriditis salmonella, and Giardia. Infection typically occurs through pathogen ingestion. Transmission may happen when drinking sewage-contaminated water, or eating food grown in sewage-contaminated land. Historically some of these diseases reappeared periodically, as, for example, in the epidemic of cholera in the nineteenth century London [10].

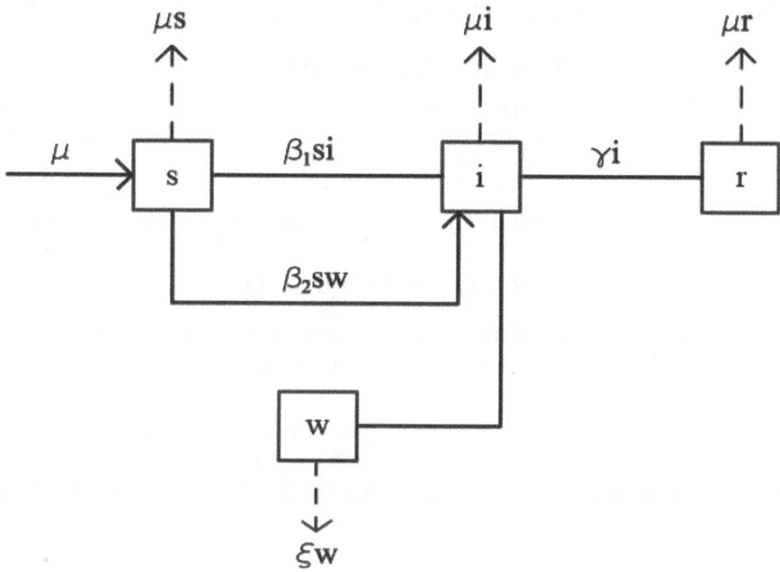

Fig. 3 Flow diagram of model (15a)–(15d); *broken lines* indicate death

Here we consider a simple SIR model with added compartment W that tracks pathogen concentration in water. We follow a recent model studied by Tien et al. [9] (see also the references in [9]). The model includes four variables: susceptible (s), infected (i) and recovered (r) individuals, and pathogens (w). In the nondimensional form where $s+i+r=1$, the four variables satisfy the following system of equations:

$$\dot{s} = \mu - \beta_1 si - \beta_2 sw - \mu s, \tag{15a}$$

$$\dot{i} = \beta_1 si + \beta_2 sw - \gamma i - \mu i, \tag{15b}$$

$$\dot{w} = \xi(i - w), \tag{15c}$$

$$\dot{r} = \gamma i - \mu r, \tag{15d}$$

where β_1 is the water reservoir–person contact rate, β_2 is the person-person contact rate, $1/\gamma$ is the infectious period, $1/\xi$ is the pathogen lifetime in the water reservoir, and μ is the birth/death rate. The phase space for the system (15a)–(15d) is

$$\Omega = \{(s,i,w,r,);\ s \geq 0,\ i \geq 0,\ 0 \leq w \leq i,\ r \geq 0,\ s+i+r=1\}. \tag{16}$$

Figure 3 illustrates the flow diagram represented by model (15a)–(15d).

Some waterborne diseases are seasonal. An example is the epidemic of cholera in the nineteenth century London [10]. To model these cases we take β_1, β_2 and ξ to be periodic functions in t with period ω. Thus, we shall consider the system

$$\dot{s} = \mu - \beta_1(t)si - \beta_2(t)sw - \mu s\,, \tag{17a}$$
$$\dot{i} = \beta_1(t)si + \beta_2(t)sw - \gamma i - \mu i\,, \tag{17b}$$
$$\dot{w} = \xi(t)(i - w)\,, \tag{17c}$$
$$\dot{r} = \gamma i - \mu r, \tag{17d}$$

where $\beta_1(t), \beta_2(t), \xi(t)$ are continuous ω-periodic functions, with initial conditions.

$$(s(0),\ i(0),\ w(0),\ r(0)) \in \Omega. \tag{18}$$

It easily follows that the solution $(s(t),\ i(t),\ w(t),\ r(t))$ remains in Ω for all $t > 0$. The DFE for both systems (15a)–(15d) and (17a)–(17d) is

$$(s,i,w,r) = (1,0,0,0).$$

The basic reproduction number for Eqs. (15a)–(15d) is easily computed to be [8]

$$\overline{R}_0 \equiv R_0(\beta_1,\beta_2) = \frac{\beta_1+\beta_2}{\gamma+\mu}\,,$$

so that

$$[R_0] = R_0(\langle\beta_1\rangle,\langle\beta_2\rangle) = \frac{\langle\beta_1\rangle+\langle\beta_2\rangle}{\gamma+\mu}\,. \tag{19}$$

Note that $\overline{R}_0 < 1$ if and only if all the eigenvalues of the matrix

$$A = \begin{pmatrix} \beta_1-\gamma-\mu & \beta_2 \\ \xi & -\xi \end{pmatrix}$$

have negative real parts. Similarly, setting

$$A(t) = \begin{pmatrix} \beta_1(t)-\gamma-\mu & \beta_2(t) \\ \xi(t) & -\xi(t) \end{pmatrix},\ \langle A\rangle = \begin{pmatrix} \langle\beta_1\rangle-\gamma-\mu & \langle\beta_2\rangle \\ \langle\xi\rangle & -\langle\xi\rangle \end{pmatrix},$$

$[R_0] < 1$ if and only if all the eigenvalues of $\langle A\rangle$ have negative real parts.

In the following sections we shall prove the assertions (13) and (14) for the system (17a)–(17d). This will be a consequence of the following theorems.

Theorem 4.1. *If $[R_0] < 1$, then the DFE for Eqs. (17a)–(17d) is globally asymptotically stable.*

Theorem 4.2. *If $[R_0] > 1$, then there exists a positive number δ such that for any initial values with $s(0) > 0$, and $0 < w(0) \leq i(0)$, the solution of Eqs. (17a)–(17d) satisfies:*

$$i(t) > \delta \text{ and } w(t) > \delta \text{ if } t \text{ is sufficiently large.}$$

Using Theorem 4.2 we shall also prove:

Theorem 4.3. *If* $[R_0] > 1$, *then there exists an* ω*-periodic solution of Eqs. (17a)–(17d) with* $i(t) > 0$, $w(t) > 0$ *for all* $t > 0$.

5 Proof of Theorem 4.1

Since $s(t) \leq 1$, we obtain from Eqs. (17a)–(17d)

$$\frac{\mathrm{d}}{\mathrm{d}t}\begin{pmatrix} i \\ w \end{pmatrix} \leq A(t)\begin{pmatrix} i \\ w \end{pmatrix}$$

where we have used the notation $(x,y)^{\mathrm{T}} \leq (x_1,y_1)^{\mathrm{T}}$ if $x \leq x_1$, $y \leq y_1$.

We introduce the solution $z = (z_1,z_2)^{\mathrm{T}}$ of

$$\frac{\mathrm{d}z}{\mathrm{d}t} = (A(t) + \delta I)z, \qquad z(0) = (i(0)+\delta, w(0)+\delta)^{\mathrm{T}}\,, \tag{20}$$

for any $\delta > 0$. Then $z > (i,w)^{\mathrm{T}}$ for small t. We claim that this inequality holds for all $t > 0$.

Indeed, otherwise there is a smallest $\bar{t}$ such that at least one of the strict inequalities is violated at $t = \bar{t}$. Suppose $z_1(\bar{t}) = i(\bar{t})$. Then $\mathrm{d}z_1(\bar{t})/\mathrm{d}t \leq \mathrm{d}i(\bar{t})/\mathrm{d}t$. We also have $z_2(\bar{t}) \geq w(\bar{t})$. Hence, at $\bar{t}$,

$$\frac{\mathrm{d}z_1}{\mathrm{d}t} \leq \frac{\mathrm{d}i}{\mathrm{d}t} \leq \beta_2 z_2 + (\beta_1 - \gamma - \mu)z_1,$$

which contradicts the first equation in Eq. (20). By a similar argument one derives a contradiction in case $z_2(\bar{t}) = w(\bar{t})$.

Taking $\delta \to 0$ we conclude that

$$i(t) \leq z_1(t),\ w(t) \leq z_2(t) \text{ for all } t > 0, \tag{21}$$

where $z = (z_1,z_2)^{\mathrm{T}}$ is the solution of

$$\frac{\mathrm{d}z}{\mathrm{d}t} = A(t)z, \qquad z|_{t=0} = (i,w)^{\mathrm{T}}|_{t=0}\,. \tag{22}$$

By integration and use of the ω-periodicity of $A(t)$ we obtain

$$z(t) = \mathrm{e}^{n\omega\langle A\rangle}\exp\left(\int_{n\omega}^{n\omega+\tau} A(s)\mathrm{d}s\right)z(0) \qquad \text{if } t = n\omega + \tau,$$

where n is any positive integer. Since $[R_0] < 1$, both eigenvalues of $\langle A \rangle$ have negative real parts, say $< -\lambda$, so that

$$|z(t)| \leq \text{const. } e^{-\lambda t}.$$

From Eq. (21) we then conclude that $(i(t), w(t)) \to 0$ as $t \to \infty$; then also $r(t) \to 0$ as $t \to \infty$, and the assertion of Theorem 4.1 follows.

6 Proof of Theorem 4.2

We shall need the following lemma.

Lemma 6.1. *Let a, b be any positive numbers.*

(i) *If*

$$\frac{dz}{dt} + az \leq b \quad \text{for } 0 < t < \Lambda$$

and $0 \leq z(0) \leq z_0$, $0 < \varepsilon < z_0 - \frac{b}{a}$, *then* $z(t) \leq \frac{b}{a} + \varepsilon$ *for all* $T_0 < t < \Lambda$ *where*

$$T_0 = \frac{1}{a} \ln \frac{z_0 - b/a}{\varepsilon}.$$

(ii) *If*

$$\frac{dz}{dt} + az \geq b \text{ for } 0 < t < \Lambda$$

and $z(0) \geq z_0 \geq 0 \quad 0 < \varepsilon < \frac{b}{a} - z_0$, *then* $z(t) \geq \frac{b}{a} - \varepsilon$ *for all* $T_1 < t < \Lambda$ *where*

$$T_1 = \frac{1}{a} \ln \frac{\frac{b}{a} - z_0}{\varepsilon}.$$

The proof follows immediately by integration.

In the sequel we shall use the bounds

$$0 < \alpha_1 \leq \beta_i(t) \leq \alpha_2 \ (i = 1, 2), \ 0 < \xi_1 \leq \xi(t) \leq \xi_2$$

Lemma 6.2. *Let* η *be any small positive number such that* $\eta < \mu/(2\gamma)$. *If*

$$i(t) < \eta \text{ for } 0 < t < \Lambda, \tag{23}$$

then

$$r(t) < \frac{2\gamma}{\mu}\,\eta \text{ for } T_0 < t < \Lambda, \tag{24}$$

where

$$T_0 = T_0(\eta) = \frac{1}{\mu}\,\ln\left(\frac{\mu}{\gamma\eta} - 1\right). \tag{25}$$

Furthermore,

$$\begin{pmatrix} i(t) \\ w(t) \end{pmatrix} \geq \mathrm{e}^{(n\omega\langle A\rangle - c\eta J)} \begin{pmatrix} i(T_0+\tau) \\ \omega(T_0+\tau) \end{pmatrix} \text{ if } t = T_0 + n\omega + \tau < \Lambda \tag{26}$$

for any positive integer n and $0 \leq \tau \leq \omega$, where c is a positive constant independent of η and J is the matrix $\begin{pmatrix} 1\,1 \\ 0\,0 \end{pmatrix}$.

Proof. From Eqs. (17d) and (23) we have

$$\frac{\mathrm{d}r}{\mathrm{d}t} + \mu r \leq \gamma\eta,\ 0 \leq t < \Lambda.$$

Suppose $r(0) < 1$. Applying Lemma 6.1 (i) with $z_0 = 1$, $\varepsilon = 1 - \frac{\gamma\eta}{\mu}$, the inequality (24) follows with T_0 as in Eqs. (17a)–(17d). The case $r(0) = 1$ follows by approximation.

Since $s = 1 - i - r$,

$$\frac{\mathrm{d}i}{\mathrm{d}t} = \beta_1(t)i + \beta_2(t)w - \gamma i - \mu i + F,\ T_0 \leq t < \Lambda,$$

where, by Eqs. (23)–(24), $|F| \leq const.\ \eta(i+w) \leq const.\ 2\eta$. Hence

$$\begin{aligned}\begin{pmatrix} \frac{\mathrm{d}i}{\mathrm{d}t} \\ \frac{\mathrm{d}w}{\mathrm{d}t} \end{pmatrix} &\geq \begin{pmatrix} \beta_1(t) - \gamma - \mu - c\eta & \beta_2(t) - c\eta \\ \xi(t) & -\xi(t) \end{pmatrix} \begin{pmatrix} i \\ w \end{pmatrix} \\ &= (A(t) - c\eta J)\begin{pmatrix} i \\ w \end{pmatrix},\end{aligned} \tag{27}$$

where c is a constant independent of η, and Eq. (26) follows by integration. □

Lemma 6.3. *If $i(t) \geq \eta$ for $0 \leq t < \Lambda$, for some $\eta > 0$, then*

$$w(t) \geq \frac{\xi_1}{2\xi_2}\eta \text{ for } \frac{1}{\xi_2}\ln 2 \leq t < \Lambda. \tag{28}$$

Proof. Since

$$\frac{\mathrm{d}w}{\mathrm{d}t} + \xi_2 w \geq \xi_1 i \geq \xi_1\eta,$$

Lemma 6.1(ii) with $z_0 = 0$, $\varepsilon = \dfrac{\xi_1 \eta}{2\xi_2}$ yields the asserted inequality. □

Lemma 6.4. *If*

$$w(t) \geq \eta \text{ for } 0 \leq t < \Lambda \tag{29}$$

for some $\eta > 0$, then

$$i(t) \geq c\eta \text{ for } t_0 \leq t < \Lambda, \tag{30}$$

where c is a positive constant independent of η, and

$$t_0 = \left(\frac{1}{\mu + 2\alpha_1} + \frac{1}{\gamma + \mu} \right) \ln 2. \tag{31}$$

Proof. From the equation for $s(t)$, Eq. (17a),

$$\frac{\mathrm{d}s}{\mathrm{d}t} + (\mu + 2\alpha_2)s \geq \mu,$$

so that, by Lemma 6.1(ii) with $z_0 = 0$, $\varepsilon = \mu/2(\mu + 2\alpha_2)$,

$$s(t) \geq \frac{\mu}{2(\mu + 2\alpha_2)} \text{ if } \bar{t}_0 \leq t < \Lambda, \tag{32}$$

where

$$\bar{t}_0 = \frac{\ln 2}{\mu + 2\alpha_2}.$$

Substituting Eqs. (28) and (32) into Eq. (17b), we get

$$\frac{\mathrm{d}i}{\mathrm{d}t} + (\gamma + \mu)i \geq \alpha_1 ws \geq \frac{\alpha_1 \mu \eta}{2(\mu + 2\alpha_2)}.$$

Applying once more Lemma 6.1(ii) with $z_0 = 0$ and

$$\varepsilon = \frac{\alpha_1 \mu \eta}{4(\gamma + \mu)(\mu + 2\alpha_2)},$$

we obtain the inequality (30) for $t_0 \leq t < \Lambda$ with t_0 as in Eq. (31). □

Lemma 6.5. *Set $B = \gamma + \mu + \xi_2$. Then*

$$i(t) \geq i(\tau)\mathrm{e}^{-B(t-\tau)}, \; w(t) \geq w(\tau)\mathrm{e}^{-B(t-\tau)}$$

for all $0 < \tau < t < \infty$.

Indeed, this follows from the inequalities

$$\frac{\mathrm{d}i}{\mathrm{d}t} \geq -(\gamma + \mu)i, \quad \frac{\mathrm{d}w}{\mathrm{d}t} \geq -\xi w.$$

Until now we have not used the assumption that $[R_0] > 1$. We shall now use this assumption. Setting $E = A - c\eta J$, c as in Eq. (27), the eigenvalues of the constant matrix E, $\lambda_i = \lambda_i(\eta)$, satisfy

$$\lambda_2 < \lambda_1, \ \lambda_1 > 0$$

if η is sufficiently small, say $\eta < \eta_0$.

We shall use the notation

$$M(t) = \max\{i(t), w(t)\}, \qquad m(t) = \min\{i(t), w(t)\}. \tag{33}$$

Lemma 6.6. *Suppose Eq. (23) holds with $0 < \eta < \min\{\eta_0, \mu/2\gamma\}$ for arbitrarily small $\eta_0 > 0$. Then*

$$M(T_0 + n\omega + \tau) \geq c_1 \mathrm{e}^{\lambda_1 n\omega}\, m(T_0 + \tau) \text{ if } T_0 + n\omega + \tau < \Lambda, \tag{34}$$

where c_1 is a positive constant independent of η and n is any positive integer.

Proof. Denote by $z(t)$ the solution of

$$\frac{\mathrm{d}z}{\mathrm{d}t} = E(t)z, \ z(T_0 + \tau) = (i(T_0 + \tau), w(T_0 + \tau)),$$

where $E(t) = A(t) - c\eta J$. By the proof of Lemma 6.2 the inequality (27) holds, so that, by a comparison argument as in Sect. 5,

$$z_1(t) \leq i(t), \ z_2(t) \leq w(t) \text{ if } T_0 + \tau < t < \Lambda. \tag{35}$$

Next,

$$z(T_0 + nw + \tau) = \exp\left(\int_{T_0+\tau}^{T_0+n\omega+\tau} E(s)\mathrm{d}s\right) z(T_0 + \tau) = \mathrm{e}^{n\omega\langle E\rangle} z(T_0 + \tau). \tag{36}$$

Let $(Q = (q_{ij})$ be a matrix such that

$$Q\langle E\rangle Q^{-1} = D, \ D = \begin{pmatrix} \lambda_1 & 0 \\ 0 & \lambda_2 \end{pmatrix}.$$

Then

$$Q\mathrm{e}^{n\omega\langle E\rangle} Q^{-1} = \mathrm{e}^{n\omega Q\langle E\rangle Q^{-1}} = \mathrm{e}^{n\omega D} = \begin{pmatrix} \mathrm{e}^{n\omega\lambda_1} & 0 \\ 0 & \mathrm{e}^{n\omega\lambda_2} \end{pmatrix}.$$

Hence, from Eq. (36),

$$Qz(T_0+n\omega+\tau)=\begin{pmatrix} e^{n\omega\lambda_1} & 0 \\ 0 & e^{nw\lambda_2} \end{pmatrix} Qz(T_0+\tau). \tag{37}$$

We next observe that (q_{11},q_{12}) is an eigenvector of $\langle E\rangle$ corresponding to λ_1. One can verify directly that, since $\lambda_1>\lambda_2$, $\lambda_1>0$, q_{11} and q_{12} have the same sign; this also follows from the general theory mentioned in the paragraph following Theorem 2.1. Since the i-th component of the right-hand side of Eq. (37) is equal to

$$e^{nw\lambda_1}[q_{i1}i(T_0+\tau)+q_{i2}w(T_0+\tau)],$$

the assertion (34) then follows from Eqs. (35) and (37). □

Lemma 6.7. *For any sufficiently small* η *(say* $0<\eta<\eta_0$*) and arbitrarily small* $m(0)$ *(*$m(0)>0$*), set*

$$T_1=T_1(\eta,m(0))=T_0(\eta)+\frac{1}{\lambda_1\omega}\left\{\left[BT_0(\eta)+\ln\frac{\eta}{c_1m(0)}\right]^+ +1\right\}, \tag{38}$$

where $T_0(\eta)$ *is defined by Eq. (25). Then there exists a first time* t_1, $0<t_1\le T_1$ *such that*

$$M(t_1)>\eta. \tag{39}$$

Proof. We proceed by contradiction, assuming that Eq. (39) is not true. Then $M(t)<\eta$ for all $0\le t\le T_1$ and, in particular, $i(t)<\eta$ for $0\le t\le T_1$. Hence we can apply Lemma 6.6 to deduce that Eq. (34) holds, and, by Lemma 6.6, we then have

$$M(T_0+n\omega)\ge c_1e^{\lambda_1 n\omega}\,m(T_0)\ge c_1e^{\lambda_1 n\omega}e^{-BT_0}m(0).$$

But the right-hand side is larger than η if

$$\lambda_1 n\omega>\left[BT_0+\ln\frac{\eta}{c_1m(0)}\right]^+,$$

i.e., $M(t)>\eta$ for $t=T_1$, which is a contradiction. □

Proof of Theorem 4.2. By Lemma 6.7 there exists a $\bar{t}_1$, $\bar{t}_1\le T_1(\eta,m(0))$ such that Eq. (39) holds, i.e., $\max(i(\bar{t}_1),w(\bar{t}_1))>\eta$. Consider first the case

$$i(\bar{t}_1)>\eta.$$

Then, by Lemma 6.5,

$$i(t+\bar{t}_1) > \eta e^{-Bt} \text{ for all } t > 0,$$

and by Lemma 6.3 we deduce that

$$w(\hat{t}_0+\bar{t}_1) > \frac{\xi_1}{2\xi_2} e^{-B\hat{t}_0} \text{ where } \hat{t}_0 = 1+\frac{1}{\xi_2}\ln 2.$$

Setting $t_1' = \hat{t}_0+\bar{t}_1$, it follows that

$$i(t_1') \geq \eta e^{-B\hat{t}_0},\ w(t_1') \geq \frac{\xi_1}{2\xi_2}\ \eta e^{-B\hat{t}_0}. \tag{40}$$

We next consider the case

$$w(\bar{t}_1) > \eta.$$

By Lemma 6.5,

$$w(t+\bar{t}_1) > \eta e^{-Bt} \text{ for all } t > 0,$$

and, by Lemma 6.4,

$$i(t_0+\bar{t}_1) > c\eta e^{-Bt_0}$$

where t_0 is defined by Eq. (31). Hence, setting $t_1'' = t_0+\bar{t}_1$ we get,

$$w(t_1'') > \eta e^{-Bt_0},\ i(t_1'') \geq c\eta e^{-Bt_0}. \tag{41}$$

We have thus proved that Eq. (39) implies that either Eq. (40) or Eq. (41) must hold, and, by Lemma 6.5, we conclude that

$$m(t_1) \geq c_2\eta,\ t_1 = \max\{t_1', t_1''\}, \tag{42}$$

and, since we can take $t_2 > 1+t_1$,

$$1 < t_1 - \bar{t}_1 < c_3, \tag{43}$$

where the constants c_2, c_3 are independent of η.

We now repeat the previous argument starting at $t = t_1$ with $m(0)$ replaced by $m(t_1)$. We then need to replace $T_1(\eta, m(0))$ by $T_1(\eta, m(t_1))$. Note that whereas $T_1(\eta, m(0)) \to \infty$ if $m(0) \to 0$, $T_1(\eta, m(t_1))$ is bounded from above by a constant $c(\eta)$ which depends only on η. We conclude that there exists a point t_2 such that

$$m(t_2) > c_2\eta,\ 1 < t_2 - t_1 < c(\eta)+c_3,$$

where the constants c_2, c_3 are the same constants as in Eqs. (42) and (43).

Proceeding similarly step-by-step, we construct a sequence $\{t_m\}$ such that

$$m(t_m) \geq c_2\eta, \; 1 < t_m - t_{m-1} \leq c(\eta) + c_3 \quad (m = 3,4,\dots).$$

Using once more Lemma 6.5, we get

$$m(t) \geq c_3\eta e^{-B(c(\eta)+c_3)} \text{ for all } t > \bar{t}_1,$$

and this completes the proof of Theorem 4.2 with sufficiently small δ. □

7 Proof of Theorem 4.3

We shall use the following Horn's Fixed Point Theorem [5]; see also [4] where this theorem is applied to epidemic models with seasonal contact rate.

Theorem 7.1 ([5]). *Let $X_0 \subset X_1 \subset X_2$ be nonempty convex sets in a Banach space X such that X_0 and X_2 are compact and X_1 is open relative to X_2. Let W be a continuous mapping $X \to X$ such that, for some positive integer m, $W^j(X_1) \subset X_2$ for $1 \leq j \leq m-1$, and $W^j(X_1) \subset X_0$ for $m \leq j \leq 2m-1$, where $W^j = \underbrace{W \circ W \circ \cdots \circ W}_{j\text{–times}}$. Then W has a fixed point in X_0.*

Proof. We take the Banach space X to be $\mathbb{R}^4$ with points $z = (s,i,w,r,)$, and introduce the convex sets

$$X_2 = \Omega \; (\Omega \text{ as in } (4.2)),$$
$$X_1 = \Omega \cap \left\{ i > \frac{1}{2}\delta, \; w > \frac{1}{2}\delta \right\},$$
$$X_0 = \Omega \cap \{ i \geq \delta, \; w \geq \delta \},$$

where δ is any sufficiently small positive number. Clearly X_0, X_1, X_2 are convex sets, X_0 and X_2 are compact, and X_1 is open relative to X_2. For any z_0 in X we define $z(t)$ to be the solution of (4.3) with $z(0) = z_0$, and introduce the mapping

$$W : z_0 \to z(\omega).$$

Then W maps X into X and, clearly, $W^j(X_1) \subset X_2$ for all $j = 1,2,3,\dots$. By Theorem 4.2, $W^j(X_1) \subset X_0$ if $j\omega > T_0\left(\delta, \frac{\delta}{2}\right)$ where $T_0(\eta, m(0))$ is the function defined in Eq. (38). Hence the conditions in Horn's theorem are satisfied with m sufficiently large. We conclude that W has a fixed point in X_0, which is the periodic solution asserted in Theorem 4.3. □

8 Extensions

The results of this chapter can be extended to other disease models. Consider first the extension of Theorem 4.1 to general systems (8a) and (8b). We assume that the

$$\mathscr{F}_i(t,x) \le \mathscr{F}_i(t,x^*,x^0_{m+1}(t),\dots,x^0_n(t)) \equiv \tilde{\mathscr{F}}_i(t,x^*),$$

where $x^0(t) = (0,\dots,0,x^0_{m+1}(t),\dots,x^0_n(t))$ is the DFE and $x^* = (x_1,\dots,x_m)$, and that

$$\mathscr{V}_i = \mathscr{V}_i(t,x^*).$$

Then

$$\frac{\mathrm{d}x_i}{\mathrm{d}t} \le \tilde{\mathscr{F}}_i(t,x^*) - \mathscr{V}_i(t,x^*) \quad (i = 1,\dots,m)$$

and, if the right-hand side is linear in x^*, then

$$\frac{\mathrm{d}x^*}{\mathrm{d}t} \le A(t)x^*. \tag{44}$$

Hence, if $A(t)$ depends linearly on the ω-periodic parameters of the system (8a) and (8b), then, by the proof of Theorem 4.1, $[R_0] < 1$ implies that $R_0 < 1$.

We next turn to extensions of Theorem 4.2 and set

$$M(t) = \max_{1\le j\le m}\{x_j(t)\},\ m(t) = \min_{1\le j\le m}\{x_j(t)\}.$$

As in the proof of Theorem 4.2, we need to establish three properties of the solution:

(i) There exists a j_0 $(1 \le j_0 \le m)$ such that if $x_{j_0}(t) < \eta$ for all $t < \Lambda$, then there exists a T_1 independent of η such that $x_j(t) < c_2\eta$ for all $1 \le j \le m$ and $T_1 < t < \Lambda$.

(ii) If $\sum_{i=1}^{m} x_j(t) < \eta$ for all $0 < t < \Lambda$, then there is a $T_0 = T_0(\eta, x(0))$ such that

$$M(T_0 + n\omega + \tau) \ge c_1 \mathrm{e}^{\lambda_1 n\omega} m(T_0 + \tau) \text{ for } T_0 + n\omega + \tau < \Lambda,$$

where c_1 is a positive constant, and λ_1 is the largest eigenvalue of the matrix $\langle A\rangle - c\eta J$ where $A(t)$ is the matrix in Eq. (44); here J can be any matrix with nonnegative elements.

(iii) If $x_i(t) > \eta$ for some $i = 1,\dots,m$, for $0 < t < \Lambda$ and any η positive and sufficiently small, then $x_j(t) > c_2\eta$ and $t_0 < t < \Lambda$ for all $j = 1,\dots,m$, where c_2 is a positive constant and c_2 and t_0 are independent of η.

A version of this procedure was utilized in [3] for a model of malaria with periodic coefficient. We give here two more examples.

The first example is an SEIR tuberculosis model [8]

$$\frac{\mathrm{d}E}{\mathrm{d}t} = (1-q)\beta(t)\frac{S}{N}I - (\mu + k(t))E, \tag{45a}$$

$$\frac{\mathrm{d}I}{\mathrm{d}t} = q\beta(t)\frac{S}{N}I - (\mu + d + r)I, \tag{45b}$$

$$\frac{\mathrm{d}S}{\mathrm{d}t} = \mu - \beta(t)\frac{S}{N}I - \mu S, \tag{45c}$$

$$\frac{\mathrm{d}R}{\mathrm{d}t} = rI - \mu R, \tag{45d}$$

where $N = S + E + I + R$. Here $x_1 = E$, $x_2 = I$, $x_3 = S$, $x_4 = R$ and $m = 2$, $n = 4$ in the notation of Sect. 3. Since $S \leq N$, we can apply the proof of Theorem 4.1 to deduce that if $[R_0] < 1$ then the DFE, $(1,0,0,0)$, is globally asymptotically stable, so that $R_0 < 1$. However, as shown in [8], for some choices of $\beta(t)$, $k(t)$ there holds: $[R_0] > 1 > R_0$, so that Theorem 4.2 does not hold.

The second example is a model of staged progression in disease transmission of HIV [6]:

$$\frac{\mathrm{d}I_1}{\mathrm{d}t} = \beta_1(t)\frac{S}{N}I_1 + \beta_2(t)\frac{S}{N}I_2 - (\nu_1 + d_1)I_1, \tag{46a}$$

$$\frac{\mathrm{d}I_2}{\mathrm{d}t} = \nu_1 I_1 - (\nu_2 + d_2)I_2, \tag{46b}$$

$$\frac{\mathrm{d}I_3}{\mathrm{d}t} = \nu_2 I_2 - d_3 I_3, \tag{46c}$$

$$\frac{\mathrm{d}S}{\mathrm{d}t} = \mu - \beta_1(t)\frac{S}{N}I_1 - \beta_2(t)\frac{S}{N}I_2 - \mu S, \tag{46d}$$

with $DFE = (1,0,0,0)$; here $N = S + I_1 + I_2 + I_3$. As shown in [12], for some choice of $\beta_1(t)$, $\beta_2(t)$ and the parameters ν_i d_j there holds: $R_0 > 1 > [R_0]$, so that Theorem 4.1 does not hold.

In order to prove Theorem 4.2 for the system (46a)–(46d) we proceed with steps (i), (ii) and (iii) described above. The first step is to show that if $I_1(t) \leq \eta$ for $0 < t < \Lambda$ then $I_2(t) \leq c_1\eta$ and $I_3(t) \leq c_2\eta$ for $T_0 < t < \Lambda$ and positive constants c_i. This follows from the differential equations for I_2 and I_3, using Lemma 6.1 (i). The second step, (ii), follows by the same arguments as in the proof of Lemma 6.6.

We next note that if $I_1(t) > \eta$ for $0 < t < \Lambda$ then $I_2(t) \geq c_3\eta$ and $I_3(t) \geq c_4\eta$ for $t_0 < t < \Lambda$ and positive constants c_3, c_4. Also, if $I_2(t) \geq \eta$ then $I_1(t) \geq c_5\eta$ for $t_0 < t < \Lambda$ with another t_0, and $c_5 > 0$; indeed the proof is similar to the proof of Lemma 6.4. First we estimate S analogously to Eq. (32), and then use the estimate

$$\beta_2(t)\frac{S}{N}I_2 \geq \frac{c\eta}{N} \geq \frac{c\eta}{N(0)}$$

on the right-hand side of the equation for I_1, noting that the parameters in Eqs. (46a)–(46d) should be such that $\mathrm{d}N/\mathrm{d}t \leq 0$ (since HIV reduces life expectancy).

We finally need to prove that if $I_3(t) \geq \eta$ for $0 < t < \Lambda$ then $I_1(t) \geq c_6\eta$ for $t_0 < t < \Lambda$ and $c_6 > 0$. This, however, we can only establish if we modify the system (46a)–(46d) by adding a term $\beta_3(t)\frac{S}{N}I_3$ to the right-hand side of the first equation and subtracting it from the right-hand side of the last equation. We conclude that Theorem 4.2 holds for this modified system.

9 Conclusion and Discussion

The basic reproduction number R_0 provides important but limited information on the spread of an infectious disease. It informs whether initial small infection leads to endemic disease (which is the case if $R_0 > 1$) or whether the infection will die out (which is the case if $R_0 < 1$). When dealing with a disease that is seasonality(periodically) dependent, it is clearly more difficult to make the same determination on the course of an initial infection; this is also reflected mathematically by the difficulty in computing R_0. The aim of this chapter was to determine, for some disease models, a procedure to compute when $R_0 < 1$ and when $R_0 > 1$ for a disease with seasonality. For clarity, we developed this method for water-dependent diseases, such as cholera, and then explained in the Sect. 8, how this method can be extended to other models.

We also proved that, in case $R_0 > 1$, there exists a periodic endemic solution. The case of a malaria model was developed earlier, in [3].

One immediate question is whether there is just one periodic solution, and, if not, how many. The answer may be important, since knowing the course of the disease progression may help in the development of a strategy to contain it.

The inequality $R_0 > 1$ tells us that the disease will become endemic. We expect that the larger R_0 is the larger by the size of infected compartment, at any future time. But such a result has not been proved and remains an interesting open problem.

References

1. Bacaër, N.: Approximation of the basic reproduction number R_0 for waterborne diseases with seasonality. Bull. Math. Biol. **69**, 1067–1091 (2007)
2. Bacaër, N., Guernaoui, S.: The epidemic threshold of vector-borne disease with seasonality. J. Math. Biol. **53**, 421–436 (2006)
3. Dembele, B., Friedman, A., Yakubu, A.A.: Malaria model with periodic mosquito birth and death rates. J. Biol. Dyn. **3**, 430–445 (2009)
4. Greenhalgh, D., Moneim, I.A.: SIRS epidemic model and simulations using different types of seasonal contact rate. Syst. Anal. Model. Simul. **43**, 573–600 (2003)
5. Horn, W.A.: Some fixed point theorems for compact maps and flows in Banach spaces. Trans. Am. Math. Soc. **149**, 391–402 (1970)

6. Hyman, J.M., Li, S., Stanley, E.A.: The differential infectivity and staged progression models for the transmission of HIV. Math. Biosci. **155**, 77–109 (1999)
7. LaSalle, J.P.: The Stability of Dynamical Systems. SIAM, Philadelphia (1976)
8. Liu, L., Zhao, X.-Q., Zhou, Y.: A tuberculosis model with seasonality. Bull. Math. Biol. **72**, 931–952 (2010)
9. Tien, J.H., Earn, D.J.D.: Multiple transmission pathways and disease dynamics in waterborne pathogen model. Bull. Math. Biol. **72**, 1506–1533 (2010)
10. Tien, J.H., Poinar, H.N., Fisman, D.N., Earn, D.J.D.: Herald waves of cholera in 19th century London. J. R. Soc. Interface **8**, 756–760 (2011)
11. van den Driessche, P., Watmough, J.: Further notes on the basic reproduction number. Chap. 6 Mathematical Epidemiology. In: Brauer, F., van den Driessche, P., Wu, J. (eds.) Springer Lecture Notes in Mathematics, vol. 1945, pp. 159–178. Springer, Berlin (2008)
12. Wang, W., Zhao, X.-Q.: Threshold dynamics for compartmental epidemic models in periodic environments. J. Dyn. Differ. Equat. **20**, 699–717 (2008)
13. Zhang, F., Zhao, X.-Q.: A periodic epidemic model in patchy environment. J. Math. Anal. Appl. **325**, 496–516 (2007)

Periodic Incidence in a Discrete-Time *SIS* Epidemic Model

Najat Ziyadi and Abdul-Aziz Yakubu

1 Introduction

Mathematical models have continued to increase our understanding of the spread of infectious diseases and their control in both humans and animals. In most infectious diseases, the incidence coefficient or contact rate (the rate of new infections) plays a key role in ensuring that the model gives a reasonable qualitative description of the real disease dynamics. To accurately gauge the impact of infectious diseases prevention efforts, it is important to understand the relation between disease transmission and the host population dynamics. In [8–11], Castillo-Chavez and Yakubu introduced a framework for studying infectious disease dynamics in strongly fluctuating populations. In their model framework, Castillo-Chavez and Yakubu assumed that the host demographics is governed by the Ricker model and the contact rate is constant. However, periodicity in infectious disease incidence is known to occur in chickenpox, measles, pertussis, gonorrhea, mumps, influenza, and other infectious diseases.

In this chapter, we extend the SIS epidemic model framework of Castillo-Chavez and Yakubu to include periodic incidence coefficients. Using the extended model, we obtain that the SIS model of Castillo-Chavez and Yakubu can exhibit oscillatory dynamics when the contact rate is periodic and the recruitment dynamics is asymptotically constant (non-cyclic and non-chaotic demographic dynamics). Some of the

N. Ziyadi (✉)
Department of Mathematics, Morgan State University, Baltimore, MD 21251, USA
e-mail: najat.ziyadi@morgan.edu

A.-A. Yakubu
Department of Mathematics, Howard University, Washington, DC 20059, USA
e-mail: ayakubu@howard.edu

U. Ledzewicz et al., *Mathematical Methods and Models in Biomedicine*, Lecture Notes on Mathematical Modelling in the Life Sciences, DOI 10.1007/978-1-4614-4178-6_15,

mechanisms that have been used to generate oscillations in epidemiological models include external forcing, seasonality, time-dependent coefficients, periodic incidence, age-structure, time-delay, and migration [1,4,5,8–12,16–19,22,26,28–31].

Following Castillo-Chavez and Yakubu, we assume that a disease invades and subdivides the target population into two compartments: susceptibles (non-infectives) and infectives. Prior to the time of disease invasion, the host population is assumed to be asymptotically constant via either Beverton-Holt or constant recruitment functions or growing geometrically or oscillatory via the Ricker recruitment function [6–11, 13–15, 17, 18, 20, 21, 23–25, 27, 29, 30]. The transition from susceptible to infective is a function of the *T-periodic* contact rate $\beta_t = \beta_{t+T}$ and the proportion of infectives (prevalence) in the population. To simplify our analysis, we assume that the disease is nonfatal and individuals (infected and susceptible) have equal probability of surviving one generation. Mild viral infections, such as most infections from rhinoviruses (causative agents of the common cold) are examples of such nonfatal infections. Our primary focus is on the impact of periodic contact rate, asymptotically constant, geometric growth and periodic demographic dynamics on the persistence or control of infectious diseases.

For many epidemiological models, the threshold parameter is the basic reproduction number, $\mathfrak{R}_0$. In this chapter, we compute $\mathfrak{R}_0$ and used it to predict disease persistence or extinction when the host population dynamics are asymptotically constant or when the host population is growing at a geometric rate. That is, potentially, by developing strategies that reduce $\mathfrak{R}_0$ to values less than 1, we can combat infectious diseases with periodic incidence, when the host population dynamics is either asymptotically constant or under geometric growth. Castillo-Chavez and Yakubu, in [8, 10], showed that the demographic equation drives the disease dynamics in discrete-time SIS models with constant contact rate. We illustrate that when the host population is asymptotically constant, it is possible for the infective and susceptible populations to exhibit oscillatory dynamics. That is, under periodic contact rate, the demographic dynamics do not drive the disease dynamics. When the demographic dynamics are oscillatory, we explore the relationship between the demographic equation and the epidemic process as the demographic model undergoes period-doubling bifurcations.

The chapter is organized as follows. In Sect. 2, we introduce the demographic equation for the study. The equation, a deterministic discrete-time model, describes the dynamics of the host population before disease invasion [2,3]. The main model, a discrete-time SIS epidemic model with periodic contact rate, is constructed in Sect. 3 of the chapter. In Sect. 4, the basic reproductive number, $\mathfrak{R}_0$, is introduced and used to predict the (uniform) persistence or extinction of the infective population, where the population dynamics are asymptotically constant or under geometric demographic dynamics. In Sect. 5, we illustrate that it is possible for the demographic dynamics to drive both the *S-dynamics* and *I-dynamics* as it undergoes period-doubling bifurcations. The implications of our results are discussed in Sect. 6.

2 Demographic Equation

Simple deterministic discrete-time epidemic models are usually formulated under the assumption that the dynamics of the total population are governed by equations of the form

$$N(t+1) = f_\gamma(N(t)) \equiv \gamma(f(N(t)) + N(t)), \tag{1}$$

where $N(t)$ is the total population size at time t, $\gamma \in (0,1)$ is the constant "probability" of surviving per generation, and $f : \mathbb{R}_+ \to \mathbb{R}_+$ models the birth or recruitment process [2, 3, 8–11, 17, 18, 29, 30].

The set of iterates of the map

$$f_\gamma(N) = \gamma(f(N) + N)$$

is equivalent to the set of density sequences generated by Model (1). In the demographic equation, this form of regulation is known to allow for complex population dynamics such as period-doubling bifurcations and chaos [8–11, 14–18, 23–25, 29, 30].

Following Castillo-Chavez and Yakubu [8–11], we focus on the following four types of recruitment functions that are commonly found in the literature.

(a) Constant Recruitment:

$$f(N) = \frac{k(1-\gamma)}{\gamma},$$

where k is a positive constant. Under constant recruitment function, $N_\infty = \lim_{t\to\infty} f_\gamma^t(N) = k$ for all $N \geq 0$ and the total population is asymptotically constant, where

$$f_\gamma^t(N) = \underbrace{f_\gamma \circ f_\gamma \circ \cdots \circ f_\gamma(N)}_{t \text{ compositions}}$$

is the composition map evaluated at the point N.

(b) Geometric Growth Recruitment:

$$f(N) = \frac{\mu}{\gamma} N,$$

where μ is a positive constant. Let

$$\mathfrak{R}_\mathrm{D} = \frac{\mu}{1-\gamma}.$$

Under geometric growth dynamics, the total population goes extinct at a geometric rate when $\mathfrak{R}_\mathrm{D} < 1$. However, the total population explodes at a geometric rate when $\mathfrak{R}_\mathrm{D} > 1$.

(c) Beverton–Holt Recruitment:

$$f(N) = \frac{(1-\gamma)\mu kN}{\gamma((1-\gamma)k + (\mu - 1 + \gamma)N)},$$

where the constant $\mu > 1$ [7–11, 20, 27]. Under the classic Beverton–Holt recruitment function, zero is an unstable fixed point. $N_\infty = k > 0$ for all $N > 0$ and, as in the case of constant recruitment, the total population is asymptotically constant.

(d) Ricker Recruitment:

$$f(N) = \frac{(1-\gamma)}{\gamma} N e^{k-N},$$

where k is a positive constant [8–11,23–25,27,29,30]. Under the classic Ricker recruitment, zero is an unstable fixed point. If $0 < k < \frac{2}{1-\gamma}$, then $N_\infty = k$ for all $N > 0$ and, as in the case of constant recruitment, the total population is asymptotically constant. However, as k increases past $\frac{2}{1-\gamma}$, the positive fixed point, k, undergoes period-doubling bifurcations.

The total population is *uniformly persistent* if there exists a constant $\eta > 0$ such that $\lim_{t\to\infty} f_\gamma^t(N) \geq \eta$ for every positive initial population size; $N > 0$. The total population is *persistent* if $\lim_{t\to\infty} f_\gamma^t(N) > 0$ for every positive initial population size. Consequently, uniform persistence implies persistence of the population. We note that, the total population is uniformly persistent whenever the recruitment function is constant or geometric growth or Beverton–Holt or Ricker function.

3 SIS Epidemic Model with Periodic Infection

To introduce the discrete-time susceptible-infective-susceptible (*SIS*) epidemic model, we assume that a nonfatal disease invades and subdivides the target population into two compartments: susceptibles (noninfectives) and infectives. Prior to the time of disease invasion, the population dynamics are assumed to be governed by Eq. (1), where the recruitment function is either constant or geometric growth or Beverton–Holt or Ricker function. Let $S(t)$ denote the population of susceptibles; $I(t)$ denotes the population of the infected, assumed infectious, $N(t) \equiv S(t) + I(t)$ denotes the total population size at generation t, N_∞ denotes the demographic positive steady state or attracting population and $\overline{N}_0$ the initial point on a globally attracting cycle, when it exists.

We assume that there is no immunity from the disease and there is no disease-induced mortality. There is no vertical transmission, and both susceptible and infected individuals reproduce into the susceptible class. Although the SIS model has no disease-induced mortality, the population does experience death via density dependence. Susceptible and infected individuals survive (respectively, die) with constant probability γ (respectively, $(1-\gamma)$) per generation.

Let $\phi : [0,\infty) \to [0,1]$ be a monotone concave probability function with $\phi(0) = 1, \phi'(N) < 0$ and $\phi''(N) \geq 0$ for all $N \in [0,\infty)$. At generation t, we assume the susceptible individuals become infected with nonlinear probability $(1 - \phi(\beta_t I(t)/N(t)))$ per generation and infected individuals recover with constant probability $\sigma \in (0,1)$, where the transmission constant β_t is $T-periodic$. That is, $\beta_{t+T} = \beta_t$. For each $t \in \{0,1,\ldots,T-1\}$, β_t is positive and models the impact of prevalence on ϕ. When infections are modeled as Poisson processes, then the "escape" function, $\phi(\beta_t I(t)/N(t)) = \mathrm{e}^{-(\beta_t I(t))/N(t)}$ [8–11, 17, 18, 29, 30].

The periodically forced frequency-dependent discrete-time SIS model implicitly assumes the ordering of events. At the end of each generation, susceptibles become infected while infected recover; both susceptibles and infected reproduce into the susceptible class; a fraction of each class is removed. This important assumptions distinguish our discrete-time model from a similar continuous-time differential equation model. Typically, continuous-time differential equation models with similar well-defined distinct temporal phases are non-autonomous. Taking into account the temporal ordering of events, we derive our model in the following three steps.

1. Disease transmission and recovery

$$\left.\begin{aligned} S_1(t) &= \phi\left(\tfrac{\beta_t I(t)}{N(t)}\right) S(t) + \sigma I(t), \\ I_1(t) &= (1-\sigma)I(t) + \left(1 - \phi\left(\tfrac{\beta_t I(t)}{N(t)}\right)\right) S(t). \end{aligned}\right\} \tag{2}$$

   That is, after disease transmission and recovery, $S_1(t)$ denotes the susceptible individuals and $I_1(t)$ denotes the infected.

2. Reproduction (both S and I reproduce into S)

$$\left.\begin{aligned} S_2(t) &= S_1(t) + f(S_1(t) + I_1(t)), \\ I_2(t) &= I_1(t). \end{aligned}\right\}$$

   That is,

$$\left.\begin{aligned} S_2(t) &= \phi\left(\tfrac{\beta_t I(t)}{N(t)}\right) S(t) + \sigma I(t) + f(N(t)), \\ I_2(t) &= (1-\sigma)I(t) + \left(1 - \phi\left(\tfrac{\beta_t I(t)}{N(t)}\right) S(t)\right). \end{aligned}\right\} \tag{3}$$

   That is, after disease transmission, recovery from the disease and reproduction, $S_2(t)$ denotes the susceptible individuals and $I_2(t)$ denotes the infected.

3. Death/survival

$$\left.\begin{aligned} S_3(t) &= \gamma S_2(t) = \gamma\left(\phi\left(\tfrac{\beta_t I(t)}{N(t)}\right) S(t) + \sigma I(t) + f(N(t))\right), \\ I_3(t) &= \gamma I_2(t) = \gamma\left((1-\sigma)I(t) + \left(1 - \phi\left(\tfrac{\beta_t I(t)}{N(t)}\right)\right) S(t)\right). \end{aligned}\right\} \tag{4}$$

   After disease transmission, recovery, reproduction and survival (death), $S_3(t)$ denotes the susceptible individuals and $I_3(t)$ denotes the infected.

Table 1 Model (5) parameters and functions

Parameter	Description
γ	Survival "probability" of susceptibles and infectives per generation
β_t	$T-Periodic$ transmission constant ($\beta_t = \beta_{t+T}$)
ϕ	Frequency-dependent escape "probability" function
σ	Recovery "probability" of infective individuals per generation
g	Density-dependent escape per-capita growth function
f	Recruitment function

Our assumptions and notation lead to the following frequency-dependent *SIS* epidemic model:

$$\left.\begin{aligned} S(t+1) &= \gamma\Big(\phi\left(\tfrac{\beta_t I(t)}{N(t)}\right)S(t)+\sigma I(t)+f(N(t))\Big), \\ I(t+1) &= \gamma\Big((1-\sigma)I(t)+\Big(1-\phi\left(\tfrac{\beta_t I(t)}{N(t)}\right)\Big)S(t)\Big), \end{aligned}\right\} \tag{5}$$

where $0<\gamma<1$, $\beta_t=\beta_{t+T}$ and $\sigma\in(0,1)$. In Model (5), we assume that events happen in the following order: disease transmission, recovery, reproduction, and survival (death). However, in real biological systems, these three events may happen in different orders. For example, when reproduction happens before disease transmission and survival (death) happens after disease transmission, proceeding as in the derivation of Model (5), we obtain the following system:

$$\left.\begin{aligned} S(t+1) &= \gamma\Big(\phi\left(\tfrac{\beta_t I(t)}{N(t)+f(N(t))}\right)(S(t)+f(N(t)))+\sigma I(t)\Big), \\ I(t+1) &= \gamma\Big((1-\sigma)I(t)+\Big(1-\phi\left(\tfrac{\beta_t I(t)}{N(t)+f(N(t))}\right)\Big)(S(t)+f(N(t)))\Big). \end{aligned}\right\} \tag{6}$$

Clearly, Model (5) is different from Model (6). Cyclic permutations of the three distinct temporal phases lead to models that are topologically conjugate to Model (5). However, noncyclic permutations of the three temporal phases may lead to models that are not topologically conjugate to Model (5). For simplicity, we will analyze Model (5). We summarize the model parameters and functions in Table 1.

Below, we summarize some of the underlying assumptions in model (5).

(a) There is no acquired immunity.
(b) There is no latent period (or it is very short).
(c) There is no disease induced mortality.
(d) Transmission is frequency dependent rather than density dependent.

Theoretical and empirical investigations have been done on comparing these assumptions. In both continuous-time and discrete-time epidemic models, it is known that these assumptions are not universally applicable [4, 12, 16–18, 29, 30].

Model (5) is a deterministic frequency-dependent SIS epidemic model and has no "probability" of transmission. The assumption of deterministic dynamics is valid in a large population, where stochasticity is unimportant. This assumption places a constraint on the applicability of the model. Stochastic transmission (including a Poisson process) in a small population (close to extinction), for example, would not be described by the SIS model.

From $N(t+1) = S(t+1) + I(t+1)$, we obtain that the total host population dynamics in the presence of the disease are described by Model (1). In the absence of the disease, the susceptible host population dynamics are described by the following single species model:

$$S(t+1) = f_\gamma(S(t)) = \gamma(S(t) + f(S(t))).$$

When the recruitment function is either constant or geometric growth or Beverton–Holt or Ricker function, then $f(S) > 0$ whenever $S > 0$. Hence, in Model (5),

$$S(0) > 0 \text{ and } I(0) = 0 \text{ implies that } S(t) > 0 \text{ and } I(t) = 0 \text{ for } t = 1, 2, \ldots.$$

In the present chapter, we use Model (5) to study the relationship between the demographic and disease dynamics in a periodically forced frequency-dependent discrete-time SIS model. When the transmission rate is constant, then Model (5) reduces to that of Castillo-Chavez and Yakubu [8–11]. Others have studied discrete-time SIS models with constant transmission and periodic survival rates [2, 3, 17, 18, 29, 30].

4 Disease Extinction Versus Disease Persistence

Classical theory of disease epidemics usually involves computation of an epidemic threshold parameter, the basic reproductive number $\mathfrak{R}_0$ [2–4]. In this section, we introduce $\mathfrak{R}_0$ for Model (5), the SIS model with periodic disease transmission. In this section, we study Model (5), where the recruitment function is either constant or described by the Beverton–Holt model.

4.1 Asymptotically Constant Recruitment Functions

When the recruitment function is either $f(N) = \frac{k(1-\gamma)}{\gamma}$ (constant recruitment) or $f(N) = ((1-\gamma)\mu kN)/(\gamma((1-\gamma)k + (\mu - 1 + \gamma)N))$ (Beverton–Holt model), then the total population is asymptotically constant. That is, long-term demographic effects disappear in Model (1) and $\lim_{t \to \infty} N(t) = N_\infty > 0$. In this case, letting $S(t) = N_\infty - I(t) \geq 0$ in the $I-equation$ of Model (5) gives rise to the following one-dimensional equivalent non-autonomous "limiting system" [6, 18, 32].

$$I(t+1) = \gamma\left((1-\sigma)I(t) + \left(1 - \phi\left(\frac{\beta_t I(t)}{N_\infty}\right)\right)(N_\infty - I(t))\right), \tag{7}$$

where $0 \leq I(t) \leq N_\infty$ for all $t \in \{0,1,2,\ldots\}$. Mathematical theorems on the qualitative dynamics equivalence of autonomous and nonautonomous systems, such as System (5) and Eq. (7), have been established by Best et al. [6], Franke and Yakubu [18] and Zhao [32].

On the closed interval $[0,N_\infty]$, let

$$F_{N_\infty,t}(I,\beta_t) = \gamma\left((1-\sigma)I + \left(1-\phi\left(\frac{\beta_t I}{N_\infty}\right)\right)(N_\infty - I)\right).$$

When the total population is asymptotically constant, then the set of sequences generated by

$$I(t+1) = F_{N_\infty,t}(I(t),\beta_t) \tag{8}$$

is the set of density sequences generated by the infective population. When $F_{N_\infty,t}(\cdot,\beta_t)$ has a unique positive fixed point and a unique critical point, we denote them by $I_{N_\infty,t}$ and $C_{N_\infty,t}$, respectively.

Lemma 4.1. *$F_{N_\infty,t}(I)$ satisfies the following conditions.*

(a) If $0 \leq I(0) \leq N_\infty$, then $F_{N_\infty,t}(I) < N_\infty$.
(b) $F'_{N_\infty,t}(0) = \gamma((1-\sigma) - \beta_t\phi'(0))$ and $F'_{N_\infty,t}(N_\infty) > -1$.
(c) $F_{N_\infty,t}(I)$ is concave down on $[0,N_\infty]$.
(d) $F_{N_\infty,t}(I) < F'_{N_\infty,t}(0)I$ on $(0,N_\infty]$.
(e) If $F'_{N_\infty,t}(0) > 1$, then $F_{N_\infty,t}$ has a unique positive fixed point $I_{N_\infty,t}$ in $[0,N_\infty]$.

Proof. (a) Since

$$\begin{aligned}
F_{N_\infty,t}(I) &= \gamma\left((1-\sigma)I + \left(1-\phi\left(\frac{\beta_t I}{N_\infty}\right)\right)(N_\infty - I)\right) \\
&\leq \gamma((1-\sigma)I + (N_\infty - I)) \\
&\leq \gamma(I + N_\infty - I) = \gamma N_\infty < N_\infty.
\end{aligned}$$

(b)

$$F'_{N_\infty,t}(I) = \gamma\left((1-\sigma) - \left(1-\phi\left(\frac{\beta_t I}{N_\infty}\right)\right) - \frac{\beta_t}{N_\infty}\phi'\left(\frac{\beta_t I}{N_\infty}\right)(N_\infty - I)\right)$$

$$\begin{aligned}
F'_{N_\infty,t}(0) &= \gamma((1-\sigma) - (1-\phi(0)) - \beta_t\phi'(0)) \\
&= \gamma((1-\sigma) - \beta_t\phi'(0)).
\end{aligned}$$

$$\begin{aligned}
F'_{N_\infty,t}(N_\infty) &= \gamma\left((1-\sigma) - \left(1-\phi\left(\frac{\beta_t N_\infty}{N_\infty}\right)\right) - \frac{\beta_t}{N_\infty}\phi'\left(\frac{\beta_t N_\infty}{N_\infty}\right)(N_\infty - N_\infty)\right) \\
&= \gamma\left((1-\sigma) - \left(1-\phi\left(\frac{\beta_t N_\infty}{N_\infty}\right)\right)\right) > -\gamma > -1.
\end{aligned}$$

(c)
$$F''_{N_\infty,t}(I) = \gamma\left(-\left(\frac{\beta_t}{N_\infty}\right)^2 \phi''\left(\frac{\beta_t I}{N_\infty}\right)(N-I) + 2\frac{\beta_t}{N_\infty}\phi'\left(\frac{\beta_t I}{N_\infty}\right)\right).$$

Since $\phi' < 0$ and $\phi'' \geq 0$ on $[0,\infty)$, we have

$$F''_{N_\infty,t}(I) < 0 \text{ on } [0, N_\infty].$$

(d) $F_{N_\infty,t}(0) = 0$ implies that $y = F'_{N_\infty,t}(0)I$ is the tangent line to the graph of $F_{N_\infty,t}(I)$ at $I = 0$. Since $F_{N_\infty,t}$ is concave down on $[0, N_\infty]$, its graph is below the tangent line at the origin on $[0, N_\infty]$. Hence,

$$F_{N_\infty,t}(I) < F'_{N_\infty,t}(0)I \text{ on } (0, N_\infty].$$

(e) $F_{N_\infty,t}(N_\infty) = \gamma((1-\sigma)N_\infty) < N_\infty$. Since $F'_{N_\infty,t}(0) > 1$, the graph of $F_{N_\infty,t}(I)$ starts out higher than the diagonal and must cross it before $I = N_\infty$. The concavity property of $F_{N_\infty,t}(I)$ (see (c)) implies that there is a unique positive fixed point. □

Let

$$\mathfrak{R}_{0,t} = \frac{-\gamma\beta_t\phi'(0)}{1-\gamma(1-\sigma)}.$$

In $\mathfrak{R}_{0,t}$, $\frac{1}{(1-\gamma(1-\sigma))}$ is the product of the average death adjusted infectious period in generations; γ is the proportion that can be invaded by the disease (survival first then infection) and at time t, $-\beta_t\phi'(0)$ is the maximum rate of infection of new recruits and susceptible individuals per infective. Thus, at time t, $\mathfrak{R}_{0,t}$ gives the average number of secondary infections due to small initial infective individuals over their life-time.

We note that $F'_{N_\infty,t}(0) > 1$ (respectively, $F'_{N_\infty,t}(0) < 1$), is equivalent to $\mathfrak{R}_{0,t} > 1$ (respectively, $\mathfrak{R}_{0,t} < 1$).

The threshold parameter (basic reproduction number),

$$\mathfrak{R}_0 = \prod_{t=0}^{T-1}\left(F'_{N_\infty,t}(0)\right),$$

determines the long-term behavior of the disease in Model (5) , where the total population is asymptotically constant. That is, we obtain that $\mathfrak{R}_0 < 1$ implies disease extinction whereas $\mathfrak{R}_0 > 1$ implies disease persistence. We collect these results in Theorem 4.1.

Theorem 4.1. *In Model* (5), *let* $N(t) = N_\infty$ *and* $N_\infty \geq I(0) > 0$.

(a) If $\mathfrak{R}_0 < 1$, *then* $\lim_{t\to\infty} I(t) = 0$. *That is, the disease goes extinct.*

(b) If $\mathfrak{R}_0 > 1$, *then* $\varliminf_{t\to\infty} F_{N_\infty,t}\circ\cdots\circ F_{N_\infty,1}\circ F_{N_\infty,0}(I) \geq \eta > 0$ *for some* $\eta > 0$. *That is, the infected population is uniformly persistent.*

Proof. (*a*) Since

$$\mathfrak{R}_0 = \prod_{t=0}^{T-1} \left(F'_{N_\infty,t}(0)\right) < 1,$$

Lemma 4.1 gives

$$I(1) = F_{N_\infty,0}(I(0)) \leq F'_{N_\infty,0}(0)I(0).$$

Thus,

$$I(2) = F_{N_\infty,1}(I(1)) \leq F'_{N_\infty,1}(0)I(1) \leq F'_{N_\infty,1}(0)F'_{N_\infty,0}(0)I(0)$$

and inductively

$$I(t) \leq \left(\prod_{t=0}^{t-1} F'_{N_\infty,t}(0)\right) I(0) \leq (\mathfrak{R}_0)^{\lfloor t/T \rfloor} \max_{j\in\{1,2,\cdots,T-1\}} \left\{\prod_{t=0}^{j-1} F'_{N_\infty,t}(0)\right\} I(0).$$

Since the sequence $\{I(t)\}$ is dominated by the decreasing sequence

$$(\mathfrak{R}_0)^{\lfloor t/T \rfloor} \max_{j\in\{1,2,\cdots,T-1\}} \left\{\prod_{t=0}^{j-1} F'_{N_\infty,t}(0)\right\} I(0),$$

it converges to zero. Hence $\lim_{t\to\infty} I(t) = 0$. The proof of Theorem 4.1 (b) is similar to that of Theorem 4.6 in [18] and is omitted. □

In Model (5), the demographic dynamics does not drive the disease dynamics. Now, we use constant recruitment and Theorem 4.1 in Example 4.1 to illustrate an infective population on a 2-cycle attractor, where the demographic dynamics is noncyclic.

Example 4.1. In Model (5), let

$$f(N) = \frac{k(1-\gamma)}{\gamma} \text{ and } \phi\left(\frac{\beta_t I}{N}\right) = e^{-\frac{\beta_t I}{N}},$$

where

$$k = 5,\ \beta_t = a + b \times (1 + (-1)^t),\ a = 0.1,\ b \in [42, 50],\ \gamma = 0.1 \text{ and } \sigma = 0.01.$$

In Example 4.1, the total population is asymptotically constant, $N_\infty = 5$, the interaction between susceptible and infected is modeled as a Poisson process, and disease transmission is $2-$ *periodic* ($\beta_{t+2} = \beta_t$). When $b = 42$, then $\beta_0 = 84.1$, $\beta_1 = 0.1$, $\mathfrak{R}_0 = 0.927 < 1$ and the disease goes extinct (Theorem 4.1 and Fig. 1). However, when $b = 45.38$ then $\mathfrak{R}_0 = 1.0001 > 1$ and the infective population persists on a positive 2-cycle attractor (Theorem 4.1 and Fig. 1). That is, increasing the periodic infection rate in Example 4.1 shifts the system from disease extinction phase to disease persistence on a 2-cycle attractor, where the total population persists on a fixed point attractor.

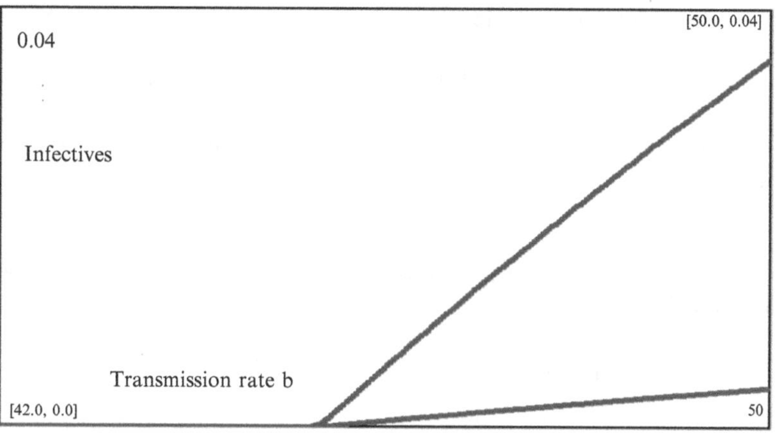

Fig. 1 In Example 4.1, the total infective population shifts from extinction to persistence as b is varied between 42 and 50. On the horizontal axis, $42 \leq b \leq 50$ and on the vertical axis, $0 \leq i \leq 0.04$

4.2 Geometric Growth Recruitment Function

When the recruitment function is a geometric growth function, we use proportions to study Model (5). That is, we introduce the new variables

$$s(t) = \frac{S(t)}{N(t)},$$

and

$$i(t) = \frac{I(t)}{N(t)}.$$

In the new variables, Model (5) with $f(N(t)) = \frac{\mu}{\gamma}N(t)$ becomes

$$\left.\begin{aligned} s(t+1) &= \tfrac{\gamma}{(\gamma+\mu)}\left(\phi\left(\beta_t i(t)\right)s(t) + \sigma i(t)\right) + \tfrac{\mu}{(\gamma+\mu)}, \\ i(t+1) &= \tfrac{\gamma}{(\gamma+\mu)}\left((1-\sigma)i(t) + \left(1-\phi\left(\beta_t i(t)\right)\right)s(t)\right). \end{aligned}\right\} \tag{9}$$

We note that,

$$s(t) + i(t) = 1$$

for all t in Model (9) implies all solutions live on the positive invariant line

$$\{(s,i) \in [0,\infty) \times [0,\infty) \mid s+i=1\}.$$

Using the substitution $s = 1-i$, the i-equation in Model (9) reduces to the "one-dimensional" equation

$$i(t+1) = \frac{\gamma}{(\gamma+\mu)}\left((1-\sigma)i(t) + \left(1-\phi\left(\beta_t i(t)\right)\right)(1-i(t))\right)$$

where $i(t) \leq 1$ for all $t \in \{0,1,2,\ldots\}$.

On the closed interval $[0,1]$, let

$$\widehat{F}_{1,t}(i) = \frac{\gamma}{(\gamma+\mu)}\left((1-\sigma)i + (1-\phi(\beta_t i))(1-i)\right).$$

When the total population is under geometric growth, then the set of sequences generated by

$$i(t+1) = \widehat{F}_{1,t}(i(t)) \tag{10}$$

is the set of density sequences generated by the "proportion" of infectives population. We note that

$$\widehat{F}'_{1,t}(0) = \frac{\gamma}{(\gamma+\mu)}\left(1-\sigma-\beta_t\phi'(0)\right).$$

Let

$$\Re_{0,t} = \frac{-\gamma\beta_t\phi'(0)}{\Re_{\mathrm{D}}(1-\gamma)+\gamma\sigma}.$$

Notice that when $\Re_{\mathrm{D}} = 1$, then the demographic effects disappear and $\Re_0$ reduces to

$$\Re_{0,t} = \frac{-\gamma\beta_t\phi'(0)}{1-\gamma(1-\sigma)}.$$

We note that

$$\widehat{F}'_{1,t}(0) = \frac{\gamma}{(\gamma+\mu)}\left(1-\sigma-\beta_t\phi'(0)\right) > 1,$$

respectively,

$$\widehat{F}'_{1,t}(0) = \frac{\gamma}{(\gamma+\mu)}\left(1-\sigma-\beta_t\phi'(0)\right) < 1),$$

is equivalent to $\Re_{0,t} > 1$, respectively, $\Re_{0,t} < 1$.

The threshold parameter (basic reproduction number),

$$\Re_0 = \prod_{t=0}^{T-1}\left(\widehat{F}'_{1,t}(0)\right),$$

determines the long-term behavior of the disease in Model (9) . As in the case of bounded asymptotic growth (Theorem 4.1), we obtain that if $\Re_0 > 1$, then the proportion of infectives in the total population persists uniformly while if $\Re_0 < 1$ the proportion of infectives in the total population decreases to zero regardless of initial population sizes. We collect these results in the following theorem.

Theorem 4.2. *Consider Model* (9).

(a) If $\Re_{\mathrm{D}} > 1$ and $\Re_0 < 1$, then $\lim_{t\to\infty}(s(t),i(t)) = (1,0)$.

That is, the proportion of infectives in the total population goes to zero while the total population is increasing at a geometric rate.

(b) *If* $\mathfrak{R}_D > 1$ *and* $\mathfrak{R}_0 > 1$, *then the proportion of infectives persists uniformly.*

(c) *If* $\mathfrak{R}_D < 1$ *and* $\mathfrak{R}_0 < 1$, *then* $\lim_{t\to\infty}(s(t),i(t)) = (1,0)$.

That is, the proportion of infectives in the total population goes to zero while the total population is decreasing at a geometric rate.

(d) *If* $\mathfrak{R}_D < 1$ *and* $\mathfrak{R}_0 > 1$, *then the proportion of infectives persists uniformly.*

That is, the proportion of infectives in the total population persists uniformly while the total population is decreasing at a geometric rate.

Proof. Since $f(N) = \frac{\mu}{\gamma}N$, N increases geometrically when $\mathfrak{R}_D > 1$ while it decreases geometrically when $\mathfrak{R}_D < 1$. To establish the result, we prove that if in addition $\mathfrak{R}_0 < 1$, then all solutions $(s(t),i(t))$ of Model (9) converge to the disease-free equilibrium point $(1,0)$ as $t \to \infty$. However, if $\mathfrak{R}_0 > 1$ we proceed exactly as in the proof of Theorem 4.1 and use Theorem 4.6 in [18] to prove that as $t \to \infty$ the proportion of infective population persists uniformly. □

5 Ricker Recruitment Function

If new recruits are governed by the Ricker model,

$$f(N) = \frac{(1-\gamma)}{\gamma} N e^{k-N},$$

and $0 < k < \frac{2}{1-\gamma}$, then $N_\infty = k$ for all $N > 0$ and the total population is asymptotically constant. If, in addition $\mathfrak{R}_0 > 1$, then by Theorem 4.1, the infective population persists uniformly. However, if $\mathfrak{R}_0 < 1$, then the infective population goes extinct (Theorem 4.1).

As k increases past $\frac{2}{1-\gamma}$, the demographic dynamics becomes periodic and Theorem 1 no longer applies. In [15], Elaydi and Yakubu proved that non-trivial periodic orbits are not globally attracting in autonomous models such as Model (1). Next, we use the following example, Example 5.1, to illustrate that when the infective population is nonzero and the demographic N-dynamics undergoes period-doubling bifurcations, then the N-dynamics can "drive" both the S-dynamics and I-dynamics in Model (5) (see Figs. 2–4).

Example 5.1. In Model (5), let

$$f(N) = \frac{(1-\gamma)}{\gamma} N e^{k-N} \text{ and } \phi\left(\frac{\beta_t I}{N}\right) = e^{-\frac{\beta_t I}{N}},$$

where

$$k \in [1,6],\ \beta_t = a + b \times (1 + (-1)^t),\ a = 20,\ b = 10,\ \gamma = 0.1 \text{ and } \sigma = 0.01.$$

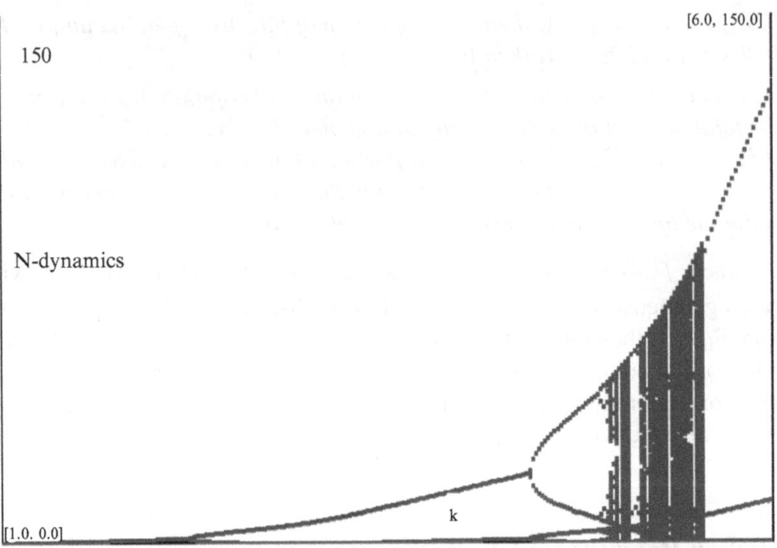

Fig. 2 In Example 5.1, the total population undergoes period-doubling bifurcation route to chaos as k is varied between 1 and 6. On the horizontal axis, $1 \leq k \leq 6$ and on the vertical axis, $0 \leq N \leq 150$

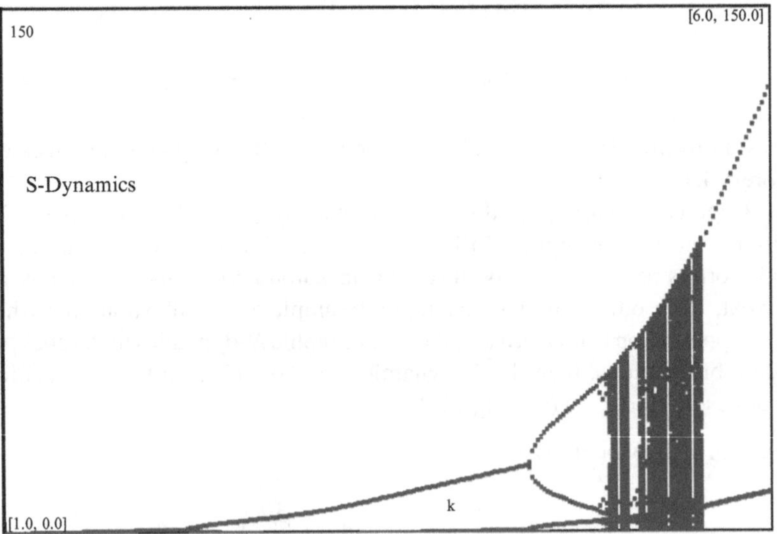

Fig. 3 In Example 5.1, the susceptible population dynamics follows the total population dynamics and undergoes period-doubling bifurcation route to chaos as k is varied between 1 and 6. On the horizontal axis, $1 \leq k \leq 6$ and on the vertical axis, $0 \leq S \leq 150$

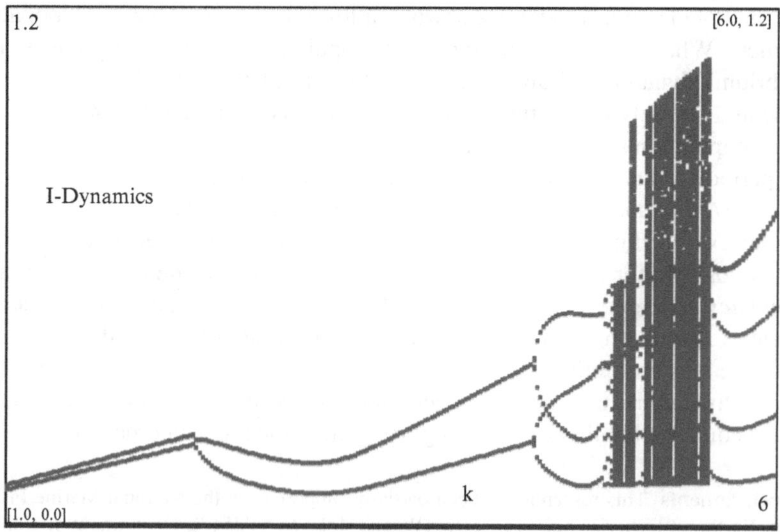

Fig. 4 In Example 5.1, the infected population dynamics follows the total population dynamics and undergoes period-doubling bifurcation route to chaos as k is varied between 1 and 6. On the horizontal axis, $1 \leq k \leq 6$ and on the vertical axis, $0 \leq I \leq 1.2$

In Example 5.1, the recruitment function is the Ricker model and, as in Example 4.1, infection is modeled as a Poisson process [8–11]. When $0 < k < \frac{2}{1-\gamma} = \frac{20}{9}$, then $N_\infty = k$ and $\Re_0 = 7.416 > 1$. As guaranteed by Theorem 4.1, the disease persists (see Fig. 3). As k is varied between 2 and 6, both the S-dynamics and I-dynamics follow the N-dynamics as it undergoes period-doubling bifurcations route to chaos. In Figs. 2–4, when $5.8 \leq k \leq 6$, by zooming on the graphs, we obtain that the S-dynamics and I-dynamics follow the N-dynamics on a period-4 cycle attractor. However, $N = S + I$ implies that the amplitudes of oscillations in the S-dynamics and I-dynamics are different from that of the N-dynamics whenever the disease is endemic.

6 Conclusion and Discussion

In this chapter, we used an extension of the discrete-time SIS epidemic model framework of Castillo-Chavez and Yakubu to study the relationship between periodic infectious disease incidence and host population dynamics. When the host population is asymptotically constant or growing at a geometric rate we computed $\Re_0$ and used it to predict disease persistence or extinction. In this case, we obtained that the transmission rates as well as survival and recovery rates are critical model parameters for the persistence or control of the disease.

As in [8–11], our model framework allows for oscillatory host population dynamics. What is $\mathfrak{R}_0$ when the host population exhibits oscillatory (non-equilibrium) dynamics? Can $\mathfrak{R}_0$ be used to control the spread of an infectious disease in a strongly fluctuating host population? What is the relationship between the host population attractor and the infective population attractor?

In periodic environments, Franke and Yakubu obtained that the demographic dynamics (*N-dynamics*) does not always drive the disease dynamics (*I-dynamics*) [18]. Our extended model results support this prediction. Furthermore, we showed that it is possible for *I-dynamics* and *S-dynamics* to follow the *N-dynamics* as the *N-dynamics* undergoes period doubling bifurcations route to chaos. In this case, our numerical explorations seem to suggest that the differences between the *N-dynamics* and *I-dynamics* are limited to their variation in "amplitude." Qualitative proofs of these results that include the case where individuals (infected and susceptibles) do not have equal probability of surviving one generation are welcome.

Acknowledgments This research has been partially supported by the National Marine Fisheries Service, Northeast Fisheries Science Center (Woods Hole, MA 02543), Department of Homeland Security, DIMACS and CCICADA of Rutgers University, Mathematical Biosciences Institute of the Ohio State University and National Science Foundation under grants DMS 0931642 and 0832782.

References

1. Alexander, M.E., Moghadas, S.M.: Periodicity in an epidemic model with a generalized nonlinear incidence. Math. Biosci. **189**, 75–96 (2004)
2. Allen, L.: Some discrete-time SI, SIR and SIS models. Math. Biosci. **124**, 83–105 (1994)
3. Allen, L., Burgin, A.M.: Comparison of deterministic and stochastic SIS and SIR models in discrete-time. Math. Biosci. **163**, 1–33 (2000)
4. Anderson, R.M., May, R.M.: Infectious Diseases of Humans: Dynamics and Control. Oxford University, Oxford (1992)
5. Bailey, N.T.J.: The simple stochastic epidemic: a complete solution in terms of known functions. Biometrika **50**, 235–240 (1963)
6. Best, J., Castillo-Chavez, C., Yakubu, A.-A.: Hierarchical competition in discrete time models with dispersal. Fields Institute Communications **36**, 59–86 (2003)
7. Beverton, R.J.H., Holt, S.J.: On the dynamics of exploited fish populations, vol. 2, p. 19. H.M. Stationery Off., London, Fish. Invest. (1957)
8. Castillo-Chavez, C., Yakubu, A.: Discrete-time S-I-S models with complex dynamics. Nonlin. Anal. TMA **47**, 4753–4762 (2001)
9. Castillo-Chavez, C., Yakubu, A.: Dispersal, disease and life-history evolution. Math. Biosc. **173**, 35–53 (2001)
10. Castillo-Chavez, C., Yakubu, A.: Epidemics on attractors. Contem. Math. **284**, 23–42 (2001)
11. Castillo-Chavez, C., Yakubu, A.A.: Intraspecific competition, dispersal and disease dynamics in discrete-time patchy environments. In: Castillo-Chavez, C., Blower, S., van den Driessche, P., Kirschner, D., Yakubu, A.-A. (eds.) Mathematical Approaches for Emerging and Reemerging Infectious Diseases: An Introduction to Models, Methods and Theory. Springer, New York (2002)
12. Cooke, K.L., Yorke, J.A.: Some equations modelling growth processes of gonorrhea epidemics. Math. Biosc. **16**, 75–101 (1973)

13. Cull, P.: Local and global stability for population models. Biol. Cybern. **54**, 141–149 (1986)
14. Elaydi, S.N.: Discrete Chaos. Chapman Hall/CRC, Boca Raton, FL (2000)
15. Elaydi, S.N., Yakubu, A.-A.: Global stability of cycles: Lotka-Volterra competition model With stocking. J. Diff. Equat. Appl. **8**, 537–549 (2002)
16. Ewald, P.W.: Evolution of Infectious Disease. Oxford University, Oxford, NY (1994)
17. Franke, J.E., Yakubu, A.-A.:Discrete-time SIS epidemic model in a seasonal environment. SIAM J. Appl. Math. **66**, 1563–1587 (2006)
18. Franke, J.E., Yakubu, A.-A.: Periodically forced discrete-time SIS epidemic with disease-induced mortality. Math. Biosci. Eng. **8**, 385–408 (2011)
19. Hadeler, K.P., van den Driessche, P.: Backward bifurcation in epidemic control. Math. Biosc. **146**, 15–35 (1997)
20. Hassell, M.P.: The dynamics of competition and predation. Studies in Biol, vol. 72. The Camelot Press Ltd, Southampton (1976)
21. Hassell, M.P., Lawton, J.H., May, R.M.: Patterns of dynamical behavior in single species populations.
J. Anim. Ecol. **45**, 471–486 (1976)
22. Hethcote, H.W., van den Driessche, P.: Some epidemiological models with nonlinear incidence. J. Math. Biol. **29**, 271–287 (1991)
23. May, R.M.: Simple mathematical models with very complicated dynamics. Nature **261**, 459–469 (1977)
24. May, R.M.: Stability and Complexity in Model Ecosystems. Princeton University Press, Princeton (1974)
25. May, R.M., Oster, G.F.: Bifurcations and dynamic complexity in simple ecological models. Am. Nat. **110**, 573–579 (1976)
26. McClusky, C.C., Muldowney, J.C.: Bendixson-Dulac criteria for difference equations. J. Dyn. Diff. Equat. **10**, 567–575 (1998)
27. Nicholson, A.J.: Compensatory reactions of populations to stresses, and their evolutionary significance. Aust. J. Zool. **2**, 1–65 (1954)
28. van den Driessche, P., Watmough, J.: A simple SIS epidemic model with a backward bifurcation. J. Math. Biol. **40**, 525–540 (2000)
29. Yakubu, A.-A.: Introduction to discrete-time epidemic models. DIMACS Ser. Discrete Math. Theor. Comput. Sci. **75**, 83–109 (2010)
30. Yakubu, A.-A., Ziyadi, N.: Discrete-time exploited fish epidemic models. Afrika Mathematika, **22**, 177–199 (2011)
31. Yorke, J.A., London, W.P.: Recurrent outbreaks of measles, chickenpox and mumps II. Am. J. Epidemiol. **98**, 453–468 (1973)
32. Zhao, X.-Q.: Asymptotic behavior for asymptotically periodic semiflows with applications. Comm. Appl. Nonl. Anal. **3**, 43–66 (1996)

Zeitfracht Medien GmbH
Ferdinand-Jühlke-Straße 7
99095 Erfurt, Deutschland
produktsicherheit@kolibri360.de